T0183088

Springer Texts in Statistics

Series Editors

G. Allen, Department of Statistics, Rice University, Houston, TX, USA

R. De Veaux, Department of Mathematics and Statistics, Williams College, Williamstown, MA, USA

R. Nugent, Department of Statistics, Carnegie Mellon University, Pittsburgh, PA, USA

Springer Texts in Statistics (STS) includes advanced textbooks from 3rd- to 4th-year undergraduate courses to 1st- to 2nd-year graduate courses. Exercise sets should be included. The series editors are currently Genevera I. Allen, Richard D. De Veaux, and Rebecca Nugent. Stephen Fienberg, George Casella, and Ingram Olkin were editors of the series for many years.

More information about this series at http://www.springer.com/series/417

Kostas Triantafyllopoulos

Bayesian Inference of State Space Models

Kalman Filtering and Beyond

 Springer

Kostas Triantafyllopoulos (iD)
School of Mathematics
University of Sheffield
Sheffield, UK

ISSN 1431-875X   ISSN 2197-4136   (electronic)
Springer Texts in Statistics
ISBN 978-3-030-76126-4   ISBN 978-3-030-76124-0   (eBook)
https://doi.org/10.1007/978-3-030-76124-0

Mathematics Subject Classification: 62F15, 62M10, 62M20, 91B84, 93E03, 93E11, 62P05, 62P20, 62P30

This Springer imprint is published by the registered company Springer Nature Switzerland AG.
The registered company address is: Gewerbestrasse 11, 6330 Cham, Switzerland

Preface

The discovery and development of state space models and the Kalman filter have their roots in the dynamical and control systems research communities. The Kalman filter was proposed initially as an alternative to the Wiener-Kolmogorov estimation theory after Kalman applied state variables to the filtering problem. Soon after its conception, the Kalman filter was associated with significant real-data application such as the Apollo project at the Ames Research Centre at NASA. The Kalman filter was soon discovered by the statistics and econometrics research communities, which were able to exploit the flexibility of the state space formulation in the quest to describe dynamic systems and to use the filter as a powerful forecasting framework. Bayesian inference and forecasting were key areas of development in the 1980s and 1990s. By that time, the state space models and the Kalman filter were found to be extremely popular in a wide range of applied science, including statistics, engineering, economics, environmetrics and biology, to name but a few. Partly due to the availability of computer advancement, over the last 20 years, there has been an increasing interest in multivariate linear Gaussian, non-linear and non-Gaussian modelling and related state space methods. This is facilitated by deploying Bayesian computation, in particular sequential Monte Carlo and Markov chain Monte Carlo methods, but also by other estimation procedures such as the unscented Kalman filter and related algorithms.

The principle idea of this book is to bring together the above models with their statistical inference and make them available to a broad audience. Teaching time series and state space modelling for over 15 years at the University of Sheffield has motivated me to take up this book project. During these years, I have met and collaborated with a number of applied scientists sharing a keen interest in state space models. Some of these models can be generated by a natural phenomenon such as met in engineering or environmetrics and some can be generated by a socio-economic structure such as in finance and its applications. The book is written from a statistician's perspective, and it uses a number of data sets to illustrate the methods from a wide range of disciplines. The first six chapters discuss state space methods and Bayesian inference for general use; the last two chapters of the book focus on finance and dynamical systems. On one hand, economics and finance have been a steady flow of motivation and development of state space methods, and on the other hand, dynamical systems is the key subject area where the story of the Kalman filter began. Furthermore, the text aims to widespread state space methods by developing suitable software to enable the reader to apply the algorithms using the data sets of the book, but also using their own data sets. We have created the package BTSA (Bayesian Time Series Analysis), available via the contributed package CRAN website within the environment for statistical computing R (https://www.r-project.org). The package includes most functions and data sets used in Chaps. 1–6; moreover, R commands incorporated in the text is believed to help understanding and implementing the algorithms once the BTSA package is installed.

The book assumes a basic technical background of linear algebra, probability, and statistics. This is about second year university level and is reviewed in Chap. 2. The textbook can be used as a one-semester course on linear Gaussian state space methods by covering Chaps. 1–4, perhaps with some inclusion of multivariate state space models in Chap. 5. Alternatively, the book can cover the basic theory of linear models in Chap. 3 and then move on to non-linear and non-Gaussian models in Chap. 6, perhaps by including some parts of Chaps. 7 and 8. The textbook is aimed at students at the higher end of undergraduate or graduate level and it is also aimed at scientists and doctoral students for self-study.

Sheffield, UK   Kostas Triantafyllopoulos
June 2021

Acknowledgements

This book would not be possible without the encouragement and input of Nick Bingham. I am indebted to him for our discussions in numerous occasions in Sheffield and in London. I am grateful to several colleagues and friends as well as graduate students for their feedback and support. In particular, I thank Dave Applebaum, Peter Young, Guy Nason, Clive Anderson, Andrew Harvey, Alan Zinober, Osman Tokhi, Giovanni Montana, Tata Subba Rao, Maurice Priestley, Attilio Meucci, John Fry, Jeremy Oakley, and Daniel Molinari. Special thanks are due to Jeff Harrison who introduced me to state space models and taught me Bayesian forecasting.

I am grateful to the Springer team and in particular to Joerg Sixt and Remi Lodh who offered me great support during the long period of this project. They have been very patient and have well accommodated my pace of work.

Finally, I would like to thank my family who has been very supportive and has encouraged me to complete the project.

Contents

Acronyms

AR	Autoregressive (model, parameters)
ARIMA	Autoregressive integrated moving average (model)
c.d.f.	Cumulative distribution function
DGLM	Dynamic generalised linear model
EKF	Extended Kalman filter
EM	Expectation maximisation (algorithm)
FFBS	Forward filtering backward sampling (algorithm)
i.i.d.	Independent and identically distributed (random variables)
GARCH	Generalised autoregressive conditional heteroskedastic models
GLM	Generalised linear model
MCMC	Markov chain Monte Carlo
MAD	Mean absolute deviation
MGARCH	Multivariate generalised autoregressive conditional heteroskedastic models
MIMO	Multiple input multiple output (dynamic system)
MSE	Mean squared error
MSSE	Mean squared standardised error
MSOP	Multivariate scaled observation precision model
OLS	Ordinary least squares
p.d.f.	Probability density function
PID	Proportional integral derivative (controller)
PF-I	Particle filter algorithm I (standard particle filter)
PF-II	Particle filter algorithm II (Liu and West particle filter)
p.m.f.	Probability mass function
RLS	Recursive least squares
SISO	Single input single output (dynamic system)
SOP	Scaled observational precision model
TRMS	Twin rotor multi-input multi-output system
UKF	Unscented Kalman filter
WAR	Whishart autoregressive process
WN	White noise (process)

Chapter 1
State Space Models

This chapter introduces time series, the *state space* model (although a more formal treatment is given in Chap. 3) and Kalman filtering. We start by defining and giving some examples of time series data. Section 1.2 discusses a data driven problem from hydro-dynamics, which motivates our first encounter with the state space model. To appreciate the wealth of applications that the state space model holds, Sect. 1.3 provides several examples from environmetrics, navigation, economics and physics. Section 1.4 gives a historical account of the Kalman filter and the chapter concludes by giving a brief content description.

1.1 Introduction

1.1.1 Time Series

In many subject areas, such as engineering, economics, biology, environmetrics, data collected are frequently collected over time. Such data, known as time series, may arise as the result of the data collection process (we may collect data in a daily, or hourly frequency) and interest is focused in understanding the dynamics of such data as well as forecasting future time series values. Examples of time series include: daily prices of financial assets, monthly marriage figures, quarterly product sales, daily temperatures in a particular city, annual precipitation levels of a lake and so forth. In all these examples, the time in which the data are collected is important, as it introduces a particular correlation or dependence structure between measurements or observations. In order to understand such data, it is necessary to consider statistical models that take into account time in the aforementioned dependence structure. The study of such models is known as *time series analysis* and has been discussed in many textbooks, see e.g. Brockwell and Davis (1991), Shumway and Stoffer (2017) and Lindsey (2004).

K. Triantafyllopoulos, *Bayesian Inference of State Space Models*, Springer Texts in Statistics, https://doi.org/10.1007/978-3-030-76124-0_1

To establish notation, we refer to time as t and to the time series observation at t as y_t or as $y(t)$ (see below). In most situations t will belong to a discrete set (such as the natural numbers $t = 1, 2, 3 \ldots$) and in this respect the collection of observations y_t, for all $t = 1, 2, \ldots$, denoted by $\{y_t, t = 1, 2, \ldots\}$ or simply $\{y_t\}$ defines a *discrete-time time series*. For example, t may represent days, months or years (as in examples above). In some cases t belongs to a continuous set (e.g. the closed interval $[0, 1]$); then the time series at t is denoted by $y(t)$ (to make explicit that t is continuous) and so we may write $\{y(t), t \in [0, 1]\}$ or $\{y(t)\}$ for a *continuous-time time series*. In this book we will primarily study discrete-time time series, unless otherwise stated. The objectives of time series analysis are

1. to build statistical models (known as time series models) that describe and understand the dynamics of observed time series data,
2. based on an observed collection of data, to forecast future values of the time series.

Both (1) and (2) adopt a time series model, but while (1) aims at finding a model that describes a collection of data, (2) adopts the model in order to forecast future data.

A general class of time series models and one that this book is focused on, known as *state space models*, suggests that at each time t the observations y_t are related to the *states* at time t, which in turn are related to the states at $t - 1$. Thus, the observation is a function of the states, and the way in which the states move in their space, hence the name state space (time series) models.

1.1.2   Examples of Time Series Data

Time series data typically comprise trend, seasonal components and their combination. This section describes three examples of time series data, which are revisited later in Chap. 4.

- **Trend (Aluminium prices).** Figure 1.1 depicts spot prices of aluminium over the time period 4 January 2005 to 31 October 2005. Aluminium, which is a non-ferrous metal, trades daily at the London metal exchange, for information of which the reader is referred to http://www.lme.com/. The data of Fig. 1.1, which exclude bank holidays and weekends, show initially some random fluctuation, followed by an upward linear trend (March–April), followed by a linear fall (May–June), followed by some random fluctuation (June–July), followed by some increasing and then decreasing trend (August–September), and finally followed by a linear trend (October).
- **Seasonal (Quarter temperatures at Sheffield).**
  Figure 1.2 shows averaged temperatures collected for each quarter at Weston Park, Sheffield, UK, for unspecified years. This data set shows clear evidence of seasonality (or cyclic variation), as the values of the 1st Quarter of each year are

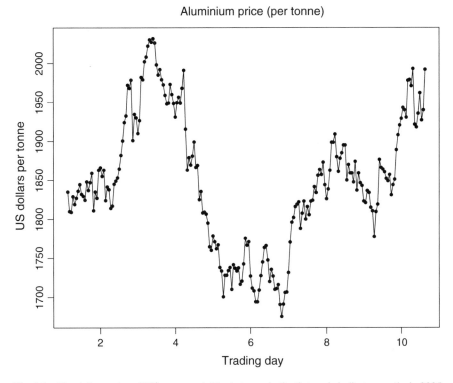

Fig. 1.1 Aluminium prices (US$ per tonne). The integers in the time axis indicate months in 2005

similar (lower values in the figure) and of course they are such due to the effect of the winter months in the first quarter. Likewise, the third quarter of each year is responsible for the higher temperature values, being influenced by the summer months.

1.2   Water Tank Dynamics and the State Space Model

In this section we describe a mechanism that generates a linear state space model. We consider a simple experiment used to measure the level of water in a tank. Figure 1.3 shows a water tank, which allows a constant flow of water entering the tank from the left side with constant flow rate 6 litres per min and leaving the tank on the bottom right side at flow rate 5 litres per min. The objective of the experiment is to measure the level L of the water in the course of time. The first observation we make is that the level is not constant as the water coming to the tank creates disturbance affecting the level of the water. Secondly, the water leaving the tank has lower flow rate than the flow rate entering the tank. Assuming water mass per litre is

Quarterly averaged temperatures in Sheffield

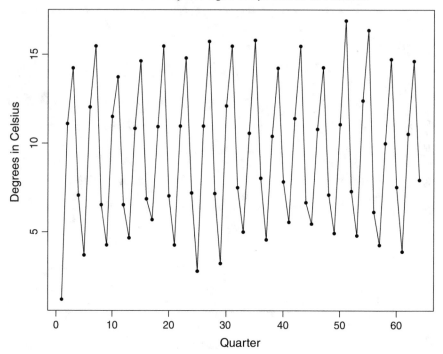

Fig. 1.2 Quarterly mean temperature in Sheffield

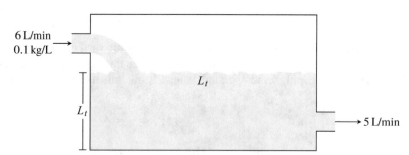

Fig. 1.3 Water tank dynamics

constant, this means that in the long run we expect the level to increase by 1 litre per min. If there was no disturbance due to water entering the tank, the volume in the tank at time t is $\ell L_t = \ell L_0 + t$, where ℓ is the length of the base of the tank and L_0 is the initial level of the water (assumed to be higher than the higher point of the right leave point of the tank). The above equation can be written as $\ell L_t = \ell L_{t-1} + 1$, starting from $\ell L_1 = \ell L_0 + 1$. With the disturbance generated by the water falling into the tank we postulate that $\ell L_t \approx \ell L_{t-1} + 1$. This in words says that the volume of the tank at time t (in min) is close to the volume of the tank at time $t - 1$ plus 1 litre. This can be modelled by introducing a sequence of random variables ζ_t, so that

$$L_t = L_{t-1} + \frac{1}{\ell} + \zeta_t, \tag{1.1}$$

where ζ_t is a i.i.d. sequential of random variables with zero mean and some variance Z. For this model we have $E(L_t) = E(L_{t-1}) + 1/\ell$ and $Var(L_t) \geq Var(L_{t-1})$.

Consider now that an observer is measuring the level L_t. This may be achieved automatically by placing a float in the tank. This will create an extra source of uncertainty on the level of the water. A possible model for the observed level y_t suggests that y_t is expected to be equal to the actual (unobserved) level L_t inflated by noise ϵ_t, caused by the float. Hence the measurement equation is

$$y_t = L_t + \epsilon_t, \tag{1.2}$$

where ϵ_t is an i.i.d. sequence of random variables with zero mean and variance σ^2.

Initially a value for L_0 (the level as time $t = 0$ should be set); alternatively it is possible to consider that L_0 is a random variable, if we wish to specify an initial distribution for it (this specification will be a useful consideration, if the engineer is not certain about the initial value L_0). Measurement model (1.2) together the evolution model (1.1) and the initial specification of L_0 define a *state space model*. The unobserved signal L_t is the state at time t and the observation y_t is a linear function of L_t. Again collecting water float measurement data $y_1, \ldots, y_n$ we can estimate the unknown water tank float at each time L_t.

Before we proceed with the definition of the general state space model, we need to establish some notation. Suppose that x represents a p-dimensional column vector, i.e.

$$x = \begin{bmatrix} x_1 \\ x_2 \\ \vdots \\ x_p \end{bmatrix},$$

where x_i is the ith element of the vector, $i = 1, 2, \ldots, p$. With $\top$ denoting transposition, we write $x^\top = [x_1, x_2, \ldots, x_p]$ for a row vector; note that with this notation x can be compactly written as $x = [x_1, x_2, \ldots, x_p]^\top$. A $p \times p$ matrix $\mathbf{X}$ is a tabular display, which places p column vectors $[x_{11}, x_{21}, \ldots, x_{p1}]^\top$, $[x_{12}, x_{22}, \ldots, x_{p2}]^\top, \ldots, [x_{1p}, x_{2p}, \ldots, x_{pp}]^\top$ one after the other, i.e. assigning the element x_{ij} at position (i, j). As is evident from the above, we use boldface to distinguish a matrix from a scalar or a vector. In a probabilistic setting, we deal with random variables (for univariate variables) and random vectors (for multivariate variables), more information on which can be found in Chap. 2. For a p-dimensional column vector y, the notation $y \sim N(\mu, \mathbf{V})$ implies that y follows the multivariate Gaussian distribution with mean vector μ and with covariance matrix $\mathbf{V}$. More details about matrix algebra and statistics relevant to the contents of this book are provided in Chap. 2.

In general, if the p-dimensional state column vector at time t is denoted by β_t, then a linear state space model may be described by equations

$$y_t = x_t^\top \beta_t + \epsilon_t \quad \text{and} \quad \beta_t = \mathbf{F}_t \beta_{t-1} + \zeta_t, \tag{1.3}$$

where x_t is a p-dimensional column design vector, $\mathbf{F}_t$ is a $p \times p$ transition matrix and the error sequences ϵ_t and ζ_t may be assumed to be independent, with zero mean; in many applications $x_t = x$ and $\mathbf{F}_t = \mathbf{F}$ are time-invariant, but the general case allows for more flexibility. It is assumed that the sequences $\{\epsilon_t\}$ and $\{\zeta_t\}$ are independent (i.e. ϵ_t is independent of ϵ_s and ζ_t is independent of ζ_s, for any $t \neq s$) and that ϵ_t and ζ_t are independent of β_0, for any t. The model is completely determined if distributions for the innovations ϵ_t and ζ_t as well as distribution of the initial state β_0 are specified.

If Gaussian distributions are assumed for ϵ_t, ζ_t and for β_0, model (1.3) is known as Gaussian linear state space model, and it can be described by

$$y_t \mid \beta_t \sim N(x_t^\top \beta_t, \sigma^2) \quad \text{and} \quad \beta_t \mid \beta_{t-1} \sim N(\mathbf{F}_t \beta_{t-1}, \mathbf{Z}), \tag{1.4}$$

where $y_t \mid \beta_t$ denotes the conditional distribution of y_t given β_t (a formal definition is given in Chap. 2), σ^2 is the variance of ϵ_t and $\mathbf{Z}$ is the covariance matrix of ζ_t.

Expression (1.4) allows us to extend the Gaussian linear state space model (1.4) to non-linear and non-Gaussian state space models. Indeed, we will say that the time series $\{y_t\}$ is generated by a general (including linear Gaussian and non-linear and non-Gaussian) state space model, if y_t can be described by the following distributions

$$p(y_t \mid \beta_t), \quad p(\beta_t \mid \beta_{t-1}) \quad \text{and} \quad p(\beta_0). \tag{1.5}$$

This, very general, model setting assumes a distribution of y_t, given some states, a distribution of β_t, given the previous state β_{t-1} and an initial state distribution $p(\beta_0)$. It is implicitly assumed that given β_t, y_t is conditionally independent of past observations and states $y_{t-1}, y_{t-2}, \ldots$ and of $\beta_{t-1}, \beta_{t-2}, \ldots$, but also of future observations and states $y_{t+1}, y_{t+2}, \ldots$ and $\beta_{t+1}, \beta_{t+2}, \ldots$; in other words, the present state β_t holds all information from past and future data and states relevant to the understanding and knowledge of y_t. Likewise, given β_{t-1}, β_t is conditionally independent of $\beta_{t-2}, \beta_{t-3}, \ldots$. Simply put, we say that given the present (state at time t), the past and the future are conditionally independent. More information about this independence structure can be found in West and Harrison (1997). It is worth noting that the above description fits the purposes of state space models for discrete and roughly equally spaced observed data. An example of a continuous-time state space model is given in Sect. 1.3.4 and further discussed in the context of dynamical systems in Sect. 8.4. In the sequel we give some illustrative examples showing the application of state space models to real-life situations.

1.3 Examples of State Space Models

This section describes some situations motivated by real-life problems, which can be modelled with a state space model. It illustrates some of the many subject areas that state space models find application.

1.3.1 Forecasting Air-Pollution Levels

Air pollution consists of the introduction of chemicals, particulate matter and biological materials into the atmosphere, causing severe damage to the environment. Many of the main air pollutants are contributing to the greenhouse effect, which is considered to be the main human made factor that affects climate change.

Nitric oxide (NO), one of the most prominent air pollutants, is emitted from high temperature combustion, and also produced naturally during thunderstorms by electrical discharge. Figure 1.4 shows NO levels (in mg/m^3) together with levels of % humidity, mean daily temperature (in °C), and wind speed (in m/s); the measurements of these variables are collected daily covering January to December 2001 and they are provided by one of the sensors sites of the air-pollution networks of Athens.

One of the objectives is to be able to use the covariates (here denoted by x_{1t}—humidity, x_{2t}—temperature and x_{3t}—wind speed) in order to forecast future values of the NO levels (denoted by y_t). Such information may be vital in issuing warning messages to the community, should the pollution levels be expected to rise, e.g. by preventing old people, and in particular those with respiratory related health problems, access particular areas of the city. Another objective may be to establish pollution trends and dynamics, so as to assess whether anti-pollution measures work and assist policy makers.

A first model is a simple regression model of x_{1t}, x_{2t}, x_{3t} on the response variable y_t, given by

$$y_t = \beta_0 + \beta_1 x_{1t} + \beta_2 x_{2t} + \beta_3 x_{3t} + \epsilon_t, \tag{1.6}$$

where ϵ_t is an independent sequence, following a Gaussian distribution with zero mean and some variance σ^2, i.e. $\epsilon_t \sim N(0, \sigma^2)$. The coefficient β_0 is the intercept and $\beta_1, \beta_2, \beta_3$ are the coefficients of the covariates x_{1t}, x_{2t}, x_{3t}. We can fit this model by using standard regression methods, see e.g. Bingham and Fry (2010).

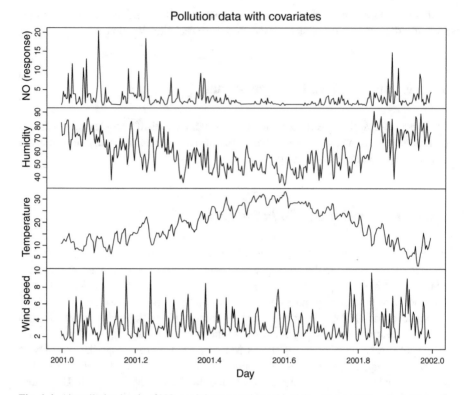

Fig. 1.4 Air-pollution levels of NO_2 and three covariates (humidity, temperature and wind speed)

If we define the state vector

$$\beta_t = \begin{bmatrix} \beta_0 \\ \beta_1 \\ \beta_2 \\ \beta_3 \end{bmatrix},$$

then model (1.6) can be put in state space form by writing

$$y_t = [1, x_{1t}, x_{2t}, x_{3t}] \begin{bmatrix} \beta_0 \\ \beta_1 \\ \beta_2 \\ \beta_3 \end{bmatrix} + \epsilon_t = x_t^\top \beta_t + \epsilon_t, \tag{1.7}$$

and

$$\beta_t = \beta_{t-1}, \tag{1.8}$$

for all t, with probability 1, i.e. $\beta_t = \beta_{t-1} = \cdots = \beta_1 = \beta$.

It can be argued that model (1.6) is insufficient, because although the response variable and covariate are time-varying, the β_i coefficients are time-invariant (for $i = 0, 1, 2, 3$). Sometimes such a model is referred to as *static* regression model, because its coefficients do not depend on time. However, as it is evidenced by Fig. 1.4, the humidity and temperature covariates (x_{1t} and x_{2t}) are changing over time and thus the coefficients β_i should be time-varying.

A simple regression model with time-varying coefficients, known also as *dynamic regression* model, is to adopt (1.7), but to replace (1.8) by

$$\beta_t = \beta_{t-1} + \zeta_t. \tag{1.9}$$

where ζ_t is an independent sequence and ζ_t follows a four-dimensional Gaussian distribution with zero mean vector and some covariance matrix. Equation (1.9) implies that $\beta_t \approx \beta_{t-1}$, but the shock ζ_t allows for some local variation in the β_t coefficients. Equations (1.7) and (1.9) define a Gaussian linear state space model (assuming that a Gaussian distribution is set for β_0).

1.3.2 Tracking a Ship

We consider the classical bearings-only tracking problem of tracking a ship being observed from a constant (not moving) observation position. State space models have been proposed for tracking various objects since the late 70s, see e.g. Aidala (1979); for a recent discussion of this problem see Fearnhead (2002) and Särkkä (2013). Figure 1.5 depicts, with solid points, the position of the ship (in $x - y$ coordinates), at each time t. At each time t, and for each of the points mentioned above, angular data of the ship's position is obtained. The objective is, using this data, to project future positions and thus to track the movement of the ship.

From Fig. 1.5 we have that $\tan(\theta_t) = y_t/x_t$, where x_t is the x coordinate of the ship at time t, y_t is the y coordinate of the ship at time t and θ_t is the respective angle. From this we have that $\theta_t = \arctan(y_t/x_t)$ and it can be postulated that the angular data we observe are inflated by noise, thus the observation model is

$$z_t = \arctan\left(\frac{y_t}{x_t}\right) + \epsilon_t. \tag{1.10}$$

In this model we only can observe θ_t inflated by noise, i.e. we observe z_t. Thus θ_t and x_t, y_t are assumed unobserved or hidden processes.

In order to provide a model for z_t, we need to set a model for the dynamics or unobserved coordinates x_t and y_t. First we postulate that the derivative of x_t at t, denoted by $\dot{x}_t$, is close to the derivative at $t - 1$, i.e. $\dot{x}_t \approx \dot{x}_{t-1}$. This is interpreted as describing a model when the ship will not have abrupt moves in the x coordinate.

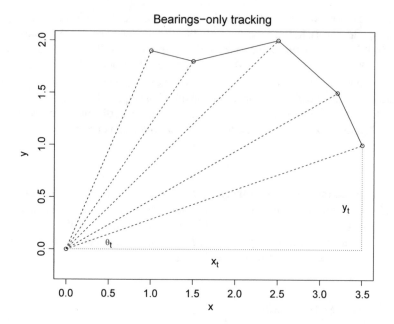

Fig. 1.5 Bearings-only tracking

Similarly we postulate that $\dot{y}_t \approx \dot{y}_{t-1}$. Finally, we can see that the x position at time t can be determined by $x_t \approx x_{t-1} + \dot{x}_{t-1}$, which originates from

$$\dot{x}_{t-1} \approx \dot{x}_t \approx \frac{\Delta x_t}{\Delta t},$$

where $\Delta x_t = x_t - x_{t-1}$ and $\Delta t = t - (t-1) = 1$. A similar formula applies for the y-coordinate, that is $y_t \approx y_{t-1} + \dot{y}_{t-1}$. It is worth noting that time t is considered to be discrete (as data is obtained at discrete times) and in this sense the derivatives (with respect to time) are merely described as ratios $\Delta x_t = x_t - x_{t-1}$ and Δy_t.

Putting the above together, we can define a state vector

$$\beta_t = \begin{bmatrix} x_t \\ y_t \\ \dot{x}_t \\ \dot{y}_t \end{bmatrix},$$

which with the above evolutions of $x_t, y_t, \dot{x}_t, \dot{y}_t$ implies

$$\beta_t = \begin{bmatrix} x_t \\ y_t \\ \dot{x}_t \\ \dot{y}_t \end{bmatrix} = \begin{bmatrix} 1 & 0 & 1 & 0 \\ 0 & 1 & 0 & 1 \\ 0 & 0 & 1 & 0 \\ 0 & 0 & 0 & 1 \end{bmatrix} \begin{bmatrix} x_{t-1} \\ y_{t-1} \\ \dot{x}_{t-1} \\ \dot{y}_{t-1} \end{bmatrix} + \begin{bmatrix} \zeta_{1t} \\ \zeta_{2t} \\ \zeta_{3t} \\ \zeta_{4t} \end{bmatrix},$$

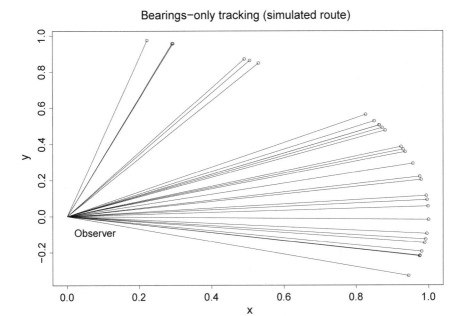

Fig. 1.6 Bearings-only tracking (simulated route)

or

$$\beta_t = \mathbf{F}\beta_{t-1} + \zeta_t. \tag{1.11}$$

Finally, from the definition of β_t, the observation model (1.10) can be written as

$$z_t = \arctan\left(\frac{[0, 1, 0, 0]\beta_t}{[1, 0, 0, 0]\beta_t}\right) + \epsilon_t. \tag{1.12}$$

Model (1.12)–(1.11) is a non-linear state space model. The evolution model (1.11) is linear, but the observation model (1.12) is non-linear. If Gaussian distributions are placed on the innovations ϵ_t, ζ_t and on the initial state β_0, then the above model is a conditionally Gaussian non-linear state space model. Following a similar approach as in Berzuini and Gilks (2001), Fig. 1.6 shows a simulated smooth trajectory of the ship in the $x - y$ plane.

1.3.3 Stochastic Volatility

The subject of dynamic change of financial asset prices has been a core interest of finance over a very long time. Such assets can be share prices and other stocks trading in global stock markets, or indices such as the Standard and Poor 500 index or international exchange rates, such as the exchange rate of the British pound with the US dollar and so forth. Unfortunately, forecasting asset prices is nearly impossible because prices are exposed to excessive variability, which severely affects forecast performance.

Suppose that p_t denotes the price at time t of an asset (or the value of an index or exchange rate) and define the logarithmic returns, known also as log-returns, as $y_t = \log p_t - \log p_{t-1}$. We note that if p_t is similar to p_{t-1}, then $y_t \approx 0$, and y_t will be high when p_t is large compared to p_{t-1} or y_t will be low when p_t will be low compared to p_{t-1}. The conditional variance of the returns y_t, known as volatility, is a measure of the variability of the share prices and hence a measure of the risk associated in forecasting the returns. For example, the 2008 credit crisis resulted in high volatility in most assets, which in turn reflected the increased uncertainty associated with investment decisions. It is now well known that volatility estimation plays a crucial role in constructing portfolios of assets in risk management.

Figure 1.7 shows log-returns of 1776 trading days of IBM share prices. We observe that the returns fluctuate around zero, but at the start of the time series the returns are particularly volatile. This simple picture illustrates that returns should fluctuate around zero and that their variance or volatility is time-varying. Thus, a plausible model is to assume that

$$y_t = \exp(h_t/2)\epsilon_t, \tag{1.13}$$

where ϵ_t is independent of ϵ_s $(t \neq s)$, $\epsilon_t \sim N(0, 1)$ and h_t is some unobserved component, which follows an autoregressive process

$$h_t - \mu = \phi(h_{t-1} - \mu) + \omega_t, \quad \omega_t \sim N(0, \sigma_\omega^2), \tag{1.14}$$

where μ is the mean of h_t, ϕ is an autoregressive coefficient and some variance σ_ω^2. An initial Gaussian distribution is set for h_0. The unobserved process h_t is the logarithm of the volatility $\text{Var}(y_t \mid h_t) = \exp(h_t)$.

Model (1.13)–(1.14) basically postulates that given h_t, y_t follows a Gaussian distribution with zero mean and variance, which is the volatility $\exp(h_t)$, where h_t follows an autoregressive process with mean μ. This model is known as a *stochastic volatility model*, as it provides a framework for the estimation of the volatility $\exp(h_t)$, via the stochastic process h_t; for more details on stochastic volatility the reader is referred to Tsay (2002, §10.7).

We can put the above model in state space form as follows. Define the bivariate state vector $\beta_t = [h_t, 1]^\top$ to be state at time t. Now Eq. (1.13) can be written as

$$y_t \mid \beta_t \sim N[0, \exp([1, 0]\beta_t)], \tag{1.15}$$

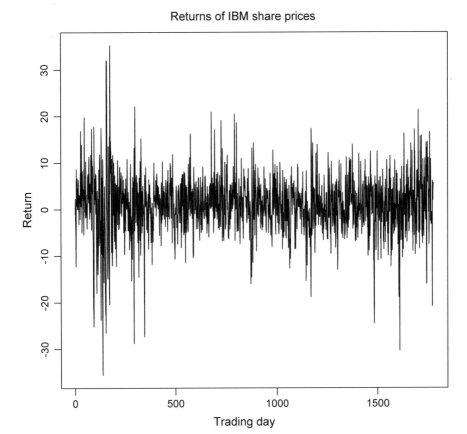

Fig. 1.7 Returns of IBM share prices

and Eq. (1.14) can be written as

$$\beta_t = \begin{bmatrix} h_t \\ 1 \end{bmatrix} = \begin{bmatrix} \phi & \mu(1-\phi) \\ 0 & 1 \end{bmatrix} \begin{bmatrix} h_{t-1} \\ 1 \end{bmatrix} + \begin{bmatrix} \omega_t \\ 0 \end{bmatrix} = \mathbf{F}\beta_{t-1} + \zeta_t \qquad (1.16)$$

Equations (1.15)–(1.16) define a Gaussian non-linear state space model, where the observation model (1.15) is non-linear in β_t, while the transition model (1.16) is linear.

1.3.4   Hookean Spring Force Dynamics

Consider a simple mechanical system describing the motion of an object in one-dimension (translational mode). Suppose that the object has mass m and it moves

Fig. 1.8 Spring single-mass
system, including a spring
and damping

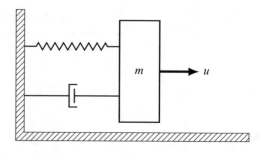

horizontally on a plane and attached to a wall on a spring, sometimes referred to
as Hookean spring (see Fig. 1.8). If $y(t)$ is the position of the object at time t (here
t is continuous), then according to Newton's laws of motion $y(t)$ is driven by the
differential equation

$$m\frac{d^2y(t)}{dt^2} + k_1\frac{dy(t)}{dt} + k_2y(t) = u(t), \tag{1.17}$$

where $dy(t)/dt = \dot{y}(t)$ is the velocity of the object at t (the first derivative of the
position $y(t)$), $d^2y(t)/dt^2$ is the acceleration (the second derivative of $y(t)$), $u(t)$ is
an applied force at t the constants k_1 is the damping or viscous friction constant and
k_2 is the spring constant. The linear restoring force is $-k_2y(t)$ and the friction force
is $-k_1\dot{y}(t)$. This model and its derivation are discussed in Anand (1984, pp. 18–19)
and illustrated in Fig. 1.8.

We can represent the above differential equation in state space form by defining
the state vector

$$x(t) = \begin{bmatrix} y(t) \\ \dot{y}(t) \end{bmatrix}$$

and writing down the position $y(t)$ as

$$y(t) = [1, 0]\begin{bmatrix} y(t) \\ \dot{y}(t) \end{bmatrix} = Hx(t), \tag{1.18}$$

with transition equation driven by (1.17)

$$\dot{x}(t) = \begin{bmatrix} 0 & 1 \\ -\frac{k_2}{m} & -\frac{k_1}{m} \end{bmatrix}x(t) + \begin{bmatrix} 0 \\ \frac{1}{m} \end{bmatrix}u(t). \tag{1.19}$$

It is easy to observe that (1.18) and (1.19) give the differential equation (1.17).

Now suppose that an experiment is conducted whereby measurements of the position $y(t)$, for some points of time t are collected. In this case it is expected that $y(t)$ will be inflated by noise, due to

1. measurement error;
2. the differential equation (1.17) not perfectly describing the system.

As a result it is natural to keep the transition (1.19) and to adopt an observation equation

$$y(t) = Hx(t) + \epsilon(t), \qquad (1.20)$$

where $\epsilon(t)$ is an error term measuring discrepancies between the observed $y(t)$ and the theoretical signal $Hx(t)$ driven by (1.17). The transition (1.19) is basically a deterministic transition equation, which describes the dynamical system, while uncertainty is passed onto $y(t)$ via the innovation or error term $\epsilon(t)$. Model (1.19)–(1.20) is a continuous-time state space model, which is considered in more detail in Sect. 8.4.

Having discussed some detailed numerical examples we now give a historical account of the development of the Kalman filter.

1.4   A Short History of the Kalman Filter

Least Squares: Gauss and Plackett

The state space model can be regarded as an extremely useful extension of regression models. Therefore, it is natural to think that estimation of the state space model shares some common ground with least squares regression (Bingham & Fry, 2010). Least squares regression was discovered independently by Adrien-Marie Legendre (Legendre, 1805) and by Carl Friedrich Gauss (Gauss, 1809). Although Gauss claimed he had discovered the method much earlier than 1809, it seems that Legendre is credited with the discovery of least squares, while the probabilistic treatment of the errors (today usually referred to as residuals) is due to Gauss. The debate of the priority of least squares has been one of the most famous in the history of mathematics and statistics and it is discussed in detail in Stigler (1986). Even in Gauss's first published account in 1809 we find his original idea of assigning a probability distribution to the errors; he refined his theory substantially in a series of papers, which appear collectively under the title "Theoria combinationis observationum erroribus minimis obnoxiae" in 1821, (Gauss, 1821/23/26). In those papers he presents a systematic and probabilistic treatment of least squares methods with application to astronomy and geodesy. It is here that Gauss effectively assigns the exponential $\exp(-x^2)$ (normal distribution) to describe the evolution of the errors x. Gauss goes on to assign a uniform prior to the parameters and to obtain a normal posterior distribution of the parameters. In doing this he gives the first account of Bayesian estimation in least squares. As Büuhler (1981, p. 140) observes "Least squares were Gauss's indispensable theoretical tool in experimental research; increasingly, he came to see it as the most important witness to the connection between mathematics and nature". Based on this view, Gauss was ahead of his time and had the capacity of a modern applied mathematician and statistician.

After Gauss several scientists contributed on the development of least squares, such as Pierre-Simon Laplace (on least squares computation), George Udny Yule (on relating least squares to correlation), Sir Francis Galton and Karl Pearson (on regression and correlation); reviews of the developments of least squares can be found in Stigler (1986) and in Aldrich (1998). Although the least squares discovery was the first mathematical attempt to build an optimal model based on observed data, regression remained moderately explored in the nineteenth century and in the beginning of the twentieth century; Robin Lewis Plackett in 1950 rediscovered least squares and established regression analysis as we know it now (Plackett, 1950, 1991). Plackett, ahead of his time, derived a recursive estimation approach, known as recursive residuals, which enables the recursive calculation of the least squares solution. Plackett's path-breaking work had two important implications: on the applied side large data sets could be handled with notable flexibility (by performing a small number of recursive calculations), and on the theoretical side his work set a framework for the derivation of more complex estimation procedures.

The Filtering Problem: Kolmogorov and Wiener

The modern axiomatic treatment of probability theory is attributed to Russian mathematician Andrey Nikolaevich Kolmogorov (1903–1987). Kolmogorov's research interests span a wide range of mathematical topics, including probability theory, topology, intuitionistic logic, turbulence, classical mechanics and computational complexity. Kolmogorov worked on stochastic processes and in 1941 he gave a solution to the problem of optimal estimation of a discrete-time stationary stochastic process (Kolmogorov, 1941).

At the same time independently of Kolmogorov, Norbert Wiener (1894–1964) was working on the equivalent problem for continuous-time stationary stochastic processes. In 1942 he completed his work on smoothing stationary time series, known today as Wiener filter, but his work was only published post-war in 1949 (Wiener, 1949). In 1942 a classified report with Wiener's method appeared with the nickname "the yellow peril" because of the colour of the cover and the difficulty of the subject. Wiener derived the solution of the least squares errors in a continuous-time stationary process as a function of the autocorrelation functions of the signal and the noise. Together with Kolmogorov, Wiener is credited with introducing the term "filtering" in time series or stochastic processes. Filtering is used for the operation of removing noise from a process inflated by noise. Today, the term "denoising" is also used, for such a filtering operation.

The Kalman Filter: Kalman

Rudolf Emil Kalman was born in 19 May 1930 in Budapest. Kalman received his bachelor's and master's degrees from MIT in electrical engineering, in 1953 and 1954, respectively. In 1955 Kalman obtained a position at Columbia University as a graduate student and lecturer. He then developed strong research interests in systems theory and systems dynamics. He was interested in algebraic methods and he is credited for finding the algebraic expression of observability and controllability of a dynamical system, see e.g. Grewal and Andrews (2015).

Although the state space form of a time-varying system was known since 1956 (Halcombe Laning & Battin, 1956), it was not connected with the filtering problem mentioned above; see also Zadeh and Desoer (1963). In 1958 Kalman had the idea to use state variables and the linear state space model for the Wiener filtering problem. Although the Wiener filter is applied to continuous-time processes, Kalman first considered discrete-time processes. In order to be able to derive the equivalent to the Wiener filter using the state space form in discrete time, Kalman equated expectation with projection, after he read Loève's book on probability theory (Loève, 1955). The use of projections has been a central element in the derivation of the new filter, known as the Kalman filter, which was published in a mechanical engineering journal in 1960 (Kalman, 1960). Kalman's new filter had two important elements: (a) it used the state state space form (which is regarded today as a very advantageous consideration as it can describe a large number of physical phenomena) and (b) it did not require the assumption of stationarity (which is significant as most real-life systems are not stationary). Just as Plackett's recursive estimation was ahead of his time, so was Kalman's method, as it was able to deal with non-stationary processes.

Kalman considers the state space model (1.3) where the states β_t represent some physical entities of the system (usually they represent an ideal theoretic state of the system—sometimes driven by differential or difference equations— while the observations y_t represent noisy versions of the states, which are subject to measurement error). Then Kalman provides an optimal filter, provided that the design vector x_t and the transition matrix $\mathbf{F}_t$ as well as the variance of ϵ_t and the covariance matrix of ζ_t are known and specified by the user. Kalman suggested that these model components may be known by the system or by past experiments and he did not study how they may be specified or estimated from the data. Later on when the Kalman filter was applied in economics (see the discussion below), the states were considered as unobserved or hidden components, not necessarily assigned to physical entities, which assist on understanding the generating process of the observations. In this approach the problem of specifying or estimating the components x_t and $\mathbf{F}_t$ as well as the variance and covariance components mentioned above, becomes more crucial as there is weak or no information provided by the physical/socio-economic system (other than the observed data) to help the identification of these components. This difference between Kalman's state space modelling framework and later applications of the Kalman filter often is not made clear. In cases where the model components x_t and $\mathbf{F}_t$ are unknown and subject to estimation, the Kalman filter is not optimal in the sense that Kalman has proved, because the additional estimation of such components introduce certain limitations and biases. On one hand this has led in certain cases to misuse of the Kalman filter, and in other cases it has motivated the development and extension of Kalman-type algorithms, such as the expectation maximisation algorithm or sequential Monte Carlo methods. Both of these approaches are discussed in detail in Chaps. 4 and 5 in this book.

Kalman's method was not received without controversy. He was not able to publish his seminal paper in an electrical or systems engineering journal, and people who attended his talks found it hard to realise the potential of the Kalman filter. However, Stanley F. Schmidt from the Ames Research Centre of NASA was able to see its importance when Kalman explained it to him in a visit in 1960. Schmidt was the first to implement the Kalman filter for the trajectory estimation and control problems for the Apollo project. He then soon discovered what is now known as the "extended Kalman filter", which is basically a linearisation of observation and state functions to deal with non-linearities and is discussed in Chap. 5. In the 1960s apart from the work of Schmidt and his colleagues, several people were involved in application and developments related to the Kalman filter, most notably Kalman and Bucy (1961), Raunch et al. (1965), Schweppe (1965) and Young (1968, 1969). Since then the Kalman filter and its generalisations have been applied widely in science and in real-life applications, e.g. in navigation systems, in digital filters, systems engineering, time series forecasting and control.

Kalman is currently emeritus professor in three universities. Kalman received the 2008 Charles Stark Draper Prize "for the development and dissemination of the optimal digital technique (known as the Kalman Filter) that is pervasively used to control a vast array of consumer, health, commercial and defence products".

Bayesian Forecasting: Jones, Harrison and Stevens

Although the state space model has been known to engineers since 1956 (Halcombe Laning & Battin, 1956), statisticians and economists did not seem to exploit it. John Fraser Muth (1930–2005), an eminent economist known as the father of the rational expectations revolution in economics, introduced in 1960 the local level model (see Sect. 1.2 above), and gave its relationship with the popular exponentially weighted moving average forecasting procedure Muth (1960). In the 60s the closest effort of statistical filtering of the state space model was the attempt of Richard H. Jones in 1966 (Jones, 1966). Jones introduced Kalman filtering to the statistics and econometrics scientific communities and he applied it in order to estimate the parameters of Muth's exponential smoothing model. He is also the first to provide a derivation of the Kalman filter based on conditional Gaussian distribution theory as opposed to projections adopted by Kalman. Jeff P. Harrison almost independently was arriving at the same line of research, with some preliminary results in his exponential smoothing 1967 paper (Harrison, 1967), which was generalised later by Godolphin and Harrison (1975). However, these studies did not realise their potential, perhaps, due to the technological and computational limitations of the 1960s.

Jeff Harrison met C. F. Stevens at Imperial Chemical Industries (ICI) where they worked together on a series of papers. They first observed that many different types of time series (especially non-stationary time series, exhibiting trend and seasonal variation) could be described by state variables and indeed by state space models. This important observation led to the adoption of the Kalman filter (as this was developed for non-stationary processes) and had a significant advantage over competitive time series methods, based upon the assumption of stationarity, e.g. Box-Jenkins methods (Box et al., 2008). In 1971 Harrison and Stevens published their first account of state space modelling for time series data (Harrison & Stevens, 1971). They adopted a derivation of the Kalman filter similar to that of Jones, but made the important observation that the state variables could incorporate many familiar time series models; this came under the title *Bayesian forecasting*, because the Bayes theorem was used to provide the forecast distribution. This new approach offered an array of new possibilities utilising Bayesian statistics, which at the time was undergoing rapid development. Some of these possibilities were explored in their seminal paper in 1976 (Harrison & Stevens, 1976) which was read before the Royal Statistical Society. It is worth mentioning that in the early 70s several economists used the Kalman filter for inventory forecasting and control (see the introduction of Morrison and Pike (1977) for references). In particular, in 1977, independently of Harrison and Stevens, Morrison and Pike (1977) rediscovered the Kalman filter as a general estimation procedure of non-stationary time series forecasting.

In the 80s many scientists realised the potential of state space modelling using the Kalman filter. Among many authors, A.C. Harvey and his co-authors published articles in 1980 and 1981 that demonstrate state space models and the Kalman filter (Harvey et al., 1980; Harvey, 1981). In 1984 Harvey showed how the Kalman filter can provide a unified estimation methodology for stationary and non-stationary time series (Harvey, 1984); he was one of the first to point out the correspondence of state space models and the Box-Jenkins autoregressive moving average models. Soon after in 1989 Harvey published his book on structural models and the Kalman filter (Harvey, 1989), which is known to have educated generations of statisticians, since 1989. The growth of state space models is reflected by textbooks and monographs, such as Anderson and Moore (1979), West and Harrison (1997) [first published in 1989] and later Grewal and Andrews (2015) and Durbin and Koopman (2012).

1.5 Layout of the Book

The aim of the book is to introduce state space models and the Kalman filter and to illustrate its wide application in science and commerce. The book aims to present a concise yet self-contained treatment of state space models with applications in many different disciplines. Throughout the book we adopt the Bayesian paradigm for estimation and forecasting, as this is judged to be the modern statistical approach which benefits from the high-performance computational power. However, in places we discuss frequentist-based inference, such as least squares and maximum likelihood estimation. The layout of the book is as follows.

Chapter 2 provides the mathematical background related to Kalman filter, including matrix algebra, analysis, probability and statistics. In this chapter we summarise the main results which are necessary for the reader to develop the technical arguments that follow in Chap. 3.

Chapter 3 introduces the state space model and develops filtering and smoothing estimation methodologies, including the Kalman filter, as well as forecasting. We motivate the state space model by considering it as an extension of the linear regression model. The chapter describes a package of the programming language for statistical computing R, which is used throughout the book. Two proofs of the Kalman filter are given, each of which provides different insight into the structure of the filter. Throughout the chapter examples illustrate the applicability of the estimation algorithms.

Chapter 4 discusses further topics on the implementation of linear state space models. We start by discussing an array of useful state space forms of real-life situations, such as time series comprising trend, seasonal, autoregressive and time-varying regression components. These are discussed with particular data in mind and the Kalman filter is applied to provide forecasting solutions. Then maximum likelihood and related concepts are described to provide a way of estimating or specifying parameters of the models. Error analysis and diagnostic checks as well as prior specification are considered in some detail. The chapter concludes by discussing model monitoring and intervention analysis.

Chapters 5 and 6 discuss models that go beyond the Kalman filter. The aim of these chapters is to give an account of modern statistical methodologies for state space models and therefore to illustrate recent trends of research in this area. These include covariance estimation of multivariate state space models, non-linear and non-Gaussian state space models, and sequential Monte Carlo estimation methods.

Chapters 7 and 8 discuss the application of state space methods in finance and in dynamical systems. While finance and economics paid particular attention to the Kalman filter, as noted earlier, systems engineering is where the Kalman filter was discovered. As a result these major areas of application of the Kalman filter and related methods are considered in Chaps. 7 and 8. The selection of illustrative examples presented in those two chapters aim to showcase the applicability and contribution of the state space models to these areas.

Chapter 2
Matrix Algebra, Probability and Statistics

This chapter offers notation and the necessary background for the development of state space methods that follows. The background lies on three basic areas, matrix algebra, probability and statistics. We start with matrix algebra in Sect. 2.1; we develop in some detail vector and matrix differentiation, while other elements of more standard matrix algebra are just mentioned. Section 2.3 discusses probability and distribution theory relevant to the needs of the book. This section aims to establish notation and to remind to the reader the notion of distribution theory relevant to filtering and state space modelling. Several examples of discrete and continuous distributions, which will be used in later chapters, are presented. Section 2.4 discusses the ideas behind statistics necessary for the chapters that follow. The maximum likelihood principle is introduced and the expectation maximisation (EM) algorithm is discussed in detail. The chapter concludes with an introduction to Bayesian inference.

2.1 Vectors, Matrices and Basic Operations

We assume that the reader is familiar with calculus of one and several variables, with matrix algebra and with basic probability and statistics. For detailed coverage of calculous of one and several variables the reader is referred to Spivak (1995) and to Lang (1987), respectively. Matrix algebra and matrix analysis are discussed in Magnus and Neudecker (1988), Harville (1997) and Horn and Johnson (2013). Below we establish some basic notation and highlight a few important concepts. We assume that the reader is familiar with matrix addition and multiplication, inverse of a matrix, symmetric and positive definite matrices, trace of a square matrix and the determinant of a square matrix.

Scalars and vectors are denoted by small letters, e.g. a, b, c, while matrices are denoted by boldface capital letters, e.g. $\mathbf{A}, \mathbf{B}, \mathbf{C}$. Sets are denoted by capital letters,

© The Author(s), under exclusive license to Springer Nature Switzerland AG 2021
K. Triantafyllopoulos, *Bayesian Inference of State Space Models*, Springer Texts in Statistics, https://doi.org/10.1007/978-3-030-76124-0_2

e.g. $A = \{a, b, c\}$ and $\mathbb{R}$ denotes the set of real numbers. An $m \times n$ matrix $\mathbf{A}$ is usually denoted by $\mathbf{A} = (a_{ij})_{i=1,...,m;\, j=1,...,n}$, where a_{ij} is the ij-th element of $\mathbf{A}$, m is the number of rows and n is the number of columns; sometimes we simply write $\mathbf{A} = (a_{ij})$, if the range of i, j is implied. If $m = n$ $\mathbf{A}$ is usually referred to as a square matrix. Unless otherwise stated, we will assume that $\mathbf{A}$ is a matrix with elements from the real field. A matrix with zero elements ($a_{ij} = 0$, for all i, j) is referred to as the zero-matrix and is denoted by $\mathbf{0}$; a square matrix with $a_{ii} = 1$ and $a_{ij} = 0$, for $i \neq j$ and $m = n$ is known as the identity matrix and is denoted by $\mathbf{I}$. A diagonal matrix is a matrix with diagonal elements $a_{11}, \ldots, a_{nn}$ and $a_{ij} = 0$, for $i \neq j$; usually we write $\mathbf{A} = \mathrm{diag}(a_{11}, \ldots, a_{nn})$ or $\mathbf{A} = \mathrm{diag}(a)$, where a is a column vector $a = [a_{11}, \ldots, a_{nn}]^\top$ and sometimes we may write $\mathrm{diag}(\mathbf{A})$ to denote a diagonal matrix with diagonal elements the diagonal elements of a matrix $\mathbf{A}$. A symmetric matrix $\mathbf{A} = (a_{ij})_{i,j=1,2,...,n}$ is a $n \times n$ square matrix with $a_{ij} = a_{ji}$, for all $i, j = 1, 2, \ldots, n$. A non-negative definite matrix is a matrix $\mathbf{A}$ with $x^\top \mathbf{A} x \geq 0$, for any vector x. If $x^\top \mathbf{A} x > 0$, for all non-zero vectors x, then $\mathbf{A}$ is called positive definite. In statistics symmetric and positive definite matrices play a significant role as they represent the variance and covariance structure of random vectors (see Sect. 2.3).

The inverse of an $n \times n$ non-singular matrix $\mathbf{A}$ is denoted by $\mathbf{A}^{-1}$, the trace of an $n \times n$ matrix $\mathbf{A}$ is denoted by $\mathrm{trace}(\mathbf{A})$, the determinant of $\mathbf{A}$ is demoted by $|\mathbf{A}|$. In the sections below we discuss in some detail vector and matrix differentiation, optimisation and limit of matrices. Below we list some of the properties of matrix operations.

1. **Integer powers of a matrix.** Consider a square matrix $\mathbf{A}$. The power of order k of $\mathbf{A}$, denoted as $\mathbf{A}^k$, is defined to be the product

$$\mathbf{A}^k = \underbrace{\mathbf{A}\mathbf{A}\cdots\mathbf{A}}_{k-\text{times}} = \mathbf{A}^{k-1}\mathbf{A} = \mathbf{A}\mathbf{A}^{k-1},$$

   for some integer $k > 0$, while for $k = 0$ we define $\mathbf{A}^0 = \mathbf{I}$.
2. **Determinant of a matrix.** Associated with any $n \times n$ matrix $\mathbf{A}$ there is a scaler, known as the determinant of $\mathbf{A}$ and denoted by $|\mathbf{A}|$. For a full definition of the determinant the reader is referred to Harville (1997, page 177). Here we just mention two of the properties of determinants.

   a. For any constant c, it is $|c\mathbf{A}| = c^n|\mathbf{A}|$;
   b. For two $n \times n$ matrices $\mathbf{A}$ and $\mathbf{B}$, it is $|\mathbf{AB}| = |\mathbf{A}||\mathbf{B}|$ (in words: the determinant of the product of two matrices is equal to the product of the determinant of the two matrices).

3. **Trace of a matrix.** Associated with any square matrix $\mathbf{A} = (a_{ij})$ is a scalar, called the trace of $\mathbf{A}$, which is denoted by $\mathrm{trace}(\mathbf{A})$ and defined as

$$\mathrm{trace}(\mathbf{A}) = a_{11} + a_{22} + \cdots + a_{nn} = \sum_{i=1}^{n} a_{ii}.$$

We give two basic properties of the trace of a matrix

a. For any $n \times n$ matrices $\mathbf{A} = (a_{ij})$ and $\mathbf{B} = (b_{ij})$, we have

$$\text{trace}(\mathbf{A} + \mathbf{B}) = \text{trace}(\mathbf{A}) + \text{trace}(\mathbf{B}).$$

b. For two $n \times n$ matrices $\mathbf{A}$ and $\mathbf{B}$ we have

$$\text{trace}(\mathbf{AB}) = \text{trace}(\mathbf{BA}).$$

4. **The vec, vech and the Kronecker product.**
For an $m \times n$ matrix $\mathbf{A}$ the vec operator rearranges the elements of $\mathbf{A}$ into a vector by stacking the columns of $\mathbf{A}$ one after the other. For example, if $\mathbf{A} = (a_{ij}) = [a_1, a_2, \ldots, a_n]$, where $a_i = [a_{1i}, a_{2i}, \ldots, a_{mi}]^\top$ represents the i-th column of $\mathbf{A}$, for $i = 1, 2, , \ldots, n$, then

$$\text{vec}(\mathbf{A}) = \begin{bmatrix} a_1 \\ a_2 \\ \vdots \\ a_n \end{bmatrix}$$

which is an $mm \times 1$ vector.
The Kronecker product of an $m \times n$ matrix $\mathbf{A}$ and a $p \times q$ matrix $\mathbf{B}$ is the $mp \times nq$ matrix defined by

$$\mathbf{A} \otimes \mathbf{B} = (a_{ij}\mathbf{B}).$$

For example, for $n = m = p = q = 2$ we have

$$\mathbf{A} \otimes \mathbf{B} = \begin{bmatrix} a_{11}\mathbf{B} & a_{12}\mathbf{B} \\ a_{21}\mathbf{B} & a_{22}\mathbf{B} \end{bmatrix} = \begin{bmatrix} a_{11}b_{11} & a_{11}b_{12} & a_{12}b_{11} & a_{12}b_{12} \\ a_{11}b_{21} & a_{11}b_{22} & a_{12}b_{21} & a_{12}b_{22} \\ a_{21}b_{11} & a_{21}b_{12} & a_{22}b_{11} & a_{22}b_{12} \\ a_{21}b_{21} & a_{21}b_{22} & a_{22}b_{21} & a_{22}b_{22} \end{bmatrix}.$$

A few properties of the vec and Kronecker operations are listed below

a. If $\mathbf{A}$, $\mathbf{B}$ and $\mathbf{C}$ are matrices so that the product $\mathbf{ABC}$ is defined, then

$$\text{vec}(\mathbf{ABC}) = (\mathbf{C}^\top \otimes \mathbf{A})\text{vec}(\mathbf{B}).$$

When matrix $\mathbf{A}$ is symmetric, then since $a_{ij} = a_{ji}$ not all elements of the vector $\text{vec}(\mathbf{A})$ are distinct. In such a case we can obtain a vectored rearrangement of

A by stacking columns of **A** with distinct elements only and this is denoted as vetch(**A**). For example, for $m = n = 2$ we have

$$\text{vech} \begin{bmatrix} a_{11} & a_{12} \\ a_{12} & a_{22} \end{bmatrix} = \begin{bmatrix} a_{11} \\ a_{12} \\ a_{22} \end{bmatrix}.$$

2.2   Vector and Matrix Differentiation

2.2.1   Background and Notation

In this section we describe the notion of vector and matrix partial derivatives, necessary for the development of this book. We will restrict our attention to two cases (1) partial derivatives of scalar function of a vector x and (2) partial derivatives of scalar function of a symmetric matrix **X**. We denote with $\mathbb{R}^n$ the cartesian product $\mathbb{R}^n = \mathbb{R} \times \mathbb{R} \times \cdots \times \mathbb{R}$, so that $\mathbb{R}^n$ contains vectors $x = [x_1, x_2, \ldots, x_n]^\top$, with $x_i \in \mathbb{R}$, for $i = 1, 2, \ldots, n$. For a more detailed treatment of matrix calculus the reader is referred to Magnus and Neudecker (1988) and Harville (1997).

1. It is assumed that a function $f(\cdot)$ of the vector x is continuously differentiable at an interior vector point c in the domain D (a subset of $\mathbb{R}^n$); for a precise definition of differentiability the reader is referred to Magnus and Neudecker (1988) and Harville (1997). We shall write $f(x)$ to denote the value of $f(\cdot)$ at point x and usually it will be assumed that $D = \mathbb{R}^n$. The partial derivative of $f(x)$ (defined at an interior point of a subset of D) with respect to x_i, denoted by $\partial f(x)/\partial x_i$, is the scalar derivative of $f(x) = f(x_1, \ldots, x_i, \ldots, x_n)$ when this is viewed as a function of x_i alone and all $x_j \neq x_i$ are viewed as constants, where $x = [x_1, \ldots, x_n]^\top$. The partial derivative of $f(x)$ with respect to x, denoted by $\partial f(x)/\partial x$ is defined to be the vector with elements the respective partial derivatives $\partial f(x)/\partial x_i$, i.e.

$$\frac{\partial f(x)}{\partial x} = \begin{bmatrix} \partial f(x)/\partial x_1 \\ \partial f(x)/\partial x_2 \\ \vdots \\ \partial f(x)/\partial x_n \end{bmatrix}.$$

Some special derivatives used in this book are covered in Sect. 2.2.2.

In the same way we can define second and higher order partial derivatives of $f(x)$. The second partial derivative of $f(x)$, denoted as $\partial^2 f(x)/\partial x \partial x^\top$ is the transpose of the first partial derivative of $\partial f(x)/\partial x$, i.e.

$$\frac{\partial^2 f(x)}{\partial x \partial x^\top} = \begin{bmatrix} \partial^2 f(x)/\partial x_1^2 & \partial^2 f(x)/\partial x_1 \partial x_2 & \cdots & \partial^2 f(x)/\partial x_1 \partial x_n \\ \partial^2 f(x)/\partial x_2 \partial x_1 & \partial^2 f(x)/\partial x_2^2 & \cdots & \partial^2 f(x)/\partial x_2 \partial x_n \\ \vdots & \vdots & \ddots & \vdots \\ \partial^2 f(x)/\partial x_n \partial x_1 & \partial^2 f(x)/\partial x_n \partial x_2 & \cdots & \partial^2 f(x)/\partial x_n^2 \end{bmatrix},$$

assuming that all $\partial^2 f(x)/\partial x_i \partial x_j$ are defined, for $i, j = 1, \ldots, n$.

2. For the purposes of partial derivatives of a function of matrices, we consider $f(\cdot)$ to be a continuously differentiable function on an $n \times n$ matrix $\mathbf{X}$ and we write $f(\mathbf{X})$ for the value of $f(\cdot)$ at $\mathbf{X}$; we note that the value of $f(\mathbf{X})$ is a real number. The domain of such a function is a subset D of all matrices with real elements, usually denoted by $\mathbb{R}^{n \times n}$. This is basically a special case of (1) as $\mathbf{X}$ can be rearranged as a long vector (by stacking all column vectors of $\mathbf{X}$). However, for the purposes of matrix calculation and convenience in the presentation we retain the matrix notation. Since $\mathbf{X}$ has n^2 elements $f(\mathbf{X})$ can be viewed as a function of n^2 variables. If, however, some of the elements of $\mathbf{X}$ are not distinct, e.g. when $\mathbf{X}$ is a symmetric matrix, then we need to consider only the distinct elements in the differentiation of $f(\mathbf{X})$.

Writing $\mathbf{X} = (x_{ij})$ for an unrestricted matrix of variables, we define the partial derivative matrix of $f(\mathbf{X})$ as the matrix with ij-th element the partial derivative of $f(\mathbf{X})$ with respect to x_{ij}, defined at an interior point of D, i.e.

$$\frac{\partial f(\mathbf{X})}{\partial \mathbf{X}} = \begin{bmatrix} \partial f(\mathbf{X})/\partial x_{11} & \partial f(\mathbf{X})/\partial x_{12} & \cdots & \partial f(\mathbf{X})/\partial x_{1n} \\ \partial f(\mathbf{X})/\partial x_{21} & \partial f(\mathbf{X})/\partial x_{22} & \cdots & \partial f(\mathbf{X})/\partial x_{2n} \\ \vdots & \vdots & \ddots & \vdots \\ \partial f(\mathbf{X})/\partial x_{n1} & \partial f(\mathbf{X})/\partial x_{n2} & \cdots & \partial f(\mathbf{X})/\partial x_{nn} \end{bmatrix}.$$

If $\mathbf{X}$ is symmetric, this logic does not work, since there are only $n(n + 1)/2$ distinct elements in $\mathbf{X}$ (out of the possible n^2). In such a case $\mathbf{X}$ is rearranged into a $[n(n + 1)/2] \times 1$ column vector $x = \text{vech}(\mathbf{X})$, as discussed in the previous section. Now the partial derivative of $f(\mathbf{X})$ can be formed as the vector partial derivative with respect to x, which then is rearranged as a symmetric matrix to form $\partial f(\mathbf{X})/\partial \mathbf{X}$.

Dealing with case (2) we will also need to define a matrix-valued function $\mathbf{F}(\cdot) = (f_{kl})_{k=1,2,\ldots,q;l=1,2,\ldots,s}$ defined on a matrix of variables $\mathbf{X} = (x_{ij})$ (unrestricted or restricted). In such a case we write $\mathbf{F}(\mathbf{X})$ for the value of $\mathbf{F}(\cdot)$ on matrix $\mathbf{X}$; the domain D of $\mathbf{F}(\cdot)$ is all $n \times n$ real-valued matrices (if $\mathbf{X}$ is unrestricted) or D can be defined as above for symmetric matrices. Each constituent function $f_{kl}(\mathbf{X})$ of the matrix $\mathbf{F}(\mathbf{X})$ is a scalar-valued function on the matrix $\mathbf{X}$ and it is assumed to be continuously differentiable at an interior point of D. Then the partial derivative of $\mathbf{F}(\mathbf{X})$ with respect to x_{ij} is defined to be the matrix

$$\frac{\partial \mathbf{F}(\mathbf{X})}{\partial x_{ij}} = \begin{bmatrix} \partial f_{11}(\mathbf{X})/\partial x_{ij} & \partial f_{12}(\mathbf{X})/\partial x_{ij} & \cdots & \partial f_{1s}(\mathbf{X})/\partial x_{ij} \\ \partial f_{21}(\mathbf{X})/\partial x_{ij} & \partial f_{22}(\mathbf{X})/\partial x_{ij} & \cdots & \partial f_{2s}(\mathbf{X})/\partial x_{ij} \\ \vdots & \vdots & \ddots & \vdots \\ \partial f_{q1}(\mathbf{X})/\partial x_{ij} & \partial f_{q2}(\mathbf{X})/\partial x_{ij} & \cdots & \partial f_{qs}(\mathbf{X})/\partial x_{ij} \end{bmatrix}.$$

2.2.2 Differentiation of Linear and Quadratic Forms

We consider here the partial derivatives of a linear form

$$f(x) = a^\top x = \sum_{i=1}^{n} a_i x_i$$

and a quadratic form

$$g(x) = x^\top \mathbf{A} x = \sum_{i=1}^{n} \sum_{k=1}^{n} a_{ik} x_i x_k,$$

where $a = [a_1, a_2, \ldots, a_n]^\top$ is a vector of constants, $\mathbf{A} = (a_{ij})$ is an $n \times n$ matrix of constants and $x = [x_1, x_2, \ldots, x_n]^\top$.

The partial derivative of $f(x)$ with respect to x_j is

$$\frac{\partial f(x)}{\partial x_j} = \frac{\partial}{\partial x_j} \sum_{i=1}^{n} a_i x_i = \sum_{i=1}^{n} a_i \frac{\partial x_i}{\partial x_j} = a_j, \tag{2.1}$$

since

$$\frac{\partial x_i}{\partial x_j} = \begin{cases} 1 & \text{if } i = j \\ 0 & \text{if } i \neq j \end{cases}.$$

Recasting (2.1) in matrix form, we obtain

$$\frac{\partial (a^\top x)}{\partial x} = a. \tag{2.2}$$

Moving on now to the quadratic form $g(x)$ we have

$$
\begin{aligned}
\frac{\partial g(x)}{\partial x_j} &= \frac{\partial}{\partial x_j} \sum_{i=1}^{n} \sum_{k=1}^{n} a_{ik} x_i x_k \\
&= \frac{\partial}{\partial x_j} \left(\sum_{i=1}^{n} a_{i1} x_i x_1 + \cdots + \sum_{i=1}^{n} a_{i,j-1} x_i x_{j-1} + \sum_{i=1}^{n} a_{i,j+1} x_i x_{j+1} \right. \\
&\qquad \left. + \cdots + \sum_{i=1}^{n} a_{in} x_i x_n + \sum_{i \neq j} a_{ij} x_i x_j + a_{jj} x_j^2 \right) \\
&= a_{j1} x_1 + \ldots a_{j,j-1} x_{j-1} + a_{j,j+1} x_{j+1} + \cdots + a_{jn} x_n \\
&\qquad + \sum_{i \neq j} a_{ij} x_i x_j + 2 a_{jj} x_j \\
&= \sum_{k=1}^{n} a_{jk} x_k + \sum_{i=1}^{n} a_{ij} x_i .
\end{aligned}
$$

Recasting this in matrix form we obtain

$$
\frac{\partial (x^{\top} \mathbf{A} x)}{\partial x} = (\mathbf{A} + \mathbf{A}^{\top}) x, \tag{2.3}
$$

since the j-th element of the vector $\mathbf{A}x$ is $\sum_{k=1}^{n} a_{jk} x_k$ and the j-th element of the vector $\mathbf{A}^{\top} x$ is $\sum_{i=1}^{n} a_{ij} x_i$.

Next we compute the second partial derivative matrix of the quadratic form $g(x)$. As mentioned in the previous section, this will be the $n \times n$ matrix with ij-th element $\partial^2 g(x) / \partial x_l x_j$, for $l, j = 1, \ldots, n$.

We have

$$
\begin{aligned}
\frac{\partial^2 (x^{\top} \mathbf{A} x)}{\partial x_l \partial x_j} &= \frac{\partial}{\partial x_j} \left[\frac{\partial (x^{\top} \mathbf{A} x)}{\partial x_l} \right] \\
&= \frac{\partial}{\partial x_j} \left(\sum_{i=1}^{n} a_{il} x_i + \sum_{k=1}^{n} a_{lk} x_k \right) \\
&= \sum_{i=1}^{n} a_{il} \frac{\partial x_i}{\partial x_j} + \sum_{k=1}^{n} a_{lk} \frac{\partial x_k}{\partial x_j} = a_{jl} + a_{lj}
\end{aligned}
$$

and recasting this in matrix form we obtain

$$
\frac{\partial^2 (x^{\top} \mathbf{A} x)}{\partial x \partial x^{\top}} = \mathbf{A} + \mathbf{A}^{\top} . \tag{2.4}
$$

If $\mathbf{A}$ is a symmetric matrix, then the partial derivatives (2.3) and (2.4) simplify to

$$\frac{\partial(x^{\top}\mathbf{A}x)}{\partial x} = 2\mathbf{A}x \quad \text{and} \quad \frac{\partial^2(x^{\top}\mathbf{A}x)}{\partial x \partial x^{\top}} = 2\mathbf{A}. \tag{2.5}$$

This equation is applied in Sect. 3.1.1 in order to derive the least squares solution in regression.

2.2.3 Differentiation of Determinant and Trace

Below we provide formulae for the partial derivatives of the logarithm of the determinant of a symmetric matrix and of the trace; both of these are used in the maximisation step of the EM algorithm in Sect. 4.3.1.

Differentiation of the Logarithm of the Determinant Consider $\mathbf{X}$ to be an $n \times n$ non-singular symmetric matrix and $|\mathbf{X}|$ to demote the determinant of $\mathbf{X}$. First we derive a formula for $\partial \log |\mathbf{X}|/\partial \mathbf{X}$.

Very closely related to the determinant is the cofactor matrix. Let $\mathbf{X}_{ij}$ be an $(n-1) \times (n-1)$ submatrix of $\mathbf{X}$ (a submatrix of $\mathbf{X}$ is a matrix whose elements are also elements of $\mathbf{X}$) obtained by striking out the row and column that contain the element x_{ij}, i.e. the i-th row and the j-th column. The signed determinant $(-1)^{i+j}|\mathbf{X}_{ij}|$ is called the cofactor of x_{ij} and is denoted by ξ_{ij}. A fundamental property of the cofactors is that

$$|\mathbf{X}| = x_{i1}\xi_{i1} + x_{i2}\xi_{i2} + \cdots + x_{in}\xi_{in} = \sum_{j=1}^{n} x_{ij}\xi_{ij} \tag{2.6}$$

and this can be used to compute the determinant of a matrix.

The $n \times n$ matrix consisting of elements ξ_{ji} in its ij-th element (i.e. the transpose of the matrix whose ij-th element is the cofactor ξ_{ij} is called the *adjoint* matrix and is denoted by adj($\mathbf{X}$). A fundamental property of adj($\mathbf{X}$) relates this matrix to the inverse matrix $\mathbf{X}^{-1}$ and it is given below

$$\mathbf{X}^{-1} = \frac{1}{|\mathbf{X}|}\text{adj}(\mathbf{X}). \tag{2.7}$$

For the proof of this result the reader is referred to Harville (1997, page 192).

From Eq. (2.6) we have that

$$\frac{\partial |\mathbf{X}|}{\partial x_{ij}} = \sum_{k=1}^{n} \xi_{ik}\frac{\partial x_{ik}}{\partial x_{ij}}$$

and since $\mathbf{X}$ is symmetric, so is the adjoint matrix $\text{adj}(\mathbf{X}) = (\xi_{ij})$. Hence from Eq. (2.7) it is

$$\frac{\partial |\mathbf{X}|}{\partial x_{ij}} = \text{trace} \left[\text{adj}(\mathbf{X}) \frac{\partial \mathbf{X}}{\partial x_{ij}} \right] = \text{trace} \left(|\mathbf{X}|\mathbf{X}^{-1} \frac{\partial \mathbf{X}}{\partial x_{ij}} \right) = |\mathbf{X}|\text{trace} \left(\mathbf{X}^{-1} \frac{\partial \mathbf{X}}{\partial x_{ij}} \right).$$

Define u_i to be the $n \times 1$ vector with a unit in the i-th position and elsewhere zeros, i.e.

$$u_i = [0, \ldots, 0, 1, 0, \ldots, 0]^\top.$$

We can see that since $\mathbf{X}$ is symmetric (i.e. $x_{ij} = x_{ji}$), then

$$\frac{\partial \mathbf{X}}{\partial x_{ij}} = \begin{cases} u_i u_i^\top, & i = j \\ u_i u_j^\top + u_j u_i^\top, & j < i \end{cases}, \tag{2.8}$$

i.e. a matrix having a unit at the ii-th position and elsewhere zero, if $i = j$ and a matrix having two units at positions ij and ji and elsewhere zeros, if $j < i$.

Now using the chain rule of differentiation for scalar functions

$$\begin{aligned} \frac{\partial \log |\mathbf{X}|}{\partial x_{ij}} &= \frac{1}{|\mathbf{X}|} \frac{\partial |\mathbf{X}|}{\partial x_{ij}} = \text{trace} \left(\mathbf{X}^{-1} \frac{\partial \mathbf{X}}{\partial x_{ij}} \right) \\ &= \begin{cases} \text{trace} \left(\mathbf{X}^{-1} u_i u_i^\top \right), & i = j \\ \text{trace} \left(\mathbf{X}^{-1} u_i u_j^\top \right) + \text{trace} \left(\mathbf{X}^{-1} u_j u_i^\top \right), & j < i \end{cases} \\ &= \begin{cases} u_i^\top \mathbf{X}^{-1} u_i, & i = j \\ u_j^\top \mathbf{X}^{-1} u_i + u_i^\top \mathbf{X}^{-1} u_j, & j < i \end{cases} \\ &= \begin{cases} x_{ii}^{(-1)}, & i = j \\ 2x_{ij}^{(-1)}, & j < i \end{cases}, \end{aligned}$$

where $\mathbf{X}^{-1} = (x_{ij}^{(-1)})$. Recasting this to matrix form we obtain

$$\frac{\partial \log |\mathbf{X}|}{\partial \mathbf{X}} = 2\mathbf{X}^{-1} - \text{diag}(\mathbf{X}^{-1}), \tag{2.9}$$

where $\text{diag}(\mathbf{X}^{-1})$ denotes a diagonal matrix with diagonal elements $x_{11}^{(-1)}$, $x_{22}^{(-1)}, \ldots, x_{nn}^{(-1)}$.

Differentiation of the Trace In this part we aim to provide a formula for $\partial \text{trace}(\mathbf{A}\mathbf{X}^{-1})/\partial \mathbf{X}$, where $\mathbf{A}$ is an $n \times n$ matrix of constants and $\mathbf{X} = (x_{ij})$ is as before an $n \times n$ non-singular symmetric matrix of variables.

Consider $\mathbf{F}(\mathbf{X})$ a matrix-valued function of $\mathbf{X}$. Then for each x_{ij} element of $\mathbf{X}$ it is

$$\frac{\partial \text{trace}[\mathbf{F}(\mathbf{X})]}{\partial x_{ij}} = \text{trace}\left[\frac{\partial \mathbf{F}(\mathbf{X})}{\partial x_{ij}}\right]. \tag{2.10}$$

To show this, we write $f_{ij}(\mathbf{X})$ to be the ij-th element function (in $\mathbf{X}$) of $\mathbf{F}(\mathbf{X})$, i.e. $\mathbf{F}(\mathbf{X}) = (f_{ij}(\mathbf{X}))$. Then

$$\frac{\partial \text{trace}[\mathbf{F}(\mathbf{X})]}{\partial x_{ij}} = \frac{\partial f_{11}(\mathbf{X})}{\partial x_{ij}} + \frac{\partial f_{22}(\mathbf{X})}{\partial x_{ij}} + \cdots + \frac{\partial f_{nn}(\mathbf{X})}{\partial x_{ij}} = \text{trace}\left[\frac{\partial \mathbf{F}(\mathbf{X})}{\partial x_{ij}}\right],$$

as required.

Next we show that if $\mathbf{F}(\mathbf{X})$ is an $n \times n$ matrix of functions, then with the definition of $\mathbf{A} = (a_{ij})$ above, we have

$$\frac{\partial \mathbf{A}\mathbf{F}(\mathbf{X})}{\partial x_{ij}} = \mathbf{A}\frac{\partial \mathbf{F}(\mathbf{X})}{\partial x_{ij}}. \tag{2.11}$$

This follows by simply writing $\mathbf{A}\mathbf{F}(\mathbf{X}) = \left[\sum_{k=1}^{n} a_{ik} f_{kj}(\mathbf{X})\right]$ and consequently

$$\frac{\partial \mathbf{A}\mathbf{F}(\mathbf{X})}{\partial x_{ij}} = \left[\sum_{k=1}^{n} a_{ik} \frac{\partial f_{kj}(\mathbf{X})}{\partial x_{ij}}\right] = \mathbf{A}\frac{\partial \mathbf{F}(\mathbf{X})}{\partial x_{ij}}.$$

Also, for two matrix-valued functions $\mathbf{F}(\mathbf{X})$ and $\mathbf{G}(\mathbf{X})$ defined on the same domain D, so that the matrix product $\mathbf{F}(\mathbf{X})\mathbf{G}(\mathbf{X})$ is defined, we have the usual multiplicative law of differentiation

$$\frac{\partial \mathbf{F}(\mathbf{X})\mathbf{G}(\mathbf{X})}{\partial x_{ij}} = \frac{\partial \mathbf{F}(\mathbf{X})}{\partial x_{ij}}\mathbf{G}(\mathbf{X}) + \mathbf{F}(\mathbf{X})\frac{\partial \mathbf{G}(\mathbf{X})}{\partial x_{ij}}. \tag{2.12}$$

To show this first write $\mathbf{F}(\mathbf{X}) = (f_{ij}(\mathbf{X}))$ and $\mathbf{G}(\mathbf{X}) = (g_{ij}(\mathbf{X}))$ so that $\mathbf{F}(\mathbf{X})\mathbf{G}(\mathbf{X}) = \left(\sum_{k=1}^{n} f_{ik}(\mathbf{X})g_{kj}(\mathbf{X})\right)$. Then

$$\frac{\partial \mathbf{F}(\mathbf{X})\mathbf{G}(\mathbf{X})}{\partial x_{ij}} = \left(\sum_{k=1}^{n} \frac{\partial f_{ik}(\mathbf{X})}{\partial x_{ij}}g_{kj}(\mathbf{X}) + \sum_{k=1}^{n} f_{ik}(\mathbf{X})\frac{\partial g_{kj}(\mathbf{X})}{\partial x_{ij}}\right)$$

$$= \frac{\partial \mathbf{F}(\mathbf{X})}{\partial x_{ij}}\mathbf{G}(\mathbf{X}) + \mathbf{F}(\mathbf{X})\frac{\partial \mathbf{G}(\mathbf{X})}{\partial x_{ij}}.$$

Now we use result (2.12) to show

$$\frac{\partial \mathbf{X}^{-1}}{\partial x_{ij}} = -\mathbf{X}^{-1}\frac{\partial \mathbf{X}}{\partial x_{ij}}\mathbf{X}^{-1}. \tag{2.13}$$

We start by the usual equation $\mathbf{X}\mathbf{X}^{-1} = \mathbf{I}$ and take partial derivatives in both sides

$$
\frac{\partial \mathbf{X}}{\partial x_{ij}}\mathbf{X}^{-1} + \mathbf{X}\frac{\partial \mathbf{X}^{-1}}{\partial x_{ij}} = 0, \quad \text{or} \quad \frac{\partial \mathbf{X}^{-1}}{\partial x_{ij}} = -\mathbf{X}^{-1}\frac{\partial \mathbf{X}}{\partial x_{ij}}\mathbf{X}^{-1}.
$$

Finally, combining (2.10), (2.11), (2.13) together with (2.8) we obtain

$$
\begin{aligned}
\frac{\partial \operatorname{trace}(\mathbf{A}\mathbf{X}^{-1})}{\partial x_{ij}} &= \operatorname{trace}\left(\frac{\partial \mathbf{A}\mathbf{X}^{-1}}{\partial x_{ij}}\right) \\
&= \operatorname{trace}\left[\mathbf{A}\left(-\mathbf{X}^{-1}\frac{\partial \mathbf{X}}{\partial x_{ij}}\mathbf{X}^{-1}\right)\right] \\
&= \begin{cases} -\operatorname{trace}(\mathbf{A}\mathbf{X}^{-1}u_i u_i^\top \mathbf{X}^{-1}), & i = j \\ -\operatorname{trace}[\mathbf{A}\mathbf{X}^{-1}(u_i u_j^\top + u_j u_i^\top)\mathbf{X}^{-1}], & j < i \end{cases} \\
&= \begin{cases} -u_i^\top \mathbf{X}^{-1}\mathbf{A}\mathbf{X}^{-1}u_i, & i = j \\ -u_j^\top \mathbf{X}^{-1}\mathbf{A}\mathbf{X}^{-1}u_i - u_i^\top \mathbf{X}^{-1}\mathbf{A}\mathbf{X}^{-1}u_j, & j < i \end{cases} \\
&= \begin{cases} -c_{ii}, & i = j \\ -c_{ji} - c_{ij}, & j < i \end{cases},
\end{aligned}
$$

where $\mathbf{X}^{-1}\mathbf{A}\mathbf{X}^{-1} = \mathbf{C} = (c_{ij})$. This equation can be written in matrix form as

$$
\frac{\partial \operatorname{trace}(\mathbf{A}\mathbf{X}^{-1})}{\partial \mathbf{X}} = -\mathbf{C} - \mathbf{C}^\top + \operatorname{diag}(\mathbf{C}).
$$

When $\mathbf{A}$ is symmetric, the above formula simplifies to

$$
\frac{\partial \operatorname{trace}(\mathbf{A}\mathbf{X}^{-1})}{\partial \mathbf{X}} = -2\mathbf{X}^{-1}\mathbf{A}\mathbf{X}^{-1} + \operatorname{diag}(\mathbf{X}^{-1}\mathbf{A}\mathbf{X}^{-1}). \tag{2.14}
$$

2.2.4 Optimisation, Integration and Limits

Optimisation Many statistical problems involve finding the maximum or minimum of a function $f(x)$ of a vector $x = [x_1, x_2, \ldots, x_n]^\top$; for example, in Sect. 3.1.1 we are interested in maximising the likelihood function in the context of ordinary regression. Assuming that $f(x)$ is continuously differentiable in the domain $D \in \mathbb{R}^n$, then the possible maximum/minimum is identified by solving the equation

$$
\frac{\partial f(x)}{\partial x} = 0.
$$

The solution $x = x^*$ of this equation gives what is known as stationary point. If the second partial derivative evaluated at this point is a positive definite matrix (negative definite matrix), then x^* is the required minimum (maximum).

For example, consider the function

$$f(x) = x^\top A x + 2a^\top x + c,$$

where A is an $n \times n$ positive definite matrix of knowns, a and c are n-dimensional vectors of knowns and x is an n-dimensional vector of variables.

From derivatives (2.5) and (2.2) we have

$$\frac{\partial f(x)}{\partial x} = 2Ax + 2a$$

and equating this to zero we obtain $x^* = -A^{-1}a$. Here we note that since A is positive definite it is also non-singular, hence A^{-1} exists.

From (2.5) the second partial derivative matrix is

$$\frac{\partial^2 f(x)}{\partial x \partial x^\top} = 2A,$$

which is a positive definite matrix, hence x^* is a minimum. The minimum value of $f(x)$ is

$$f(x^*) = (-Aa)^\top A(-A^{-1}a) + 2a^\top(-A^{-1}a) + c = -a^\top A^{-1}a + c.$$

Integration In this section we give a brief discussion on integration of scalar functions of argument a vector or a matrix. Suppose that $f(x)$ is a scalar function of argument a vector $x = [x_1, \ldots, x_n]^\top$. We assume that $f(x)$ is continuous in each interior point of D (a subset of $\mathbb{R}^n$). Then the multiple integral of $f(x)$ over D is denoted by

$$\int_D f(x)\,dx = \int\int\cdots\int_D f(x_1, x_2, \ldots, x_n)\,dx_1\,dx_2\cdots dx_n.$$

It is outside the scope of this book to discuss a technical development of multiple integrals. For a formal definition and calculus of several variables, the reader is referred to Lang (1987); an excellent treatment of calculus of one variable is given in Spivak (1995). Similarly, multiple integrals of a scalar function $f(X)$ of a matrix X may be considered and the notation

$$\int_D f(X)\,dX$$

will be adopted, where D is the domain of $f(\mathbf{X})$. Note that when $\mathbf{X}$ is a symmetric matrix, the integral needs to be amended to include only the variables that are distinct. Following standard notation we still keep the above notation, but it is understood that integration is carried over the $n(n+1)/2$ distinct elements of $\mathbf{X}$.

Limit of a Sequence of Matrices A sequence of square matrices is a collection of matrices $\mathbf{A}_1, \mathbf{A}_2, \ldots$, indexed by $t = 1, 2, \ldots$. We write $\mathbf{A}_t$ for the t-th term of the sequence. In order to form the limit of $\mathbf{A}_t$ as $t \to \infty$, we need first to introduce the notion of *norm* or distance for matrices. The norm of a matrix (usually referred to as matrix norm) is denoted by $\| \cdot \|$ and is a function of any $m \times n$ matrix $\mathbf{A}$ that satisfies the following conditions

1. $\| \mathbf{A} \| \geq 0$;
2. $\| \mathbf{A} \| = 0$, if and only if $\mathbf{A} = \mathbf{0}$;
3. $c\mathbf{A} = |c| \| \mathbf{A} \|$, where c is a scalar and $|c|$ denotes the modulus of c;
4. $\| \mathbf{A} + \mathbf{B} \| \leq \| \mathbf{A} \| + \| \mathbf{B} \|$, for any $m \times n$ matrices $\| \mathbf{A} \|$ and $\| \mathbf{B} \|$.

Some matrix norms also satisfy the property

$$\| \mathbf{AB} \| \leq \| \mathbf{A} \| \| \mathbf{B} \|,$$

for any $n \times n$ matrices $\mathbf{A}$ and $\mathbf{B}$.

For example, the Euclidean norm for any vector a (a matrix with $n = 1$) is defined as

$$\| a \| = \left(a_1^2 + a_2^2 + \cdots + a_m^2 \right)^{1/2} = \left(a^\top a \right)^{1/2},$$

where $a = [a_1, a_2, \ldots, a_m]^\top$.

The Frobenius norm is a generalisation of the Euclidean norm to matrices and is defined for any $m \times n$ matrix $\mathbf{A} = (a_{ij})$ as

$$\| \mathbf{A} \| = \left(\sum_{i=1}^{m} \sum_{j=1}^{n} a_{ij}^2 \right)^{1/2} = \left[\operatorname{trace}(\mathbf{A}^\top \mathbf{A}) \right]^{1/2}.$$

Having defined the matrix norm, we can define the distance between two matrices $\mathbf{A}$ and $\mathbf{B}$ as $d = \| \mathbf{A} - \mathbf{B} \|$. Note that in case of $n = 1$, this definition gives the well known Euclidean distance between two vectors $a = [a_1, a_2, \ldots, a_m]^\top$ and $b = [b_1, b_2, \ldots, b_m]^\top$

$$d = \sqrt{(a_1 - b_1)^2 + (a_2 - b_2)^2 + \cdots + (a_m - b_m)^2}.$$

The distance of two vectors or matrices are used to measure convergence of iterative algorithms, e.g. in the context of the EM algorithm discussed in Sect. 2.4.2 below and in Sect. 4.3.1.

The sequence of matrices $\mathbf{A}_1, \mathbf{A}_2, \ldots$, sometimes written as $\{\mathbf{A}_t\}$ is said to converge to a limiting matrix $\mathbf{A}$, if and only if for any $\varepsilon > 0$, there exists some $t_0 > 0$ such that for any $t > t_0$ it is

$$\| \mathbf{A}_t - \mathbf{A} \| < \varepsilon.$$

This definition extends immediately the definition of a limit of sequence of real numbers by replacing scalars by matrices and moduli by norms; for more details on limits of one variable the reader is referred to Spivak (1995). Matrix norms are covered in detail in the classic text of matrix analysis (Horn & Johnson, 2013).

2.3   Probability and Distribution Theory

2.3.1   *Random Vectors and Probability Distributions*

Probability and statistics study uncertainty. Uncertainty arises when possible outcomes of an experiment, or a phenomenon in general, are uncertain. For example, the future movements of shares in the stock market are uncertain; likewise the possible increase or decrease in global temperature, which may be accounted for an indicator of climate change, are uncertain. Probability provides the mathematical notion for describing such phenomena, by introducing random outcomes responsible for the underlying uncertainty. Statistics adopts probability models with the aim to make inference based on observed data.

It is assumed that the reader is familiar with the basic notions of probability, with random vectors and with probability distributions and with their properties. Below we provide a brief summary of some of the basic definitions and properties necessary for establishing notation.

1. A p-dimensional random vector, demoted by X is a vector whose elements are random variables; we shall write

$$X = \begin{bmatrix} X_1 \\ X_2 \\ \vdots \\ X_p \end{bmatrix}.$$

Clearly when $p = 1$, X is reduced to a random variable. If the values of X_i are discrete (the domain of X_i is a discrete set), then X is known as a *discrete* random vector, while if the domain of X_i is a continuous set, then X is known as a *continuous* random vector.

2. Associated with X is its joint cumulative distribution function (c.d.f.) defined as

$$F_X(x) = F_X(x_1, x_2, \ldots, x_p) = P(X_1 \leq x_1, X_2 \leq x_2, \ldots, X_p \leq x_p),$$

which is the probability of the event $A = \{X_1 \leq x_1\} \cap \{X_2 \leq x_2\} \cap \cdots \cap \{X_p \leq x_p\}$, a subset of $\mathbb{R}^p$.

If X is discrete we define the joint probability mass function (p.m.f.) of X as

$$p_X(x) = p_X(x_1, x_2, \ldots, x_p) = P(X_1 = x_1, X_2 = x_2, \ldots, X_p = x_p), \tag{2.15}$$

the probability of the event $A = \{X_1 = x_1\} \cap \{X_2 = x_2\} \cap \cdots \cap \{X_p = x_p\}$, where $A \cap B$ denotes the intersection of the events A and B.

If X is continuous, we define the joint probability density function (p.d.f.) of X is

$$p_X(x_1, x_2, \ldots, x_p) = \frac{\partial^p F_{X_1 X_2 \cdots X_p}(x_1, x_2, \ldots, x_p)}{\partial x_1 \partial x_2 \cdots \partial x_p}. \tag{2.16}$$

It follows that

$$F_X(x) = \int_{-\infty}^{x_1} \int_{-\infty}^{x_2} \cdots \int_{-\infty}^{x_p} p_X(u_1, u_2, \ldots, u_p) \, du_1 \, du_2 \cdots du_p \tag{2.17}$$

and a similar formula applies for discrete random vectors, if we replace the integrals by sums. When $p = 1$ and X is continuous the above formulae simplify to

$$p_X(x) = \frac{d F_X(x)}{dx} \quad \text{and} \quad F_X(x) = \int_{-\infty}^{x} p_X(u) \, du.$$

Equations (2.15)–(2.17) indicate that knowledge of the p.m.f. or p.d.f. implies knowledge of the c.d.f. and vice versa.

3. The marginal distribution of X_i is defined by integrating out the rest X_j ($j \neq i$) random variables, i.e.

$$p_{X_i}(x_i) = \int_{-\infty}^{\infty} p_X(x_1, x_2, \ldots, x_p) \, dx_1 \cdots dx_{i-1} \, dx_{i+1} \cdots dx_p$$

and similarly we define the marginal c.d.f. of X_i. If X is discrete there is an analogous definition of the marginal p.m.f. where the above integral is replaced by a sum.

We define the conditional p.d.f. (p.m.f. if discrete) as the ratio of the joint p.d.f. (p.m.f. if discrete) and the marginal p.d.f (p.m.f. if discrete), e.g. if $X =$

$[X_1, X_2]^\top$ is continuous, then the conditional p.d.f. of X_1, given $X_2 = x_2$ is defined as

$$p_{X_1|X_2}(x_1 \mid x_2) = \frac{p_X(x_1, x_2)}{p_{X_2}(x_2)}.$$

The random variables $X_1, X_2, \ldots, X_p$ are said to be independent, if their joint p.d.f. (or p.m.f. if they are discrete) can be written as a product of their marginals, i.e.

$$p_X(x_1, x_2, \ldots, x_p) = p_{X_1}(x_1) p_{X_2}(x_2) \cdots p_{X_p}(x_p).$$

If the random variables $X_1, X_2, \ldots, X_p$ are not independent, then they are said to be dependent.

4. If X is continuous then the expectation of X is defined as

$$E(X) = \int_{\mathbb{R}^p} x p_X(x_1, x_2, \ldots, x_p) \, dx_1 \, dx_2 \cdots dx_p$$

and if X is discrete the above integral is replaced by a sum.

For any continuous random vectors X, Y with conditional p.d.f. $p_{X|Y}(x \mid y)$ we define the conditional expectation of X, given $Y = y$ as the expectation of X, with respect to the conditional density $X \mid Y = y$, i.e.

$$E(X \mid Y = y) = \int_{R_x} x p_{X|Y}(x \mid y) \, dx,$$

for any value y of Y, where R_x is the domain of $p_{X|Y}(x \mid y)$.

5. The covariance matrix of X is

$$\text{Var}(X) = E[X - E(X)][X - E(X)]^\top$$

and the covariance of two random vectors X and Y is

$$\text{Cov}(X, Y) = E[X - E(X)][Y - E(Y)]^\top.$$

The following two properties are referred to as the *tower properties* and are useful when we can calculate expectations of variances of $X \mid Y$, but not directly of X. For any random vectors X and Y we have

a. $E(X) = E[E(X \mid Y)]$;
b. $\text{Var}(X) = E[\text{Var}(X \mid Y)] + \text{Var}[E(X \mid Y)]$.

Property (a) is used to derive the independence of the residuals in Theorem 4.2 (Sect. 4.4).

2.3.2 *Common Discrete Distributions*

In this section we discuss four discrete distributions, the Poisson, the binomial, the negative binomial and the multinomial distribution. Some of these distributions are used to formulate discrete-response state space models, such as the dynamic generalised linear models of Sect. 6.2. The binomial distribution is also discussed in Sect. 2.4.3. For simplicity in the notation we drop the subscript X in the p.m.f. $p_X(x)$, i.e. we write $p(x)$ where reference to the random variable X is implied.

Poisson Distribution Let X be a discrete random variable taking values $0, 1, 2, 3, \ldots$. If its p.m.f. is given by

$$p(x) = P(X = x) = \exp(-\lambda)\frac{\lambda^x}{x!}, \quad x = 0, 1, 2, \ldots.$$

then X is said to follow the Poisson distribution, written $X \sim \text{Pois}(\lambda)$, for $\lambda > 0$.

The mean and the variance of X is

$$\text{E}(X) = \text{Var}(X) = \lambda.$$

Figure 2.1 shows the p.m.f. of the Poisson distribution, for $\lambda = 2$.

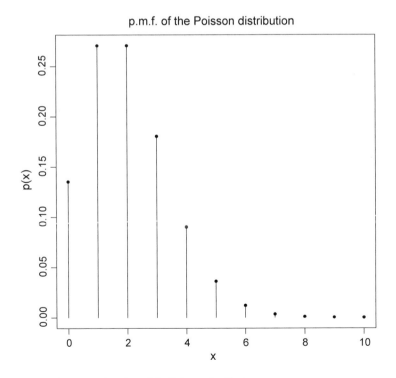

Fig. 2.1 Probability mass function of the Poisson distribution

Binomial Distribution The random variable X with p.m.f.

$$p(x) = P(X = x) = \binom{n}{x}\pi^x(1-\pi)^{n-x}, \quad x = 0, 1, 2, \ldots, n$$

is said to follow the binomial distribution of size n and probability π; we write $X \sim \mathrm{Binom}(n, \pi)$. The binomial coefficient is defined by

$$\binom{n}{x} = \frac{n!}{x!(n-x)!}.$$

The distribution is generated as the probability of the sum of n independent trials, each following the Bernoulli distribution with p.m.f. $\pi^x(1-\pi)^{1-x}$, where $x = 0, 1$ (here $P(X = 0) = \pi$ and $P(X = 1) = 1 - \pi$).

The mean and variance of the binomial distribution are

$$\mathrm{E}(X) = n\pi \quad \text{and} \quad \mathrm{Var}(X) = n\pi(1-\pi).$$

Figure 2.2 shows the p.m.f. of the binomial distribution, for $n = 10$ and $\pi = 0.6$.

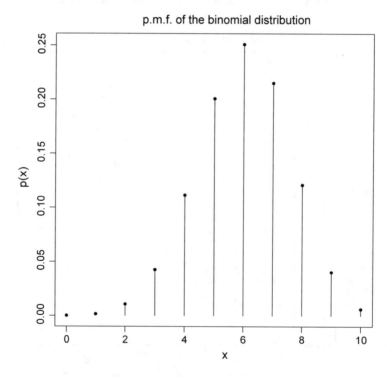

Fig. 2.2 Probability mass function of the binomial distribution

Negative Binomial Distribution The negative binomial distribution is generated as the distribution of the number of successes in independent Bernoulli trials before a fixed number of failures λ occur. More specifically, X is said to follow the negative binomial distribution, if its p.m.f. is

$$p(x) = P(X = x) = \binom{x + \lambda - 1}{x}(1 - \pi)^{\lambda}\pi^{x}, \quad x = 0, 1, 2, 3, \ldots,$$

where π is the probability of success of each Bernoulli trial. We shall write $X \sim \text{NegBinom}(\lambda, \pi)$ to indicate that X follows the negative binomial distribution with number of failures λ and probability of success π. More generally the negative binomial p.m.f. $p(x)$ is defined for real-valued $\lambda > 0$, although in this case the above genesis of the distribution does not apply.

The mean and variance of X are

$$E(X) = \frac{\pi\lambda}{1 - \pi} \quad \text{and} \quad \text{Var}(X) = \frac{\pi\lambda}{(1 - \pi)^2}.$$

We observe that

$$\text{Var}(X) = \frac{E(X)}{1 - \pi} > E(X),$$

and so the negative binomial distribution is suitable for describing over-dispersed data, i.e. data which variance is larger than its mean.

Figure 2.3 shows the p.m.f. of the negative binomial distribution, with $\pi = 0.6$ and $\lambda = 8$.

Multinomial Distribution The multinomial is a generalisation of the binomial distribution to accommodate for several (more than two) categories. Suppose that we have $k \geq 2$ categories and in each category i we have a fixed probability of success π_i, with $\pi_1 + \pi_2 + \cdots + \pi_k = 1$. Let n be the total number of trials or counts and let X_i be the random variable that records the number of success in category i, for $i = 1, 2, \ldots, k$ and $X_1 + X_2 + \cdots + X_k = n$. With these definitions the random vector $X = [X_1, X_2, \ldots, X_k]^\top$ is said to follow the multinomial distribution, if its joint p.m.f. is given by

$$p(x) = P(X_1 = x_1, X_2 = x_2, \ldots, X_k = x_k) = \frac{n!}{x_1!x_2!\cdots x_k!}\pi_1^{x_1}\pi_2^{x_2}\cdots\pi_k^{x_k}.$$

By means of notation we use $X \sim \text{Multin}(n, \pi)$, where $\pi = [\pi_1, \pi_2, \ldots, \pi_k]^\top$.

We observe that if $k = 2$ this distribution is reduced to the binomial distribution, since $X_1 + X_2 = n$ or $X_2 = n - X_1$ and $\pi_1 + \pi_2 = 1$ or $\pi_2 = 1 - \pi_1$. We note that X is only random in $k - 1$ variables, because we can always write $X_k = n - \sum_{i=1}^{k-1} X_i$

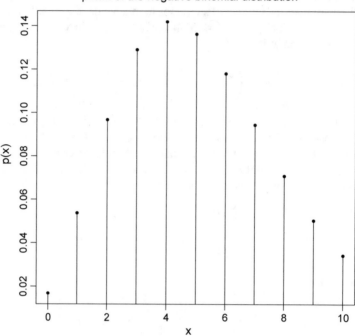

Fig. 2.3 Probability mass function of the negative binomial distribution

and $\pi_k = 1 - \sum_{i=1}^{k-1} \pi_i$ (e.g. in the binomial distribution $k = 2$ there is one random variable X_1.

The mean vector and the covariance matrix of X are

$$E(X) = n \begin{bmatrix} \pi_1 \\ \pi_2 \\ \vdots \\ \pi_k \end{bmatrix}$$

and

$$\text{Var}(X) = n \begin{bmatrix} \pi_1(1 - \pi_1) & -\pi_1\pi_2 & \cdots & -\pi_1\pi_k \\ -\pi_1\pi_2 & \pi_2(1 - \pi_2) & \cdots & -\pi_2\pi_k \\ \vdots & \vdots & \ddots & \vdots \\ -\pi_1\pi_k & -\pi_2\pi_k & \cdots & \pi_k(1 - \pi_k) \end{bmatrix}.$$

2.3.3 *Common Continuous Distributions*

In this subsection we discuss the main continuous distributions, namely Gaussian or normal, gamma and inverse gamma, beta, continuous uniform and the Student t distributions. Most of Chaps. 3 and 4 make use of the multivariate Gaussian distribution (Sect. 2.4.3 includes an example on Bayesian regression using the Gaussian distribution); the gamma and Student t distributions are discussed in Sect. 4.3.3.

Gaussian or Normal Distribution First we discuss the univariate version of the Gaussian distribution. Let X be a continuous random variable. If its p.d.f. is

$$p(x) = \frac{1}{\sqrt{2\pi}\sigma} \exp\left[-\frac{(x-\mu)^2}{2\sigma^2}\right], \quad x \in \mathbb{R},$$

where $\sigma^2 > 0$, then X is said to follow the Gaussian distribution with mean μ and variance σ^2; we write $X \sim N(\mu, \sigma^2)$.

Figure 2.4 shows the p.d.f. and the c.d.f. of the Gaussian distribution with $\mu = 0$ and $\sigma^2 = 1$, i.e. $X \sim N(0, 1)$.

Moving on to the multivariate Gaussian distribution, let $X = [X_1, \ldots, X_p]^\top$ denote a random vector. X is said to follow the multivariate Gaussian or normal distribution, with notation $X \sim N(\mu, \mathbf{V})$, if its p.d.f. is given by

$$p(x) = \frac{1}{(2\pi)^{p/2}|\mathbf{V}|^{1/2}} \exp\left[-\frac{1}{2}(x-\mu)^\top \mathbf{V}^{-1}(x-\mu)\right], \quad x \in \mathbb{R}^p,$$

where $\mu = [\mu_1, \mu_2, \ldots, \mu_p]^\top$ is the mean vector of X and $\mathbf{V}$ is the covariance matrix of X. We note that if $p = 1$, this is reduced to the above univariate Gaussian distribution with $\mathbf{V} = \sigma^2$.

Figure 2.5 shows the p.d.f. $p(x, y)$ of the Gaussian distribution above with $x_1 = x$, $x_2 = y$,

$$\mu = \begin{bmatrix} 0 \\ 0 \end{bmatrix} \quad \text{and} \quad \mathbf{V} = \begin{bmatrix} 1 & 0.8 \\ 0.8 & 1 \end{bmatrix}.$$

Gamma Distribution The positive random variable $X > 0$ is said to follow the gamma distribution with parameters $\alpha, \beta > 0$, if its p.d.f. is given by

$$p(x) = \frac{\beta^\alpha}{\Gamma(\alpha)} x^{\alpha-1} \exp(-\beta x), \quad x > 0,$$

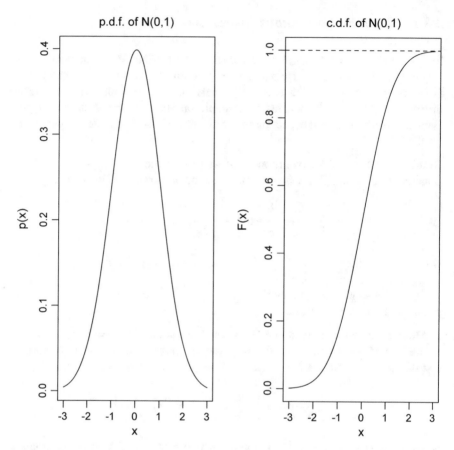

Fig. 2.4 Probability density function and cumulative distribution function of a normal distribution $N(0, 1)$

where $\Gamma(\alpha)$ denotes the gamma function of argument α defined by

$$\Gamma(\alpha) = \int_0^\infty u^{\alpha-1} \exp(-u) \, du.$$

If $\alpha = \nu/2$ and $\beta = 1/2$, the distribution is known as the chi-square distribution with ν degrees of freedom, where ν is a positive real number.

The mean and the variance of X are

$$E(X) = \frac{\alpha}{\beta} \quad \text{and} \quad \text{Var}(X) = \frac{\alpha}{\beta^2}.$$

Figure 2.6 shows the p.d.f.'s of three gamma distributions, namely $G(1, 1)$, $G(2, 1)$ and $G(5, 1)$.

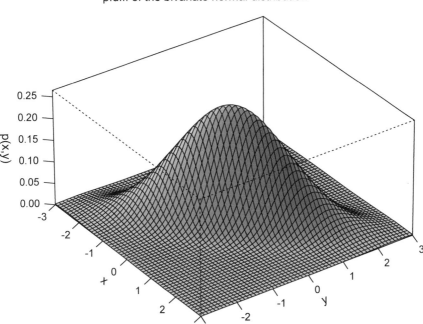

Fig. 2.5 Probability density function of the bivariate normal distribution

Closely related to the gamma distribution is the inverse gamma distribution. Let $X \sim G(\alpha, \beta)$, then the random variable $Y = 1/X$ is said to follow the inverse gamma distribution, with p.d.f.

$$p(y) = \frac{\beta^\alpha}{\Gamma(\alpha)} \frac{1}{y^{\alpha+1}} \exp\left(-\frac{\beta}{y}\right), \quad y > 0.$$

We shall write that $Y \sim IG(\alpha, \beta)$.

The mean and the variance of Y is

$$E(Y) = \frac{\beta}{\alpha - 1},$$

for $\alpha > 1$ and

$$\text{Var}(Y) = \frac{\beta^2}{(\alpha - 1)^2(\alpha - 2)},$$

for $\alpha > 2$.

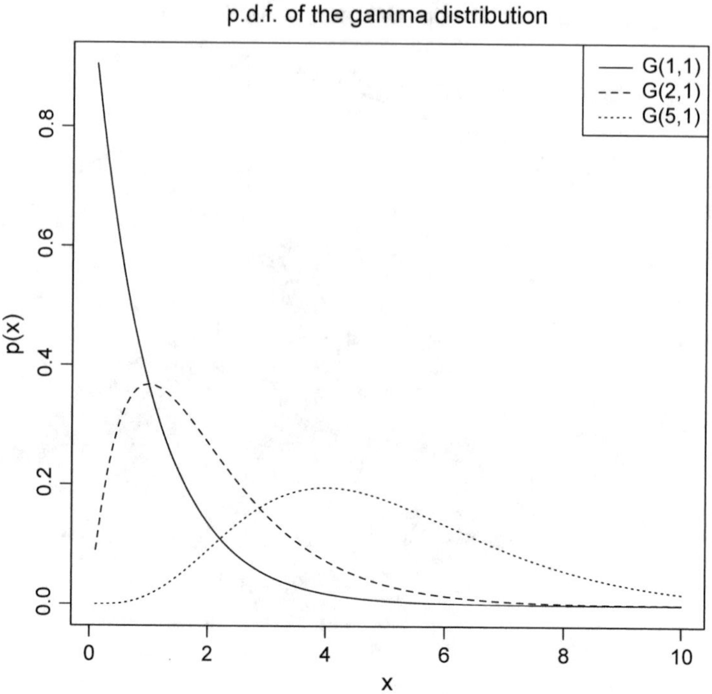

p.d.f. of the gamma distribution

Fig. 2.6 Probability density function of the gamma distribution

Beta Distribution The random variable $0 \leq X \leq 1$ is said to follow the beta distribution, if its p.d.f. is given by

$$p(x) = \frac{\Gamma(\alpha + \beta)}{\Gamma(\alpha)\Gamma(\beta)} x^{\alpha-1} (1-x)^{\beta-1}, \quad 0 \leq x \leq 1,$$

where $\alpha, \beta > 0$. By means of notation we write $X \sim \text{Beta}(\alpha, \beta)$.

The mean and the variance of X are

$$\text{E}(X) = \frac{\alpha}{\alpha + \beta} \quad \text{and} \quad \text{Var}(X) = \frac{\alpha\beta}{(\alpha + \beta)^2 (\alpha + \beta + 1)}.$$

Figure 2.7 shows the p.d.f.'s of three beta distributions, namely $\text{Beta}(0.5, 0.5)$, $\text{Beta}(2, 2.5)$ and $\text{Beta}(5, 1)$.

Uniform Distribution The random variable X follows the continuous uniform distribution, if its p.d.f. is given by

$$p(x) = \begin{cases} \frac{1}{b-a}, & a \leq x \leq b \\ 0, & \text{otherwise} \end{cases}$$

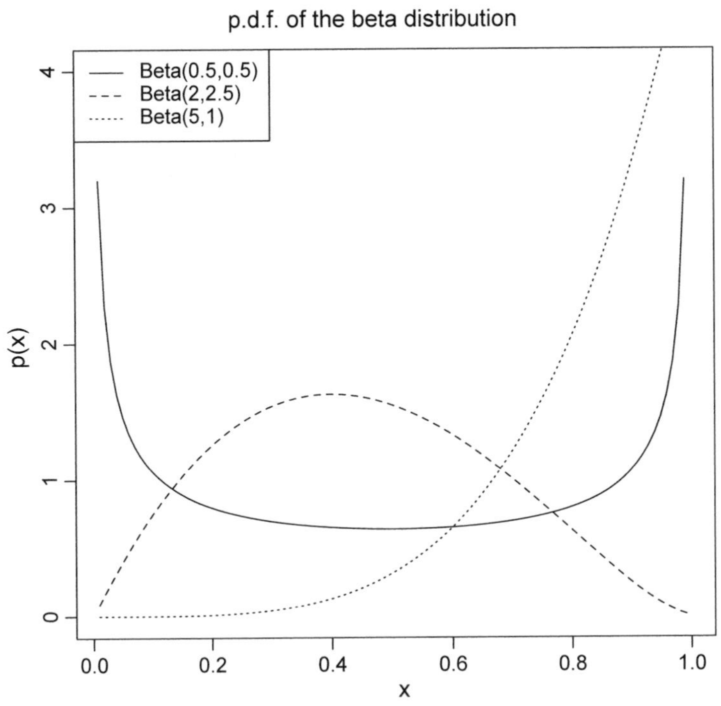

p.d.f. of the beta distribution

Fig. 2.7 Probability density function of the beta distribution

for any $a < b$. We will use the notation $X \sim U(a, b)$ to indicate that X follows the uniform distribution with parameters a and b.

The mean of X is

$$E(X) = \frac{a + b}{2},$$

and the variance of X is

$$\text{Var}(X) = \frac{(b - a)^2}{12}.$$

Figure 2.8 shows the p.d.f. of $U(-1, 1)$.

Student t Distribution We start with the univariate case. The random variable X is said to follow the Student t distribution, if its p.d.f. is

$$p(x) = \frac{\nu^{\nu/2} \Gamma\left(\frac{\nu+1}{2}\right)}{\sqrt{\pi} \Gamma\left(\frac{\nu}{2}\right) \sigma} \left[\nu + \frac{(x - \mu)^2}{\sigma^2}\right]^{-(\nu+1)/2}, \quad x \in \mathbb{R},$$

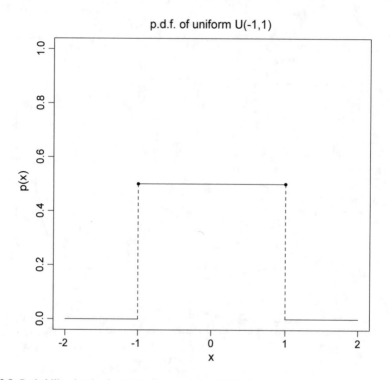

Fig. 2.8 Probability density function of the uniform distribution

where $v > 0$, μ and $\sigma^2 > 0$ are the parameters of the distribution. By means of notation we write $X \sim t(v, \mu, \sigma^2)$. The parameter $v > 0$ is known as the degrees of freedom.

The mean of X is

$$E(X) = \mu,$$

for $v > 1$ and the variance of X is

$$\text{Var}(X) = \frac{v\sigma^2}{v - 2},$$

for $v > 2$. If $0 < v \leq 2$ the variance is not defined; an example of a distribution with undefined mean and variance is the Cauchy distribution, obtained as the Student t distribution for $v = 1$.

Figure 2.9 shows the p.d.f. of Student t distribution with 3 and 10 degrees of freedom as compared with a Gaussian $N(0, 1)$ distribution. We observe that the Student t distribution has heaver tails than the Gaussian and that the smaller the degrees of freedom the heavier the tails are. In fact we can see that as $v \to \infty$, the

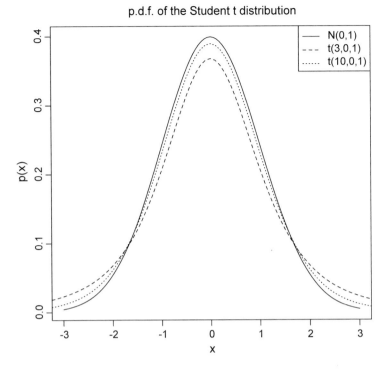

p.d.f. of the Student t distribution

Fig. 2.9 Probability density function of the Student t distribution

p.d.f. of the Student t distribution converges to that of a Gaussian distribution (see Exercise 2.10).

Now consider a random vector $X = [X_1, X_2, \ldots, X_p]^{\top}$. X is said to follow the multivariate Student t distribution with $\nu > 0$ degrees of freedom, mode or location vector μ and scale covariance matrix $\mathbf{V}$, if its density is

$$p(x) = \frac{\nu^{\nu/2}\Gamma\left(\frac{\nu+p}{2}\right)}{\pi^{p/2}\Gamma\left(\frac{\nu}{2}\right)|\mathbf{V}|^{1/2}}\left[\nu + (x - \mu)^{\top}\mathbf{V}^{-1}(x - \mu)\right]^{-(\nu+p)/2}.$$

In terms of notation we write $X \sim t(\nu, \mu, \mathbf{V})$.

The mean vector of X is

$$E(X) = \mu,$$

for $\nu > 1$ and the covariance matrix of X is

$$\text{Var}(X) = \frac{\nu}{\nu - 2}\mathbf{V},$$

for $\nu > 2$.

We observe that for $p = 1$ and $\mathbf{V} = \sigma^2$, we obtain the univariate Student t distribution, with p.d.f. given above.

As in the univariate case, the multivariate Student t distribution has heavier tails than the multivariate Gaussian distribution $N(\mu, \mathbf{V})$, controlled by the degrees of freedom v (the smaller v the heavier the tails). As $v \to \infty$ the Student t p.d.f. converges to the p.d.f. of the normal $N(\mu, \mathbf{V})$.

2.4   Statistics

2.4.1   Principle Set-Up and Objectives

Statistics is concerned with quantifying and managing uncertainty in observational studies, may these be controlled experiments, or socio-economic or physical phenomena. With observational study broadly we mean studies based on observed outcomes, which are uncertain. A good introductory treatment of Statistics can be found in Trosset (2009).

Uncertainty arises in all walks of life. In climate change, for example, the potential increase of the global temperature is hugely uncertain, as is the extent to this increase attributed to human-made actions. In a physical system the output signal may be uncertain due to measurement error and in economics fluctuations of asset prices are uncertain due to market and investors' movements.

Considering a vector $x = [x_1, x_2, \ldots, x_n]^\top$ of n observed or recorded measurements we can postulate that uncertainty around the values of x is generated by x being a particular realisation of a random vector $X = [X_1, X_2, \ldots, X_n]^\top$. Thus, it is natural to think that the probability distribution of X will summarise the uncertainty around x and thus it may be used to describe the generating process of x and to forecast future values of x. This distribution will typically depend upon some parameter vector θ, which is responsible for the shape of the distribution of X. We write $p(x \mid \theta)$ to indicate the p.d.f. of X (if X is continuous) or the p.m.f. of X (if X is discrete). Statistical inference is concerned about estimating θ, based on the observed data $x_1, x_2, \ldots, x_n$ and hence provide forecasts of future values of x, based on this observed data. Within the above model-based framework two main estimation approaches may be deployed

1. Maximum likelihood inference (frequentist statistics); and
2. Posterior inference (Bayesian statistics).

In (1) we consider that θ is a fixed, but known quantity and estimation procedures largely involve maximum likelihood analysis, see e.g. Sect. 2.4.2 below. In (2) θ is assumed to be random and estimation procedures involve the derivation of the conditional distribution of θ, given the data x; Sect. 2.4.3 discusses Bayesian inference and gives two examples for illustration purposes.

2.4.2 Maximum Likelihood Estimation: The EM Algorithm

The likelihood function has a central role on the development of estimators in statistical inference. The likelihood function, a function of the unknown parameter or parameter vector θ, is just the joint distribution of the random sample $X_1, X_2, \ldots, X_n$, given θ, i.e.

$$L(\theta; x) = p(x_1, x_2, \ldots, x_n \mid \theta),$$

where $x = [x_1, x_2, \ldots, x_n]^\top$. The likelihood function is a (deterministic) function of θ, which gives a measure of how *likely* is θ for a given sample $x_1, x_2, \ldots, x_n$. If θ is not very likely then $L(\theta; x)$ should be small, while if it is likely the value of $L(\theta; x)$ should be large. Thus, say for two values θ_1 and θ_2 of θ the inequality

$$L(\theta_1; x) > L(\theta_2; x)$$

will favour θ_1 as being more likely than θ_2, for this sample. This naturally leads us to choose the value θ that maximises the likelihood function, hence the *maximum likelihood* principle. Usually, for computational efficiency, we choose to maximise the logarithm of the likelihood, known as log-likelihood function, written as

$$\ell(\theta; x) = \log L(\theta; x).$$

In many situations it is hard to find the maximum of the likelihood function. The expectation maximisation (EM) algorithm, originally developed in Dempster et al. (1977) and further discussed in many textbooks, see e.g. Fahrmeir and Tutz (2001), is a popular choice of an *indirect* maximisation algorithm of the likelihood $L(\theta; x)$. The EM algorithm is an iterative algorithm, which consists of two steps: in the E-step the conditional expectation of the log-likelihood function is computed, given the past sample values and the current estimate of θ, and in the M-step a new estimate of θ is computed that maximises the expected log-likelihood function from the E-step. Below we provide description and the foundations of the EM algorithm, which is then used in Sect. 4.3.1 to derive the EM algorithm for state space models.

Our aim is to maximise $L(\theta; x)$ with respect to θ. Consider an unobserved discrete random vector Z and denote with R_z the domain of the p.m.f. of Z. We can express $L(\theta; x)$ in terms of Z as

$$L(\theta; x) = p(x \mid \theta) = \sum_{z \in R_z} p(x, z \mid \theta).$$

Notice that this is the marginal distribution of X, after we sum out Z. Then the EM algorithm is

1. Initialise $\hat{\theta}^{(0)}$;
2. E-step: calculate the conditional expectation

$$Q(\theta \mid \hat{\theta}^{(k)}) = \mathrm{E}[\log L(\theta; x, z) \mid x, \hat{\theta}^{(k)}];$$

3. M-step: find the maximum

$$\hat{\theta}^{(k+1)} = \arg \max_{\theta} Q(\theta \mid \hat{\theta}^{(k)});$$

4. Update $k \to k + 1$ and repeat steps (2)-(3) until convergence (see below).

As $k \to \infty$, $\hat{\theta}^{(k)}$ converges to the maximum likelihood estimate $\hat{\theta}$ of $L(\theta; x)$. In practice the algorithm terminates when $\hat{\theta}^{(k+1)}$ and $\hat{\theta}^{(k)}$ are close to each other, hence convergence is assumed. The rule to check for convergence is if the Euclidean norm of $\hat{\theta}^{(k+1)} - \hat{\theta}^{(k)}$ is smaller to a pre specified tolerance, i.e.

$$\| \hat{\theta}^{(k+1)} - \hat{\theta}^{(k)} \| < \text{Tol}.$$

Usually we set $\text{Tol} = 0.001$ or smaller.

Before we prove the correctness of the algorithm we give the Gibbs inequality. For any two discrete distributions with p.m.f.'s $p(x)$ and $q(x)$, respectively, we have

$$\sum_{x \in R_x} p(x) \log p(x) \geq \sum_{x \in R_x} p(x) \log q(x), \tag{2.18}$$

with equality if and only if $p(x) = q(x)$, for all $x \in R_x$, where R_x is the domain of $p(x)$.

To prove this we use the well known inequality $\log x \leq x - 1$, for any $x > 0$.

$$\sum_{x \in R_x} p(x) \log \frac{q(x)}{p(x)} \leq \sum_{x \in R_x} p(x) \left[\frac{q(x)}{p(x)} - 1 \right]$$

$$= \sum_{x \in R_x} q(x) - \sum_{x \in R_x} p(x)$$

$$= \sum_{x \in R_x} q(x) - 1 \leq 0$$

and so $\sum_{x \in R_x} p(x) \log p(x) \geq \sum_{x \in R_x} p(x) \log q(x)$.

Now we prove the correctness of the EM algorithm. From the definition of the joint distribution $p(x, z \mid \theta) = p(z \mid x, \theta) p(x \mid \theta)$ we have

$$p(x \mid \theta) = \frac{p(x, z \mid \theta)}{p(z \mid x, \theta)}$$

and so

$$\log p(x \mid \theta) = \log p(x, z \mid \theta) - \log p(z \mid x, \theta)$$

$$= \underbrace{\sum_{z \in R_z} p(z \mid x, \hat{\theta}^{(k)}) \log p(x, z \mid \theta)}_{\text{expectation of } \log p(x,z|\theta)} - \underbrace{\sum_{z \in R_z} p(z \mid x, \hat{\theta}^{(k)}) \log p(z \mid x, \theta)}_{\text{expectation of } \log p(z|x,\theta)}$$

$$= Q(\theta \mid \hat{\theta}^{(k)}) + G(\theta \mid \hat{\theta}^{(k)}).$$

Then

$$\log p(x \mid \theta) - \log p(x \mid \hat{\theta}^{(k)}) = Q(\theta \mid \hat{\theta}^{(k)}) - Q(\hat{\theta}^{(k)} \mid \hat{\theta}^{(k)})$$

$$+ G(\theta \mid \hat{\theta}^{(k)}) - G(\hat{\theta}^{(k)} \mid \hat{\theta}^{(k)}).$$

Now applying the Gibbs inequality (2.18) with $p(x) = p(z \mid x, \hat{\theta}^{(k)})$ and $q(x) = p(z \mid x, \theta)$, we obtain

$$G(\theta \mid \hat{\theta}^{(k)}) \geq G(\hat{\theta}^{(k)} \mid \hat{\theta}^{(k)}).$$

Hence

$$\log p(x \mid \theta) - \log p(x \mid \hat{\theta}^{(k)}) \geq Q(\theta \mid \hat{\theta}^{(k)}) - Q(\hat{\theta}^{(k)} \mid \hat{\theta}^{(k)}).$$

From this inequality, an improvement in Q ($Q(\theta \mid \hat{\theta}^{(k)}) \geq Q(\hat{\theta}^{(k)} \mid \hat{\theta}^{(k)})$) results in an improvement in $L(\theta; x) = \log p(x \mid \theta)$ ($\log p(x \mid \theta) \geq p(x \mid \hat{\theta}^{(k)})$).

As we obtain in the next iteration $\hat{\theta}^{(k+1)}$ as the maximum of $Q(\theta \mid \hat{\theta}^{(k)})$, an improvement in $L(\theta; x)$ is achieved. As a result, as $k \to \infty$, $\hat{\theta}^{(k)}$ converges to the maximum likelihood estimate $\hat{\theta}$.

2.4.3 Bayesian Inference

Bayesian statistics has been developed extensively over the past three decades, in particular the advance of computationally intensive estimation methods have made the Bayesian paradigm a very attractive estimation approach. Excellent expositions of Bayesian methods can be found in O'Hagan and Forster (2004), Leonard and Hsu (1999), Gamerman and Lopes (2006) and Robert (2007), among others.

In Bayesian inference unknown parameters are assumed to form a random vector θ. The distribution $p(\theta)$ of θ *prior* to observing data is known as the prior distribution of θ and reflects the prior belief we may have before observing the data. Apart from this prior belief, we are going to adopt the model formulation of Sect. 2.4.1, i.e. that the data generating process is described by the distribution $p(x \mid \theta)$. In other words the conditional distribution $p(x \mid \theta)$ is a model statement which asserts that if we know the parameter vector θ, then we know the distribution of the data $p(x \mid \theta) = p(x_1, \ldots, x_n \mid \theta)$. In our notation, $p(x \mid \theta) = L(\theta; x)$ is just the likelihood function of θ. The principle building block in Bayesian inference is the Bayes theorem, which states that the conditional distribution of θ given the data x is given by

$$p(\theta \mid x) = \frac{p(x \mid \theta)p(\theta)}{p(x)},$$

or simply

$$p(\theta \mid x) \propto L(\theta; x)p(\theta), \tag{2.19}$$

where $1/p(x)$ is known as the proportionality constant. The term $p(\theta \mid x)$ is the *posterior* distribution of θ, because it is the distribution of θ after (*a posteriori*) the data x is observed. Sometimes $p(\theta \mid x)$ is referred to as the *updated belief* or *posterior belief* and the likelihood $L(\theta; x)$ referred to as the *evidence* the data provides. The marginal distribution $p(x)$ of X is the forecast distribution of X, i.e.

$$p(x) = \int_{\Theta} p(x, \theta)\, d\theta = \int_{\Theta} p(x \mid \theta)p(\theta)\, d\theta,$$

where Θ is the domain of θ (if X is discrete the integral above is replaced by a relevant sum). In many situations the computation of $p(x)$ is not available in closed form, since integration of high dimensional integrals becomes very quickly intractable. In such cases simulation-based inference is usually adopted, in particular sequential Monte Carlo and Markov chain Monte Carlo methods; the former is discussed in this book in Sect. 6.7 and the latter is discussed in Sect. 5.7 (introduction to MCMC and Gibbs sampling) and in Sect. 6.8 (Metropolis-Hastings algorithm). Book-length treatment of MCMC methods can be found in Gamerman and Lopes (2006) and Robert (2007). Below two examples are provided to illustrate the utility and application of Bayesian analysis.

Example 2.1 (Bayesian Regression) Consider the linear model

$$y_i = x_i^\top \beta + \epsilon_i,$$

where y_i denotes a scaler response variable, x_i is a p-dimensional vector containing p covariates or explanatory variables and ϵ_i denotes an i.i.d. random variable, usually assumed to follow the normal distribution $N(0, \sigma^2)$, for some variance

σ^2 and for $i = 1, \ldots, n$. This model is revisited in Sect. 3.1, where regression is discussed from a least squares and maximum likelihood standpoints. A detailed discussion of Bayesian analysis for the linear model is given in Press (1989, Chapter 5) and in O'Hagan and Forster (2004, Chapter 11). This model is usually cast in matrix form as

$$y = \mathbf{X}\beta + \epsilon,$$

where $y = [y_1, y_2, \ldots, y_n]^\top$, $\epsilon = [\epsilon_1, \epsilon_2, \ldots, \epsilon_n]^\top$ and

$$\mathbf{X} = \begin{bmatrix} x_1^\top \\ x_2^\top \\ \vdots \\ x_n^\top \end{bmatrix}.$$

Thus, conditionally on β, y follows a normal distribution

$$y \mid \beta \sim N(\mathbf{X}\beta, \sigma^2 \mathbf{I}). \tag{2.20}$$

Let us consider that the prior distribution of β is normal, given by

$$\beta \sim N(\tilde{\beta}, \mathbf{R}), \tag{2.21}$$

where the prior expectation $\tilde{\beta}$ and the prior covariance matrix $\mathbf{R}$ of β are assumed known.

From (2.20) and (2.21) by applying Bayes theorem (2.19) the posterior distribution of β is

$$p(\beta \mid y) \propto p(y \mid \beta)p(\beta)$$

$$\propto \exp\left\{-\frac{1}{2}\left[(y - \mathbf{X}\beta)^\top(y - \mathbf{X}\beta)/\sigma^2 + (\beta - \tilde{\beta})^\top \mathbf{R}^{-1}(\beta - \tilde{\beta})\right]\right\}$$

$$\propto \exp\left[-\frac{1}{2}(\beta - \hat{\beta})^\top \mathbf{P}^{-1}(\beta - \hat{\beta})\right],$$

where

$$\hat{\beta} = \mathbf{P}(\mathbf{X}^\top y/\sigma^2 + \mathbf{R}^{-1}\tilde{\beta}) \quad \text{and} \quad \mathbf{P}^{-1} = \mathbf{R}^{-1} + \mathbf{X}^\top \mathbf{X}/\sigma^2. \tag{2.22}$$

Hence the posterior distribution of β is $\beta \mid y \sim N(\hat{\beta}, \mathbf{P})$. Therefore, credible intervals and other *a posteriori* features can be extracted from this distribution.

We note that if $\mathbf{R}^{-1} = 0$, then $\mathbf{P} = \sigma^2(\mathbf{X}^\top \mathbf{X})^{-1}$ (assuming that $\mathbf{X}$ is of full rank p) and so the posterior mean of β is equal to $\hat{\beta} = (\mathbf{X}^\top \mathbf{X})^{-1}\mathbf{X}^\top y$, which is the least squares and maximum likelihood estimates of β; see the relevant discussion in

Sect. 3.1. In this case ($\mathbf{R}^{-1} = \mathbf{0}$) the elements of $\mathbf{R}$ converge to infinity and prior (2.21) is an improper uniform distribution; for a related discussion see O'Hagan and Forster (2004, Chapter 11). In practice we may set a large covariance matrix, say $\mathbf{R} = 1000\mathbf{I}$, so that $\mathbf{R}^{-1} \approx \mathbf{0}$. This is known as *weakly informative prior specification* and is discussed in Sect. 4.5 and in O'Hagan and Forster (2004) and Robert (2007).

By writing $\mathbf{R}^{-1} = \mathbf{P}^{-1} - \mathbf{X}^\top \mathbf{X}/\sigma^2$ from (2.22) and substituting it in $\hat{\beta}$ we obtain

$$\hat{\beta} = \tilde{\beta} + \frac{\mathbf{P}\mathbf{X}^\top}{\sigma^2}(y - \mathbf{X}\tilde{\beta}), \tag{2.23}$$

or in words that the posterior mean $\hat{\beta}$ of β is equal to the prior mean $\tilde{\beta}$ plus a *gain* factor $K = \mathbf{P}\mathbf{X}^\top/\sigma^2$ times the residual $e = y - \mathbf{X}\tilde{\beta}$. Here $\mathbf{X}\tilde{\beta}$ is the prediction of y (hence their difference is the residual e). Equation (2.23) is important, as we will see in Sect. 3.2 it is generalised to provide the celebrated Kalman filter within the context of state space and time series modelling.

Example 2.2 (Beta-Binomial Model) Consider that $\theta = \pi$ denotes a probability of success of independent trails, such that

$$X \mid \pi \sim \text{Binom}(n, \pi)$$

follows a binomial distribution with probability of success π and size n. Suppose that a prior distribution for π is the beta distribution

$$\pi \sim \text{Beta}(2, 3).$$

By applying Bayes rule (2.19) we have that the posterior distribution of π given $X = x$ is proportional to

$$p(\pi \mid X = x) \propto p(x \mid \pi)p(\pi)$$
$$\propto \pi^{x+2-1}(1 - \pi)^{n+3-x-1},$$

which is proportional to a beta distribution with parameters $x + 2$ and $n + 3 - x$, i.e.

$$\pi \mid X = x \sim \text{Beta}(x + 2, n + 3 - x).$$

For example, if $n = 5$ and x is observed to be equal to 4, then

$$E(\pi \mid X = 4) = \frac{6}{10}.$$

In this case we observe that the posterior mean of π is larger than the prior mean of π, which is $E(\pi) = 2/5$.

The forecast distribution of X can be computed as

$$p(x) = \int_0^1 p(x \mid \pi)p(\pi)\,d\pi$$

$$= \frac{\Gamma(5)}{\Gamma(2)\Gamma(3)}\binom{n}{x}\int_0^1 \pi^{x+2-1}(1-\pi)^{n+3-x-1}\,d\pi$$

$$= 10\binom{n}{x}\frac{\Gamma(x+2)\Gamma(n+3-x)}{\Gamma(n+5)}$$

$$= \frac{10(x+1)(x+2)(n-x+1)(n-x+2)(n-x+3)}{(n+1)(n+2)(n+3)(n+3)(n+4)(n+5)},$$

where we have used $\Gamma(x) = x\Gamma(x-1) = x!$, for x integer. Also from the posterior beta distribution $\pi \mid X = x \sim \text{Beta}(x+2, n++3-x)$, with $\int_0^1 p(\pi \mid x)\,d\pi = 1$ or

$$\int_0^1 \frac{\Gamma(x+2+n+3-x)}{\Gamma(x+2)\Gamma(n+3-x)}\pi^{x+2-1}(1-\pi)^{n+3-x-1}\,d\pi = 1,$$

it follows that

$$\int_0^1 \pi^{x+2-1}(1-\pi)^{n+3-x-1}\,d\pi = \frac{\Gamma(x+2)\Gamma(n+3-x)}{\Gamma(n+5)}.$$

As it is evident, in this example, it is harder to compute $p(x)$ than it is to compute the posterior $p(\pi \mid x)$. The above two examples illustrate the use of *conjugate* prior distributions, having the property that the posterior distribution has the same form as the prior, only differing on their respective parameters (e.g. in Example 2.1 both the prior and the posterior of β are normal distributions and in Example 2.2 both the prior and the posterior are beta distributions). The conjugate prior distributions have been criticised as being limited, see e.g. Robert (2007, Chapter 3), but they are certainly a worthy consideration in recursive estimation of time series, because they facilitate a sequential prior to posterior updating over time.

2.5   Exercises

1. Show that for any non-singular $n \times n$ matrices $\mathbf{A}$ and $\mathbf{B}$ the following is true

$$(\mathbf{A} + \mathbf{B})^{-1} = \mathbf{B}^{-1}(\mathbf{A}^{-1} + \mathbf{B}^{-1})^{-1}\mathbf{A}^{-1}.$$

2. If $\mathbf{I}$ is the $n \times n$ identity matrix, show that $\text{trace}(c\mathbf{I}) = cn$. Hence show that $\text{trace}(\mathbf{0}) = 0$.

3. Show that trace($\mathbf{ABC}$) = trace($\mathbf{CAB}$) = trace($\mathbf{BCA}$).
4. Let $\mathbf{X}$ be a symmetric matrix of variables. Evaluate the partial derivatives

$$\frac{\partial\,\mathrm{trace}(\mathbf{AX}^k\mathbf{B})}{\partial\mathbf{X}}, \quad \frac{\partial\log|\mathbf{AX}^k\mathbf{B}|}{\partial\mathbf{X}} \quad \text{and} \quad \frac{\partial\,\mathrm{trace}(\mathbf{AX}^{-1}\mathbf{B})}{\partial\mathbf{X}},$$

where k is a positive integer and $\mathbf{A}, \mathbf{B}$ are any matrices with constants.

5. For a $n \times n$ matrix $\mathbf{A}$ and for any non-zero n-dimensional vector x show that

$$\|\mathbf{A}\| = |x^\top\mathbf{A}x|,$$

is a matrix norm, where $|\cdot|$ denotes modulus.
If $\lim_{k\to\infty}\mathbf{A}^k = \mathbf{0}$, use the above formula to show

$$\sum_{k=0}^{\infty}\mathbf{A}^k = (\mathbf{I}-\mathbf{A})^{-1}.$$

6. The random variable X follows the geometric distribution, if its p.m.f. is

$$p(x) = P(X = x) = (1 - \pi)^x\pi, \quad x = 0, 1, 2, \ldots$$

This distribution is generated as the number of failures until the first success in independent Bernoulli trials each having probability of success π.
Show that, for integer x, the c.d.f. of X is

$$F(x) = 1 - (1 - \pi)^{x+1}.$$

Show that the expectation and the variance of X are

$$E(X) = \frac{1 - \pi}{\pi} \quad \text{and} \quad \mathrm{Var}(X) = \frac{1 - \pi}{\pi^2}.$$

Hint: for $E(X)$: differentiate both sides of

$$\sum_{x=0}^{\infty}(1 - \pi)^x\pi = 1,$$

with respect to π. A similar argument can be applied for the variance.

7. For any $n, x > 0$ show

$$\frac{\Gamma(n + x)}{x!\Gamma(n)} = \binom{n + x - 1}{x}.$$

8. Let X be a random variable that follows the negative binomial distribution $X \sim$ NegBinom(λ, π), for some known $\lambda > 0$ and probability π. Show that

$$\binom{\lambda + x - 1}{x} = \binom{x + \lambda - 1}{\lambda - 1}$$

hence the p.m.f. of X can be written as

$$p(x) = P(X = x) = \binom{\lambda + x - 1}{x}(1 - \pi)^\lambda \pi^x, \quad x = 0, 1, 2, \ldots$$

9. The random variable X follows the Laplace distribution, if its p.d.f. is given by

$$p(x) = \frac{1}{2b} \exp\left(-\frac{|x - \mu|}{b}\right), \quad x \in \mathbb{R},$$

where μ is a location parameter and $b > 0$ is a scale parameter. Show that the c.d.f. of X is

$$F(x) = \begin{cases} \frac{1}{2} \exp\left(\frac{x - \mu}{b}\right), & x < \mu \\ 1 - \frac{1}{2} \exp\left(-\frac{x - \mu}{b}\right), & x \geq \mu \end{cases}$$

10. The random variable X follows the Pareto distribution with shape $\alpha > 0$ and scale $\beta > 0$, if its p.d.f. is

$$p(x) = \begin{cases} \frac{\alpha \beta^\alpha}{x^{\alpha+1}}, & x \geq \beta \\ 0, & x < \beta \end{cases}$$

Show that the c.d.f. of X is

$$F(x) = \begin{cases} 1 - \left(\frac{\beta}{x}\right)^\alpha, & x \geq \beta \\ 0, & x < \beta \end{cases}$$

Show that the expectation of X is

$$E(X) = \begin{cases} \infty, & \alpha \leq 1 \\ \frac{\alpha \beta}{\alpha - 1}, & \alpha > 1 \end{cases}$$

11. Let $p(x)$ be the p.d.f. of the normal distribution $N(0, 1)$ and $q_\nu(x)$ be the p.d.f. of the Student t distribution with ν degrees of freedom $t(\nu, 0, 1)$. Using the limit

$$\lim_{n \to \infty} \frac{\Gamma(n + \alpha)}{\Gamma(n)n^\alpha} = 1,$$

for any $\alpha > 0$, show that

$$\lim_{\nu \to \infty} q_\nu(x) = p(x),$$

i.e. the Student t p.d.f. converges to the $N(0, 1)$ p.d.f. as $\nu \to \infty$.
Hint: Make use of the well known limit

$$\lim_{n \to \infty} \left(1 + \frac{\alpha}{n}\right)^n = \exp(\alpha),$$

for any $\alpha \in \mathbb{R}$.

12. Let X and Y be any continuous random vectors. Show

$$\mathrm{Cov}(X, Y) = \mathrm{E}(XY^\top) - \mathrm{E}(X)\mathrm{E}(Y)^\top.$$

13. Let X, Y and Z be any continuous random vectors. Show

$$\mathrm{Cov}(X, Y) = \mathrm{E}[\mathrm{Cov}(X, Y \mid Z)] + \mathrm{Cov}[\mathrm{E}(X \mid Z), \mathrm{E}(Y \mid Z)].$$

14. Suppose that the random variables X and Y follow a joint Gaussian distribution

$$\begin{bmatrix} X \\ Y \end{bmatrix} \sim N\left\{ \begin{bmatrix} \mu_x \\ \mu_y \end{bmatrix}, \begin{bmatrix} \sigma_x^2 & \sigma_{xy} \\ \sigma_{xy} & \sigma_y^2 \end{bmatrix} \right\},$$

where μ_x is the expectation of X, μ_y is the expectation of Y, σ_x^2 is the variance of X, σ_y^2 is the variance of Y and σ_{xy} is the covariance of X and Y.
Show that the joint p.d.f. of X and Y can be written as

$$p_{XY}(x, y) = \frac{1}{2\pi \sigma_x \sigma_y \sqrt{1 - \rho^2}}$$
$$\times \exp\left[-\frac{\sigma_y^2(x - \mu_x)^2 - 2\rho\sigma_x\sigma_y(x - \mu_x)(y - \mu_y) + \sigma_x^2(y - \mu_y)^2}{2\sigma_x^2\sigma_y^2(1 - \rho^2)} \right].$$

Show that the marginal distributions of X and Y are

$$X \sim N(\mu_x, \sigma^2) \quad \text{and} \quad Y \sim N(\mu_x, \sigma^2).$$

Show that if X and Y are independent random variables, then $\mathrm{Cov}(X, Y) = 0$. The converse claim is not generally true. However, if X, Y are jointly normally distributed, then this is true. Prove this claim, i.e. if the joint distribution of X, Y is Gaussian as above, then $\mathrm{Cov}(X, Y) = 0$ implies that X and Y are independent.

15. Suppose that the random vector $X = [X_1, X_2, \ldots, X_k]^\top$ follows the multi-nomial distribution with size $n = \sum_{i=1}^{k} X_i$ and probability vector $\pi = [\pi_1, \pi_2, \ldots, \pi_k]^\top$, i.e.

$$X \sim \text{Multin}(n, \pi).$$

Show that the marginal distribution of X_1 is the binomial distribution

$$X_1 \sim \text{Binom}(n, \pi_1)$$

and the conditional distribution of the random vector $[X_2, \ldots, X_k]^\top$, given $X_1 = x_1$ is

$$[X_2, \ldots, X_k]^\top \mid X_1 = x_1 \sim \text{Multin}\left(n - x_1, \left[\frac{p_2}{1 - p_1}, \ldots, \frac{p_k}{1 - p_1}\right]^\top\right).$$

16. Three random variables have joint p.d.f.

$$p_{XYZ}(x, y, z) = \begin{cases} \frac{c}{xy} \exp(-yz), & 1 < x, y < 2, \quad z > 0 \\ 0, & \text{otherwise} \end{cases}$$

a. Show that the constant c is $c = \frac{2}{\log 2}$.
b. Find the marginal joint p.d.f. of $[X, Y]^\top$, and the marginal p.d.f.'s of X and Y.
   Show that X and Y are independent, but X, Y and Z are not independent.
c. Calculate the mean and variance of X and Y. Also calculate the mean vector of $W = [X, Y]^\top$ and the covariance matrix of W.
d. For $1 < x, y < 2$, find the joint distribution function $F_{XY}(x, y)$ of $[X, Y]^\top$ and calculate the probability $P(X \leq 3/2, Y \leq 4/3)$.

17. Consider an independent random sample $X_1, X_2, \ldots, X_n$, where X_i is generated from a Poisson distribution, with rate λ, i.e. $X_i \sim \text{Pois}(\lambda)$. Show that the maximum likelihood estimate of λ is

$$\hat{\lambda} = \frac{1}{n} \sum_{i=1}^{n} x_i,$$

where $x_1, x_2, \ldots, x_n$ are the observed values of $X_1, X_2, \ldots, X_n$.

18. Consider an independent random sample $X_1, X_2, \ldots, X_n$, where X_i is generated from the binomial distribution $X_i \sim \text{Binom}(m, \pi)$, for some m. Show that

the maximum likelihood estimate of π is

$$\hat{\pi} = \frac{1}{mn} \sum_{i=1}^{n} x_i,$$

where $x_1, x_2, \ldots, x_n$ are the observed values of $X_1, X_2, \ldots, X_n$.

19. Consider an independent random sample $X_1, X_2, \ldots, X_n$, where X_i is generated from the geometric distribution with probability of success π (see Exercise 2.7). Show that the maximum likelihood estimate of π is

$$\hat{\theta} = \frac{n}{n + \sum_{i=1}^{n} x_i},$$

where $x_1, x_2, \ldots, x_n$ are the observed values of $X_1, X_2, \ldots, X_n$.

20. Consider an independent random sample $X_1, X_2, \ldots, X_n$, where X_i is generated from the inverse Gaussian distribution, with p.d.f.

$$p_{X_i}(x_i) = \sqrt{\frac{\theta}{2\pi x_i^3}} \exp\left(-\frac{\theta(x_i - \mu)^2}{2\mu^2 x_i}\right), \quad x_i > 0, \quad \mu, \theta > 0.$$

Assuming that μ is known, show that the maximum likelihood estimate of θ is

$$\hat{\theta} = \frac{n\mu^2}{\sum_{i=1}^{n}(x_i - \mu)^2 x_i^{-1}},$$

where $x_1, x_2, \ldots, x_n$ are the observed values of $X_1, X_2, \ldots, X_n$.

21. The distribution of data X, given parameter $\lambda > 0$, is Poisson

$$X \mid \lambda \sim \text{Pois}(\lambda).$$

If, for some α and β, the prior of λ is gamma

$$\lambda \sim G(\alpha, \beta),$$

then

a. show that the posterior distribution of λ is gamma, i.e.

$$\lambda \mid X = x \sim G(\alpha + x, \beta + 1),$$

where x is the observed value of X;

b. show that the forecast distribution of X is negative binomial

$$X \sim \text{NegBinom}\left(\alpha, \frac{1}{\beta + 1}\right).$$

22. Suppose that, given probability of success π, the distribution of data X is the geometric distribution of Exercise 2.7. If the prior of π is beta

$$\pi \sim \text{Beta}(\alpha, \beta),$$

for some α and β, then

a. show that the posterior distribution of π is beta

$$\pi \mid X = x \sim \text{Beta}(\alpha + 1, \beta + x),$$

where x is the observed value of X;
b. show that the forecast distribution of X is

$$p(x) = \frac{\alpha\beta(\beta + 1) \cdots (\beta + x)}{(\alpha + \beta)(\alpha + \beta + 1) \cdots (\alpha + \beta + x + 1)}.$$

23. Given $\lambda > 0$, the data X follows the exponential distribution with p.d.f.

$$p(x \mid \lambda) = \begin{cases} \lambda \exp(-\lambda x), & x \geq 0 \\ 0, & x < 0 \end{cases}$$

If the prior of λ is gamma

$$\lambda \sim G(\alpha, \beta),$$

for some α and β, then

a. show that the posterior distribution of λ is gamma

$$\lambda \mid X = x \sim G(\alpha + 1, \beta + x),$$

where x is the observed value of X;
b. show that the forecast distribution of X is

$$p(x) = \begin{cases} \frac{\alpha\beta^\alpha}{(\beta+x)^{\alpha+1}}, & x \geq 0 \\ 0, & x < 0 \end{cases}$$

Chapter 3
The Kalman Filter

This chapter introduces the linear *state space* model and discusses filtering, smoothing and forecasting. Section 3.1 motivates the state space model as a natural extension of the usual multiple regression model, which adopts ordinary least squares and maximum likelihood estimation methods. In Sect. 3.1.1, we give a brief review of these estimation methods, and for completeness purposes we provide the full derivations. These derivations provide technical motivation for the following. Section 3.1.2 discusses recursive estimation of regression with the emphasis placed on local estimation, leading to recursive least squares. The definition of the state space model follows in Sect. 3.1.3. Filtering is discussed in Sect. 3.2, where two derivations of the Kalman filter are given. Smoothing and forecasting are discussed in the next two sections, and the chapter concludes with coverage of observability and the steady state of linear time-invariant state space models.

3.1 From Regression to the State Space Model

3.1.1 Ordinary Least Squares

In this section we briefly describe regression methods for linear models, a detailed account of which is given in Bingham and Fry (2010); see also Sect. 2.4.3. Suppose that observations $y_1, y_2, \ldots, y_n$ become available over time $t = 1, 2, \ldots, n$, for some positive integer n. In some situations, the time index t may represent a finite discrete set $\{a_1, \ldots, a_n\}$, which is equivalent to $\{1, 2, \ldots, n\}$, i.e. there is a one-to-one mapping from elements of $\{a_1, \ldots, a_n\}$ to $\{1, 2, \ldots, n\}$. For example, a_i may represent years, or quarters, or any other, suitably defined discrete objects (not necessarily representing time). It is further assumed that $\{a_1, \ldots, a_n\}$ is, at least approximately, an equally spaced set, i.e. that $a_t - a_{t-1}$ is the same for all

© The Author(s), under exclusive license to Springer Nature Switzerland AG 2021
K. Triantafyllopoulos, *Bayesian Inference of State Space Models*, Springer Texts in Statistics, https://doi.org/10.1007/978-3-030-76124-0_3

$t = 2, 3, \ldots, n$. For convenience, in what follows we will work with a time index $\{1, 2, \ldots, n\}$, but the above discussion motivates the more general case.

In the context of regression, suppose that we have p variables (also indexed by time), x_{it} ($i = 1, \ldots, p$), which form a column vector $x_t = (x_{1t}, \ldots, x_{pt})^\top$, where $\top$ denotes transposition. We wish to express a relationship between y_t and x_t, and thus we form the linear regression model

$$y_t = x_{1t}\beta_1 + \cdots + x_{pt}\beta_p + \epsilon_t = x_t^\top \beta + \epsilon_t, \tag{3.1}$$

where $\beta = (\beta_1, \ldots, \beta_p)^\top$ denotes a p-variate vector of regression coefficients and ϵ_t is the error or innovation term of the above model, accounting for the distance between y_t and $x_t^\top \beta$. According to the Gauss–Markov conditions (Bingham & Fry, 2010), $\{\epsilon_t\}$ is a sequence of independent random variables (ϵ_t is independent of ϵ_s, for any $t \neq s$) with zero mean and common variance σ^2; $\{\epsilon_t\}$ is referred to as a *white noise* process, see e.g. Brockwell and Davis (1991), and we write $\epsilon_t \sim \mathrm{WN}(0, \sigma^2)$. The purpose of regression is to estimate β, based on a realised collection of observations $y_{1:n} = \{y_1, \ldots, y_n\}$. This estimation can be carried out in several equivalent ways (in Sect. 2.4.3, we describe the Bayesian estimation), but the standard one is by minimising the sum of squares

$$S(\beta) = \sum_{t=1}^{n} (y_t - x_t^\top \beta)^2. \tag{3.2}$$

To facilitate the calculations, it is useful to write model (3.1) in matrix form as

$$y = \begin{bmatrix} y_1 \\ y_2 \\ \vdots \\ y_n \end{bmatrix} = \begin{bmatrix} x_1^\top \\ x_2^\top \\ \vdots \\ x_n^\top \end{bmatrix} \beta + \begin{bmatrix} \epsilon_1 \\ \epsilon_2 \\ \vdots \\ \epsilon_n \end{bmatrix} = \mathbf{X}\beta + \epsilon.$$

Then, assuming that $\mathbf{X}$ is of full rank, the well known solution that minimises $S(\beta)$ gives the estimator of β as

$$\hat{\beta} = (\mathbf{X}^\top \mathbf{X})^{-1} \mathbf{X}^\top y = \left(\sum_{t=1}^{n} x_t x_t^\top \right)^{-1} \sum_{t=1}^{n} x_t y_t. \tag{3.3}$$

For completeness purposes, we give the proof next. We wish to minimise

$$S(\beta) = \epsilon^\top \epsilon = (y - \mathbf{X}\beta)^\top (y - \mathbf{X}\beta) = y^\top y - 2\beta^\top \mathbf{X}^\top y + \beta^\top \mathbf{X}^\top \mathbf{X}\beta.$$

Using (2.3) and (2.5), the first partial derivative of $S(\beta)$ with respect to β is

$$\frac{\partial S(\beta)}{\partial \beta} = -2\mathbf{X}^\top y + 2\mathbf{X}^\top \mathbf{X}\beta,$$

which for $\beta = \hat{\beta}$ satisfies the equation

$$\frac{\partial S(\hat{\beta})}{\partial \hat{\beta}} = 0 \Rightarrow \hat{\beta} = (\mathbf{X}^\top \mathbf{X})^{-1}\mathbf{X}^\top y,$$

given that $\mathbf{X}$ is of full rank, so that the inverse of $\mathbf{X}^\top \mathbf{X}$ exists.

Thus $\hat{\beta}$ is a stationary vector, and we can use Eq. (2.5) to verify that its second partial derivative is a positive definite matrix, i.e.

$$\frac{\partial^2 S(\beta)}{\partial \beta \partial \beta^T} = 2\mathbf{X}^\top \mathbf{X}.$$

Thus, the sum of squares $S(\beta)$ is minimised at $\hat{\beta}$.

So far, no assumption has been made about the distribution of the sequence $\{\epsilon_t\}$. If one is prepared to accept that, additionally to the white noise assumption of $\{\epsilon_t\}$, ϵ_t $(t = 1, \ldots, n)$ follows a normal distribution with zero mean and variance σ^2, i.e. $\epsilon_t \sim N(0, \sigma^2)$, then the maximum likelihood estimation of β and σ^2 is available and given as follows. From the model definition (3.1) and the white noise assumption $\epsilon_t \sim N(0, \sigma^2)$, we can write down the distribution of y_t given the parameters β and σ^2 as

$$y_t \mid \beta, \sigma^2 \sim N(x_t^\top \beta, \sigma^2).$$

Since $\{\epsilon_t\}$ is a white noise, it follows that, given β and σ^2, $y_1, \ldots, y_n$ are independent.

The likelihood function of β and σ^2 is the joint distribution of $y_1, \ldots, y_n$, given these parameters, which by using the above two facts is

$$
\begin{aligned}
L(\beta, \sigma^2; y_{1:n}) &= p(y_1, \ldots, y_n \mid \beta, \sigma^2) \\
&= \prod_{t=1}^{n} p(y_t \mid \beta, \sigma^2) \quad \text{(from the independence of } y_1, \ldots, y_n) \\
&= \prod_{t=1}^{n} \frac{1}{\sqrt{2\pi}\sigma} \exp\left[-\frac{(y_t - x_t^\top \beta)^2}{2\sigma^2} \right] \\
&= \frac{1}{(2\pi)^{n/2}\sigma^n} \exp\left[-\frac{1}{2\sigma^2} \sum_{t=1}^{n} (y_t - x_t^\top \beta)^2 \right],
\end{aligned}
$$

and the log-likelihood function of β and σ^2 is

$$\ell(\beta, \sigma^2; y_{1:n}) = \log L(\beta, \sigma^2; y_{1:n})$$

$$= -n \log \sqrt{2\pi} - \frac{n}{2} \log \sigma^2 - \frac{1}{2\sigma^2} \sum_{t=1}^{n} (y_t - x_t^\top \beta)^2. \tag{3.4}$$

In order to find $\hat{\beta}$ and $\hat{\sigma}^2$ that maximise (3.4), first we maximise it with respect to β and then with respect to σ^2.

The partial first derivative of $\ell(\cdot)$ with respect to β is

$$\frac{\partial \ell(\beta, \sigma^2; y_{1:n})}{\partial \beta} = \frac{1}{\sigma^2} \sum_{t=1}^{n} x_t (y_t - x_t^\top \beta),$$

from which by equating it to zero, we obtain $\hat{\beta}$ exactly as in (3.3).

For σ^2, we have

$$\frac{\partial \ell(\beta, \sigma^2; y_{1:n})}{\partial \sigma^2} = -\frac{n}{2\sigma^2} + \frac{1}{2\sigma^4} \sum_{t=1}^{n} (y_t - x_t^\top \beta)^2,$$

which after evaluating it at $\beta = \hat{\beta}$, and solving the equation $\partial \ell(\hat{\beta}, \hat{\sigma}^2; y_{1:n})/\partial \hat{\sigma}^2 = 0$, yields

$$\hat{\sigma}^2 = \frac{1}{n} \sum_{t=1}^{n} (y_t - x_t^\top \hat{\beta})^2. \tag{3.5}$$

It is of some interest to express the estimator (3.5) in terms of $\mathbf{X}$ and y only. By expanding (3.5) and recalling the definitions of y and $\mathbf{X}$ above, we can readily see that

$$\hat{\sigma}^2 = n^{-1} \left(\sum_{t=1}^{n} y_t^2 - \sum_{t=1}^{n} y_t x_t^\top \hat{\beta} - \hat{\beta}^\top \sum_{t=1}^{n} x_t y_t + \hat{\beta}^\top \sum_{t=1}^{n} x_t x_t^\top \hat{\beta} \right)$$

$$= n^{-1} (y^\top y - y^\top \mathbf{X} \hat{\beta} - \hat{\beta} \mathbf{X}^\top y + \hat{\beta}^\top \mathbf{X}^\top \mathbf{X} \hat{\beta})$$

$$= n^{-1} y^\top (\mathbf{I} - \mathbf{X}(\mathbf{X}^\top \mathbf{X})^{-1} \mathbf{X}^\top) y, \tag{3.6}$$

after using $\hat{\beta} = (\mathbf{X}^\top \mathbf{X})^{-1} \mathbf{X}^\top y$. We note that in the above maximum likelihood estimators $\hat{\sigma}^2$, it is usual to replace n by $n - p$, which makes $\hat{\sigma}^2$ an unbiased estimator.

The above notion of the estimation of β is known as ordinary least squares (OLS), because all observations $y_1, \ldots, y_n$ have on average the same contribution on $\hat{\beta}$. This means that for large n, distant observations (such as y_1 or y_2) have the same weight on $\hat{\beta}$ as the more recent observations (such as y_n or y_{n-1}). This is usually not desirable when the data exhibit a time-dependence structure or when a localised estimation of β is required. For example, considering share prices of the stock market as y_t, the estimate of the price of today is likely to depend more on the related price of yesterday, but not as much on the price of last month. Thus, we wish to put more weight at more recent observations in the estimation of $\hat{\beta}$.

3.1.2 Recursive Least Squares

Motivated by the discussion above, we consider model (3.1), but here we replace $S(\beta)$ in (3.2) by the weighted or discounted sum of squares

$$S(\beta) = \sum_{j=0}^{n-1} \delta^j (y_{n-j} - x_{n-j}^\top \beta)^2, \tag{3.7}$$

where δ is a discount or forgetting factor (assumed known), satisfying $0 < \delta \leq 1$, and the weights δ^j have a discounting effect for $j = 0, 1, \ldots$. First of all, we note that $\delta = 1$ returns the sum of squares of OLS above, but if $\delta < 1$, the above sum puts more weight to the observations y_n, y_{n-1} and places less emphasis upon distant observations. This can be seen if we expand the above sum as

$$S(\beta) = (y_n - x_n^\top \beta)^2 + \delta(y_{n-1} - x_{n-1}^\top \beta)^2 + \cdots + \delta^{n-1}(y_1 - x_1^\top \beta)^2$$

and we note that $\delta^j \approx 0$, for sufficiently large j.

The *memory* of this model is defined by the geometric series

$$\sum_{j=0}^{n-1} \delta^j = 1 + \delta + \delta^2 + \cdots + \delta^{n-1} = \frac{1 - \delta^n}{1 - \delta},$$

since, at each occasion j, we forget at a rate of δ^j. If $0 < \delta < 1$, the above sum converges to $(1 - \delta)^{-1}$, as n converges to infinity. Thus, for $\delta = 1$ (OLS), the memory is equal to infinity (in this case we do not forget any data in the estimation of β), but if say $\delta = 0.5$, then the memory is equal to 2 (which we can think of as using only the two most recent observations in the calculation of the new $\hat{\beta}$).

The calculation of $\hat{\beta}$ is the result of the minimisation of (3.7), which can be obtained by direct differentiation. However, the expression of $\hat{\beta}$ may follow immediately from OLS if we rewrite the model in compact form as

$$
y = \begin{bmatrix} \delta^{(n-1)/2} y_1 \\ \vdots \\ \delta^{1/2} y_{n-1} \\ y_n \end{bmatrix} = \begin{bmatrix} \delta^{(n-1)/2} x_1^\top \\ \vdots \\ \delta^{1/2} x_2^\top \\ x_n^\top \end{bmatrix} \beta + \begin{bmatrix} \epsilon_1 \\ \vdots \\ \epsilon_{n-1} \\ \epsilon_n \end{bmatrix} = \mathbf{X}\beta + \epsilon.
$$

Then it is easy to see that the ordinary sum of squares for the above model coincides with the weighted sum of squares (3.7). Thus, the estimator of β now becomes

$$
\hat{\beta} = (\mathbf{X}^\top \mathbf{X})^{-1} \mathbf{X}^\top y = \left(\sum_{j=0}^{n-1} \delta^j x_{n-j} x_{n-j}^\top \right)^{-1} \sum_{j=0}^{n-1} \delta^j x_{n-j} y_{n-j}. \tag{3.8}
$$

Again we observe that for $\delta = 1$, we obtain the OLS solution as a special case. For $\delta < 1$, we put larger weight on y_n, y_{n-1} and x_n, x_{n-1} than on more distant y_i, x_i, e.g. y_1, x_1.

Next, we express $\hat{\beta}_n$ and $\hat{\sigma}^2$ recursively, leading to recursive least squares (RLS). It is important to note that in time series applications, one is interested in computing $\hat{\beta}_t$, for each time point $t = 1, \ldots, n$, for example, to enable real-time forecasting. In the classical application of linear models and regression (see e.g. Bingham and Fry (2010)), one would have to compute the inverse of $\mathbf{X}^\top \mathbf{X}$ for each time t, but here we develop $\hat{\beta}_t$ from the previously computed $\hat{\beta}_{t-1}$, for $t = 1, \ldots, n$.

We start with some definitions. For each t, we define

$$
\mathbf{H}_t = \sum_{i=0}^{t-1} \delta^i x_{t-i} x_{t-i}^\top = x_t x_t^\top + \delta \mathbf{H}_{t-1},
$$

$$
h_t = \sum_{i=0}^{t-1} \delta^i x_{t-i} y_{t-i} = x_t y_t + \delta h_{t-1}.
$$

Also, define

$$
e_t = y_t - x_t^\top \hat{\beta}_{t-1}
$$

to be the one-step ahead forecast error or residual (the difference of the forecast $x_t^\top \hat{\beta}_{t-1}$ from the observed value y_t) and

$$
K_t = \mathbf{H}_t^{-1} x_t.
$$

The next theorem establishes recursive estimation of β and σ^2.

Theorem 3.1 *With the above model definitions, at each time $t = 1, \ldots, n$, we have the following:*

1. *The least squares and maximum likelihood estimator of β is*
   $\hat{\beta}_t = \hat{\beta}_{t-1} + K_t e_t$;
2. *The maximum likelihood estimator $\hat{\sigma}_t^2$ of σ^2 is given by the recursions $n_t \hat{\sigma}_t^2 = n_{t-1} \hat{\sigma}_{t-1}^2 + r_t e_t$ and $n_t = n_{t-1} + 1$, with $n_0 = 0$;*
3. *The recursive updating of $\mathbf{P}_t = \mathbf{H}_t^{-1}$ is*

$$\mathbf{P}_t = \frac{1}{\delta}\left(\mathbf{I} - \frac{\mathbf{P}_{t-1} x_t x_t^\top}{\delta + x_t^\top \mathbf{P}_{t-1} x_t}\right)\mathbf{P}_{t-1},$$

where $K_t = \mathbf{P}_t x_t$, $e_t = y_t - x_t^\top \hat{\beta}_{t-1}$ is the one-step prediction error, $r_t = y_t - x_t^\top \hat{\beta}_t$ is the posterior or residual error and initial values for $\hat{\beta}_0$, $\mathbf{P}_0$ and $\hat{\sigma}_0^2$ are assumed known.

Proof First we prove (1). Using the least squares estimator $\hat{\beta}_t = \mathbf{H}_t^{-1} h_t$, we have

$$
\begin{aligned}
\hat{\beta}_t - \hat{\beta}_{t-1} &= \mathbf{H}_t^{-1} h_t - \mathbf{H}_{t-1}^{-1} h_{t-1} \\
&= \mathbf{H}_t^{-1}(\delta h_{t-1} + x_t y_t) - \mathbf{H}_{t-1}^{-1} h_{t-1} \\
&= (\delta \mathbf{H}_t^{-1} - \mathbf{H}_{t-1}^{-1}) h_{t-1} + \mathbf{H}_t^{-1} x_t y_t \\
&= \mathbf{H}_t^{-1} x_t y_t - \mathbf{H}_t^{-1} x_t x_t^\top \mathbf{H}_{t-1}^{-1} h_{t-1} \\
&= \mathbf{H}_t^{-1} x_t (y_t - x_t^\top \mathbf{H}_{t-1}^{-1} h_{t-1}) \\
&= \mathbf{H}_t^{-1} x_t (y_t - x_t^\top \hat{\beta}_{t-1}) \\
&= K_t e_t.
\end{aligned}
$$

Proceeding with the proof of (2), we use the maximum likelihood estimator $\hat{\sigma}_t^2 = n_t^{-1} y^\top (\mathbf{I} - \mathbf{X}(\mathbf{X}^\top \mathbf{X})^{-1}\mathbf{X}^\top)y$, given in (3.6), with $t = n_t = n_{t-1} + 1$, for $n_0 = 0$. Then,

$$
\begin{aligned}
n_t \hat{\sigma}_t^2 &= y^\top y - y^\top \mathbf{X}\hat{\beta}_t \\
&= \sum_{i=0}^{t-1} y_{t-i}^2 - h_t^\top \hat{\beta}_t \\
&= \sum_{i=0}^{t-2} y_{t-1-i}^2 - h_{t-1}^\top \hat{\beta}_{t-1} + y_t^2 - h_t^\top \hat{\beta}_t + h_{t-1}^\top \hat{\beta}_{t-1} \\
&= n_{t-1}\hat{\sigma}_{t-1}^2 + y_t^2 - y_t x_t^\top \hat{\beta}_t - h_{t-1}^\top K_t e_t \\
&= n_{t-1}\hat{\sigma}_{t-1}^2 + (y_t - h_t^\top K_t)e_t
\end{aligned}
$$

$$= n_{t-1}\hat{\sigma}_{t-1}^2 + (y_t - x_t^\top \hat{\beta}_t)e_t$$

$$= n_{t-1}\hat{\sigma}_{t-1}^2 + r_t e_t.$$

For the proof of (3), first we establish that the matrix $\mathbf{P}_{t-1}x_t x_t^\top \mathbf{P}_t$ is symmetric. Indeed, from the definition of $\mathbf{P}_t = \mathbf{H}_t^{-1}$, $\mathbf{P}_t$ and $\mathbf{P}_{t-1}$ are symmetric, since $\mathbf{H}_t$ and $\mathbf{H}_{t-1}$ are both symmetric and

$$\mathbf{P}_{t-1}x_t x_t^\top \mathbf{P}_t = \mathbf{P}_{t-1}(\mathbf{P}_t^{-1} - \delta \mathbf{P}_{t-1}^{-1})\mathbf{P}_t = \mathbf{P}_{t-1} - \delta \mathbf{P}_t$$

$$= \mathbf{P}_t(\mathbf{P}_t^{-1} - \delta \mathbf{P}_{t-1}^{-1})\mathbf{P}_{t-1} = \mathbf{P}_t x_t x_t^\top \mathbf{P}_{t-1},$$

since $\mathbf{H}_t - \delta \mathbf{H}_{t-1} = x_t x_t^\top$.

Again, from the definition $\mathbf{P}_t = \mathbf{H}_t^{-1}$ and the recursion $\mathbf{H}_t = x_t x_t^\top + \delta \mathbf{H}_{t-1}$, we obtain

$$\mathbf{P}_t = (\delta \mathbf{P}_{t-1}^{-1} + x_t x_t^\top)^{-1} = [(\delta \mathbf{I} + x_t x_t^\top)\mathbf{P}_{t-1}^{-1}]^{-1} = \mathbf{P}_{t-1}(\delta \mathbf{I} + x_t x_t^\top \mathbf{P}_{t-1})^{-1}$$

so $\mathbf{P}_{t-1} = \mathbf{P}_t(\delta \mathbf{I} + x_t x_t^\top \mathbf{P}_{t-1})$ so $\mathbf{P}_t x_t = (\delta + x_t^\top \mathbf{P}_{t-1}x_t)^{-1}\mathbf{P}_{t-1}x_t$

so $\mathbf{P}_{t-1} - \delta \mathbf{P}_t = (\delta + x_t^\top \mathbf{P}_{t-1}x_t)^{-1}\mathbf{P}_{t-1}x_t x_t^\top \mathbf{P}_{t-1}$

so $$\mathbf{P}_t = \frac{1}{\delta}\left(\mathbf{I} - \frac{\mathbf{P}_{t-1}x_t x_t^\top}{\delta + x_t^\top \mathbf{P}_{t-1}x_t}\right)\mathbf{P}_{t-1},$$

as required.   □

Theorem 3.1 provides key results. If one specifies prior values for $\hat{\beta}_0$, $\mathbf{P}_0$ and $\hat{\sigma}_0^2$, then for the computation of the estimators $\hat{\beta}_t$ and $\hat{\sigma}_t^2$, one needs only to save the quantities at time $t - 1$ $\hat{\beta}_{t-1}$ and $\hat{\sigma}_{t-1}^2$. Moreover, there is no matrix inversion needed for this algorithm to run. Indeed, the recursion of $\mathbf{P}_t$ in (3) allows the computation of $K_t = \mathbf{H}_t^{-1}x_t = \mathbf{P}_t x_t$, which appears in (1) and (2) of the theorem, without performing any matrix inversion. If we write the updating of the maximum likelihood of σ^2 recursively, we obtain

$$n_n \hat{\sigma}_n^2 = n_{n-1}\hat{\sigma}_{n-1}^2 + r_n e_n = n_{n-2}\hat{\sigma}_{n-2} + r_{n-1}e_{n-1} + r_n e_n = \cdots$$

$$= n_0 \hat{\sigma}_0^2 + \sum_{t=1}^n r_t e_t,$$

and so with $n_0 = 0$, $\hat{\sigma}_n^2$ is the average of the product of the residual and prediction errors, i.e.

$$\hat{\sigma}_n^2 = \frac{1}{n}\sum_{t=1}^n r_t e_t.$$

Theorem 3.1 gives the recursive least squares (RLS) algorithm, which has been used extensively in signal processing, see e.g. Haykin (2001) or Cowan and Grant (1985, Section 3.2). For $\delta = 1$, the RLS algorithm provides a recursive application of OLS, which is useful when the inversion of $\mathbf{X}^\top \mathbf{X}$ is not possible or it is likely to introduce computational instabilities. Recursive versions of the OLS is introduced in 1950 by Plackett (1950) and further explored in many studies; for a recent exposition, the reader is referred to Young (2011). The RLS algorithm is summarised below.

Recursive least squares

1. Initial values $\hat{\beta}_0$, $\mathbf{P}_0$, n_0, $\hat{\sigma}_0^2$ and δ;
2. For each time $t \geq 1$,

$$\mathbf{P}_t = \frac{1}{\delta}\left(\mathbf{I} - \frac{\mathbf{P}_{t-1}x_t x_t^\top}{\delta + x_t^\top \mathbf{P}_{t-1}x_t}\right)\mathbf{P}_{t-1};$$

3. For each t, the estimator of β_t is $\hat{\beta}_t = \hat{\beta}_{t-1} + K_t e_t$, where $e_t = y_t - x_t^\top \hat{\beta}_{t-1}$ and $K_t = \mathbf{P}_t x_t$;
4. For each t, estimator $\hat{\sigma}_t^2$ of $\hat{\sigma}^2$ is given by $n_t \hat{\sigma}^2 = n_{t-1}\hat{\sigma}_{t-1}^2 + r_t e_t$, where $n_t = n_{t-1} + 1$ and $r_t = y_t - x_t^\top \hat{\beta}_t$.

3.1.3 The State Space Model

Considering the RLS model of the previous section, we observe that β appears to be time-invariant, but the introduction of the discount factor δ introduces forgetting factor, and so it makes β (or rather the estimate of it) adaptable to local fluctuations of the data. As a result, it is natural to think as β evolving over time. The globally time-invariant β of the usual regression of Sect. 3.1.1 can be written as $\beta_n = \beta_{n-1} = \cdots = \beta_1 = \beta_0 = \beta$ or $\beta_t = \beta_{t-1}$, for each t, and $\beta_0 = \beta$. This allows us to motivate an 'almost' time-invariant setting, as $\beta_t \approx \beta_{t-1}$, which again can be described with an evolution equation $\beta_t = \beta_{t-1} + \zeta_t$, where ζ_t is a white noise process. Here, since ζ_t is a random variable, β_t is also random and is known as a *state* vector. We can then propose a model

$$y_t = x_t^\top \beta_t + \epsilon_t \quad \text{(observation model)}, \tag{3.9a}$$

$$\beta_t = \beta_{t-1} + \zeta_t \quad \text{(transition model)}, \tag{3.9b}$$

where we may assume that ϵ_t and ζ_t are independent. As we will see later, under some specification of the covariance matrix of ζ_t, this model proposes exactly the same estimation recursions as Theorem 3.1 for the RLS. Thus, with respect to estimation, we can establish that the above model is equivalent to the RLS, in the sense of producing the same estimators $\hat{\beta}_t$ and $\hat{\sigma}_t^2$. Model (3.9a)–(3.9b) defines a linear *state space* model. We note that the observations y_t are related linearly to the states β_t via the observation model and β_t are linearly linked to β_{t-1}. The transition model describes the transition of the states from $t-1$ to t, and in this case, β_t is a random walk. Model (3.9a)–(3.9b) is known as dynamic or time-varying regression, and it has been proposed for much data in economics and in other areas, see e.g. Pankratz (1991), West and Harrison (1997, Chapter 3) and Commandeur and Koopman (2007). If we set $x_t = 1$, for all t, i.e.

$$y_t = \beta_t + \epsilon_t \quad \text{and} \quad \beta_t = \beta_{t-1} + \zeta_t,$$

then the above model is known as *local level model* or *random walk plus noise model*, and it has played a key role on the development of state space models, see Harrison (1967), Harrison and Stevens (1976) and Harvey (1989) among others. For this model, we observe that $\beta_t = E(y_t \mid \beta_t)$, defined to be the level of the time series, is local, expressed by the random walk transition of β_t, hence the name 'local level'.

The general linear state space model considers model (3.9a)–(3.9b) but replaces the random walk of β_t by a more general Markov chain, i.e.

$$y_t = x_t^\top \beta_t + \epsilon_t \quad \text{(observation model)}, \tag{3.10a}$$

$$\beta_t = \mathbf{F}_t \beta_{t-1} + \zeta_t \quad \text{(transition model)}, \tag{3.10b}$$

where x_t is a $p \times 1$ vector, β_t is a $p \times 1$ vector and $\mathbf{F}_t$ is a $p \times p$ matrix. The Markov property of (3.10b) implies that given β_{t-1}, the distribution of β_t does not depend on past states $\beta_{t-2}, \beta_{t-3} \ldots$. In other words, given the present β_t, the future β_{t+k} and the past β_{t-j} are conditionally independent, for any k, j. The vector x_t is referred to as the *design* vector, the matrix $\mathbf{F}_t$ is known as the *transition* or *state* matrix and the independent white noise sequences ϵ_t and ζ_t are known as *innovations*; sometimes ϵ_t is referred to as *observation innovation* and ζ_t as *state innovation*.

In the above model, it may be of interest setting out the distributions of the innovations ϵ_t and ζ_t as univariate and multivariate Gaussian, i.e.

$$\epsilon_t \sim N(0, \sigma^2) \quad \text{and} \quad \zeta_t \sim N(0, \mathbf{Z}), \tag{3.11}$$

where σ^2 is an observation variance and $\mathbf{Z}$ is a $p \times p$ transition covariance matrix. The variance σ^2 and the covariance matrix $\mathbf{Z}$ may be time-varying, i.e. σ_t^2 and $\mathbf{Z}_t$, and this consideration may be useful in some situations, e.g. in finance, σ_t^2 may represent volatility. Some of these considerations will be explored in later chapters, but for now we will operate with a time-invariant observation variance $\sigma_t^2 = \sigma^2$, while we will allow the transition covariance matrix $\mathbf{Z}_t$ to be time-varying. An excellent introduction to state space models can be found in Durbin (2004);

book-length treatments of state space models include Jazwinski (1970), Anderson and Moore (1979), Harvey (1989), West and Harrison (1997), Commandeur and Koopman (2007) and Petris et al. (2009).

Model (3.10a)–(3.10b) is fully specified if a prior or initial Gaussian distribution for β_0 is set, i.e.

$$\beta_0 \sim N(\hat{\beta}_{0|0}, \mathbf{P}_{0|0}), \tag{3.12}$$

for some mean vector $\hat{\beta}_{0|0}$ and some covariance matrix $\mathbf{P}_{0|0}$.

In the following three sections, we discuss the estimation of β_t forward in time (known as filtering), estimation backward in time (known as smoothing) and forecasting. Given a working data set $y_{1:t} = \{y_1, \ldots, y_t\}$, filtering refers to the estimation of β_t, given $y_{1:t}$, smoothing refers to the prediction of β_t and y_t, given $y_{1:n}$ for $t = 1, 2, \ldots, n$, and forecasting refers to the prediction of β_{t+k} and y_{t+k}, given $y_{1:t}$, for some positive integer k.

3.2 Filtering

3.2.1 A First Derivation of the Kalman Filter

We discuss two forms (with two separate proofs) of the filtering algorithm, known as the Kalman filter. As it is common in important results, other proofs have been provided; some of these proofs have opened paths for interesting statistical estimation theories and applications. The interested reader should consult Duncan and Horn (1972), Hartigan (1969) and Eubank (2006). A review of the Kalman filter can be found in Meinhold and Singpurwalla (1983). An account of the Kalman filter from an econometrics perspective can be found in Pollock (2003). An alternative proof of the filter for stationary processes is proposed in Priestley and Rao (1975). The algorithm, which is given below, proposes recursive application of the conditional or posterior distribution of β_t (conditioned upon data $y_{1:t}$) from the respective conditional or posterior distribution of β_{t-1} (conditioned upon data $y_{1:t-1}$), for all time t starting at $t = 1$.

Theorem 3.2 (Kalman Filter) *Consider the state space model* (3.10a)–(3.10b) *together with the error or innovations distribution* (3.11) *and the initial distribution* (3.12). *Then, for each time $t = 1, \ldots, n$, the following apply:*

1. *The forecast distribution of β_t at time $t - 1$ is $\beta_t \mid y_{1:t-1} \sim N(\hat{\beta}_{t|t-1}, \mathbf{P}_{t|t-1})$, where $\hat{\beta}_{t|t-1} = \mathbf{F}_t \hat{\beta}_{t-1|t-1}$ and $\mathbf{P}_{t|t-1} = \mathbf{F}_t \mathbf{P}_{t-1|t-1} \mathbf{F}_t^\top + \mathbf{Z}_t$.*
2. *The posterior distribution of β_t at time t is $\beta_t \mid y_{1:t} \sim N(\hat{\beta}_{t|t}, \mathbf{P}_{t|t})$, where $\hat{\beta}_{t|t} = \hat{\beta}_{t|t-1} + K_t e_t$, $\hat{y}_{t|t-1} = x_t^\top \hat{\beta}_{t|t-1}$, $e_t = y_t - \hat{y}_{t|t-1}$, $q_{t|t-1} = x_t^\top \mathbf{P}_{t|t-1} x_t + \sigma^2$, $K_t = \mathbf{P}_{t|t-1} x_t / q_{t|t-1}$ and $\mathbf{P}_{t|t} = \mathbf{P}_{t|t-1} - q_{t|t-1} K_t K_t^\top$.*

Proof The proof is inductive. We note that distribution (2) is valid for $t = 0$, since this is just the assumed initial distribution (3.12). Let us assume that (2) is valid for $t - 1$, so that $\beta_{t-1} \mid y_{1:t-1} \sim N(\hat{\beta}_{t-1|t-1}, \mathbf{P}_{t-1|t-1})$. Then from transition equation (3.10b), we have

$$E(\beta_t \mid y_{1:t-1}) = \mathbf{F}_t E(\beta_{t-1} \mid y_{1:t-1}) = \mathbf{F}_t \hat{\beta}_{t-1|t-1} = \hat{\beta}_{t|t-1}$$

and

$$Var(\beta_t \mid y_{1:t-1}) = \mathbf{F}_t Var(\beta_{t-1} \mid y_{1:t-1})\mathbf{F}_t^\top + \mathbf{Z}_t = \mathbf{F}_t \mathbf{P}_{t-1|t-1}\mathbf{F}_t^\top + \mathbf{Z}_t = \mathbf{P}_{t|t-1},$$

since β_{t-1} and ζ_t are independent as β_{t-1} can be written as a linear combination of $\zeta_{t-1}, \zeta_{t-2}, \ldots, \zeta_1$ and ζ_s is independent of ζ_t, for any $s \neq t$. The above establishes the mean vector and the covariance matrix of $\beta_t \mid y_{1:t-1}$. Since β_t is a linear combination of β_{t-1} and ζ_t, both of which follow Gaussian distributions, it follows that $\beta_t \mid y_{1:t-1}$ follows a Gaussian distribution with mean vector $\hat{\beta}_{t|t-1}$ and covariance matrix $\mathbf{P}_{t|t-1}$, as given above. This establishes the distribution for part (1).

Proceeding now to (2), we first note that the distribution of $y_t \mid y_{1:t-1}$ is Gaussian with mean $\hat{y}_{t|t-1} = x_t^\top \hat{\beta}_{t|t-1}$ and variance $q_{t|t-1} = x_t^\top \mathbf{P}_{t|t-1} x_t + \sigma^2$. This follows from the observation model (3.10a) and the distribution of β_t in part (1). Consequently, we form the joint distribution of β_t and y_t given $y_{1:t-1}$, as

$$\begin{bmatrix} \beta_t \\ y_t \end{bmatrix} \mid y_{1:t-1} \sim N \left\{ \begin{bmatrix} \hat{\beta}_{t|t-1} \\ \hat{y}_{t|t-1} \end{bmatrix}, \begin{bmatrix} \mathbf{P}_{t|t-1} & c_t \\ c_t^\top & q_{t|t-1} \end{bmatrix} \right\},$$

where c_t is the covariance of β_t and y_t (given $y_{1:t-1}$), which is

$$c_t = Cov(\beta_t, y_t \mid y_{1:t-1}) = Cov(\beta_t, x_t^\top \beta_t + \epsilon_t \mid y_{1:t-1}) = \mathbf{P}_{t|t-1} x_t,$$

since β_t is independent of ϵ_t, as β_t can be written as a linear combination of $\zeta_t, \zeta_{t-1}, \ldots, \zeta_1$ and ϵ_t is independent of ζ_s, for any t, s.

Now form the joint distribution of $\beta_t - K_t y_t$ and y_t as

$$\begin{bmatrix} \beta_t - K_t y_t \\ y_t \end{bmatrix} \mid y_{1:t-1} \sim N \left\{ \begin{bmatrix} \hat{\beta}_{t|t-1} - K_t \hat{y}_{t|t-1} \\ \hat{y}_{t|t-1} \end{bmatrix}, \right.$$
$$\left. \begin{bmatrix} \mathbf{P}_{t|t-1} - q_{t|t-1} K_t K_t^\top & 0 \\ 0^\top & q_{t|t-1} \end{bmatrix} \right\},$$

since

$$Var(\beta_t - K_t y_t \mid y_{1:t-1}) = \mathbf{P}_{t|t-1} - q_{t|t-1} K_t K_t^\top - 2 K_t x_t^\top \mathbf{P}_{t|t-1}$$
$$= \mathbf{P}_{t|t-1} - q_{t|t-1} K_t K_t^\top$$

and

$$\mathrm{Cov}(\beta_t - K_t y_t, y_t \mid y_{1:t-1}) = \mathbf{P}_{t|t-1} x_t - K_t q_{t|t-1} = 0,$$

from the definition of $K_t = \mathbf{P}_{t|t-1} x_t / q_{t|t-1}$. Thus, given $y_{1:t-1}$, $\beta_t - K_t y_t$ and y_t are conditionally independent (since they are uncorrelated having a joint Gaussian distribution), and so by denoting by $p(\cdot)$ the respective probability density functions, we have

$$p(\beta_t - K_t y_t \mid y_{1:t}) = p(\beta_t - K_t \mid y_{1:t-1}),$$

which implies that the distribution of $\beta_t \mid y_{1:t}$ is Gaussian. Independence also implies

$$\mathrm{E}(\beta_t - K_t y_t \mid y_{1:t}) = \mathrm{E}(\beta_t - K_t y_t \mid y_{1:t-1}) \Rightarrow \hat{\beta}_{t|t} = \mathrm{E}(\beta_t \mid y_{1:t}) = \hat{\beta}_{t|t-1} + K_t e_t$$

and

$$\mathrm{Var}(\beta_t - K_t y_t \mid y_{1:t}) = \mathrm{Var}(\beta_t - K_t y_t \mid y_{1:t-1})$$
$$\Rightarrow \mathbf{P}_{t|t} = \mathrm{Var}(\beta_t \mid y_{1:t}) = \mathbf{P}_{t|t-1} - q_{t|t-1} K_t K_t^{\top}.$$

This establishes $\beta_t \mid y_{1:t} \sim N(\hat{\beta}_{t|t}, \mathbf{P}_{t|t})$. □

Some comments are in order.

- We start with some notational clarifications: $\hat{\beta}_{t|t-1}$ is the mean $\mathrm{E}(\beta_t \mid y_{1:t-1})$, i.e. the forecast of β_t at time t, given data up to time $t-1$. When y_t is observed, the data set is updated to $y_{1:t}$, and then $\hat{\beta}_{t|t}$ is the mean $\mathrm{E}(\beta_t \mid y_{1:t})$. In other words, the subscript $t \mid t-1$ indicates forecast at time t, given information $y_{1:t-1}$, and the subscript $t \mid t$ indicates filtered estimate at time t, given information $y_{1:t}$. In a similar vein, $\hat{y}_{t|t-1}$ is the one-step ahead forecast of y_t (given information up to $t-1$), and e_t is the one-step prediction error (the difference of the above prediction from the observed value of y_t). Forecasting will be covered in more detail in Sect. 3.4.
- The distribution $p(y_i \mid \beta_i)$ is the likelihood of β_i based on the single observation y_t; this is required because Theorem 3.2 suggests a recursive application sequentially over time (see also the last comment below). The distribution $p(\beta_t \mid y_{1:t-1})$ is the prior distribution at time t, meaning it is the distribution of β_t, given the past $y_{1:t-1}$ and prior of observing y_t at time t. The posterior distribution of β_t at time t is the distribution of β_t after (a posteriori) y_t and the past $y_{1:t-1}$ are observed. The Kalman filter relies upon the specification of the distribution of an initial state vector β_0; this distribution is referred to as *prior distribution* because it is the distribution of β_0 prior of observing any data. Note the difference between the prior distribution of β_0 and the prior distribution of β_t at time t.

- The vector K_t is known as the *Kalman gain*; its name originates from the fact that after multiplying it by e_t and adding it to the forecast of β_t, we obtain the posterior estimate of β_t ($\hat{\beta}_{t|t} = \hat{\beta}_{t|t-1} + K_t e_t$).
- The above theorem provides an algorithm that works recursively: with starting distribution (3.12), we can obtain $\hat{\beta}_{1|0}, \mathbf{P}_{1|0}$, and by observing y_1, we obtain $\hat{\beta}_{1|1}, \mathbf{P}_{1|1}$, and this completes a full cycle of the algorithm. Then we compute $\hat{\beta}_{2|1}, \mathbf{P}_{2|1}$, and upon observing y_2, we compute $\hat{\beta}_{2|2}, \mathbf{P}_{2|2}$ and so forth. The power of the algorithm lies in the fact that at each time t, in order to compute $\hat{\beta}_{t|t}$ and $\mathbf{P}_{t|t}$, we only need to store the current observation y_t and the respective posterior estimates at $t-1$, i.e. $\hat{\beta}_{t-1|t-1}$ and $\mathbf{P}_{t-1|t-1}$.

A Note on Statistical Computing

For computation purposes, we use the programming environment R (https:// www.r-project.org). One of the most important features of R is the numerous packages available for free. These can be accessed via the Contributed Packages website: http://cran.r-project.org/web/packages/. A package can be installed via the `Packages` tab in the R console and then loaded in the current working directory from the same tab. We have developed the package BTSA (Bayesian Time Series Analysis), which performs Bayesian computation for state space models and includes many of the R functions used in the text. The package has a manual, which is downloaded for free in the link http://cran.r-project.org/web/packages/BTSA/BTSA.pdf.

Example 3.1 (Annual Temperatures of Central England)   We consider a historical time series, consisting of average annual temperatures (°C) in central England for the decades 1663–1674 to 1993–2002. Annual values are calculated from monthly data by totaling the monthly values and dividing by 12. This series is based on a famous compilation in 1974 by meteorologist Gordon Manley of monthly mean temperatures for the UK West-Central Midlands, 1659–1973, see Manley (1974). The data for the earliest years, before introduction of the thermometer, are based on careful research in documentary sources such as old diaries. Manley draws attention to the fact that the data values before 1723 are much less reliable than the later values. Manley's original series has now been expanded to give daily values from 1723 to the present. The series is one of the longest consistent records of temperature in existence for anywhere in the world.

There are many questions of interest, particularly in connection with climate change, including whether there are any regularities in temperature fluctuations, whether there is evidence of a consistent rise in temperature going beyond natural fluctuations and so forth. Part of the data are shown by the solid points in Fig. 3.1.

In this example we are interested in setting up a state space model for this data set and applying the Kalman filter, e.g. in order to forecast the temperature values of future dates or to describe changes in the level or the mean of the temperatures over time. By looking at the dynamics of the data, we observe that it fluctuates around some level, but there is clear evidence of local evolution, e.g. the temperatures up to 1750 seem to have much more uncertainty around the level and the temperatures after 1900 seem to have a slight increasing trend. This local evolution suggests that a local level model may be a reasonable representation of the data. Thus we use the model

$$y_t = \beta_t + \epsilon_t \qquad \text{(observation model)}, \qquad (3.13a)$$

$$\beta_t = \beta_{t-1} + \zeta_t \qquad \text{(transition model)}, \qquad (3.13b)$$

where y_t denotes the temperature at time t, β_t is the level of the series and the remaining terms of the model are as in model (3.10a)–(3.10b). Here we have $p = 1$, so that β_t is scalar, and we have used $\beta_0 \sim N(9, 1000)$, $\sigma^2 = 1$ and $Z = 10$. The value of $\hat{\beta}_{0|0} = 9$ is motivated by the fact that the mean annual temperature is believed to fluctuate around $9\,°C$. The variance $P_{0|0} = 1000$ suggests a large uncertainty in this initial belief of the mean temperature. The values of σ^2 and Z are chosen here somewhat arbitrarily, but they can be estimated by maximum likelihood methods; for more details on this approach, see Chap. 4. The following commands in R were used to read the data and to fit the above model:

```
> # read data
> temp <- read.table("temp.txt")
> temp <- temp[,2]

> # fit local level model
> fit <- bts.filter(temp, x0=1, F0=1, obsvar=1, Z0=10,
+ beta0=9, P0=1000, DISO=FALSE, VAREST=FALSE)
```

Figure 3.1 plots the first 100 points of the data (solid points) with its forecasts, which are the estimated states $\hat{\beta}_{t|t-1}$. From Fig. 3.1, we observe that the forecasts generally follow closely the time series values y_t. The estimated level $\hat{\beta}_{t|t}$ fluctuates around its mean 9.190. However, there are some poor forecasts, notably for $t = 82$ (year 1740), the forecast mean is $\hat{y}_{82|1:81} = 9.252$, while the observation for that year is $y_{82} = 6.84$. This causes the forecast at $t = 83$ (Year 1741) to be extremely poor ($\hat{y}_{83|1:82} = 7.042$, with observation $y_{83} = 9.3$). Figure 3.1 is produced using the R commands:

```
> # define time series objects to be plotted
> tempts <- ts(temp[1:100], start=1659, frequency=1)
> pred <- ts(fit$FittedMean[1:100], start=1659, frequency=1)
> # Time series plot
> ts.plot(pred,lty=2, col=2, main=expression("Annual central
+ England temperatures with forecasts"),ylab="Degrees in Celsius",
+ xlab="Year", ylim=c(6.7,10.5) )
> points(tempts, pch=20)
> points(pred, pch=4, col=2)
> legend("bottomleft",c("Observations","Forecast mean"),
+ pch=c(20, 4), col=c(1,2))
```

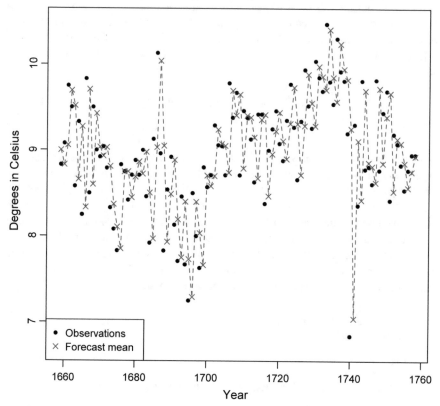

Fig. 3.1 Temperature values of central England (solid points) and their forecasts (dashed lines/ticks)

It is of some interest to calculate the forecast variance $q_t = x_t^\top P_{t|t-1} x_t + \sigma^2 = P_{t|t-1} + \sigma^2$. This is provided in the `bts.filter` command of the `btw` package as

```
> q <- fit$FittedVar
```

The first value of `q` is $q_{1|0} = 1011$, the second drops to $q_{2|1} = 11.99901$, the third is $q_{3|2} = 11.91666$ and then $q_{t|t-1} = 11.91608$ is time-invariant for any $t \geq 4$. Thus after $t = 3$, the forecast variance q_t converges to 11.91608. The same convergence result applies for the filtered variance $P_{t|t}$, i.e. $P_{t|t} = 0.91608$, for $t \geq 4$. This desirable phenomenon, of stable filtered variances, is common in a wide class of state space models and is discussed in some detail in Sect. 3.5 and also in Sect. 4.5.

3.2.2 A Second Derivation of the Kalman Filter

In some situations, the Gaussian assumption in the innovation terms ϵ_t and ζ_t of the state space model (3.10a)–(3.10b) may be dropped. In such a case, ϵ_t and ζ_t are only partially specified via their mean and variances, but their distribution is unspecified. This may be the result of lack of support for the data and the transition equation to follow Gaussian distributions, or it may reflect the modeller's reluctance to specify a Gaussian distribution for ϵ_t and ζ_t. For such cases, the recursions of $\hat{\beta}_{t|t-1}, \mathbf{P}_{t|t-1}, \hat{\beta}_{t|t}, \mathbf{P}_{t|t}$ of Theorem 3.2 still hold. In other words, by relaxing the assumption of normality of the innovations of the state space model, we lose the specification of the conditional and predictive distributions, but we retain the first two moments as a means of filtering. This remarkable result is presented in the next theorem.

Theorem 3.3 *In the state space model (3.10a)–(3.10b), suppose that the distribution assumption of the innovations ϵ_t and ζ_t is dropped, i.e. we only assume that $E(\epsilon_t) = 0$, $Var(\epsilon_t) = \sigma^2$, $E(\zeta_t) = 0$ and $Var(\zeta_t) = \mathbf{Z}_t$, with σ^2 and $\mathbf{Z}_t$ as defined in (3.10a)–(3.10b). We assume that the elements of the design vector x_t of (3.10a) are linearly independent. If the estimator $\hat{\beta}_{t|t}$ of β_t is required to be linear in y_t and to minimise the sum of squares $S = \sum_{t=1}^{n}(y_t - x_t^\top \beta_t)^2$ subject to the minimisation of the cost function $E[(\beta_t - \hat{\beta}_{t|t})^\top(\beta_t - \hat{\beta}_{t|t})]$, then it is given by $\hat{\beta}_{t|t} = \hat{\beta}_{t|t-1} + K_t e_t$, where $\hat{\beta}_{t|t-1}$, K_t and e_t are as in Theorem 3.2. The minimum covariance matrix of β_t is given by the recursion $\mathbf{P}_{t|t} = \mathbf{P}_{t|t-1} - q_{t|t-1} K_t K_t^\top$, where $q_{t|t-1}$ is defined as in Theorem 3.2.*

Proof First note that given $\hat{\beta}_{t-1|t-1}$ and $\mathbf{P}_{t-1|t-1}$ at time $t - 1$, the recursions for $\hat{\beta}_{t|t-1}$ and $\mathbf{P}_{t|t-1}$ follow directly from part (1) of Theorem 3.2.

Suppose we wish to obtain an estimator of β_t that is linear in y_t, that is $\hat{\beta}_{t|t} = a_t + K_t y_t$, for some a_t and K_t (to be specified later). Then we can write

$$\hat{\beta}_{t|t} = a_t^* + K_t e_t, \tag{3.14}$$

with $a_t^* = a_t - K_t \hat{y}_{t|t-1}$, $e_t = y_t - \hat{y}_{t|t-1}$ and $\hat{y}_{t|t-1}$ as in Theorem 3.2. We will show that for some K_t, if $\hat{\beta}_{t|t}$ is required to minimise the sum of squares

$$S = \sum_{t=1}^{n}(y_t - x_t^\top \beta_t)^2, \tag{3.15}$$

then $a_t^* = \mathbf{F}_t \hat{\beta}_{t-1|t-1} = \hat{\beta}_{t|t-1}$. To prove this, first define

$$
y = \begin{bmatrix} y_1 \\ y_2 \\ \vdots \\ y_n \end{bmatrix}, \quad
\mathbf{X} = \begin{bmatrix} x_1^\top & 0 & \cdots & 0 \\ 0 & x_2^\top & \cdots & 0 \\ \vdots & \vdots & \ddots & \vdots \\ 0 & 0 & \cdots & x_n^\top \end{bmatrix}, \quad
\beta = \begin{bmatrix} \beta_1 \\ \beta_2 \\ \vdots \\ \beta_n \end{bmatrix},
$$

$$
\epsilon = \begin{bmatrix} \epsilon_1 \\ \epsilon_2 \\ \vdots \\ \epsilon_n \end{bmatrix}, \quad
e = \begin{bmatrix} e_1 \\ e_2 \\ \vdots \\ e_n \end{bmatrix}, \quad
\mathbf{K} = \begin{bmatrix} K_1 & 0 & \cdots & 0 \\ 0 & K_2 & \cdots & 0 \\ \vdots & \vdots & \ddots & \vdots \\ 0 & 0 & \cdots & K_n \end{bmatrix}.
$$

Then we can write the observation model (3.10a) as $y = \mathbf{X}\beta + \epsilon$ and the sum of squares (3.15) as

$$
S(\beta) = (y - \mathbf{X}\beta)^\top (y - \mathbf{X}\beta),
$$

and from the assumed linear form (3.14), we have $\hat{\beta} = a^* + \mathbf{K}e$, where $a^{*\top} = [(a_1^*)^\top, \ldots, (a_n^*)^\top]$. We will show that $a^* = b^*$, where $b^* = \mathrm{E}(\beta) = \beta^{*\top} = [\hat{\beta}_{1|0}^\top, \ldots, \hat{\beta}_{n|n-1}^\top]$. With the above $\hat{\beta}$, the sum of squares can be written as

$$
\begin{aligned}
S \equiv S(\hat{\beta}) &= (y - \mathbf{X}a^* - \mathbf{X}\mathbf{K}e)^\top (y - \mathbf{X}a^* - \mathbf{X}\mathbf{K}e) \\
&= (y - \mathbf{X}a^*)^\top (y - \mathbf{X}a^*) - 2(y - \mathbf{X}a^*)^\top \mathbf{X}\mathbf{K}e \\
&\quad + e^\top \mathbf{K}^\top \mathbf{X}^\top \mathbf{X}\mathbf{K}e,
\end{aligned}
$$

which is minimised when $\mathrm{E}(y - \mathbf{X}a^*) = 0$. To see this, first set $z = y - \mathbf{X}a^*$, and then notice that the function $f(z) = z^\top z - 2z^\top c_1 + c_2$ is equal to S, if $c_1 = \mathbf{X}\mathbf{K}e$ and $c_2 = e^\top \mathbf{K}^\top \mathbf{X}^\top \mathbf{X}\mathbf{K}e$ (constants not depending on a^*). Now, using vector differentiation, we have

$$
\frac{d f(z)}{d z} = 2\mathbf{I}z - 2c_1,
$$

which is equal to zero when $z = c_1$. The second derivative of $f(z)$ is equal to

$$
\frac{d^2 f(z)}{d z d z^T} = 2\mathbf{I},
$$

which is a positive definite matrix, and hence the stationary vector $z = c_1$ minimises $f(z)$. Thus, the value of a^* that minimises S satisfies the equation $y - \mathbf{X}a^* = \mathbf{X}\mathbf{K}e$, or $\mathrm{E}(y - \mathbf{X}a^*) = \mathbf{X}\mathbf{K}\mathrm{E}(e) = 0$, since $\mathrm{E}(e) = 0$.

This implies $\mathrm{E}(y) = \mathbf{X}a^* = \mathbf{X}\mathrm{E}(\beta) = \mathbf{X}b^*$. With the definition of $\mathbf{X}$, a^* and b^*, this in turn implies $x_t^\top c = \sum_{i=1}^p x_{ti} c_i = 0$, where $x_t = (x_{1t}, \ldots, x_{tp})^\top$ and $c =$

$(c_1, \ldots, c_p)^\top$ and $c = a_t^* - b_t^*$. Since the elements of x_t are linearly independent, it follows that $c = 0$, and hence $a_t^* = b_t^*$, for all $t = 1, \ldots, n$. Thus, $a_t^* = \hat{\beta}_{t|t-1}$, and from (3.14), we have

$$\hat{\beta}_{t|t} = \hat{\beta}_{t|t-1} + K_t e_t, \tag{3.16}$$

for some value of K_t to be defined. From the definition of $\mathbf{P}_{t|t}$, we have that

$$
\begin{aligned}
\mathbf{P}_{t|t} &= \mathrm{E}[(\beta_t - \hat{\beta}_{t|t})(\beta_t - \hat{\beta}_{t|t})^\top] \\
&= \mathrm{E}\{[\beta_t - \hat{\beta}_{t|t-1} - K_t(x_t^\top \beta_t + \epsilon_t - x_t^\top \hat{\beta}_{t|t-1})] \\
&\quad \times [\beta_t - \hat{\beta}_{t|t-1} - K_t(x_t^\top \beta_t + \epsilon_t - x_t^\top \hat{\beta}_{t|t-1})]\}^\top \\
&= \mathrm{E}[(\mathbf{I} - K_t x_t^\top)(\beta_t - \hat{\beta}_{t|t-1})(\beta_t - \hat{\beta}_{t|t-1})^\top (\mathbf{I} - x_t K_t^\top) \\
&\quad + \epsilon_t^2 K_t K_t^\top] \\
&= (\mathbf{I} - K_t x_t^\top) \mathbf{P}_{t|t-1}(\mathbf{I} - x_t K_t^\top) + \sigma^2 K_t K_t^\top \\
&= \mathbf{P}_{t|t-1} - K_t x_t^\top \mathbf{P}_{t|t-1} - \mathbf{P}_{t|t-1} x_t K_t^\top + q_{t|t-1} K_t K_t^\top, \tag{3.17}
\end{aligned}
$$

where $q_{t|t-1} = x_t^\top \mathbf{P}_{t|t-1} x_t + \sigma^2$, since β_t and ϵ_t are independent, as β_t can be written as a linear combination of the innovation vectors $\zeta_t, \ldots, \zeta_1$ and ϵ_t and ζ_s are independent for any t, s.

The covariance matrix $\mathbf{P}_{t|t}$ in (3.17) is given as a function of K_t and so the minimum $\mathbf{P}_{t|t}$ is achieved by minimising the cost function

$$\mathrm{E}[(\beta_t - \hat{\beta}_{t|t})^\top (\beta_t - \hat{\beta}_{t|t})] = \mathrm{trace}\{\mathrm{E}[(\beta_t - \hat{\beta}_{t|t})(\beta_t - \hat{\beta}_{t|t})^\top]\} = \mathrm{trace}(\mathbf{P}_{t|t}),$$

with respect to K_t.

By applying the trace to (3.17), we get

$$
\begin{aligned}
\mathrm{trace}(\mathbf{P}_{t|t}) &= \mathrm{trace}(\mathbf{P}_{t|t-1}) - \mathrm{trace}(K_t x_t^\top \mathbf{P}_{t|t-1}) - \mathrm{trace}(\mathbf{P}_{t|t-1} x_t K_t^\top) \\
&\quad + \mathrm{trace}(q_{t|t-1} K_t K_t^\top) \\
&= \mathrm{trace}(\mathbf{P}_{t|t-1}) - \mathrm{trace}(x_t^\top \mathbf{P}_{t|t-1} K_t) - \mathrm{trace}(K_t \mathbf{P}_{t|t-1} x_t) \\
&\quad + q_{t|t-1} \mathrm{trace}(K_t K_t^\top) \\
&= \mathrm{trace}(\mathbf{P}_{t|t-1}) - 2 x_t^\top \mathbf{P}_{t|t-1} K_t + q_{t|t-1} \mathrm{trace}(K_t K_t^\top),
\end{aligned}
$$

and thus K_t is the solution of the matrix equation

$$\frac{\partial \mathrm{trace}(\mathbf{P}_{t|t})}{\partial K_t} = -2\mathbf{P}_{t|t-1} x_t + 2 q_{t|t-1} K_t = 0,$$

where $\partial \text{trace}(\mathbf{P}_{t|t})/\partial K_t$ denotes the partial derivative of the trace of $\mathbf{P}_{t|t}$ with respect to K_t. Solving the above equation, we obtain $K_t = \mathbf{P}_{t|t-1} x_t / q_{t|t-1}$. The quantity K_t is optimal in the sense that among all linear estimators $\hat{\beta}_{t|t}$, (3.16) minimises $\text{E}[(\beta_t - \hat{\beta}_{t|t})^\top (\beta_t - \hat{\beta}_{t|t})]$. With $K_t = \mathbf{P}_{t|t-1} x_t / q_{t|t-1}$, from (3.17), the minimum covariance matrix $\mathbf{P}_{t|t}$ becomes $\mathbf{P}_{t|t} = \mathbf{P}_{t|t-1} - q_{t|t-1} K_t K_t^\top$. $\square$

Theorem 3.3 provides important results. It validates the recursions of $\hat{\beta}_{t|t}$ and $\mathbf{P}_{t|t}$, even in situations where the modeller is reluctant to specify Gaussian distributions for the innovations ϵ_t and ζ_t. Even if one is happy to accept the Gaussian assumption of the innovations, Theorem 3.3 shows that $\hat{\beta}_{1|1}, \ldots \hat{\beta}_{n|n}$ minimise the sum of squares subject to the minimisation of the cost function $\text{E}[(\beta_t - \hat{\beta}_{t|t})^\top (\beta_t - \hat{\beta}_{t|t})]$, for each t. This is equivalent in the minimisation of

$$\sum_{t=1}^{n}(y_t - x_t^\top \beta_t)^2 + \mu \sum_{t=2}^{n}(\beta_t - \mathbf{F}_t \beta_{t-1})^\top (\beta_t - \mathbf{F}_t \beta_{t-1}),$$

for some $\mu \geq 0$. Derivation of the optimal $\hat{\beta}_t$ under this approach is known as flexible least squares (FLS) and it is derived in Kalaba and Tesfatsion (1988) and further developed in Tesfatsion and Kalaba (1989). This approach returns exactly the Kalman filter recursion of $\hat{\beta}_{t|t}$, as it is shown in Montana et al. (2009). The parameter μ controls the magnitude of the distance $\beta_t - \mathbf{F}_t \beta_{t-1}$ and in Montana et al. (2009) it is shown that μ is related to the covariance matrix of ζ_t via $\mathbf{Z}_t = \mu^{-1}\mathbf{I}$. However, it turns out that the Kalman filter setting is more general than that of the FLS, as it enables a more general specification for $\mathbf{Z}_t$, i.e. components of ζ_t not having equal variances μ^{-1} and zero covariances.

In the rest of this book, unless otherwise stated, we will be assuming the Gaussian distributions for the innovations, but we should bear in mind that this assumption may be relaxed as discussed above. The following is a summary of the Kalman filtering algorithm under the assumption of Gaussian distributions of the innovations.

Kalman Filter

1. Prior distribution at $t = 0$: $\beta_0 \sim N(\hat{\beta}_{0|0}, \mathbf{P}_{0|0})$;
2. Posterior distribution of β_{t-1} at time $t - 1$:
   $\beta_{t-1} \mid y_{1:t-1} \sim N(\hat{\beta}_{t-1|t-1}, \mathbf{P}_{t-1|t-1})$;
3. Prior distribution of β_t at time $t - 1$:
   $\beta_t \mid y_{1:t-1} \sim N(\hat{\beta}_{t|t-1}, \mathbf{P}_{t|t-1})$,
   where $\hat{\beta}_{t|t-1} = \mathbf{F}_t \hat{\beta}_{t-1|t-1}$ and $\mathbf{P}_{t|t-1} = \mathbf{F}_t \mathbf{P}_{t-1|t-1} \mathbf{F}_t^\top + \mathbf{Z}_t$;
4. Posterior distribution at time t:

(continued)

$\beta_t \mid y_{1:t} \sim N(\hat{\beta}_{t|t}, \mathbf{P}_{t|t})$, where

$$\hat{\beta}_{t|t} = \hat{\beta}_{t|t-1} + K_t e_t, \quad \mathbf{P}_{t|t} = \mathbf{P}_{t|t-1} - q_{t|t-1} K_t K_t^\top,$$

$$\hat{y}_{t|t-1} = x_t^\top \hat{\beta}_{t|t-1}, \quad e_t = y_t - \hat{y}_{t|t-1}, \quad q_{t|t-1} = x_t^\top \mathbf{P}_{t|t-1} x_t + \sigma^2,$$

$$K_t = \mathbf{P}_{t|t-1} x_t / q_{t|t-1}.$$

3.3 Smoothing

In this section we consider estimation of the states β_t and of the observations y_t, given information $y_{1:n} = \{y_1, \ldots, y_n\}$, for $t = 1, 2, \ldots, n$; this is known as the *fixed-interval smoothing problem*, because n is fixed. The distributions of $\beta_t \mid y_{1:n}$ and $y_t \mid y_{1:n}$ are called the smoothed state and observation distributions, as they are the distributions smoothed by data $y_{1:n}$; they are provided in the next theorem. Smoothing is an important development of the theory and application of state space models and has been discussed, among others, in Anderson and Moore (1979), Catlin (1989), De Jong (1989), Harvey (1989), Koopman (1993) and Durbin and Koopman (2012). The classic fixed-interval smoothing algorithm that follows was derived in 1965 by Raunch et al. (1965). We also prove the recurrence updating of the lag-one covariance smoother discussed in Shumway and Stoffer (2017).

3.3.1 Fixed-Interval Smoothing

Theorem 3.4 (Fixed-Interval Smoothing) *In the state space model* (3.10a)– (3.10b) *with information* $y_{1:n} = \{y_1, \ldots, y_n\}$, *the following smoothing distributions apply:*

1. *For each* $t = 1, \ldots, n$, *the smoothed state distribution is* $\beta_t \mid y_{1:n} \sim N(\hat{\beta}_{t|n}, \mathbf{P}_{t|n})$, *where* $\hat{\beta}_{t|n} = \hat{\beta}_{t|t} + \mathbf{L}_t(\hat{\beta}_{t+1|n} - \hat{\beta}_{t+1|t})$ *and* $\mathbf{P}_{t|n} = \mathbf{P}_{t|t} + \mathbf{L}_t(\mathbf{P}_{t+1|n} - \mathbf{P}_{t+1|t})\mathbf{L}_t^\top$, *with* $\mathbf{L}_t = \mathbf{P}_{t|t} \mathbf{F}_{t+1}^\top \mathbf{P}_{t+1|t}^{-1}$ *and* $\hat{\beta}_{t|t}, \hat{\beta}_{t+1|t}, \mathbf{P}_{t|t}, \mathbf{P}_{t+1|t}$ *being calculated via the Kalman filter (Theorem 3.2).*
2. *For each* $t = 1, \ldots, n - 1$, *the smoothed observation distribution is* $y_t \mid y_{1:n} \sim N(\hat{y}_{t|n}, q_{t|n})$, *where* $\hat{y}_{t|n} = x_t^\top \hat{\beta}_{t|n}$ *and* $q_{t|n} = x_t^\top \mathbf{P}_{t|n} x_t + \sigma^2$.

Proof First we prove part (1) of the theorem. The proof proceeds by induction (backwards in time) for $t = 1, 2, \ldots, n$. For $t = n$, the stated distribution of $\beta_t \mid y_{1:n}$ is just the posterior distribution $\beta_n \mid y_{1:n}$ of Theorem 3.2. Suppose that the theorem is true for $t + 1$ ($t \leq n - 1$), i.e. the smoothed state distribution of β_{t+1} is $\beta_{t+1} \mid y_{1:n} \sim N(\hat{\beta}_{t+1|n}, \mathbf{P}_{t+1|n})$, for some known $\hat{\beta}_{t+1|n}, \mathbf{P}_{t+1|n}$. The smoothed state distribution of β_t is defined as the marginal distribution of (β_t, β_{t+1}) if β_{t+1} is integrated out, i.e.

$$p(\beta_t \mid y_{1:n}) = \int_{\mathbb{R}^p} p(\beta_t, \beta_{t+1} \mid y_{1:n}) \, d\beta_{t+1}$$

$$= \int_{\mathbb{R}^p} p(\beta_t \mid \beta_{t+1}, y_{1:n}) p(\beta_{t+1} \mid y_{1:n}) \, d\beta_{t+1}. \qquad (3.18)$$

The second term in the integral of (3.18) is the Gaussian distribution above, assumed by induction. In the first term, we apply the Bayes theorem

$$p(\beta_t \mid \beta_{t+1}, y_{1:n}) = \frac{p(y \mid \beta_t, \beta_{t+1}, y_{1:t}) p(\beta_t \mid \beta_{t+1}, y_{1:t})}{p(y \mid \beta_{t+1}, y_{1:t})},$$

where $y = (y_{t+1}, \ldots, y_n)$. Now, given β_{t+1}, y is independent of β_t. Intuitively, this works since β_t precedes β_{t+1} and so β_t is 'embedded' in β_{t+1}. More mathematically, we can show the independence by noting that $y \mid \beta_t, \beta_{t+1}$ is a Gaussian distribution (from the observation model (3.10a)) and noting that $\text{Cov}(y, \beta_t \mid \beta_{t+1}, y_{1:t}) = 0$, because each element of y can be written as a linear combination of β_{t+1}. This establishes the independence and so the two terms with y in the above equation cancel out. Thus, applying the Bayes theorem for the remaining term, we obtain

$$p(\beta_t \mid \beta_{t+1}, y_t) \propto p(\beta_{t+1} \mid \beta_t, y_{1:t}) p(\beta_t \mid y_{1:t}).$$

Now we know from the transition model (3.10b) that $\beta_{t+1} \mid \beta_t, y_t \sim N(\mathbf{F}_{t+1}\beta_t, \mathbf{Z}_{t+1})$ and from the Kalman filter (Theorem 3.2) that $\beta_t \mid y_{1:t} \sim N(\hat{\beta}_{t\mid t}, \mathbf{P}_{t\mid t})$. Then we form the joint distribution of β_t and β_{t+1}, given $y_{1:t}$ as

$$\begin{bmatrix} \beta_t \\ \beta_{t+1} \end{bmatrix} \mid y_{1:t} \sim N \left\{ \begin{bmatrix} \hat{\beta}_{t\mid t} \\ \hat{\beta}_{t+1\mid t} \end{bmatrix}, \begin{bmatrix} \mathbf{P}_{t\mid t} & c_t \\ c_t^\top & \mathbf{P}_{t+1\mid t} \end{bmatrix} \right\},$$

where the covariance c_t is

$$c_t = \text{Cov}(\beta_t, \beta_{t+1} \mid y_{1:t}) = \text{Cov}(\beta_t, \mathbf{F}_{t+1}\beta_t + \zeta_{t+1} \mid y_{1:t})$$

$$= \text{Var}(\beta_t \mid y_{1:t}) \mathbf{F}_{t+1}^\top = \mathbf{P}_{t\mid t} \mathbf{F}_{t+1}^\top.$$

As a result, in a similar way as in Theorem 3.2, we obtain the conditional distribution of $\beta_t \mid \beta_{t+1}, y_{1:t}$ as $\beta_t \mid \beta_{t+1}, y_{1:t} \sim N(\hat{\beta}_{t\mid n}^*, \mathbf{P}_{t\mid n}^*)$, with

$$\hat{\beta}_{t\mid n}^* = \hat{\beta}_{t\mid t} + \mathbf{L}_t(\beta_{t+1} - \hat{\beta}_{t+1\mid t}),$$

$$\mathbf{P}_{t\mid n}^* = \mathbf{P}_{t\mid t} - \mathbf{L}_t \mathbf{P}_{t+1\mid t} \mathbf{L}_t^\top,$$

where $\mathbf{L}_t$ is as defined in the theorem. From these two equations, the mean vector and the covariance matrix of $\beta_t \mid y_{1:n}$ are calculated using the tower properties (see Sect. 2.3.1) as

$$\hat{\beta}_{t|n} = E(\beta_t \mid y_{1:n}) = E[E(\beta_t \mid \beta_{t+1}, y_{1:n}) \mid y_{1:n}]$$
$$= E[\hat{\beta}_{t|t} + \mathbf{L}_t(\beta_{t+1} - \hat{\beta}_{t+1|t}) \mid y_{1:n}]$$
$$= \hat{\beta}_{t|t} + \mathbf{L}_t(\hat{\beta}_{t+1|n} - \hat{\beta}_{t+1|t})$$

and

$$\mathbf{P}_{t|n} = E[\text{Var}(\beta_t \mid \beta_{t+1}, y_{1:n}) \mid y_{1:n}] + \text{Var}[E(\beta_t \mid \beta_{t+1}, y_{1:n}) \mid y_{1:n}]$$
$$= E(\mathbf{P}_{t|t} - \mathbf{L}_t\mathbf{P}_{t+1|t}\mathbf{L}_t^\top \mid y_{1:n})$$
$$+\text{Var}[\hat{\beta}_{t|t} + \mathbf{L}_t(\beta_{t+1} - \hat{\beta}_{t+1|t}) \mid y_{1:n}]$$
$$= \mathbf{P}_{t|t} - \mathbf{L}_t\mathbf{P}_{t+1|t}\mathbf{L}_t^\top + \mathbf{L}_t\mathbf{P}_{t+1|n}\mathbf{L}_t^\top$$
$$= \mathbf{P}_{t|t} + \mathbf{L}_t(\mathbf{P}_{t+1|n} - \mathbf{P}_{t+1|t})\mathbf{L}_t^\top.$$

The distribution of $\beta_t \mid y_{1:n}$ is Gaussian because from (3.18) this is defined as the marginal of a joint Gaussian distribution (of β_t and β_{t+1}). This settles part (1).

Part (2) follows immediately from the observation model (3.10a) and part (1). The mean and variance of y_t, given $y_{1:n}$, are

$$\hat{y}_{t|n} = E(y_t \mid y_{1:n}) = E(x_t^\top \beta_t + \epsilon_t \mid y_{1:n}) = x_t^\top \hat{\beta}_{t|n}$$

and

$$q_{t|n} = \text{Var}(y_t \mid y_{1:n}) = \text{Var}(x_t^\top \beta_t + \epsilon_t \mid y_{1:n}) = x_t^\top \mathbf{P}_{t|n}x_t + \sigma^2,$$

since β_t is independent of ϵ_t. Since the distributions of ϵ_t and of β_t, given $y_{1:n}$, are both Gaussian, it follows that the distribution of $y_t \mid y_{1:n}$ is also Gaussian, and this completes the proof. □

We note that a $(1 - \alpha)\%$ smoothing interval for y_t, based on information $y_{1:n}$, is obtained as $\hat{y}_{t|n} \pm z_{1-\alpha/2}\sqrt{q_{t|n}}$, where $z_{1-\alpha/2}$ denotes the $(1 - \alpha/2)\%$-quantile of the standard Gaussian distribution $N(0, 1)$. The quantity $\hat{y}_{t|n}$ of the second part of the theorem is known as the *smoothed forecast mean*, or just *smoothed forecast* at time t, based on information $y_{1:n} = \{y_1, \ldots, y_n\}$.

The above theorem proposes an algorithm that works recursively backward in time. At time n, we obtain $\beta_n \mid y_{1:n} \sim N(\hat{\beta}_{n|n}, \mathbf{P}_{n|n})$ via the Kalman filter. Then we obtain backward in time the distributions $\beta_{n-1} \mid y_{1:n}$, $\beta_{n-2} \mid y_{1:n}$ and so on, up to $\beta_1 \mid y_{1:n}$, which are used to obtain the smoothed distributions $y_t \mid y_{1:n}$. The algorithm is summarised below.

Fixed-Interval Smoothing

For any $t = 1, 2, \ldots, n$:

1. Initial state distribution at $t = n$: $\beta_n \mid y_{1:n} \sim N(\hat{\beta}_{n|n}, \mathbf{P}_{n|n})$, where $\hat{\beta}_{n|n}$ and $\mathbf{P}_{n|n}$ are computed by the Kalman filter (Theorem 3.2).
2. Smoothed state distribution, for $t = n - 1, n - 2, \ldots, 1$:
$\beta_t \mid y_{1:n} \sim N(\hat{\beta}_{t|n}, \mathbf{P}_{t|n})$, where

$$\hat{\beta}_{t|n} = \hat{\beta}_{t|t} + \mathbf{L}_t(\hat{\beta}_{t+1|n} - \hat{\beta}_{t+1|t}),$$

$$\mathbf{P}_{t|n} = \mathbf{P}_{t|t} + \mathbf{L}_t(\mathbf{P}_{t+1|n} - \mathbf{P}_{t+1|t})\mathbf{L}_t^\top,$$

with $\mathbf{L}_t = \mathbf{P}_{t|t}\mathbf{F}_{t+1}^\top\mathbf{P}_{t+1|t}^{-1}$ and $\hat{\beta}_{t|t}, \hat{\beta}_{t+1|t}, \mathbf{P}_{t|t}, \mathbf{P}_{t+1|t}$ being calculated by the Kalman filter.
3. Smoothed observation distribution, for $t = 1, 2, \ldots, n - 1$:
$y_t \mid y_{1:n} \sim N(\hat{y}_{t|n}, q_{t|n})$,
where $\hat{y}_{t|n} = x_t^\top \hat{\beta}_{t|n}$ and $q_{t|n} = x_t^\top \mathbf{P}_{t|n} x_t + \sigma^2$.

Example 3.2 (Annual Temperature Example Continued) In Example 3.1, we computed forecasts $\hat{y}_{t|t-1}$ of observations y_t, for the annual temperature data for central England. In this example, we compute the smooth estimates $y_{t|n} = x_t^\top \hat{\beta}_{t|n}$ of y_t, for the adopted local level model (3.13a)–(3.13b) of Example 3.1, where $t = 1, \ldots, 343$ and $n = 344$ (corresponding to the years 1659, 1660, \ldots, 2002). Figure 3.2 plots the first 100 values of the smooth estimates $\hat{y}_{t|n}$ ($t = 1, \ldots, n-1$) (dashed line with stars), together with the forecasts $\hat{y}_{t|t-1}$ ($t = 1, \ldots, n$) (dotted line with ticks) and the observed data (solid points). This plot is produced by the R commands

```
> # temp and pred is defined previously
> fit.s <- bts.smooth(temp, x0=1, F0=1, obsvar=1, Z0=10,
+ beta0=9, P0=1000, DISO=FALSE)
> smooth1 <- ts(fit.s$SmoothMean[1:100],  start=1659, frequency=1)
> ts.plot(smooth1, lty=2, col=4,main=expression("Temperatures
+ with forecasts and smooth estimates"),xlab="Year",
+ ylab="Degrees Celsius",ylim=c(6.7,10.5))
> #lines(smooth1, lty=2)
> points(tempts, pch=20)
> points(pred, pch=4, col=2)
> points(smooth1, pch=8, col=4)
> legend("bottomleft",c("Observations","Forecast mean",
+ "Smooth estimate"), pch=c(20, 4, 8), col=c(1,2, 4))
> abline(v=1740, lty=3)
```

It is expected that the smooth estimates of y_t will be considerably closer to the observations y_t than the forecasts, since they are using all observations $y_1, \ldots, y_{344}$.

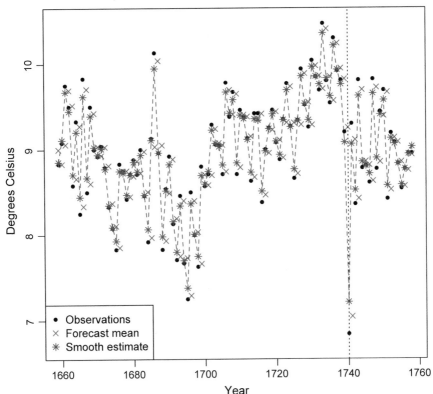

Fig. 3.2 Smoothed estimates (dashed lines/stars), forecasts (dotted lines/ticks) and actual observations (solid points) for the temperature data

At a first glance, a plot of these means shows little difference to the forecast means $\hat{y}_{t|t-1}$. However, a more careful examination reveals that the smoothed forecasts are much more accurate than the forecasts, which are based on filtering, and lag behind the data by one time point. This effect is much more prominent for the 'extreme' observations. For example, the observation in year 1740 is 6.84 °C. The smoothed estimate is 7.212 °C, but the forecast is 9.252 °C. This happens because the forecast is prior to the observation of $t = 82$ (corresponding to year 1740), and thus it takes into account only past observations $y_{81}, y_{80} \ldots$, which are much higher than y_{82}; for example, $y_{81} = 9.2$ °C and $y_{80} = 9.81$ °C. In contrast, for $\hat{y}_{82|n}$ the smooth estimate of y_{82}, all years are taken into account, resulting in much more accurate estimation. This is clearly depicted by the vertical line in Fig. 3.2, from which we see that in year 1740 the actual observation and the smoothed estimate are close, but the respective filtered forecast is far from these; the forecast that appears to be close to the observation refers to the next time point, year 1741.

A second observation is that the forecast of $y_{83} = 9.3\,°C$ is equal to $7.04\,°C$, which again is very inaccurate. This is due to the fact that this forecast is affected by the low value of $y_{82} = 6.84\,°C$; we can observe that most forecasts lag behind the original observations by one time point. This effect is more prominent in extreme observations, such as y_{82} as discussed above, and it is a characteristic of most forecasting systems. The smoothed forecasts avoid this lagging and produce more accurate forecasts, but the price to pay is that they require the availability of all data. Thus, smoothing is not suitable for real-time forecasting, which requires at each time to forecast the observation at the next time point when the entire data are not available. Smoothing is then more useful when all data are available, and used routinely, in order to describe trends or to fit a model to historical data; sometimes it is used as a pre-processing tool for a wider data analysis. In some studies, forecasting is referred to as *out of sample* estimation (as the estimation is carried out for observations not belonging to the sample), and smoothing is referred to as *in-sample* estimation (as the entire sample is used for the estimation).

3.3.2 The Lag-One Covariance Smoother

In this section, the lag-one covariance smoother is discussed, which gives a sequential updating of the lag-one covariance matrix backward in time. This is useful in the derivation and computation of the expectation maximisation (EM) algorithm (used in maximum likelihood estimation) as discussed in detail in Sect. 4.3.1.

Theorem 3.5 (Lag-One Covariance Smoothing) *In the state space model* (3.10a)–(3.10b) *with information* $y_{1:n} = \{y_1, \ldots, y_n\}$, *define the lag-one covariance matrix* $\mathbf{P}_{t,t-1|n} = Cov(\beta_t, \beta_{t-1} \mid y_{i:n})$. *This can be computed backwards in time* $t = n, n-1, \ldots, 3, 2$, *according to*

1. $\mathbf{P}_{n,n-1|n} = (\mathbf{I} - K_n x_n^\top)\mathbf{F}_n\mathbf{P}_{n-1|n-1}$, *where* K_n *is the Kalman gain and* $\mathbf{P}_{n-1|n-1}$ *is computed by the Kalman filter (Theorem 3.2);*
2. $\mathbf{P}_{t-1,t-2|n} = \mathbf{P}_{t-1|t-1} + \mathbf{L}_{t-1}(\mathbf{P}_{t,t-1|n} - \mathbf{F}_t\mathbf{P}_{t-1|t-1})\mathbf{L}_{t-2}^\top$, *for* $t = 2, \ldots, n$, *where* $\mathbf{P}_{t-1|t-1}$ *is computed by the Kalman filter (Theorem 3.2) and* $\mathbf{L}_{t-1}$ *and* $\mathbf{L}_{t-2}$ *are computed by the fixed-interval smoothing algorithm (Theorem 3.4).*

Proof First we prove (1). Write down the joint distribution of β_n, β_{n-1} and y_n, given information $y_{1:n-1}$:

$$\begin{bmatrix} \beta_n \\ \beta_{n-1} \\ y_n \end{bmatrix} \mid y_{1:n-1} \sim N \left\{ \begin{bmatrix} \hat{\beta}_{n|n-1} \\ \hat{\beta}_{n-1|n-1} \\ \hat{y}_{n|n-1} \end{bmatrix}, \begin{bmatrix} \mathbf{P}_{n|n-1} & \mathbf{F}_n\mathbf{P}_{n-1|n-1} & \mathbf{P}_{n|n-1}x_n \\ \mathbf{P}_{n-1|n-1}\mathbf{F}_n^\top & \mathbf{P}_{n-1|n-1} & \mathbf{P}_{n-1|n-1}\mathbf{F}_n^\top x_n \\ x_n^\top\mathbf{P}_{n|n-1} & x_n^\top\mathbf{F}_n\mathbf{P}_{n-1|n-1} & q_{n|n-1} \end{bmatrix} \right\},$$

where $\mathrm{Cov}(\beta_n, \beta_{n-1} \mid y_{1:n-1}) = \mathrm{Cov}(F_n\beta_{n-1} + \zeta_n, \beta_{n-1} \mid y_{1:n-1}) = F_n P_{n-1|n-1}$, $\mathrm{Cov}(\beta_n, y_n \mid y_{1:n-1}) = \mathrm{Cov}(\beta_n, x_n^\top \beta_n + \epsilon_n \mid y_{1:n-1}) = P_{n|n-1}x_n$, and $\mathrm{Cov}(\beta_{n-1}, y_n \mid y_{1:n-1}) = \mathrm{Cov}(\beta_{n-1}, x_n^\top \beta_n + \epsilon_n \mid y_1{}^{n}{}_{-1}) = \mathrm{Cov}(\beta_{n-1}, x_n^\top F_n\beta_{n-1} + x_n^\top \zeta_n + \epsilon_n \mid y_{1:n-1}) = P_{n-1|n-1}F_n^\top x_n$.

By applying the conditional result of Theorem 3.2, we have that the covariance matrix of the state vector $(\beta_n^\top, \beta_{n-1}^\top)^\top$, given information $y_{1:n-1}$ and y_n (or equivalently given $y_{1:n}$), is

$$\mathrm{Var}\left(\begin{bmatrix} \beta_n \\ \beta_{n-1} \end{bmatrix} \mid y_{1:n}\right) = \begin{bmatrix} P_{n|n} & P_{n,n-1|n} \\ P_{n,n-1|n}^\top & P_{n-1|n} \end{bmatrix}$$

$$= \begin{bmatrix} P_{n|n-1} & F_n P_{n-1|n-1} \\ P_{n-1|n-1}F_n^\top & P_{n-1|n-1} \end{bmatrix} - q_{n|n-1}^{-1} \begin{bmatrix} P_{n|n-1}x_n \\ P_{n-1|n-1}F_n^\top x_n \end{bmatrix}$$

$$\times \left[x_n^\top P_{n|n-1}, x_n^\top F_n P_{n-1|n-1} \right],$$

from which we obtain

$$P_{n,n-1|n} = F_n P_{n-1|n-1} - q_{n|n-1}^{-1} P_{n|n-1}x_n x_n^\top F_n P_{n-1|n-1}$$

$$= (I - K_n x_n^\top)F_n P_{n-1|n-1},$$

as required.

Proceeding now to (2), first we define

$$\hat{\beta}_{t|s}^* = \beta_t - \hat{\beta}_{t|s}, \tag{3.19}$$

for any $t, s = 1, \dots, n$.

First notice that $\hat{\beta}_{t|s}^*$ and $y = [y_1, \dots, y_s]^\top$ are independent. From the tower property, it follows that $\hat{\beta}_{t|s}^*$ and y are uncorrelated, i.e.

$$E(\hat{\beta}_{t|s}^* y^\top) = E[E(\hat{\beta}_{t|s}^* y^\top \mid y) \mid y] = E[\{E(\beta_t \mid y) - \hat{\beta}_{t|s}\}y^\top \mid y] = 0$$

$$= E(\hat{\beta}_{t|s}^*)E(y)^\top.$$

Since the distribution of β_t given y is Gaussian, $\hat{\beta}_{t|s}^*$ and y are independent.

An implication of that independence is

$$P_{t-1,t-2|n} = E(\hat{\beta}_{t-1|n}^* \hat{\beta}_{t-2|n}^{*\top} \mid y_{1:n}) = E(\hat{\beta}_{t-1|n}^* \hat{\beta}_{t-2|n}^{*\top}), \tag{3.20}$$

and so in the following we use unconditional expectations to prove the recursion of $P_{t-1,t-2|n}$.

From Eq. (3.19) and from the smoothing recursion of $\hat{\beta}_{t|n}$ of Theorem 3.4, we can see that

$$
\begin{aligned}
\hat{\beta}^*_{t-1|n} &= \beta_{t-1} - \hat{\beta}_{t-1|n} \\
&= \beta_{t-1} - [\hat{\beta}_{t-1|t-1} + \mathbf{L}_{t-1}(\hat{\beta}_{t|n} - \hat{\beta}_{t|t-1})] \\
&= \beta_{t-1} - \hat{\beta}_{t-1|t-1} + \mathbf{L}_{t-1}[\beta_t - \hat{\beta}_{t|n} - (\beta_t - \hat{\beta}_{t|t-1})] \\
&= \hat{\beta}^*_{t-1|t-1} + \mathbf{L}_{t-1}(\hat{\beta}^*_{t|n} - \hat{\beta}^*_{t|t-1}),
\end{aligned}
$$

which can be written as

$$
\begin{aligned}
\hat{\beta}^*_{t-1|n} + \mathbf{L}_{t-1}\hat{\beta}_{t|n} &= \hat{\beta}^*_{t-1|t-1} + \mathbf{L}_{t-1}(\hat{\beta}^*_{t|n} - \hat{\beta}^*_{t|t-1} + \hat{\beta}_{t|n}) \\
&= \hat{\beta}^*_{t-1|t-1} + \mathbf{L}_{t-1}\mathbf{F}_t\hat{\beta}_{t-1|t-1}, \quad\quad (3.21)
\end{aligned}
$$

since $\hat{\beta}^*_{t|n} - \hat{\beta}^*_{t|t-1} + \hat{\beta}_{t|n} = \beta_t - \hat{\beta}^*_{t|t-1} = \hat{\beta}_{t|t-1} = \mathbf{F}_t\hat{\beta}_{t-1|t-1}$.

Now we multiply $\hat{\beta}^*_{t-1|n}$ by $\hat{\beta}^T_{t-2|n}$ (which is provided by (3.21) for time $t-2$) and taking expectations

$$
\begin{aligned}
&\mathbf{P}_{t-1,t-2|n} + \mathbf{L}_{t-1}\mathrm{E}(\hat{\beta}_{t|n}\hat{\beta}^T_{t-1|n})\mathbf{L}^T_{t-2} + \mathbf{L}_{t-1}\mathrm{E}(\hat{\beta}_{t|n}\hat{\beta}^{*T}_{t-2|n}) + \mathrm{E}(\hat{\beta}^*_{t-1|n}\hat{\beta}^T_{t-1|n})\mathbf{L}^T_{t-2} \\
&\quad = \mathrm{E}(\hat{\beta}^*_{t-1|t-1}\hat{\beta}^{*T}_{t-2|t-2}) + \mathbf{L}_{t-1}\mathbf{F}_t\mathrm{E}(\hat{\beta}_{t-1|t-1}\hat{\beta}^{*T}_{t-2|t-2}) \\
&\quad\quad + \mathrm{E}(\hat{\beta}^*_{t-1|t-1}\hat{\beta}_{t-2|t-2})\mathbf{F}^T_{t-1}\mathbf{L}^T_{t-2} + \mathbf{L}_{t-1}\mathbf{F}_t\mathrm{E}(\hat{\beta}_{t-1|t-1}\hat{\beta}^T_{t-2|t-2})\mathbf{F}^T_{t-1}\mathbf{L}^T_{t-2}.
\end{aligned}
$$
$$(3.22)$$

In the rest of the proof, we calculate the expectations in (3.22). Firstly, we note that

$$
\mathrm{E}(\hat{\beta}_{t|n}\hat{\beta}^{*T}_{t-2|n}) = \mathrm{E}(\hat{\beta}^*_{t-1|n}\hat{\beta}^T_{t-1|n}) = \mathrm{E}(\hat{\beta}^*_{t-1|t-1}\hat{\beta}^T_{t-2|t-2}) = 0.
$$

This can be established by using the tower property of expectations, e.g. for the first expectation

$$
\begin{aligned}
\mathrm{E}(\hat{\beta}_{t|n}\hat{\beta}^{*T}_{t-2|n}) &= \mathrm{E}[\mathrm{E}\{\hat{\beta}_{t|n}(\beta_{t-2} - \hat{\beta}_{t-2|n}) \mid y_{1:n}\} \mid y_{1:n}] \\
&= \hat{\beta}_{t|n}[\mathrm{E}(\beta_{t-2} \mid y_{1:n}) - \hat{\beta}_{t-2|n}] = 0.
\end{aligned}
$$

Similarly, one obtains the other two expectations.

The next step is to obtain a recursion for $\mathrm{E}(\hat{\beta}^*_{t-1|t-1}\hat{\beta}^{*T}_{t-2|t-2})$. From the Kalman filter recursion of $\hat{\beta}_{t-1|t-1}$, we have

$$
\begin{aligned}
\hat{\beta}^*_{t-1|t-1} &= \beta_{t-1} - \hat{\beta}_{t-1|t-1} \\
&= \beta_{t-1} - \hat{\beta}_{t-1|t-2} - K_{t-1}(y_{t-1} - x^T_{t-1}\hat{\beta}_{t-1|t-2})
\end{aligned}
$$

$$= \hat{\beta}^*_{t-1|t-2} - K_{t-1}(x^\top_{t-1}\beta_{t-1} + \epsilon_{t-1} - x^\top_{t-1}\hat{\beta}_{t-1|t-2})$$

$$= \hat{\beta}^*_{t-1|t-2} - K_{t-1}x^\top_{t-1}\hat{\beta}^*_{t-1|t-2} - K_{t-1}\epsilon_{t-1},$$

and substituting this into $\mathrm{E}(\hat{\beta}^*_{t-1|t-1}\hat{\beta}^{*\top}_{t-2|t-2})$, we obtain

$$\mathrm{E}(\hat{\beta}^*_{t-1|t-1}\hat{\beta}^{*\top}_{t-2|t-2}) = \mathrm{E}(\hat{\beta}^*_{t-1|t-2}\hat{\beta}^{*\top}_{t-2|t-2}) - K_{t-1}x^\top_{t-1}\mathrm{E}(\hat{\beta}^*_{t-1|t-2}\hat{\beta}^{*\top}_{t-2|t-2})$$

$$- K_{t-1}\mathrm{E}(\epsilon_{t-1}\hat{\beta}^{*\top}_{t-2|t-2})$$

$$= \mathbf{P}_{t-1,t-2|t-2} - K_{t-1}x^\top_{t-1}\mathbf{P}_{t-1,t-2|t-2}, \tag{3.23}$$

since ϵ_{t-1} and $\hat{\beta}^*_{t-2|t-2}$ are independent.

From the recursion of $\hat{\beta}^*_{t-1|t-1}$ by subtracting β_{t-1} from both sides of that recursion, we see that

$$\hat{\beta}_{t-1|t-1} = \hat{\beta}_{t-1|t-2} + K_{t-1}x^\top_{t-1}\hat{\beta}^*_{t-1|t-2} + K_{t-1}\epsilon_{t-1} \tag{3.24}$$

and so

$$\mathrm{E}(\hat{\beta}_{t-1|t-2}\hat{\beta}^{*\top}_{t-2|t-2}) = K_{t-1}x^\top_{t-1}\mathrm{E}(\hat{\beta}^*_{t-1|t-2}\hat{\beta}^{*\top}_{t-2|t-2})$$

$$= K_{t-1}x^\top_{t-1}\mathbf{P}_{t-1,t-2|t-2}. \tag{3.25}$$

Also,

$$\mathrm{E}(\hat{\beta}_{t|n}\hat{\beta}^\top_{t-1|n}) = \mathrm{E}(\beta_t - \hat{\beta}^*_{t|n})(\beta_{t-1} - \hat{\beta}^{*\top}_{t-1|n})$$

$$= \mathrm{E}(\beta_t\beta^\top_{t-1}) - \mathbf{P}_{t,t-1|n}$$

$$= \mathrm{E}(\mathbf{F}_t\beta_{t-1} + \zeta_t)(\mathbf{F}_{t-1}\beta_{t-2} + \zeta_{t-1})^\top - \mathbf{P}_{t,t-1|n}$$

$$= \mathbf{F}_t\mathrm{E}(\beta_{t-1}\beta^\top_{t-2})\mathbf{F}^\top_{t-1} + \mathbf{F}_t\mathbf{Z} - \mathbf{P}_{t,t-1|n}, \tag{3.26}$$

and using (3.24),

$$\mathrm{E}(\hat{\beta}_{t-1|t-1}\hat{\beta}^\top_{t-2|t-2}) = \mathrm{E}(\hat{\beta}_{t-1|t-2} + K_{t-1}x^\top_{t-1}\hat{\beta}^*_{t-1|t-2} + K_{t-1}\epsilon_{t-1})\hat{\beta}^\top_{t-2|t-2}$$

$$= \mathrm{E}(\hat{\beta}_{t-1|t-2}\hat{\beta}^\top_{t-2|t-2})$$

$$= \mathrm{E}(\beta_{t-1} - \hat{\beta}^*_{t-1|t-2})(\beta_{t-2} - \hat{\beta}^*_{t-2|t-2})^\top$$

$$= \mathrm{E}(\hat{\beta}_{t-1}\hat{\beta}^\top_{t-2}) - \mathbf{P}_{t-1,t-2|t-2}, \tag{3.27}$$

as $\mathrm{E}(\beta_{t-1}\hat{\beta}^{*\top}_{t-2|t-2}) = \mathrm{E}(\hat{\beta}^*_{t-1|t-2}\beta^\top_{t-2}) = \mathbf{P}_{t-1,t-2|t-2}.$

We substitute (3.23), (3.25), (3.26) and (3.27) in (3.22), which after some algebra yields

$$\mathbf{P}_{t-1,t-2|n} = \mathbf{L}_{t-1}(\mathbf{P}_{t,t-1|n} - \mathbf{F}_t \mathbf{P}_{t-1|t-1})\mathbf{L}_{t-2}^{\top} + \mathbf{P}_{t-1|t-1}\mathbf{L}_{t-2}^{\top},$$

after using

$$\mathbf{P}_{t|t} = \mathbf{P}_{t|t-1} - q_t K_t K_t^{\top} = (\mathbf{I} - K_t x_t^{\top})\mathbf{P}_{t|t-1},$$

from the Kalman filter.   □

Theorem 3.5 is useful in the computation of the expectation maximisation (EM) algorithm, in Chap. 4, for the evaluation of maximum likelihood estimates.

3.4   Forecasting

With data $y_{1:t} = \{y_1, \ldots, y_t\}$, the forecasting problem involves the derivation of the distributions of β_{t+k} and y_{t+k}, known also as k-step ahead forecast state and observation distributions, respectively. The positive integer k is known as the *lead time* of the forecast and the maximum value it takes is known as the *forecast horizon*. To the following we assume that the transition matrix $\mathbf{F}_t = \mathbf{F}$ is time-invariant; this consideration is met in most practical models (e.g. for all models in Sect. 4.1) and is imposed here for simplification and elaboration purposes (the proof of the general result is analogous to the time-invariant case and is left to the reader as an exercise).

Theorem 3.6 *In the state space model* (3.10a)–(3.10b) *with transition matrix* $\mathbf{F}_t = \mathbf{F}$ *and with information* $y_{1:t} = \{y_1, \ldots, y_t\}$, *the following apply.*

1. *The k-step ahead forecast state distribution is given by* $\beta_{t+k} \mid y_{1:t} \sim N(\hat{\beta}_{t+k|t}, \mathbf{P}_{t+k|t})$, *with* $\hat{\beta}_{t+k|t} = \mathbf{F}^k \hat{\beta}_{t|t}$ *and*

$$\mathbf{P}_{t+k|t} = \mathbf{F}^k \mathbf{P}_{t|t}(\mathbf{F}^k)^{\top} + \sum_{j=0}^{k-1} \mathbf{F}^j \mathbf{Z}_{t+k-j}(\mathbf{F}^j)^{\top},$$

   where $\hat{\beta}_{t|t}$ *and* $\mathbf{P}_{t|t}$ *are computed by the Kalman filter (Theorem 3.2).*
2. *The k-step ahead forecast observation distribution is given by* $y_{t+k} \mid y_{1:t} \sim N(\hat{y}_{t+k|t}, q_{t+k|t})$, *where* $\hat{y}_{t+k|t} = x_{t+k}^{\top}\hat{\beta}_{t+k|t}$ *and* $q_{t+k|t} = x_{t+k}^{\top}\mathbf{P}_{t+k|t}x_{t+k} + \sigma^2$.

Proof Suppose that at time t with information $y_{1:t}$, we have obtained, from an application of the Kalman filter, the filtered distribution $\beta_t \mid y_{1:t} \sim N(\hat{\beta}_{t|t}, \mathbf{P}_{t|t})$.

First we derive the predictive distribution of β_{t+k} given $y_{1:t}$. To this end, we need to express β_{t+k} as a function of β_t. Applying recurrently the state equation (3.10b), we have

$$
\begin{aligned}
\beta_{t+k} &= \mathbf{F}\beta_{t+k-1} + \zeta_{t+k} \\
&= \mathbf{F}(\mathbf{F}\beta_{t+k-2} + \zeta_{t+k-1}) + \zeta_{t+k} \\
&= \cdots \\
&= \mathbf{F}^k \beta_t + \sum_{j=0}^{k-1} \mathbf{F}^j \zeta_{t+k-j}.
\end{aligned}
$$

Now, using this expression, the mean vector and the covariance matrix of $\beta_{t+k} \mid y_{1:t}$ are

$$
\hat{\beta}_{t+k|t} = \mathrm{E}(\beta_{t+k} \mid y_{1:t}) = \mathbf{F}^k \mathrm{E}(\beta_t \mid y_{1:t}) = \mathbf{F}^k \hat{\beta}_{t|t}
$$

and

$$
\begin{aligned}
\mathbf{P}_{t+k|t} &= \mathrm{Var}(\beta_{t+k} \mid y_{1:t}) \\
&= \mathbf{F}^k \mathrm{Var}(\beta_t \mid y_{1:t})(\mathbf{F}^k)^\top + \sum_{j=0}^{k-1} \mathbf{F}^j \mathrm{Var}(\zeta_{t+k-j} \mid y_{1:t})(\mathbf{F}^j)^\top,
\end{aligned}
$$

which returns the stated covariance matrix, if we note that $\mathrm{Var}(\beta_t \mid y_{1:t}) = \mathbf{P}_{t|t}$, $\mathrm{Var}(\zeta_t) = \mathbf{Z}_t$. Since β_{t+k} is a linear combination of the random vectors β_t and ζ_t's, all of which follow Gaussian distributions, the distribution of $\beta_{t+k} \mid y_{1:t}$ will also be Gaussian, and this completes the proof for β_{t+k}.

The stated form of the predictive distribution of $y_{t+k} \mid y_{1:t}$ follows immediately by writing down the observation equation $y_{t+k} = x_{t+k}^\top \beta_{t+k} + \epsilon_{t+k}$ and applying the distribution of $\beta_{t+k} \mid y_{1:t}$. This derivation is left to the reader as an exercise and is complete by parallel to the derivation of the smoothed distribution of $y_t \mid y_{1:n}$ in Theorem 3.4 of the previous section. □

Some comments are in order.

- The observation forecast distribution (the distribution of $y_{t+k} \mid y_{1:t}$) is usually referred as the k-step ahead forecast distribution or simply the forecast distribution. The state forecast distribution (the distribution of $\beta_{t+k} \mid y_{1:t}$) is referred to as the forecast state distribution.
- The mean $\hat{y}_{t+k|t}$ of the forecast distribution is a function of $k = 1, 2, 3, \ldots$ and is known as a *forecast function*. Its form is important because it can be used to classify different models aimed at forecasting. For example, the forecast function of a linear trend model is a straight line and that of a seasonal model exposes a cyclic variation (for detailed considerations and related proofs, see Sect. 4.1 of

the next chapter). For a time-invariant transition matrix $\mathbf{F}_t = \mathbf{F}$ (which is the case in many applied models), the forecast function takes the form

$$\hat{y}_{t+k|t} = x_{t+k}^{\top} \mathbf{F}^k \hat{\beta}_{t|t}. \tag{3.28}$$

• We note that if $\mathbf{Z}_t = \mathbf{Z}$ is time-invariant, the covariance matrix $\mathbf{P}_{t+k|t}$ of β_{t+k} takes the attractive form

$$\mathbf{P}_{t+k|t} = \mathbf{F}^k \mathbf{P}_t (\mathbf{F}^k)^{\top} + \sum_{j=0}^{k-1} \mathbf{F}^j \mathbf{Z} (\mathbf{F}^j)^{\top}.$$

• Recalling that in the Kalman filter (Theorem 3.2), we compute $\hat{\beta}_{t|t-1}$ and $\mathbf{P}_{t|t-1}$, we note that these quantities are the mean vector and covariance matrix of the one-step forecast state distribution of β_t, given $y_{1:t-1}$. Likewise, $\hat{y}_{t|t-1}$ and $q_{t|t-1}$ are the mean and variance of the one-step forecast observation distribution of y_t, given $y_{1:t-1}$.

• A $(1 - \alpha)\%$ forecast interval for y_{t+k}, based on information $y_{1:t}$, is obtained as $\hat{y}_{t+k|t} \pm z_{1-\alpha/2} \sqrt{q_{t+k|t}}$, where $z_{1-\alpha/2}$ denotes the $(1 - \alpha/2)\%$-quantile of the standard Gaussian distribution $N(0, 1)$.

• We note that Theorem 3.6 suggests a recursive algorithm: at each time t, we compute $\hat{\beta}_{t|t}$ and $\mathbf{P}_{t|t}$ by the Kalman filter, and then we calculate $\hat{\beta}_{t+k|t}$, $\mathbf{P}_{t+k|t}$, $\hat{y}_{t+k|t}$ and $q_{t+k|t}$. This algorithm is summarised below.

Forecasting

For any $t \geq 2$:

1. Initial state distribution at t: $\beta_t \mid y_{1:t} \sim N(\hat{\beta}_{t|t}, \mathbf{P}_{t|t})$, where $\hat{\beta}_{t|t}$ and $\mathbf{P}_{t|t}$ are calculated by the Kalman filter (Theorem 3.2).

2. k-step forecast state distribution: $\beta_{t+k} \mid y_{1:t} \sim N(\hat{\beta}_{t+k|t}, \mathbf{P}_{t+k|t})$, where

$$\hat{\beta}_{t+k|t} = \mathbf{F}^k \hat{\beta}_{t|t} \quad \text{and} \quad \mathbf{P}_{t+k|t} = \mathbf{F}^k \mathbf{P}_t (\mathbf{F}^k)^{\top} + \sum_{j=0}^{k-1} \mathbf{F}^j \mathbf{Z}_{t+k-j} (\mathbf{F}^j)^{\top}.$$

3. k-step forecast observation distribution: $y_{t+k} \mid y_{1:t} \sim N(\hat{y}_{t+k|t}, q_{t+k|t})$, where $\hat{y}_{t+k|t} = x_{t+k}^{\top} \hat{\beta}_{t+k|t}$ and $q_{t+k|t} = x_{t+k}^{\top} \mathbf{P}_{t+k|t} x_{t+k} + \sigma^2$.

Example 3.3 (Annual Temperature Example Continued) In Examples 3.1 and 3.2, we computed filtered and smoothed estimates $\hat{\beta}_{t|t}$ of the level β_t, for the annual temperature data for central England. In this example, we compute predictions for the actual temperatures y_t and the level β_t. In particular, we are interested in 1-year,

2-year and 3-year ahead predictions at $t = n = 344$ (corresponding to year 2002); in other words, we are interested in temperature predictions for the years 2003, 2004 and 2005, based on the data up to 2002. In R this is implemented by the command:

```
> # temp is defined previously
> forecast <- bts.predict(temp, x0=1, F0=1, obsvar=1, Z0=10,
+ beta0=9, P0=1000, DISO=FALSE, VAREST=FALSE, kmax=3)
```

The above command creates the object forecast with the values of $\hat{y}_{344+k|344}$ and $q_{344+k|344}$, for $k = 1, 2, 3$. These values are returned by

```
> forecast$ForMean
[[1]]
[1] 10.54656

[[2]]
[1] 10.54656

[[3]]
[1] 10.54656
```

for the forecast mean and

```
> forecast$ForVar
[[1]]
[1] 11.91608

[[2]]
[1] 21.91608

[[3]]
[1] 31.91608
```

for the forecast variance.

Thus, $\hat{y}_{345|344} = \hat{y}_{346|344} = \hat{y}_{347|344} = 10.54656$ and $q_{345|344} = 11.91608$, $q_{346|344} = 21.91608$ and $q_{347|344} = 31.91608$. In other words, all three prediction means are equal to $11.9\,°C$ (1 dp accuracy), but the respective prediction variances increase with $k = 1, 2, 3$. The increase of variance is expected, as given information $y_{1:344}$, the further we look into the future, the more uncertainty is added, e.g. $q_{344+k|344} < q_{344+j|344}$, for $k < j$, and the effect is magnified by the magnitude of the difference $j - k$. This is a property shared by any state space model. The equality of the prediction means is, however, particular to the local level model and in fact is its characteristic. Finally, based on the above R output, we can easily compute predictive intervals, e.g. a 95% 2-step ahead predictive interval for y_{346} is

$$\hat{y}_{346|344} \pm z_{1-0.05/2}\sqrt{q_{346|344}} \approx 10.547 \pm 1.96\sqrt{21.916} = (1.371, 19.723),$$

where $z_{1-0.05/2} \approx 1.96$ is calculated in R using the command qnorm(0.975). Here, we observe that the relatively large prediction variance has resulted in a too wide prediction interval. As k increases (and the forecast variance increases), the prediction intervals will widen and the forecasts will become more uncertain.

3.5   Steady State of the Kalman Filter

An important feature of the state estimator of the Kalman filter $\hat{\beta}_t$ is that after some
time its fluctuation is steady. This is important in control systems (see Chap. 8)
where the states include input signals and the observations output signals. One of
the aims in the design of a control system is to obtain a stable output signal, provided
that the input signals are stable. For a class of state space models (with time-invariant
components), the fluctuations of the estimated states become time-invariant, and
hence the states are steady; this is known as the *steady state* of the model and it can
be thought of as the system being in an 'equilibrium' state. The key to arrive at the
steady state is to show that the sequence of the posterior covariance matrices of the
states $(\mathrm{Var}(\beta_t \mid y_{1:t}) = \mathbf{P}_t)$ converges to some matrix $\mathbf{P}$. The steady state of state
space models is discussed in detail in Balakrishnan (1984, Section 4.2) and in Chan
and Guo (1991, Section 3.3). This section provides the details of the convergence
of $\mathbf{P}_t$; we start by discussing the concept of observability, which is interesting in its
own right and will be revisited in Chap. 8.

3.5.1   Observability

Consider the linear state space model (3.10a)–(3.10b), where the design vector $x_t = x$ and the transition matrix $\mathbf{F}_t = \mathbf{F}$ are assumed to be time-invariant. There are many
state space models in this class, which can generate or describe useful time series as
several examples in the next chapter (Sect. 4.1) indicate.

Informally, observability relates to whether the states of the model can be
determined from the observed data. More formally, suppose that observations
$y_1, \ldots, y_n$ are obtained for some n, and assume that the observation innovations
$\epsilon_1, \ldots, \epsilon_n$ and the state innovations $\zeta_1, \ldots, \zeta_n$ are known. With this set-up, we seek
to derive some condition to ensure that $\beta_1, \ldots, \beta_n$ are uniquely determined.

From the state equation (3.10b), write

$$\beta_t = \mathbf{F}^{t-1}\beta_1 + \sum_{i=0}^{t-2} \mathbf{F}^i \zeta_{t-i}, \tag{3.29}$$

and from the observation equation (3.10a), write

$$y_t = x^\top \mathbf{F}^{t-1}\beta_1 + \sum_{i=0}^{t-2} x^\top \mathbf{F}^i \zeta_{t-i} + \epsilon_t. \tag{3.30}$$

If we set

$$y_t' = y_t - \sum_{i=0}^{t-2} x^\top \mathbf{F}^i \zeta_{t-i} - \epsilon_t,$$

then (3.30) implies

$$y_t' = x^\top \mathbf{F}^{t-1} \beta_1, \quad t = 1, \ldots, n. \tag{3.31}$$

This is a system of n linear equations (in β_1), from which β_1 may be determined. Then any β_t, for $t \geq 2$, can be determined using (3.29).

The system of equations (3.31) has a unique solution (in β_1) if the zero solution is the unique solution of the homogeneous system of equations $x^\top \mathbf{F}^{t-1} z = 0, t = 1, \ldots, n$ (in the $p \times 1$ vector z). Since z has dimension $p \leq n$, the above is equivalent to $x^\top \mathbf{F}^{i-1} z = 0$ having unique zero solution, for $i = 1, \ldots, p$. This in turn is equivalent to the $p \times p$ matrix

$$O = \begin{bmatrix} x^\top \\ x^\top \mathbf{F} \\ \vdots \\ x^\top \mathbf{F}^{p-1} \end{bmatrix} \tag{3.32}$$

being invertible or having full rank p. If this is the case, we can determine $\beta_1, \ldots, \beta_n$ from the data $y_1, \ldots, y_n$, and we will say that the states are observable. The matrix O is known as the *observability matrix*, and when it is of full rank it suggests a one-to-one relationship between the observations and the states; in such a case, the state space model is said to be observable. Observability may be used as a guide to determine the minimum order of the state vector β_t to ensure this one-to-one correspondence between states and observations, hence to reduce redundant dimensionality.

Example 3.4 (Local Level Model) Consider the local level model

$$y_t = \beta_t + \epsilon_t \quad \text{and} \quad \beta_t = \beta_{t-1} + \zeta_t, \tag{3.33}$$

where the white noise processes $\{\epsilon_t\}$ and $\{\zeta_t\}$ are mutually independent and independent of the initial state β_0. This model is obviously observable, since the observability matrix is $O = 1$ (here $p = 1$), which of course has full rank.

Example 3.5 Consider now the following state space model, which is described in detail in Sect. 4.1.1:

$$y_t = [1, 0] \begin{bmatrix} \beta_{1t} \\ \beta_{2t} \end{bmatrix} + \epsilon_t = x^\top \beta_t + \epsilon_t,$$

$$\begin{bmatrix} \beta_{1t} \\ \beta_{2t} \end{bmatrix} = \begin{bmatrix} 1 & 1 \\ 0 & 1 \end{bmatrix} \begin{bmatrix} \beta_{1,t-1} \\ \beta_{2,t-1} \end{bmatrix} + \begin{bmatrix} \zeta_{1t} \\ \zeta_{2t} \end{bmatrix} = \mathbf{F}\beta_{t-1} + \zeta_t,$$

with the usual assumptions on the innovations ϵ_t and ζ_t.

This state space model is observable, since the observability matrix

$$O = \begin{bmatrix} 1 & 0 \\ 1 & 1 \end{bmatrix}$$

is of full rank $p = 2$.

Now suppose that in the above model, the transition matrix is

$$\mathbf{F} = \begin{bmatrix} \lambda & 0 \\ 0 & \mu \end{bmatrix},$$

for some constants λ and μ. Then the observability matrix of this new model is

$$O = \begin{bmatrix} 1 & 0 \\ \lambda & 0 \end{bmatrix},$$

which rank is $1 < 2 = p$, and hence this model is not observable.

By writing $\beta = [\beta_{1t}, \beta_{2t}]^\top$ and $\zeta_t = [\zeta_{1t}, \zeta_{2t}]^\top$, we see

$$y_t = \beta_{1t} + \epsilon_t, \quad \beta_{1t} = \lambda\beta_{1,t-1} + \zeta_{1t} \quad \text{and} \quad \beta_{2t} = \mu\beta_{2,t-1} + \zeta_{2t},$$

from which we observe that β_{2t} does not affect y_t and is completely redundant (assuming that the covariance matrix of ζ_t is a diagonal matrix, so that β_{1t} and β_{2t} are uncorrelated). Thus, it is reasonable to suggest that the above model is reduced to the model

$$y_t = \beta_t + \epsilon_t \quad \text{and} \quad \beta_t = \lambda\beta_{t-1} + \zeta_t,$$

where the dimension of the state vector β_t is reduced from 2 to 1. We can easily verify that this model is now observable. Such a reduction in state dimensions may be useful, in particular having in mind forecasting of future observations, but if interest lies in the state β_{2t}, it may be useful to keep it in the model.

3.5.2   Steady State of the Local Level Model

Consider the local level model, defined by (3.33), where $\text{Var}(\zeta_t) = Z$ is time-invariant. From the Kalman filter (Theorem 3.2), we have recursions for the posterior mean $\hat{\beta}_{t|t}$ and the posterior variance $P_{t|t}$ of β_t. With this set-up in place, we will show that the sequence of variances $\{P_{t|t}\}$ converges to a limit, which is independent of the prior $P_{0|0}$ and is given by

$$P = \lim_{t \to \infty} P_{t|t} = \frac{Z}{2} \left(\sqrt{1 + \frac{4\sigma^2}{Z}} - 1 \right).$$

First we show that this limit exists. To this end, we show that the sequence $\{P_{t|t}\}$ is bounded and monotonic.

From Theorem 3.2 with $x_t = F_t = 1$ (local level model), the posterior variance $P_{t|t}$ is

$$P_{t|t} = P_{t|t-1} - K_t^2 q_{t|t-1} = P_{t|t-1} - \frac{P_{t|t-1}^2}{q_{t|t-1}} = \frac{P_{t|t-1}\sigma^2}{q_{t|t-1}} = K_t \sigma^2, \qquad (3.34)$$

where we have used $K_t = P_{t|t-1}/q_{t|t-1}$ and $q_{t|t-1} = P_{t|t-1} + \sigma^2$ (Theorem 3.2). From (3.34), it follows that $0 \le P_{t|t} \le \sigma^2$, and hence $\{P_{t|t}\}$ is bounded, because

$$0 \le K_t = \frac{P_{t|t-1}}{P_{t|t-1} + \sigma^2} \le 1, \quad \text{with} \quad \sigma^2 \ge 0.$$

For the monotonicity, first we prove

$$P_{t|t}^{-1} = P_{t|t-1}^{-1} + \sigma^{-2}.$$

Indeed, we have

$$\left(P_{t|t-1}^{-1} + \sigma^{-2} \right)^{-1} = \frac{\sigma^2 P_{t|t-1}}{\sigma^2 + P_{t|t-1}} = K_t \sigma^2 = P_{t|t}.$$

Consequently,

$$P_{t|t}^{-1} - P_{t-1|t-1}^{-1} = P_{t|t-1}^{-1} - P_{t-1|t-2}^{-1}$$

$$= \frac{P_{t-1|t-1} - P_{t-2|t-2}}{P_{t|t-1}P_{t-1|t-2}}$$

$$= C_t \left(P_{t-1|t-1}^{-1} - P_{t-2|t-2}^{-1} \right),$$

where

$$C_t = \frac{P_{t-1|t-1}P_{t-2|t-2}}{P_{t|t-1}P_{t-1|t-2}} > 0.$$

By applying that formula repeatedly, we have

$$P_{t|t}^{-1} - P_{t-1|t-1}^{-1} = C_t C_{t-1} \cdots C_2 \left(P_{1|1}^{-1} - P_{0|0}^{-1} \right).$$

Hence, $P_{t|t}^{-1} \geq P_{t-1|t-1}^{-1}$, for all t, if $P_{1|1}^{-1} \geq P_{0|0}^{-1}$; likewise, $P_{t|t}^{-1} < P_{t-1|t-1}^{-1}$, for all t, if $P_{1|1}^{-1} < P_{0|0}^{-1}$. Thus, the sequence $\{P_{t|t}^{-1}\}$ is either increasing or decreasing (depending on the sign of $P_{1|1}^{-1} - P_{0|0}^{-1}$) and so it is monotonic.

Since $\{P_{t|t}^{-1}\}$ is monotonic, it follows that the sequence of variances $\{P_{t|t}\}$ is monotonic too, and as it is bounded, its limit $\lim_{t\to\infty} P_{t|t}$ exists. From (3.34) by taking limits and using $P_{t|t-1} = P_{t-1|t-1} + Z$, we have

$$P = \frac{(P+Z)\sigma^2}{P+Z+\sigma^2}$$

or

$$P^2 + ZP - Z\sigma^2 = 0.$$

By keeping only the non-negative solution of this quadratic equation (since $P \geq 0$ as it is the limit of a variance), we have

$$P = \frac{-Z + \sqrt{Z^2 + 4Z\sigma^2}}{2} = \frac{Z}{2}\left(\sqrt{1 + \frac{4\sigma^2}{Z}} - 1 \right),$$

as required. It is clear by the formula of P that it does not depend on the initial or prior value of $P_{0|0}$; this value of $P_{0|0}$ can affect only whether $\{P_{t|t}\}$ is increasing or decreasing.

The steady state of the Kalman filter is obtained by replacing $P_{t|t}$ in the recursion of $\hat{\beta}_{t|t}$ by its limit P, i.e.

$$\hat{\beta}_{t|t} = \hat{\beta}_{t-1|t-1} + \frac{P+Z}{P+Z+\sigma^2}(y_t - \hat{\beta}_{t-1|t-1}).$$

3.5.3 Steady State of Linear State Space Models

In this section we consider the steady state of the state space model (3.10a)–(3.10b), where the design vector $x_t = x$, the transition matrix $\mathbf{F}_t = \mathbf{F}$ and the transition covariance matrix $\mathbf{Z}_t = \mathbf{Z}$ are assumed to be time-invariant. The following theorem states the main result, that is that under mild conditions, the limit of the posterior covariance matrix $\mathbf{P}_{t|t} = \text{Var}(\beta_t \mid y_{1:t})$ exists and does not depend on the prior covariance matrix $\mathbf{P}_{0|0}$ of β_0. This is a fundamental result in state space theory; see e.g. Balakrishnan (1984, Section 4.2), Chan and Guo (1991, Section 3.3) and West and Harrison (1997).

Before we give the main theorem, first we give some preliminary discussion on the convergence of non-negative definite matrices. Write $\mathbf{A} \geq \mathbf{0}$ to denote that an $n \times n$ matrix $\mathbf{A}$ is non-negative definite. For some $n \times n$ matrices $\mathbf{A} \geq \mathbf{0}$ and $\mathbf{B} \geq \mathbf{0}$, we write $\mathbf{A} \geq \mathbf{B}$ to denote that $\mathbf{A} - \mathbf{B} \geq \mathbf{0}$. Likewise, we define $\mathbf{A} > \mathbf{0}$ if $\mathbf{A}$ is positive definite and $\mathbf{A} > \mathbf{B}$ to denote that $\mathbf{A} - \mathbf{B} > \mathbf{0}$. Based on this notion, a sequence of non-negative matrices is said to be bounded if $\mathbf{M}_1 \leq \mathbf{A}_t \leq \mathbf{M}_2$, for some matrices $\mathbf{M}_1$ and $\mathbf{M}_2$, for all t; note that all non-negative definite matrices are bounded below by $\mathbf{M}_1 = \mathbf{0}$. The sequence of non-negative definite matrices $\{\mathbf{A}_t\}$ is known to be increasing (equiv. decreasing) if there is some t_0 such that $\mathbf{A}_t \geq \mathbf{A}_{t-1}$ (equiv. $\mathbf{A}_t \leq \mathbf{A}_{t-1}$), for any $t \geq t_0$. If $\{\mathbf{A}_t\}$ is either increasing or decreasing, it is said to be monotonic. If $\{\mathbf{A}_t\}$ is bounded and monotonic, then by extending to matrices a classical result of real-valued arithmetic sequences, it follows that its limit $\mathbf{A}$ exists, i.e. $\lim_{t\to\infty} \mathbf{A}_t = \mathbf{A}$.

Theorem 3.7 *Consider the state space model (3.10a)–(3.10b) with constant design vector $x_t = x$, transition matrix $\mathbf{F}_t = \mathbf{F}$ and transition covariance matrix $\mathbf{Z}_t = \mathbf{Z}$. If this model is observable, then the sequence of posterior covariance matrices $Var(\beta_t \mid y_{1:t}) = \mathbf{P}_{t|t}$ converges to a limit $\mathbf{P}$, which does not depend on the prior covariance matrix $\mathbf{P}_{0|0}$.*

Proof This proof mimics the proof of Harrison (1997). According to (1) above, we need to prove that the sequence of posterior covariance matrices $\{\mathbf{P}_{t|t}\}$ is bounded and monotonic, and hence it is convergent. For any $t \geq p$, we can define the vector $Y_t = [y_{t-p+1}, y_{t-p+2}, \ldots, y_t]^\top$. If we expand β_t in the transition equation (3.10b), we obtain

$$\beta_t = \mathbf{F}^{p-1}\beta_{t-p+1} + \sum_{i=1}^{p-2} \mathbf{F}^i \zeta_{t-i}. \tag{3.35}$$

Then, with the definition of Y_t above and from the definition of the observability matrix O, we have

$$Y_t = \begin{bmatrix} x^\top \\ x^\top \mathbf{F} \\ \vdots \\ x^\top \mathbf{F}^{p-1} \end{bmatrix} \beta_{t-p+1} + \begin{bmatrix} \epsilon_{t-p+1} \\ x^\top \zeta_{t-p+2} + \epsilon_{t-p+2} \\ \vdots \\ \sum_{i=0}^{p-2} x^\top \mathbf{F}^i \zeta_{t-i} + \epsilon_t \end{bmatrix} = O\beta_{t-p+1} + E_t.$$

Now, since the model is observable, O has full rank, and hence $\beta_{t-p+1} = O^{-1}Y_t - O^{-1}E_t$, and substituting this into (3.35), we obtain

$$\beta_t = \mathbf{F}^{p-1}O^{-1}Y_t + \omega_t,$$

where $\omega_t = \sum_{i=0}^{p-2} \mathbf{F}^i \zeta_{t-i} - \mathbf{F}^{p-1} O^{-1} E_t$. We note that ω_t has a bounded covariance matrix $\mathbf{C} = \mathrm{Var}(E_t)$, which is a function of σ^2 and $\mathbf{Z}$.

We will show that $\mathbf{0} \le \mathbf{P}_{t|t} \le \mathbf{C}$. Indeed, as $\mathbf{P}_{t|t}$ is a covariance matrix, it is non-negative definite matrix, hence $\mathbf{P}_{t|t} \ge \mathbf{0}$. From the definition of y_t, conditioning on $y_{1:t}$ implies conditioning on y_t. Hence

$$\mathbf{P}_{t|t} = \mathrm{Var}(\beta_t \mid y_{1:t}) \le \mathrm{Var}(\beta_t \mid \mathbf{F}^{p-1} O^{-1} Y_t) = \mathrm{Var}(\omega_t) = \mathbf{C}.$$

This establishes that the sequence $\{\mathbf{P}_{t|t}\}$ is bounded.

Proceeding now to monotonicity, define the information $\tilde{y}_{d:t} = \{\beta_d, y_{d+1}, \dots, y_t\}$, which includes the state vector β_d and data y_{d+1} up to and including y_t, for any $d = 0, 1, \dots, t-1$. From the tower property (5b) (Sect. 2.3), we have

$$\mathrm{Var}(\beta_t \mid y_{1:t}) = \mathrm{E}[\mathrm{Var}(\beta_t \mid \tilde{y}_{d:t}, y_{1:t}) \mid y_{1:t}] + \mathrm{Var}[\mathrm{E}(\beta_t \mid \tilde{y}_{d:t}, y_{1:t}) \mid y_{1:t}]. \tag{3.36}$$

Note that $\hat{\beta}_{t|t}$ is a linear function of $\hat{\beta}_{d|d}, y_{d+1}, \dots, y_t$ and that given $\tilde{y}_{d:t}$, β_d is known. Then, we have

$$\mathrm{E}(\beta_t \mid \tilde{y}_{d:t}, y_{1:t}) = \mathrm{E}(\beta_t \mid \tilde{y}_{d:t}) = \sum_{i=0}^{t-d+1} b_{t-d,i} y_{t-i} + \mathbf{B}_{t-d} \beta_d, \tag{3.37}$$

for some vectors $b_{t-d,i}$ and matrix $\mathbf{B}_{t-d}$. For any $0 \le d \le t-1$, we write $\mathbf{P}_{t|t}^* = \mathrm{Var}(\beta_t \mid \tilde{y}_{0:t})$; $\mathbf{P}_{t|t}^*$ does not depend on the actual values of $\tilde{y}_{0:t}$, but only on the time t. Then from (3.37), we have

$$\mathrm{E}[\mathrm{Var}(\beta_t \mid \tilde{y}_{d:t}) \mid y_{1:t}] = \mathrm{Var}(\beta_t \mid \tilde{y}_{d:t}) = \mathrm{Var}(\beta_{t-d} \mid \tilde{y}_{0:t-d}) = \mathbf{P}_{t-d|t-d}^*.$$

All state space models can be classified in the following three cases:

1. $\mathbf{P}_{0|0} = \mathbf{P}_{0|0}^*$ so that $\mathbf{P}_{0|0} = \mathbf{0}$ and $y_{1:t} \equiv \tilde{y}_{0:t}$.
   Using (3.36), we obtain

$$\begin{aligned}
\mathbf{P}_{t|t}^* &= \mathrm{E}[\mathrm{Var}(\beta_t \mid \tilde{y}_{d:t}) \mid \tilde{y}_{0:t}] + \mathrm{Var}[\mathrm{E}(\beta_t \mid \tilde{y}_{d:t}) \mid \tilde{y}_{0:t}] \\
&= \mathbf{P}_{t-d|t-d}^* + \mathrm{Var}(\mathbf{B}_{t-d} \beta_d \mid \tilde{y}_{0:t}) \\
&\ge \mathbf{P}_{t-d|t-d}^*.
\end{aligned} \tag{3.38}$$

Thus the sequence $\{\mathbf{P}_{t|t}^*\}$ is monotonic (non-increasing), and from $\mathbf{P}_{t|t} = \mathrm{Var}(\beta_t \mid y_{1:t}) = \mathrm{Var}(\beta_t \mid \tilde{y}_{0:t}) = \mathbf{P}_{t|t}^*$, it follows that $\{\mathbf{P}_{t|t}\}$ is monotonic too. Since $\{\mathbf{P}_{t|t}\}$ is bounded and monotonic, it follows that $\lim_{t \to \infty} \mathbf{P}_{t|t} = \lim_{t \to \infty} \mathbf{P}_{t|t}^*$ exists. Equation (3.38) implies

$$\lim_{t \to \infty} \mathrm{Var}(\mathbf{B}_{t-d} \beta_d \mid \tilde{y}_{0:t}) = 0. \tag{3.39}$$

2. Let now $\mathbf{P}_{0|0} > 0$ (positive definite) and $\sigma^2 > 0$ and $\mathbf{Z} > 0$. First we see that from (3.39),

$$
\begin{aligned}
\text{Var}((\mathbf{B}_{t-d}\beta_d \mid y_{1:t}) &= \text{E}[\text{Var}(\mathbf{B}_{t-d}\beta_d \mid \tilde{y}_{0:t}) \mid y_{1:t}] \\
&\quad + \text{Var}[\text{E}(\mathbf{B}_{t-d}\beta_d \mid \tilde{y}_{0:t}) \mid y_{1:t}] \\
&\to 0 \quad \text{as } t \to \infty.
\end{aligned}
$$

This together with (3.36) and (3.38) implies

$$
\begin{aligned}
\mathbf{P}_{t|t} &= \text{Var}(\beta_t \mid y_{1:t}) \\
&= \text{E}[\text{Var}(\beta_t \mid \tilde{y}_{d:t}) \mid y_{1:t}] + \text{Var}[\text{E}(\beta_t \mid \tilde{y}_{d:t}) \mid y_{1:t}] \\
&= \mathbf{P}^*_{t-d|t-d} + \text{Var}(\mathbf{B}_{t-d}\beta_d \mid y_{1:t}) \to \mathbf{P}^*.
\end{aligned}
$$

3. Finally, let $\sigma^2 \mathbf{Z} \not\succ \mathbf{0}$ or that either $\sigma^2 = 0$ or $\mathbf{Z}$ is positive semi-definite (a non-negative definite matrix, which is not positive definite). Define $\sigma^2(z) = \sigma^2 + z$ and $\mathbf{Z}(z) = \mathbf{Z} + z\mathbf{I}$, for some $0 \le z < 1$. We have the following:

a. The limit $\lim_{t\to\infty} \mathbf{P}^*_{t|t}(0)$ exists from (1).
b. From (2), the limit $\lim_{t\to\infty} \mathbf{P}_{t|t}(z) = \mathbf{P}^*(z)$ exists for all $0 < z < 1$ because $\sigma^2(z) > 0$ and $\mathbf{Z}(z) > 0$.
c. $\{\mathbf{P}_{t|t}(z)\}$ is bounded, continuous in z and monotonic in z. Thus, as $t \to \infty$, $\mathbf{P}_{t|t}(z)$ converges uniform ally to $\mathbf{P}(z)$ and

$$
\lim_{t\to\infty} \mathbf{P}_{t|t} = \lim_{t\to\infty} \lim_{z\to\infty} \mathbf{P}_{t|t}(z) = \lim_{z\to\infty} \mathbf{P}^*(z) = \mathbf{P}^*(0) = \mathbf{P}^*.
$$

To sum up the above proves that in each case $\lim_{t\to\infty} \mathbf{P}_{t|t}$ exists and it is equal to $\lim_{t\to\infty} \mathbf{P}^*_{t|t} = \mathbf{P}^*$. This implies that the limit of $\mathbf{P}_{t|t}$ does not depend on the initial prior $\mathbf{P}_{0|0}$. □

Some comments are in order.

1. With the conditions of Theorem 3.7, the limit $\mathbf{P}$ satisfies the algebraic Riccati equation

$$
\mathbf{P} = \mathbf{FPF}^\top + \mathbf{Z} + [x^\top(\mathbf{FPF}^\top + \mathbf{Z})x + \sigma^2]^{-1}(\mathbf{FPF}^\top + \mathbf{Z})xx^\top(\mathbf{FPF}^\top + \mathbf{Z}),
$$

which is a direct consequence of the posterior covariance updating of $\mathbf{P}_{t|t}$ and the existence of $\lim_{t\to\infty} \mathbf{P}_{t|t} = \mathbf{P}$. This solution may be solved analytically in simple cases, such as in the local level model of Sect. 3.5.2, but in general one needs to resort to numerical methods for the calculation of $\mathbf{P}$.

2. With the conditions of Theorem 3.7, it follows that the following limits exist and
   do not depend on the prior covariance matrix $\mathbf{P}_{0|0}$:

   a. $\lim_{t \to \infty} \mathbf{P}_{t|t-1} = \mathbf{FPF}^{\top} + \mathbf{Z}$;
   b. $\lim_{t \to \infty} q_{t|t-1} = q = x^{\top}(\mathbf{FPF}^{\top} + \mathbf{Z})x + \sigma^2$;
   c. $\lim_{t \to \infty} K_t = K = q^{-1}(\mathbf{FPF}^{\top} + \mathbf{Z})\mathbf{F}$.

3. The steady state of the Kalman filter is

$$\hat{\beta}_{t|t} = \mathbf{F}\hat{\beta}_{t-1|t-1} + K(y_t - x^{\top}\mathbf{F}\hat{\beta}_{t-1|t-1}),$$

just by replacing $\mathbf{P}_{t|t}$ by its limit $\mathbf{P}$ in $\hat{\beta}_{t|t}$.

3.6 Exercises

1. For a time series $\{y_t\}$, we fit a state space model with $p = 1$, $x_t = 1$, $F_t = 1$,
   $\sigma^2 = 200$ and $Z_t = 10$. Given that the filtered distribution at t is $\beta_t \mid y_{1:t} \sim$
   $N(300, 40)$:

   a. write down the observation and transition equations of the state space model.
   b. what are the two-step ahead forecast distribution for $\beta_{t+2} \mid y_{1:t}$ and the
      predictive distribution for the sum $S \mid y_{1:t}$, where $S = y_{t+1} + y_{t+2}$?
   c. show that $\text{Cov}(\beta_{t+2}, S) = 110$ and hence, or otherwise, obtain the joint
      distribution of $(\beta_{t+2}, S)^{T}$.

2. Consider the time series $\{y_t\}$ generated by the state space model with $x_t = 1$,
   $F_t = \lambda$, σ^2, $Z_t = Z$, where the variances σ^2 and Z and the constant λ are all
   known. Define the time series

$$z_t = y_t - \lambda y_{t-1}.$$

   a. Write down the observation and evolution equations of the state space model
      of y_t.
   b. By obtaining the mean and the variance of z_t together with the autocovari-
      ances $\text{Cov}(z_t, z_{t-k})$, for integer k or otherwise, define the joint probability
      distribution of the time series $\{z_t\}$.

3. A simple model for the number u_t of unemployed people in a region in month t
   is as follows. It ignores discreteness and supposes that each month a proportion
   α, $(0 < \alpha < 1)$, of the unemployed find work, so that

$$u_t = (1 - \alpha)u_{t-1} + n_t,$$

where n_t is the number becoming newly unemployed during month t. In turn, n_t is assumed to vary around a constant number v according to

$$n_t = v + \eta_t,$$

where $\{\eta_t\}$ is a sequence of independent Gaussian innovations $N(0, \sigma_\eta^2)$. Because of statistical difficulties, u_t is never recorded exactly: the *recorded* number of unemployed in month t, y_t, is related to it by

$$y_t = u_t + \epsilon_t,$$

where $\{\epsilon_t\}$ is a sequence of independent Gaussian innovations $N(0, \sigma_\epsilon^2)$.

a. Show that if $\beta_t = (u_t, v)^\top$, the system can be described by

$$y_t = x_t^\top \beta_t + \epsilon_t \quad \text{and} \quad \beta_t = F_t \beta_{t-1} + \zeta_t,$$

where x_t is a constant vector, F_t is a constant matrix and ζ_t is a random vector. Give the values of x_t and F_t, and write down the mean vector and covariance matrix of ζ_t.

b. Show that, if α and v are known, the *Kalman Filter prediction equations* imply that the one-step forecast $\hat{u}_{t+1|t}$ of the number of people unemployed at time $t + 1$ given an estimate $\hat{u}_{t|t}$ of the number unemployed at time t is

$$\hat{u}_{t+1|t} = (1 - \alpha)\hat{u}_{t|t} + v.$$

c. Suppose it is known that $\alpha = 0.05$, $v = 0.2$ and $\sigma_\epsilon^2 = \sigma_\eta^2 = 0.005$, where units of measurement are millions of unemployed. At time $t = 0$, it is estimated that $u_0 = 0.7$ with estimation variance 0.01. Use this information to predict the value of U_1, and give the standard deviation of the associated prediction error.

d. If the number of unemployed at time $t = 1$ were subsequently *recorded* as $y_1 = 0.85$, would your prediction of u_1 need to be revised upwards or downwards, and by how much? Give reasons for your answer.

4. A company trades 10 products, with the i-th product projected to give a return r_{it} at time t, for $i = 1, 2, \ldots, 10$. The company believes that each of these returns r_{it} follows an autoregressive process

$$r_{it} = 0.9 r_{i,t-1} + \zeta_{it},$$

where ζ_{it} is a white noise with variance 1, $\zeta_{it} \sim N(0, 1)$, and ζ_{it} is independent of ζ_{jt}, for any $i \neq j$.

Due to a data recording error, r_{it} is not available. However, the aggregate return can be observed subject to additive noise, according to the model

$$y_t = \sum_{i=1}^{10} r_{it} + \epsilon_t,$$

where ϵ_t is a white noise with variance 1, $\epsilon_t \sim N(0, 1)$, and it is assumed that ϵ_t is independent of ζ_{it}, for any t and for any i.

a. Define the state

$$\beta_t = \sum_{i=1}^{10} r_{it}.$$

Show that y_t follows a state space model

$$y_t = x\beta_t + \epsilon_t$$
$$\beta_t = F\beta_{t-1} + \zeta_t,$$

and determine x, F, ζ_t and the variance of ζ_t.

b. A prior distribution for β_0 is set as

$$\beta_0 \sim N(0, 100).$$

If the first observation is $y_1 = 2$, perform the Kalman filter iteration for $t = 1$ and obtain the posterior distribution of

$$\beta_1 \mid \{y_1 = 2\}.$$

c. Using the result in (b), obtain a 95% predictive interval for y_2.
d. Describe briefly what is the likely effect on the posterior distribution of β_t (for large t), if the prior distribution of β_0 changes from (i) $\beta_0 \sim N(0, 1)$ to (ii) $\beta_0 \sim N(0, 1000)$.

5. It is well known that an economy's growth as measured by gross domestic product (GDP) is related to unemployment rate. Arthur Okun studied how much GDP is likely to fall, if unemployment increased by a certain level. Let y_t denote the GDP growth of the UK economy at time t, and let x_t denote the unemployment rate at time t. A simple regression between the two can reveal the likely decrease of the GDP growth, for an increase of the unemployment rate. However, a more elaborate analysis considers the following dynamic regression model:

$$y_t = \alpha + \gamma_t x_t + \epsilon_t,$$

where α is a static intercept, ϵ_t is a Gaussian white noise with variance 1 and γ_t is a time-varying slope, which follows the autoregressive model

$$\gamma_t = 0.3\gamma_{t-1} + v_t,$$

with v_t a Gaussian white noise with variance 2. It is further assumed that ϵ_t and v_s are independent for any t, s and that γ_0 is independent of v_t, for any t.

a. Define the state vector

$$\beta_t = \begin{bmatrix} \alpha \\ \gamma_t \end{bmatrix}.$$

Write the above model in state space form,

$$y_t = L_t^\top \beta_t + \kappa_t,$$
$$\beta_t = F\beta_{t-1} + \zeta_t,$$

and determine the design vector L_t and the evolution matrix F. Write down κ_t and ζ_t and obtain their distributions.

b. The above state space model is fitted to data of length n. The posterior distribution of β_n, given information $y_{1:n} = \{y_1, \ldots, y_n\}$, is

$$\beta_n \mid y_{1:n} \sim N\left\{ \begin{bmatrix} -1.5 \\ 3 \end{bmatrix}, \begin{bmatrix} 1 & 0 \\ 0 & 10 \end{bmatrix} \right\}.$$

If $x_{n+1} = 0.5$ and $y_{n+1} = 1$, then obtain the posterior distribution $p(\beta_{n+1} \mid y_{1:n+1})$ of β_{n+1}, given information $y_{1:n+1}$. Provide a 95% credible interval for γ_{n+1}, given $y_{1:n+1}$.

6. Prove theoretically the claim of Example 3.3, that the k-step ahead forecast mean in a local level model does not depend on k and that the k-step forecast variance is increasing with k. In particular, considering the local level model:

$$y_t = \beta_t + \epsilon_t \quad \text{and} \quad \beta_t = \beta_{t-1} + \zeta_t,$$

where ϵ_t and ζ_t are independent, $\epsilon_t \sim N(0, \sigma^2)$ and $\zeta_t \sim N(0, Z)$, show that the k-step ahead forecast mean and variance of y_{t+k} are

$$\hat{y}_{t+k|t} = \hat{\beta}_{t|t} \quad \text{and} \quad q_{t+k|t} = P_{t|t} + \sigma^2 + kZ,$$

for any $k = 1, 2, \ldots$. Thus, show that for fixed t, $\lim_{k \to \infty} q_{t+k|t} = \infty$. What do you learn about the local level model forecast ability? Explain why the above local level model is also referred to as *steady forecasting* model.

7. Using R, simulate 200 values of the state space model with $p = 3$. Using a simulated data $y_1, \ldots, y_{200}$, calculate in R the filtered mean vector $\hat{\beta}_{t|t}$ and the covariance matrix $\mathbf{P}_{t|t}$. Make a plot of the one-step ahead predictions $\hat{y}_{t|t-1}$ against the actual data y_t, and comment on the goodness of fit. Using R, calculate a 99% prediction interval of y_{201}, based on data $y_{1:200}$.

8. The following table gives 20 observations from a time series $\{y_t\}$:

	y_t
$t = 1 - 10$	10 11 11 9 7 9 5 8 8 10
$t = 11 - 20$	12 11 12 13 10 9 11 12 8 9

   a. Put the data in R and make a time series plot of the data.
   b. In R, fit a local level model with $\sigma^2 = 1$, $Z = 1$, $\hat{\beta}_{0|0} = 0$ and $P_{0|0} = 100$, and provide filtered and smoothed estimates of the level β_t. How do the filtered estimates compare to the smoothed estimates of the level?
   c. Produce one-step forecasts at each time point t, and comment on the goodness of fit.
   d. Repeat (b)–(c) with $Z = 10$ and comment.

9. In the context of the state space model (3.10a)–(3.10b), prove that the posterior covariance matrix $\mathbf{P}_{t|t}$ can alternatively be updated via the recursion

$$\mathbf{P}_{t|t}^{-1} = \mathbf{P}_{t|t-1}^{-1} + x_t x_t^\top / \sigma^2,$$

where $\mathbf{P}_{t|t-1}$ is given by the Kalman filter (Theorem 3.2). Thus, provide an alternative recursion for the Kalman filter.

10. In the local level model of Exercise 3.5, show that the smoothed variance $P_{t|n}$ can be approximated as

$$P_{t|n} = LZ + L^2 P_{t+1|n},$$

where L the limit of L_t (see Theorem 3.4) is equal to

$$L = \frac{(\sqrt{Z + 4\sigma^2} - \sqrt{Z})^2}{4\sigma^2}.$$

Expand $P_{t|n}$ above, and take its limit as $t \to \infty$ to show that the smoothed variance converges approximately to

$$\lim_{t \to \infty} P_{t|n} = \frac{4\sigma^2 Z(\sqrt{Z + 4\sigma^2} - \sqrt{Z})^2}{16\sigma^4 - (\sqrt{Z + 4\sigma^2} - \sqrt{Z})^4}.$$

11. Suppose that the time series $\{y_t\}$ is generated by the local level model

$$y_t = \beta_t + \epsilon_t \quad \text{and} \quad \beta_t = \beta_{t-1} + \zeta_t,$$

where ϵ_t and ζ_t are independent, $\epsilon_t \sim N(0, \sigma^2)$ and $\zeta_t \sim N(0, Z)$. Based on observed data $y_{1:n} = \{y_1, y_2, \ldots, y_n\}$, let $\hat{\beta}_{t|n}$ denote the smoothed mean of β_t at time t and $\hat{\beta}_{t|t}$ denote the posterior mean of β_t at time t.

a. Show that

$$|\hat{\beta}_{t|n} - \hat{\beta}_{t|t}| < \sum_{i=t+1}^{n} |e_i| \quad \text{and} \quad |\hat{\beta}_{t|n} - \hat{\beta}_{t+1|n}| < \sum_{i=t+1}^{n} |e_i|,$$

where e_i is the residual at time $i = t+1, \ldots, n$.

b. If

$$|e_i| < \frac{1}{n^2},$$

for each i, then show that $\hat{\beta}_{t|n}$ converges to $\hat{\beta}_{t|t}$ as $n \to \infty$.

Give reasoning to the following statement: 'As t is closer to n we expect the smoothed mean $\hat{\beta}_{t|n}$ to be closer to the posterior mean $\hat{\beta}_{t|t}$; as t is far apart from n, then $\hat{\beta}_{t|n}$ will be much more accurate compared to $\hat{\beta}_{t|t}$'.

12. Consider the state space model (3.10a)–(3.10b) with

$$x_t = \begin{bmatrix} \gamma \\ 0 \end{bmatrix}, \quad \mathbf{F}_t = \begin{bmatrix} 1 & \lambda \\ 0 & 1 \end{bmatrix} \quad \text{and} \quad \mathbf{Z}_t = \begin{bmatrix} Z_1 & 0 \\ 0 & Z_2 \end{bmatrix},$$

for some $\gamma, \lambda \in \mathbb{R}$ and $Z_1, Z_2 \geq 0$.

Show that if $\gamma \neq 0$ and $\lambda \neq 0$, then the limit of $\mathbf{P}_{t|t}$ exists, as $t \to \infty$.

13. Consider the observable state space model

$$y_t = x\beta_t + \epsilon_t \quad \text{and} \quad \beta_t = F\beta_{t-1} + \zeta_t,$$

where ϵ_t and ζ_t are independent, $\epsilon_t \sim N(0, \sigma^2)$ and $\zeta_t \sim N(0, Z)$. Assume as usual the initial distribution $\beta_0 \sim N(\hat{\beta}_{0|0}, P_{0|0})$, for some known mean $\hat{\beta}_{0|0}$ and variance $P_{0|0}$.

Show that the limit of the posterior variance $P_{t|t} = \text{Var}(\beta_t \mid y_{1:t})$ exists and is equal to

$$\lim_{t \to \infty} P_{t|t} = \frac{\sigma^2}{2x^2 F^2} \left[\sqrt{(px^2 - F^2 + 1)^2 + 4px^2 F^2} + F^2 - px^2 - 1 \right],$$

where $p = Z/\sigma^2$. Observe that for $x = F = 1$, the above result reduces to the steady state of the local level model (see Sect. 3.5.2).

Chapter 4
Model Specification and Model Performance

The previous chapter studies estimation procedures for the general state space model, assuming that the design vector x_t, the transition matrix $\mathbf{F}_t$, the observation variance σ^2 and the transition covariance matrix $\mathbf{Z}_t$, as well as the prior mean vector $\hat{\beta}_{0|0}$ and the prior covariance matrix $\mathbf{P}_{0|0}$, are all known. This chapter discusses how these components may be chosen, either estimated using the data or specified by the user and how their choice may affect the performance of the model.

The successful application of the Kalman filter and the related estimation procedures of Chap. 3 require careful specification of these components. Some of these (x_t and $\mathbf{F}_t$) are usually implied by the desired model, while others (such as σ^2 and $\mathbf{Z}_t$) are harder to specify. For the former, we give several particular models in Sect. 4.1, which reveal some of the wealth of state space models. Examples include trend, seasonal, dynamic regression, autoregression and regression with autocorrelated errors. The inverse problem that of decomposing a given state space model to simpler component state space models is discussed in Sect. 4.2.2.

Even though the above components may be implied by a particular state space model, they are likely to depend on hyperparameters. More generally, any of the model components may depend on hyperparameters, whose estimation plays a critical role in the application of the Kalman filter. In Sect. 4.3 we discuss three approaches: maximum likelihood estimation (aimed at general application), specification of $\mathbf{Z}_t$ using discount factors and estimation of σ^2 using Bayesian conjugate methods. After all these components are specified, model performance can be judged by using residual or error analysis. This is the subject of Sect. 4.4. Section 4.5 discusses prior specification, i.e. specification of $\hat{\beta}_{0|0}$ and $\mathbf{P}_{0|0}$, as well as of any other relevant prior quantities. Finally, Sect. 4.6 discusses automatic model monitoring, with the emphasis placed on outlier detection and sequential model performance.

© The Author(s), under exclusive license to Springer Nature Switzerland AG 2021   111
K. Triantafyllopoulos, *Bayesian Inference of State Space Models*, Springer Texts in Statistics, https://doi.org/10.1007/978-3-030-76124-0_4

4.1 Specification of Model Components

Suppose we wish to set up a state space model (3.10a)–(3.10b) as defined in
Sect. 3.1.3. In the implementation of this model and the application of the Kalman
filter or the other estimation algorithms, presented in Chap. 3, we need to specify the
model components x_t, $\mathbf{F}_t$, σ^2 and $\mathbf{Z}_t$ (see Sect. 3.1.3). As already noted in Chap. 3,
some of these components may be time-invariant and according to what type of
data we wish to model, x_t and $\mathbf{F}_t$ (the design vector and the transition matrix,
respectively) will take particular forms. For example, for the local level model of
Sect. 3.1.3 we have $x_t = F_t = 1$. In other situations, such as in dynamic regression,
x_t will include known covariates, to which y_t is related, and thus in model (3.9a)–
(3.9b) we know x_t and we set $\mathbf{F}_t = \mathbf{I}$ to define a random walk evolution for the
states β_t. In the next sections we describe particular forms of state space models,
which reveal particular forms of the components x_t and $\mathbf{F}_t$.

4.1.1 Trend State Space Models

Trend state space models refer to models that are capable of analysing data that are
expected or believed to follow a trend. The term "trend" here is understood as a
polynomial function in time t, e.g. a straight line, a quadratic and cubic. Data that
can be analysed using trend models may include economic time series, or other data
that exhibit growth or decline over the course of time.

Example 4.1 (Aluminium Prices Data) Figure 4.1 depicts spot prices of aluminium
over the time period 4 January 2005 to 31 October 2005. Aluminium, a non-
ferrous metal, trades daily at the London metal exchange, for information in
which the reader is referred to http://www.lme.com/. Figure 4.1 graphs aluminium
prices, excluding bank holidays and weekends, and shows initially some random
fluctuation, followed by an upward linear trend (March–April), followed by a
linear fall (May–June), followed by some random fluctuation (June–July), followed
by some increasing and then decreasing trend (August–September), and finally
followed by a linear trend (October).

Linear Growth Model
The first state space model we consider is the so-called *linear growth* or *local linear
trend* state space model. This model employs the state space model (3.10a)–(3.10b)
with design vector x_t and transition matrix $\mathbf{F}_t$, given by

$$x_t = x = \begin{bmatrix} 1 \\ 0 \end{bmatrix} \quad \text{and} \quad \mathbf{F}_t = \mathbf{F} = \begin{bmatrix} 1 & 1 \\ 0 & 1 \end{bmatrix}.$$

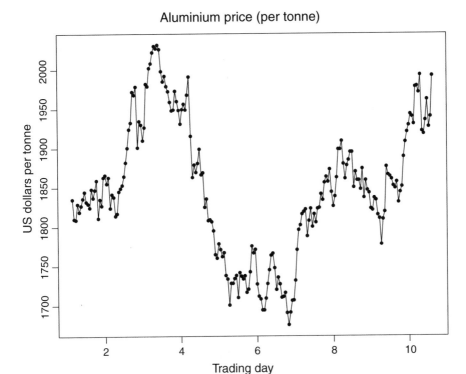

Fig. 4.1 Aluminium prices (US$ per tonne). The integers in the time axis indicate months in 2005

Thus, by writing the state vector as $\beta_t = [\beta_{1t}, \beta_{2t}]^\top$ and the transition innovations as $\zeta_t = [\zeta_{1t}, \zeta_{2t}]^\top$, the model can be written as

$$y_t = x^\top \beta_t + \epsilon_t = [1, 0] \begin{bmatrix} \beta_{1t} \\ \beta_{2t} \end{bmatrix} + \epsilon_t = \beta_{1t} + \epsilon_t, \tag{4.1a}$$

$$\beta_t = \begin{bmatrix} \beta_{1t} \\ \beta_{2t} \end{bmatrix} = \begin{bmatrix} 1 & 1 \\ 0 & 1 \end{bmatrix} \begin{bmatrix} \beta_{1,t-1} \\ \beta_{2,t-1} \end{bmatrix} + \begin{bmatrix} \zeta_{1t} \\ \zeta_{2t} \end{bmatrix},$$

$$\beta_{1t} = \beta_{1,t-1} + \beta_{2,t-1} + \zeta_{1t}, \tag{4.1b}$$

$$\beta_{2t} = \beta_{2,t-1} + \zeta_{2t}, \tag{4.1c}$$

where as usual $\epsilon_t \sim N(0, \sigma^2)$, $\zeta_t \sim N(0, \mathbf{Z}_t)$, with

$$\mathbf{Z}_t = \mathbf{Z} = \mathrm{Var} \begin{bmatrix} \zeta_{1t} \\ \zeta_{2t} \end{bmatrix} = \begin{bmatrix} z_{11,t} & z_{12,t} \\ z_{12,t} & z_{22,t} \end{bmatrix},$$

where $z_{11,t}$ is the variance of ζ_{1t}, $z_{22,t}$ is the variance of ζ_{2t} and $z_{12,t}$ is the covariance of ζ_{1t} and ζ_{2t}.

Some comments are in order. The linear growth model is defined by Eqs. (4.1a)–(4.1c): basically the observations y_t fluctuate around β_{1t}—the locally linear level of the time series—hence the name of the model. The level β_{1t} follows a linear function (in t) plus random error. To see this write down recursively β_{1t} by replacing $\beta_{1,t-1}$ from (4.1b) and $\beta_{2,t-1}$ from (4.1c) to get

$$
\begin{aligned}
\beta_{1t} &= \beta_{1,t-1} + \beta_{2,t-1} + \zeta_{1t} \\
&= \beta_{1,t-2} + 2\beta_{2,t-2} + \zeta_{1t} + \zeta_{1,t-1} + \zeta_{2,t-1} \\
&= \beta_{1,t-3} + 3\beta_{2,t-3} + \zeta_{1t} + \zeta_{1,t-1} + \zeta_{1,t-2} + \zeta_{2,t-1} + 2\zeta_{2,t-2} \\
&= \cdots \\
&= \underbrace{\beta_{1,0} + t\beta_{2,0}}_{\text{Linear function}} + \underbrace{\sum_{i=1}^{t} \zeta_{1i} + \sum_{j=1}^{t-1} j\zeta_{2,t-j}}_{\text{Random error}}.
\end{aligned}
$$

Thus, we think of y_t as fluctuating randomly around the level β_{1t}, which in itself fluctuates randomly around a linear function of t.

Example 4.2 (Aluminium Prices Data Example Continued) Using the Kalman filter we have fitted a linear growth model to the aluminium price data of Example 4.1. We have used $\sigma^2 = 1$ and

$$
\mathbf{Z}_t = \begin{bmatrix} 10 & 0 \\ 0 & 2 \end{bmatrix}.
$$

First of all we note that $\mathbf{Z}_t$ is time-invariant and we write $\mathbf{Z}_t = \mathbf{Z}$. Secondly, we have assumed that the components β_{1t} and β_{2t} are independent (hence $z_{12,t} = 0$) and the variance of ζ_{1t} is larger than that of ζ_{2t}. Also, we have chosen an initial state

$$
\beta_0 = \begin{bmatrix} \beta_{1,0} \\ \beta_{2,0} \end{bmatrix} \sim N \left\{ \begin{bmatrix} 1800 \\ 0 \end{bmatrix}, \begin{bmatrix} 1000 & 0 \\ 0 & 1000 \end{bmatrix} \right\}.
$$

From historical data is known that the mean of (y_t) fluctuates around 1800 US dollars. Thus, since β_{1t} is the level of the series $(E(y_t \mid \beta_{1t}) = \beta_{1t})$, a sensible choice is to set $\hat{\beta}_{1,0|0}$ to its historical value, i.e. 1800, while $\hat{\beta}_{2,0|0}$ can take an arbitrary value (here we have set it to $\hat{\beta}_{2,0|0} = 1$). The prior covariance matrix $\mathbf{P}_{0|0}$ is set to be proportional to the identity matrix (again consistent with a priori independence of the two components $\beta_{1,0}$ and $\beta_{2,0}$; the large variances of $\beta_{1,0}$ and $\beta_{2,0}$ reflect on the associated uncertainty on the specification of $\hat{\beta}_0$, prior to observing the data.

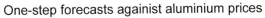

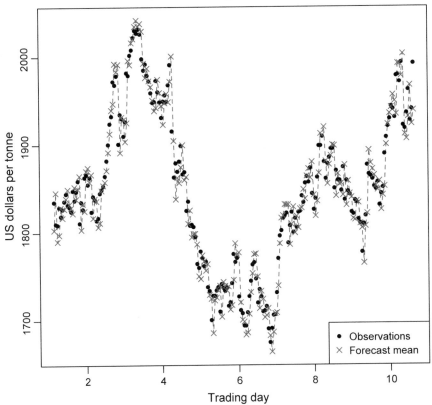

Fig. 4.2 One-step forecasts of the linear growth model (dashed lines) against the aluminium prices (solid points)

Figure 4.2 (in next page) plots one-step forecasts of the aluminium price data described above. The R code used is given below (NB: need first to load the R package gplots).

```
> # time series data
> alum1 <- read.table("alum.txt")
> alum <- alum1[,2]

> # fit linear trend
> fit <- bts.filter(alum, x0=c(1,2), F0=matrix(c(1,0,1,1),2,2),
+ obsvar=1, Z0=matrix(c(10,0,0,2),2,2), beta0=c(1800,1),
+ P0=1000*diag(2), DISO=FALSE, VAREST=FALSE )

> # define time series objects to be plotted
> alumts <- ts(alum1[,2], start=c(1,4), frequency=22)
> pred <- ts(fit$FittedMean, start=c(1,4), frequency=22)
```

```
> # time series plot
> ts.plot(pred, lty=2, col=2, main=expression("One-step forecasts
+ against aluminium prices"),
+ xlab="Trading day",ylab="US dollars per tonne")
> points(alumts, pch=20)
> points(pred, pch=4, col=2)
> legend("bottomright",c("Observations","Forecast mean"),
+ pch=c(20, 4), col=c(1,2))
```

From Fig. 4.2 we observe that the model seems to fit the data well (the one-step ahead forecasts seem to match the data well).

Returning to model (4.1a)–(4.1c), we can observe that, based on a data set $y_{1:t} = \{y_1, \ldots, y_t\}$, the forecast function $\hat{y}_{t+k|t}$ (see Eq. (3.28) in the previous chapter) is a linear function in k. First from the definition of $\mathbf{F}$ we have

$$\mathbf{F}^k = \begin{bmatrix} 1 & k \\ 0 & 1 \end{bmatrix}.$$

We prove this by induction. First note that $\mathbf{F}^1 = \mathbf{F}$ and then write

$$\mathbf{F}^k = \mathbf{F}^{k-1}\mathbf{F} = \begin{bmatrix} 1 & k-1 \\ 0 & 1 \end{bmatrix}\begin{bmatrix} 1 & 1 \\ 0 & 1 \end{bmatrix} = \begin{bmatrix} 1 & k \\ 0 & 1 \end{bmatrix}.$$

From the Kalman filter we have that $\beta_t \mid y_{1:t} \sim N(\hat{\beta}_{t|t}, \mathbf{P}_{t|t})$, with $\hat{\beta}_{t|t} = [\hat{\beta}_{1,t|t}, \hat{\beta}_{2,t|t}]^\top$. Then from the definition of the forecast function (see Theorem 3.6) we obtain

$$\hat{y}_{t+k|t} = x_{t+k}^\top \mathbf{F}^k \hat{\beta}_{t|t}$$

$$= [1, 0]\begin{bmatrix} 1 & k \\ 0 & 1 \end{bmatrix}\begin{bmatrix} \hat{\beta}_{1,t|t} \\ \hat{\beta}_{2,t|t} \end{bmatrix}$$

$$= \hat{\beta}_{1,t|t} + \hat{\beta}_{2,t|t}k,$$

which is a linear function in k (or a straight line with intercept $\hat{\beta}_{1,t|t}$ and slope $\hat{\beta}_{2,t|t}$).

Example 4.3 (Aluminium Prices Data Example Continued) Continuing on the aluminium prices Example 4.2, we can extract the posterior mean vector at $t = 210$ (corresponding to the last time point, 31 October 2005), using the R command

```
> fit$PostMean[[210]]
            [,1]
[1,] 1959.51481
[2,]   15.08425
```

Thus, the forecast function is

$$\hat{y}_{210+k|210} = 1959.51 + 15.08k, \quad k = 1, 2, \ldots$$

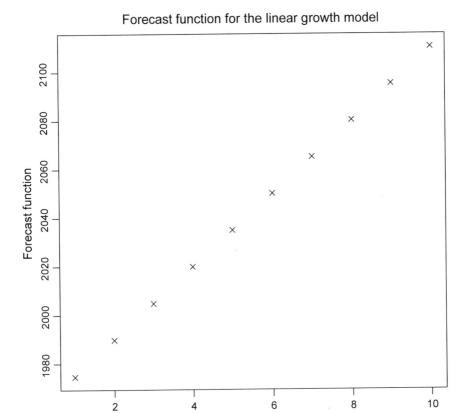

Fig. 4.3 Forecast function for the linear growth model at $t = 210$

which is plotted in Fig. 4.3, for $k = 1, 2, \ldots, 10$, by using the R commands

```
> x <- 1959.51 + (1:10) * 15.08
> plot(x, pch=4, xlab="Lead time (i)", ylab="Forecast function",
+   main=expression("Forecast function for the linear growth
+   model"))
```

We observe that for relatively low values of k, the forecasts are too high (compared to the past data), which make them unrealistic. As a result, for the aluminium prices data considered here, a linear growth model is suitable only for $k = 1$ (one-step ahead forecasting).

Polynomial Trend State Space Models
The linear growth trend model can be generalised to trend models following higher order polynomials than linear functions, similar state space models with extensions are discussed in Godolphin and Harrison (1975) and Godolphin and Stone (1980). Indeed, the most general model, known as $(p − 1)$-th order *polynomial trend model*,

employs the state space model (3.10a)–(3.10b) with design vector x_t and transition matrix $\mathbf{F}_t$, given by

$$
x_t = x = \begin{bmatrix} 1 \\ 0 \\ \vdots \\ 0 \end{bmatrix} \quad \text{and} \quad \mathbf{F}_t = \mathbf{F} = \begin{bmatrix} 1 & 1 & 0 & \cdots & 0 \\ 0 & 1 & 1 & \cdots & 0 \\ \vdots & \vdots & \vdots & \ddots & \vdots \\ 0 & 0 & 0 & \cdots & 1 \end{bmatrix}. \tag{4.2}
$$

If we write $\beta_t = [\beta_{1t}, \ldots, \beta_{pt}]^\top$ and $\zeta_t = [\zeta_{1t}, \ldots, \zeta_{pt}]^\top$, then in analogy with the linear growth model, we can write the polynomial trend model as

$$
y_t = x^\top \beta_t + \epsilon_t = \beta_{1t} + \epsilon_t, \tag{4.3a}
$$

$$
\beta_{jt} = \beta_{j,t-1} + \beta_{j+1,t-1} + \zeta_{jt}, \quad j = 1, \ldots, p-1 \tag{4.3b}
$$

$$
\beta_{pt} = \beta_{p,t-1} + \zeta_{pt}, \tag{4.3c}
$$

Usually $\mathbf{Z}_t$ is set to be diagonal, i.e.

$$
\mathbf{Z}_t = \mathrm{Var}(\zeta_t) = \mathbf{Z} = \begin{bmatrix} Z_1 & 0 & \cdots & 0 \\ 0 & Z_2 & \cdots & 0 \\ \vdots & \vdots & \ddots & \vdots \\ 0 & 0 & \cdots & Z_p \end{bmatrix},
$$

where Z_i is the variance of ζ_{it}, for $i = 1, \ldots, p$.

We can observe that model (4.3a)–(4.3c), reduces to the linear growth model of Eqs. (4.1a)–(4.1c), if $p = 2$. If $p = 3$, the 2nd-order polynomial trend model is known as *quadratic growth* or *quadratic trend*. Higher order polynomial trend models are rarely used in practice, but the general polynomial model given above offers a generic approach for their study and their understanding.

One important property of the $(p-1)$-th order polynomial trend model is that its forecast function is a $(p-1)$-order polynomial (notice that in the linear growth this is a straight line, in the quadratic growth this is a quadratic and so forth). Before we give the general proof we discuss the forecast function of the quadratic growth $(p = 3)$.

Recall first that the forecast function is given by $\hat{y}_{t+k|t} = x_{t+k}^\top \mathbf{F}^k \hat{\beta}_{t|t}$, see Eq. (3.28) in Sect. 3.4. Then, in analogy with the linear growth model, we can see that

$$
\mathbf{F}^k = \begin{bmatrix} 1 & k & k(k-1)/2 \\ 0 & 1 & k \\ 0 & 0 & 1 \end{bmatrix}.
$$

The proof of this result is done by induction (similarly as in the linear growth model); for the general proof see below (p. 119).

Then, by writing $\hat{\beta}_{t|t} = [\hat{\beta}_{1,t|t}, \hat{\beta}_{2,t|t}, \hat{\beta}_{3,t|t}]^{\top}$, the forecast function is

$$\hat{y}_{t+k|t} = x_{t+k}^{\top} F^k \hat{\beta}_{t|t}$$

$$= [1, 0, 0] \begin{bmatrix} 1 & k & k(k-1)/2 \\ 0 & 1 & k \\ 0 & 0 & 1 \end{bmatrix} \begin{bmatrix} \hat{\beta}_{1,t|t} \\ \hat{\beta}_{2,t|t} \\ \hat{\beta}_{3,t|t} \end{bmatrix}$$

$$= \hat{\beta}_{1,t|t} + \left(\hat{\beta}_{2,t|t} - \frac{\hat{\beta}_{3,t|t}}{2} \right) k + \frac{\hat{\beta}_{3,t|t}}{2} k^2,$$

which is a quadratic in k.

Returning to the general case of the $(p-1)$-th polynomial trend model, we show that its forecast function is a $(p-1)$-order polynomial in k. First we show that with the transition matrix F defined in Eq. (4.2) above, we have

$$F^k = \begin{bmatrix} 1 & \binom{k}{1} & \binom{k}{2} & \cdots & \binom{k}{p-1} \\ 0 & 1 & \binom{k}{1} & \cdots & \binom{k}{p-2} \\ 0 & 0 & 1 & \cdots & \binom{k}{p-2} \\ \vdots & \vdots & \vdots & \ddots & \vdots \\ 0 & 0 & 0 & \cdots & 1 \end{bmatrix}, \tag{4.4}$$

where

$$\binom{k}{i} = 0,$$

for $i > k$.

The proof is by induction. From the definition of F, it is trivial to check that (4.4) is true for $k = 1$. Assuming that (4.4) is true for k, we will show that it is also true for $k + 1$. We have

$$F^{k+1} = F^k F$$

$$= \begin{bmatrix} 1 & \binom{k}{1} & \binom{k}{2} & \cdots & \binom{k}{p-1} \\ 0 & 1 & \binom{k}{1} & \cdots & \binom{k}{p-2} \\ 0 & 0 & 1 & \cdots & \binom{k}{p-2} \\ \vdots & \vdots & \vdots & \ddots & \vdots \\ 0 & 0 & 0 & \cdots & 1 \end{bmatrix} \begin{bmatrix} 1 & 1 & 0 & \cdots & 0 \\ 0 & 1 & 1 & \cdots & 0 \\ 0 & 0 & 1 & \cdots & 0 \\ \vdots & \vdots & \vdots & \ddots & \vdots \\ 0 & 0 & 0 & \cdots & 1 \end{bmatrix}$$

$$
= \begin{bmatrix}
1 & 1+\binom{k}{1} & \binom{k}{1}+\binom{k}{2} & \cdots & \binom{k}{p-2}+\binom{k}{p-1} \\
0 & 1 & 1+\binom{k}{1} & \cdots & \binom{k}{p-3}+\binom{k}{p-2} \\
0 & 0 & 1 & \cdots & \binom{k}{p-4}+\binom{k}{p-3} \\
\vdots & \vdots & \vdots & \ddots & \vdots \\
0 & 0 & 0 & \cdots & 1
\end{bmatrix}
$$

$$
= \begin{bmatrix}
1 & \binom{k+1}{1} & \binom{k+1}{2} & \cdots & \binom{k+1}{p-1} \\
0 & 1 & \binom{k+1}{1} & \cdots & \binom{k+1}{p-2} \\
0 & 0 & 1 & \cdots & \binom{k+1}{p-2} \\
\vdots & \vdots & \vdots & \ddots & \vdots \\
0 & 0 & 0 & \cdots & 1
\end{bmatrix},
$$

because for each $i = 1, 2, \ldots, p$ we have

$$
\binom{k}{i-1} + \binom{k}{i} = \frac{k!}{(i-1)!(k-i)!}\left(\frac{1}{k+1-i} + \frac{1}{i}\right) = \frac{(k+1)!}{i!(k+1-i)!} = \binom{k+1}{i}.
$$

Thus, (4.4) is true for any $k \geq 1$.

Now write the posterior mean at time t as $\hat{\beta}_{t|t} = [\hat{\beta}_{t|t,1}, \ldots, \hat{\beta}_{t|t,p}]^{\top}$. With $\mathbf{F}^k$ in place, the k-step forecast function of the polynomial trend model is

$$
\hat{y}_{t+k|t} = x^{\top} \mathbf{F}^k \hat{\beta}_{t|t}
$$

$$
= [1, 0, 0, \ldots, 0] \begin{bmatrix}
1 & \binom{k}{1} & \binom{k}{2} & \cdots & \binom{k}{p-1} \\
0 & 1 & \binom{k}{1} & \cdots & \binom{k}{p-2} \\
0 & 0 & 1 & \cdots & \binom{k}{p-2} \\
\vdots & \vdots & \vdots & \ddots & \vdots \\
0 & 0 & 0 & \cdots & 1
\end{bmatrix}
\begin{bmatrix}
\hat{\beta}_{t|t,1} \\
\hat{\beta}_{t|t,2} \\
\hat{\beta}_{t|t,3} \\
\vdots \\
\hat{\beta}_{t|t,p}
\end{bmatrix}
$$

$$
= \sum_{i=1}^{p} \hat{\beta}_{t|t,i} \binom{k}{i-1} = c_{1t} + c_{2t}k + c_{3t}k^2 + \cdots + c_{pt}k^{p-1},
$$

where the coefficients c_{it} depend on $\hat{\beta}_{t|t}$, but not on k. The examples of linear growth ($p = 2$) and quadratic trend ($p = 3$) give expressions of the values of c_{it}.

4.1.2 Superposition of State Space Models

An important model building tool, known as the *superposition* of state space models, suggests that complex models are built as the sum of simple state space models.

State space superpositions are discussed in detail in West and Harrison (1997, Section 6.2.1) or in Petris (2010, Section 3.2).

Consider N time series $\{y_{it}\}$ ($i = 1, \ldots, N$), each of which following a state space model, defined by equations

$$y_{it} = x_{it}^{\top} \beta_{it} + \epsilon_{it}, \qquad \epsilon_{it} \sim N(0, \sigma_i^2), \tag{4.5a}$$

$$\beta_{it} = \mathbf{F}_{it}\beta_{i,t-1} + \zeta_{it}, \quad \zeta_{it} \sim N(0, \mathbf{Z}_{it}), \tag{4.5b}$$

with the initial state $\beta_{i,0} \sim N(\hat{\beta}_{i,0|0}, \mathbf{P}_{i,0|0})$, for some mean vector $\hat{\beta}_{i,0|0}$ and covariance matrix $\mathbf{P}_{i,0|0}$. For each i, the innovations ϵ_{it} and ζ_{it} are assumed individually and mutually independent (as in the definition of the state space model in Sect. 3.1.3). Furthermore, it is assumed that the N models are independent, i.e. ϵ_{it} is independent of ϵ_{js}, for any $i, j = 1, \ldots, N$ and for any $t, s = 1, \ldots$, and ζ_{it} is independent of ζ_{js}, for any $i, j = 1, \ldots, N$ and for any $t, s = 0, 1, \ldots$.

Then the time series $\{y_t\}$ defined as the sum of $y_{1t}, \ldots, y_{Nt}$,

$$y_t = \sum_{i=1}^{N} y_{it}, \tag{4.6}$$

follows a state space model (3.10a)–(3.10b), with state vector β_t and innovations ϵ_t and ζ_t, defined by

$$\beta_t = \begin{bmatrix} \beta_{1t} \\ \beta_{2t} \\ \vdots \\ \beta_{Nt} \end{bmatrix}, \quad \epsilon_t = \sum_{i=1}^{N} \epsilon_{it}, \quad \zeta_t = \begin{bmatrix} \zeta_{1t} \\ \zeta_{2t} \\ \vdots \\ \zeta_{Nt} \end{bmatrix}.$$

The design vector x_t and the transition matrix $\mathbf{F}_t$ of y_t are

$$x_t = \begin{bmatrix} x_{1t} \\ x_{2t} \\ \vdots \\ x_{Nt} \end{bmatrix}, \quad \mathbf{F}_t = \begin{bmatrix} \mathbf{F}_{1t} & 0 & \cdots & 0 \\ 0 & \mathbf{F}_{2t} & \cdots & 0 \\ \vdots & \vdots & \ddots & \vdots \\ 0 & 0 & \cdots & \mathbf{F}_{Nt} \end{bmatrix}.$$

Finally, from the definitions of ϵ_t and ζ_t and the independence assumption, it follows that

$$\sigma^2 = \sum_{i=1}^{n} \sigma_i^2 \quad \text{and} \quad \mathbf{Z}_t = \begin{bmatrix} \mathbf{Z}_{1t} & 0 & \cdots & 0 \\ 0 & \mathbf{Z}_{2t} & \cdots & 0 \\ \vdots & \vdots & \ddots & \vdots \\ 0 & 0 & \cdots & \mathbf{Z}_{Nt} \end{bmatrix}.$$

The proof of the above result follows by noticing that from (4.6) and the individual
N models the observation equation of y_t is

$$
y_t = \sum_{i=1}^{N}\left(x_{it}^{\mathsf{T}}\beta_{it} + \epsilon_{it}\right) = [x_{1t}^{\mathsf{T}}, x_{2t}^{\mathsf{T}}, \dots, x_{Nt}^{\mathsf{T}}]\begin{bmatrix} \beta_{1t} \\ \beta_{2t} \\ \vdots \\ \beta_{Nt} \end{bmatrix} + \sum_{i=1}^{N}\epsilon_{it} = x_t^{\mathsf{T}}\beta_t + \epsilon_t,
$$

with x_t, β_t, ϵ_t as defined earlier. Also from the transition equation of the individual
models (4.5b) $\beta_{it} = F_{it}\beta_{i,t-1} + \zeta_{it}$, we have

$$
\beta_t = \begin{bmatrix} \beta_{1t} \\ \beta_{2t} \\ \vdots \\ \beta_{Nt} \end{bmatrix} = \begin{bmatrix} F_{1t} & 0 & \cdots & 0 \\ 0 & F_{2t} & \cdots & 0 \\ \vdots & \vdots & \ddots & \vdots \\ 0 & 0 & \cdots & F_{Nt} \end{bmatrix}\begin{bmatrix} \beta_{1,t-1} \\ \beta_{2,t-1} \\ \vdots \\ \beta_{N,t-1} \end{bmatrix} + \begin{bmatrix} \zeta_{1t} \\ \zeta_{2t} \\ \vdots \\ \zeta_{Nt} \end{bmatrix}
$$

$$
= F_t\beta_{t-1} + \zeta_t,
$$

which establishes the transition equation of the model for y_t.

An important property of the superposition described above is that the k-step
forecast distribution of y_t can be obtained by adding separately the respective
forecast distributions of the N individual models. In particular, the forecast function
of y_t is the sum of the forecast functions of $y_{1t}, \dots, y_{Nt}$. This follows from (4.6) by
taking expectations:

$$
\hat{y}_{t+k|t} = E(y_{t+k} \mid y_{1:t}) = E\left(\sum_{i=1}^{N} y_{i,t+k} \mid y_{1:t}\right) = \sum_{i=1}^{N} E(y_{i,t+k} \mid y_{i,1:t})
$$

$$
= \sum_{i=1}^{N} \hat{y}_{i,t+k|t}, \tag{4.7}
$$

where $\hat{y}_{i,t+k|t}$ is the forecast function of the time series y_{it}, for $i = 1, \dots, N$. In
a similar way and by using the assumption of independence of the N individual
models, it follows that the k-step forecast variance of y_t is the sum of the forecast
variances of the N individual models, i.e. $q_{t+k|t} = \sum_{i=1}^{N} q_{i,t+k|t}$. Since y_t is
a sum of $y_{1t}, \dots, y_{Nt}$, each of which follows a Gaussian forecast distribution,
the distribution of $y_{t+k} \mid y_{1:t}$ is Gaussian too, thus we have the result $y_{t+k} \mid
y_{1:t} \sim N(\hat{y}_{t+k|t}, q_{t+k|t})$, for $k = 1, 2, \dots$. These results are very important
for forecasting; according to the above, we can build complex forecast functions
by composing forecast functions of simple individual models (such as trend and
seasonal models, see the following sections). An important consequence is that
if there is a deterioration in model performance (e.g. experiencing high forecast

errors), the modeller needs to look at the individual models and fix the problem only for those component models that are responsible.

The above principle of superposition is relevant to the so-called *structural* time series models, which comprise a superposition of several individual models, for more details of which the reader is referred to Harvey (1989) or Durbin and Koopman (2012).

Example 4.4 Suppose we have data that we believe is comprised of local level or random variation and linear trend. Such a data set could be for example, the aluminium prices of Example 4.1, where in the start of the series a local level variation seems more persistent, followed by a linear trend. In such a case we could consider for y_t (the price per tonne of aluminium), the observation equation

$$
y_t = [x_{1t}, x_{2t}^\top] \begin{bmatrix} \beta_{1t} \\ \beta_{2t} \end{bmatrix} + \epsilon_t = [1, | 1, 0] \begin{bmatrix} \beta_{1t} \\ \beta_{1,2t} \\ \beta_{2,2t} \end{bmatrix} + \epsilon_t = \beta_{1t} + \beta_{1,2t} + \epsilon_t \qquad (4.8)
$$

and transition equation

$$
\beta_t = \begin{bmatrix} \beta_{1t} \\ \beta_{1,2t} \\ \beta_{2,2t} \end{bmatrix} = \begin{bmatrix} 1 & 0 & 0 \\ 0 & 1 & 1 \\ 0 & 0 & 1 \end{bmatrix} \begin{bmatrix} \beta_{1,t-1} \\ \beta_{1,2,t-1} \\ \beta_{2,2,t-1} \end{bmatrix} + \begin{bmatrix} \zeta_{1t} \\ \zeta_{1,2t} \\ \zeta_{2,2t} \end{bmatrix} .
$$

We note that the first unit in $x_t = x = [1, 1, 0]^\top$ and the top left unit in the transition matrix

$$
\mathbf{F}_t = \mathbf{F} = \begin{bmatrix} 1 & 0 & 0 \\ 0 & 1 & 1 \\ 0 & 0 & 1 \end{bmatrix}
$$

correspond to the local level component model, with transition $\beta_{1t} = \beta_{1,t-1} + \zeta_{1t}$. The sub-vector $[1, 0]^\top$ of x and the sub-matrix

$$
\begin{bmatrix} 1 & 1 \\ 0 & 1 \end{bmatrix}
$$

of $\mathbf{F}$ correspond to the linear growth component model, with transition

$$
\begin{bmatrix} \beta_{1,2t} \\ \beta_{2,2t} \end{bmatrix} = \begin{bmatrix} 1 & 1 \\ 0 & 1 \end{bmatrix} \begin{bmatrix} \beta_{1,2,t-1} \\ \beta_{2,2,t-1} \end{bmatrix} + \begin{bmatrix} \zeta_{1,2t} \\ \zeta_{2,2t} \end{bmatrix} .
$$

Finally, from (4.8), we see that y_t is a sum of a local level (β_{1t}) and a linear growth component ($\beta_{1,2t}$).

4.1.3 Fourier Form Seasonal Models

A seasonal time series is a time series which exhibits some periodicity, i.e. it includes patterns that, subject to random error, are repeating over a period of time. The term "seasonality" is used as in many applications, this period represents a seasonal cycle, e.g. quarter, month or annum. Figure 4.4 shows averaged temperatures collected for each quarter at Weston Park, Sheffield, UK, for unspecified years. This data set shows clear evidence of seasonality, as the values of the 1st Quarter of each year are similar (lower values in the figure) and of course they are such due to the effect of the winter months in the first quarter. Likewise, the third quarter of each year is responsible for the higher temperature values, being influenced by the summer months.

There are several types of state space models that describe seasonality. One is described in the seminal paper of Harrison and Stevens (1976), and it makes use of cyclical transition matrices; for a detailed discussion of this approach the reader is referred to West and Harrison (1997, Section 8.2). A second popular approach of modelling seasonality, within the state space framework, is by representing the seasonal process by a Fourier series. An early study on this area is given in Harrison

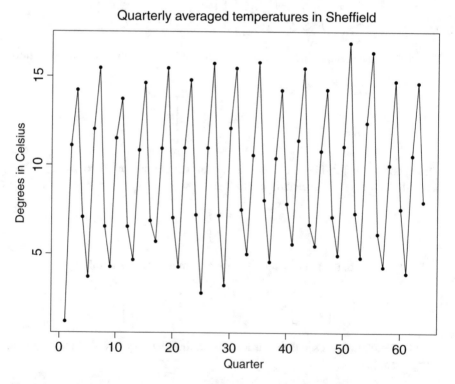

Fig. 4.4 Quarterly mean temperature in Sheffield

(1965) and more recent expositions include Harvey (1989), Harvey (2004) and West and Harrison (1997, Section 8.4). Below, we discuss this approach in some detail.

There are many textbooks discussing Fourier analysis; for an introductory treatment the reader is referred to Dyke (1999). Formally, a deterministic function (or sequence, if t is integer) $f(\cdot)$ is periodic with period T, if $f(t + T) = f(t)$, for all t. A typical example is a trigonometric function, such as the sine or the cosine, which have period 2π, i.e. $\sin(x + 2\pi) = \sin(x)$. A stochastic process $\{y_t\}$ is periodic or seasonal with cycle or period T, if the distribution of y_{t+T} is the same as the distribution of y_t. In a nutshell Fourier analysis tells us that periodic stochastic processes are represented as a weighted sum (possible an infinite sum) of sines and cosines plus random noise. So even if it is not exact, our function can probably be accurately reconstructed using a small number of sines and cosines. This sum is then recast in state space form to provide the required state space model. The technical details are as follows.

Any deterministic periodic function in L^2 (the space of all functions satisfying $\int_A |f(t)|^2 \, dt < \infty$, where A is the domain of $f(\cdot)$), such as $f(\cdot)$ above, can be represented as a Fourier series, for information of which and details of convergence see Dyke (1999) or Rudin (1976, Chapter 8). Under some conditions of convergence, see (Rudin, 1976, Chapter 8) for details, $f(t)$ can be represented by the Fourier series

$$a_0 + \sum_{i=1}^{\infty} [a_i \cos(i\omega t) + b_i \sin(i\omega t)],$$

where the coefficients a_0, a_i, b_i are known as the Fourier coefficients and they are calculated as integrals (if t is continuous) or as infinite sums (if t is discrete) of $f(t)$; for details see Dyke (1999) or Rudin (1976, Chapter 8). Thus, $f(t)$ can be approximated by the finite sum $S_N(t)$

$$S_N(t) = a_0 + \sum_{i=1}^{N} [a_i \cos(i\omega t) + b_i \sin(i\omega t)],$$

where $a_0, a_1, \ldots, a_N$ and $b_1, \ldots, b_N$ are complex numbers and $\omega = 2\pi/T$ is the frequency.

In order to be able to approximate a stochastic process $\{y_t\}$ with a Fourier form, we construct $S_N(t)$ by adding noise to it. Since $S_N(t)$ is a finite sum, y_t can be constructed as the superposition of N component state space models; these are known as *harmonic state space models* and are described below. The ith time series y_{it} follows a harmonic state space model, if it is given by Eqs. (4.5a)–(4.5b), with components

$$x_{it} = x = \begin{bmatrix} 1 \\ 0 \end{bmatrix} \quad \text{and} \quad \mathbf{F}_{it} = \mathbf{F}_i = \begin{bmatrix} \cos(i\omega) & \sin(i\omega) \\ -\sin(i\omega) & \cos(i\omega) \end{bmatrix}. \tag{4.9}$$

Here y_{it} $(i = 1, \ldots, N)$ represents the term $a_i \cos(i\omega t) + b_i \sin(i\omega t)$ plus noise. To see this first note that with x_{it} as above and from the state space model of y_{it}, we have

$$y_{it} = x_{it}^{\top} \beta_{it} + \epsilon_{it} = \beta_{i,1,t} + \epsilon_{it}, \tag{4.10}$$

with $\beta_{it} = [\beta_{i,1,t}, \beta_{i,2,t}]^{\top}$. Similarly, with F_i as above and from the transition equation (4.5b) we can write

$$\begin{bmatrix} \beta_{i,1,t} \\ \beta_{i,2,t} \end{bmatrix} = \begin{bmatrix} \cos(i\omega) & \sin(i\omega) \\ -\sin(i\omega) & \cos(i\omega) \end{bmatrix} \begin{bmatrix} \beta_{i,1,t-1} \\ \beta_{i,2,t-1} \end{bmatrix} + \begin{bmatrix} \zeta_{i,1,t} \\ \zeta_{i,2,t} \end{bmatrix},$$

where $\zeta_{it} = [\zeta_{i,1,t}, \zeta_{i,2,t}]^{\top}$. This separates out into two equations,

$$\beta_{i,1,t} = \cos(i\omega)\beta_{i,1,t-1} + \sin(i\omega)\beta_{i,2,t-1} + \zeta_{i,1,t}, \tag{4.11}$$

$$\beta_{i,2,t} = -\sin(i\omega)\beta_{i,1,t-1} + \cos(i\omega)\beta_{i,2,t-1} + \zeta_{i,1,t}. \tag{4.12}$$

By iterating these two equations and substituting (4.12) into (4.11), we obtain

$$\beta_{i,1,t} = [\cos^2(i\omega) - \sin^2(i\omega)]\beta_{i,1,t-2} + 2\sin(i\omega)\cos(i\omega)\beta_{i,2,t-2} + \text{error}$$

$$= \cos(2i\omega)\beta_{i,1,t-2} + \sin(2i\omega)\beta_{i,2,t-2} + \text{error}$$

$$= \cdots$$

$$= \cos(i\omega t)\beta_{i,1,0} + \sin(i\omega t)\beta_{i,2,0} + \text{error}, \tag{4.13}$$

where the term error indicates a function of the ζ_t's which represents some noise. From Eqs. (4.10) and (4.13) we see that y_{it} is equal to $a_i \cos(i\omega t) + b_i \sin(i\omega t)$ plus noise, where $a_i = \beta_{i,1,0}$ and $b_i = \beta_{i,2,0}$.

In order to proceed to the state space representation of the seasonal time series $\{y_t\}$, y_t is constructed as the superposition of the N harmonic state space models plus a local level model for a_0, i.e. $y_t = \sum_{i=0}^{N} y_{it}$, where y_{0t} follows a local level model and N is defined by

$$N = \begin{cases} T/2, & \text{if } T \text{ is even} \\ (T-1)/2, & \text{if } T \text{ is odd} \end{cases} \tag{4.14}$$

In case that $i\omega = \pi$ (known as the Nyquist frequency), the component state space model for y_{it} is reduced to

$$x_{Nt} = x_N = 1, \quad F_{Nt} = F_N = -1,$$

because the transition matrix defined above for $i\omega = \pi$ gives

$$\mathbf{F}_N = \begin{bmatrix} -1 & 0 \\ 0 & -1 \end{bmatrix}$$

and the second row implies that $\beta_{N,2,t}$ does not have any contribution to y_{Nt}.

Putting all these together, y_t is defined by the state space model (3.10a)–(3.10b), with design vector and transition matrix defined by

$$x_t = x = \begin{bmatrix} 1 \\ x_1 \\ x_2 \\ \vdots \\ x_N \end{bmatrix}, \quad \mathbf{F}_t = \mathbf{F} = \begin{bmatrix} 1 & 0 & 0 & \cdots & 0 \\ 0 & \mathbf{F}_1 & 0 & \cdots & 0 \\ 0 & 0 & \mathbf{F}_2 & \cdots & 0 \\ \vdots & \vdots & \vdots & \ddots & \vdots \\ 0 & 0 & 0 & \cdots & \mathbf{F}_N \end{bmatrix}, \tag{4.15}$$

where N is defined by (4.14) and, for $0 < i\omega < \pi$, $x_i, \mathbf{F}_i$ are given by equation (4.9), while for $i\omega = \pi$, $x_N = 1, F_N = -1$, for $i = 1, \ldots, N$. The top unit element of x and the top left unit element of $\mathbf{F}$ correspond to the local level state space model, used for the estimation of a_0. This local level may be replaced by some other suitable state space model, e.g. a linear growth model, as discussed in Sect. 4.1.4.

The above state space model is known as the *full-effects Fourier form* model; similar models are discussed in West and Harrison (1997, Section 8.6) and Harvey (2004). We can see that the dimension of x is equal to $p = T$ (using the local level model for the description of the level of the series). This follows as for odd T, there are $(T-1)/2$ seasonal components $x_2, x_3, \ldots, x_N$ and each x_i is a bivariate vector, so that $p = T - 1 + 1 = T$. Likewise, we can see that when T is even, there are $(T-2)/2$ bivariate vectors x_i plus a scale x_N (for the Nyquist frequency) plus the local level component amounting to $N = T - 2 + 1 + 1 = T$.

From the above it is clear that in some cases, in particular when the period T takes a large value, the full-effects model, defined above, is not practical to use, because the dimension p is too large. For example, suppose that y_t represents a time series measured at daily frequency and exhibiting annual seasonality. In this case $T = p = 365$, implying that the state vector β_t has dimension 365×1 and the transition matrix $\mathbf{F}_t$ has dimension 365×365. These dimensions appear to be too large, and although modern computing systems can handle computations with such dimensions well, it seems reasonable to make an effort to reduce the value of p. The high dimension of β_t implies higher uncertainty about the specification of the distribution of ζ_t (in particular regarding the transition covariance matrix $\mathbf{Z}_t$) and high uncertainty around the estimation of β_t. Such a dimensionality reduction may be carried out by simply replacing N by $N_1 << N$ and thus resulting in what is known a *reduced Fourier form* model. The choice of N_1 is done by experimentation (or using error analysis criteria). More details of reduced Fourier state space models can be found in West and Harrison (1997, Section 8.6.6).

Example 4.5 (Sheffield Temperature Data) We consider the data of the Sheffield temperatures $\{y_t\}$ described in the start of this section; see Fig. 4.4. These are quarterly data comprising of averaged daily temperatures (in °C). Since the data are quarterly, the period is $T = 4$ and the frequency is $\omega = 2\pi/4 = \pi/2$. For this data we propose a local level full seasonal Fourier state space model, so that y_t is defined by equations (3.10a)–(3.10b) with

$$
x_t = x = \begin{bmatrix} 1 \\ 1 \\ 0 \\ 1 \end{bmatrix} \quad \text{and} \quad \mathbf{F}_t = \mathbf{F} = \begin{bmatrix} 1 & 0 & 0 & 0 \\ 0 & 0 & 1 & 0 \\ 0 & -1 & 0 & 0 \\ 0 & 0 & 0 & -1 \end{bmatrix}.
$$

The units on the top of x and top left of $\mathbf{F}$ correspond to the local level, while the sub-vector $[1, 0, 1]^T$ and the sub-matrix

$$
\begin{bmatrix} 0 & 1 & 0 \\ -1 & 0 & 0 \\ 0 & 0 & -1 \end{bmatrix}
$$

correspond to the seasonal variation.

The observation variance and the transition covariance matrix are set to $\sigma^2 = 1$ and $\mathbf{Z}_t = \mathbf{Z} = 100\mathbf{I}$, while the initial state is set to $\beta_0 \sim N\{[0, 0, 0, 0]^T, 1000\mathbf{I}\}$.

Figure 4.5 shows the one-step forecasts $\hat{y}_{t|t-1}$ against the actual values of y_t, for $t = 1, \ldots, 64$. We observe that the estimates follow closely the data, except of the first four observations. These four observations may be considered as a training data set, which are required for the model to adapt to the data; it is remarkable to note that only four observations are needed for this training stage, after which the model seems to fit well to the data. To fit this model in R, we used the commands

```
> # load Sheffield temperatures data
> sheftemq <- scan("Sheftemq.txt")

> # define transition matrix
> F1 <- matrix(c(1))
> F2 <- matrix(c(0,-1,1,0),2,2)
> F3 <- matrix(c(-1),1)
> F0 <- blockDiagMat(F1,F2,F3)

> # fit the state space model
> fit <- bts.filter(sheftemq, x0=c(1,1,0,1), F0=F0,
+ Z0=100*diag(4), obsvar=1, beta0=rep(0,4), P0=1000*diag(4),
+ DISO=FALSE, VAREST=FALSE)

# time series plot
> pred <- ts(fit$FittedMean)
> ts.plot(pred, lty=2, col=2, main=expression("One-step forecasts
+ of the temperatures in Sheffield"),xlab="Quarter",
```

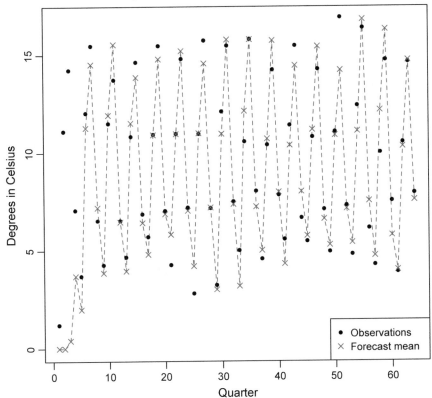

Fig. 4.5 Smoothed predictions (dashed line) against the actual average temperatures (solid line)

```
+ ylab="Degrees in Celsius")
> points(sheftemq, pch=20)
> points(pred, pch=4, col=2)
> legend("bottomright",c("Observations","Forecast mean"),
+ pch=c(20, 4), col=c(1,2))
```

We close this section by deriving the forecast function $\hat{y}_{t+k|t}$ of the seasonal time series y_t. Suppose that a collection of data $y_{1:t} = \{y_1, \ldots, y_t\}$ is obtained. Based on this information, using the Kalman filter we can obtain the posterior mean vector of β_t, $\hat{\beta}_{t|t}^T = [\hat{\beta}_{0,t|t}, \hat{\beta}_{1,t|t}^T, \ldots, \hat{\beta}_{N,t|t}^T]$, where $\hat{\beta}_{i,t|t} = [\hat{\beta}_{i,1,t|t}, \hat{\beta}_{i,2,t|t}]^T$, being the posterior mean vector of β_{it} and $\hat{\beta}_{0,t|t}$ being the posterior mean of β_{0t}. First we derive the forecast function of each individual model. To this end, first we prove that with the definition of $\mathbf{F}_i$ as above, we have

$$\mathbf{F}_i^k = \begin{bmatrix} \cos(ki\omega) & \sin(ki\omega) \\ -\sin(ki\omega) & \cos(ki\omega) \end{bmatrix}. \tag{4.16}$$

The proof is inductive. First note that (4.16) is true for $k = 2$. This is done by direct matrix multiplication, i.e.

$$\mathbf{F}_i^2 = \mathbf{F}_i \mathbf{F}_i = \begin{bmatrix} \cos(2i\omega) & \sin(2i\omega) \\ -\sin(2i\omega) & \cos(2i\omega) \end{bmatrix}.$$

Suppose now that (4.16) is true for $k-1$. Next we prove that (4.16) is true for k. We have

$$\mathbf{F}_i^k = \mathbf{F}_i^{k-1}\mathbf{F}_i = \begin{bmatrix} \cos(k-1)i\omega & \sin(k-1)i\omega \\ -\sin(k-1)i\omega & \cos(k-1)i\omega \end{bmatrix} \begin{bmatrix} \cos(i\omega) & \sin(i\omega) \\ -\sin(i\omega) & \cos(i\omega) \end{bmatrix}$$

$$= \begin{bmatrix} \cos(k-1)i\omega\cos(i\omega) - \sin(k-1)i\omega\sin(i\omega) \\ -\sin(k-1)i\omega\cos(i\omega) - \cos(k-1)i\omega\sin(i\omega) \end{bmatrix}$$

$$\begin{array}{c} \cos(k-1)i\omega\sin(i\omega) + \sin(k-1)i\omega\cos(i\omega) \\ \cos(k-1)i\omega\cos(i\omega) - \sin(k-1)i\omega\sin(i\omega) \end{array} \Big],$$

which validates (4.16), if we write down the following trigonometric formulae

$$\cos(ki\omega) = \cos[(k-1)i\omega + i\omega] = \cos(k-1)i\omega\cos(i\omega) - \sin(k-1)i\omega\sin(i\omega),$$

$$\sin(ki\omega) = \sin[(k-1)i\omega + i\omega] = \cos(k-1)i\omega\sin(i\omega) + \sin(k-1)i\omega\cos(i\omega).$$

After establishing (4.16), the forecast function of each component y_{it} is

$$y_{i,t+k|t} = x_{i,t+k}^{\top} \mathbf{F}^k \hat{\beta}_{i,t|t}$$

$$= [1, 0] \begin{bmatrix} \cos(ki\omega) & \sin(ki\omega) \\ -\sin(ki\omega) & \cos(ki\omega) \end{bmatrix} \begin{bmatrix} \hat{\beta}_{i,1,t|t} \\ \hat{\beta}_{i,2,t|t} \end{bmatrix}$$

$$= \cos(ki\omega)\hat{\beta}_{i,1,t|t} + \sin(ki\omega)\hat{\beta}_{i,2,t|t}.$$

Thus, using (4.7), the forecast function of $y_t = \sum_{i=1}^N y_{it}$ is

$$\hat{y}_{t+k|t} = \hat{y}_{0,t+k|t} + \sum_{i=1}^N \hat{y}_{i,t+k|t} = \beta_{0,t|t} + \sum_{i=1}^N \left[\cos(ki\omega)\hat{\beta}_{i,1,t|t} + \sin(ki\omega)\hat{\beta}_{i,2,t|t} \right],$$

since the forecast function of the local level model is $\hat{y}_{0,t+k|t} = \hat{\beta}_{0,t|t}$, where $E(\beta_{0t} \mid y_{1:t}) = \hat{\beta}_{0,t|t}$.

4.1.4   Trend-Seasonal Models

The previous section describes seasonal time series, i.e. time series with a repeating pattern (subject to noise). In many practical situations, this pattern is combined with some trend or other irregular pattern, examples of such data are given among others in Pole et al. (1994). An example is the classical airline passenger data, introduced in Brown (1962) and used in many textbooks, including the classic work of Box et al. (2008). Figure 4.6 plots this data, consisting of numbers of passengers (in thousands) carried by international airlines each month, from January 1949 to December 1960. A close look at the data reveals a clear annual seasonality, with increasing trend. The seasonality is expected since summer months attract usually more passengers than winter months and the trend reflects airlines' efforts over the years to attract more passengers and to grow their business. It is curious to observe that the seasonality has a time-varying amplitude, in fact after around January 1955, the amplitude of the seasonality increases considerably.

In order to propose a state space model for such data, we will assume that $\{y_t\}$, the time series of interest, comprises of trend or seasonal or any other patterns, which are additive and can be modelled individually with a separate state space model. For example, we may assume that

$$y_t = y_{1t} + y_{2t} + y_{3t},$$

where y_{1t} represents a trend, y_{2t} seasonal variation and y_{3t} local level or irregular variation. We will refer to these y_{it} as *components* and their related state space

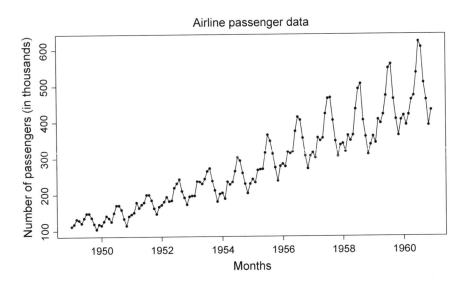

Fig. 4.6 Numbers of passengers carried in international airlines

models as *component models*. It is then natural to think that y_{1t} can be modelled with a trend state space model (as in Sect. 4.1.1), y_{2t} with a Fourier form state space model (as in Sect. 4.1.3, usually not including the local level component in (4.15)) and y_{3t} as a local level model. Consequently, the state space model for y_t is the superposition of these three components (Sect. 4.1.2). This approach of superimposing simple component models, in order to build a complex larger model is one of the advantages of the state space form, also known as *structural time series model*, see e.g. Harvey (1989). Of course, the general theory is trivially extended to a finite set of N component models, i.e. $y_t = \sum_{i=1}^{N} y_{it}$, although in practice it is of interest to keep N low and to identify meaningful components, such as those described above.

Example 4.6 (Turkey Sales Data)   In order to illustrate the trend-seasonal state space model, we consider data comprising quarterly turkey sales from autumn 1974 until summer 1982. The data which are described in more detail in Pole et al. (1994, page 270) are shown in Fig. 4.7 (solid points). We observe that the data exhibit seasonal variation of period $T = 4$ and an increasing trend. Based on the above discussion a plausible model for y_t (the number of turkey sales at time t) is the superposition of a linear trend and a full-effects Fourier form model. Thus, the model for y_t has observation equation

$$y_t = \begin{bmatrix} 1, & 0, & 1, & 0, & 1 \end{bmatrix} \begin{bmatrix} \beta_{1,1,t} \\ \beta_{1,2,t} \\ \beta_{2,1,t} \\ \beta_{2,2,t} \\ \beta_{2,3,t} \end{bmatrix} + \epsilon_t = x_t^\top \beta_t + \epsilon_t$$

and transition equation

$$\begin{bmatrix} \beta_{1,1,t} \\ \beta_{1,2,t} \\ \beta_{2,1,t} \\ \beta_{2,2,t} \\ \beta_{2,3,t} \end{bmatrix} = \begin{bmatrix} 1 & 1 & 0 & 0 & 0 \\ 0 & 1 & 0 & 0 & 0 \\ 0 & 0 & 0 & 1 & 0 \\ 0 & 0 & -1 & 0 & 0 \\ 0 & 0 & 0 & 0 & -1 \end{bmatrix} \begin{bmatrix} \beta_{1,1,t-1} \\ \beta_{1,2,t-1} \\ \beta_{2,1,t-1} \\ \beta_{2,2,t-1} \\ \beta_{2,3,t-1} \end{bmatrix} + \begin{bmatrix} \zeta_{1,1,t} \\ \zeta_{1,2,t} \\ \zeta_{2,1,t} \\ \zeta_{2,2,t} \\ \zeta_{2,3,t} \end{bmatrix},$$

where the state vector is $\beta_t = [\beta_{1,1,t}, \beta_{1,2,t}, \beta_{2,1,t}, \beta_{2,2,t}, \beta_{2,3,t}]^\top$ and $\zeta_t = [\zeta_{1,1,t}, \zeta_{1,2,t}, \zeta_{2,1,t}, \zeta_{2,2,t}, \zeta_{2,3,t}]^\top$. The horizontal lines indicate the two different components, the trend with state vector $\beta_{1t} = [\beta_{1,1,t}, \beta_{1,2,t}]^\top$ and transition matrix

$$\begin{bmatrix} 1 & 1 \\ 0 & 1 \end{bmatrix}$$

and the seasonal with state vector $\beta_{2t} = [\beta_{2,1,t}, \beta_{2,2,t}, \beta_{2,3,t}]^\top$ and transition matrix

$$\begin{bmatrix} 0 & 1 & 0 \\ -1 & 0 & 0 \\ 0 & 0 & -1 \end{bmatrix}.$$

For the analysis that follows we have set $\beta_0 \sim N(0, 1000\mathbf{I})$, where 0 indicates the 5×1 zero-vector and as usual $\mathbf{I}$ is the identity matrix; we have also set $\sigma^2 = 10$ and $\mathbf{Z}_t = \mathbf{Z} = 100\mathbf{I}$. Figure 4.7 shows the one-step forecasts $\hat{y}_{t|t-1}$ (dashed lines with ticks), the 95% forecast intervals $\hat{y}_{t|t-1} \pm z_{1-0.05/2}\sqrt{q_{t|t-1}}$ (dashed-dotted lines), the smoothed forecasts $\hat{y}_{t|n}$ (dotted lines) and the actual observations (solid points), where $z_{1-0.05/2}$ is the 95% quantile of the standard normal distribution $N(0, 1)$ and $q_{t|t-1}$ is the one-step forecast variance at time t. We observe that most of the data lie inside the forecast intervals, except for some observations at the start of

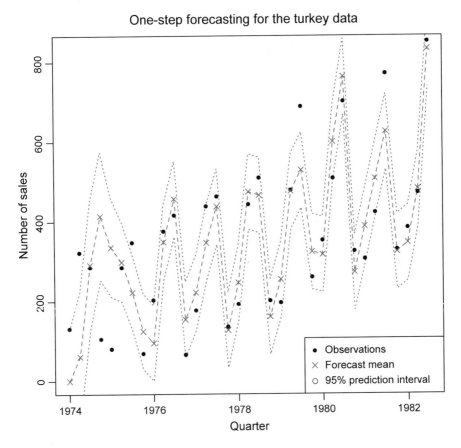

Fig. 4.7 One-step forecasting and smoothing for the turkey data

the series and a couple towards the end. Some observations are quite close to the forecast mean, while others are closer to the interval, indicating some drop in the forecast performance. The observations that are outside of the forecast intervals in the start of the data should not cause concern, because the model needs some time to learn and early points of time can be considered as a training data set. The smoothed forecast shows an overall evolution of the time series, which indicates a linear slowly increasing trend. The R commands for fitting the model and producing the above plot are given below.

```
> # read data
> turkey <- scan("turkey.txt")
> # define state space model
> x0 <- c(1,0,1,0,1)
> F1 <- matrix(c(1,0,1,1),2,2)
> F2 <- matrix(c(0,-1,1,0),2,2)
> F3 <- matrix(c(-1),1,1)
> F0 <- blockDiagMat(F1, F2, F3)
> Z0 <- 100*diag(5)
> obsvar <- 10
> beta0 <- rep(0,5)
> P0 <- 1000*diag(5)

> # fit model
> fit <- bts.filter(turkey, x0=x0, F0=F0, Z0=Z0,
+ beta0=beta0, P0=P0, DISO=FALSE, VAREST=FALSE)

> # calculate prediction intervals
> pred <- ts(fit$FittedMean,start=c(1974,1), frequency=4)
> t <- 1
> predL <- updt(list(), fit$FittedMean[[t]]-
+ 1.96*sqrt(fit$FittedVar[[t]]) )
> predU <- updt(list(), fit$FittedMean[[t]]+
+ 1.96*sqrt(fit$FittedVar[[t]]) )
> for(t in 2:35){
>    predL[[t]] <- fit$FittedMean[[t]]-
+    1.96*sqrt(fit$FittedVar[[t]])
>    predU[[t]] <- fit$FittedMean[[t]]+
+    1.96*sqrt(fit$FittedVar[[t]])
>  }
> predL <- ts(predL, start=c(1974,1), frequency=4)
> predU <- ts(predU, start=c(1974,1), frequency=4)
> turkeyts <- ts(turkey, start=c(1974,1), frequency=4)

> # time series plot
> ts.plot(pred, lty=2, main=expression("One-step forecasting
+ for the turkey data"),xlab="Quarters",
+ ylab="Number of sales",lwd=1)
> lines(predL, lwd=1,lty=3)
> lines(predU, lwd=1,lty=3)
> points(turkeyts, pch=20)
> points(pred, pch=4, col=2)
> legend("bottomright",c("Observations","Forecast mean",
+ "95% prediction interval"), pch=c(20, 4, 1), col=c(1,2,4))
```

4.1.5 Time-Varying Regression

We have already discussed regression models in Sects. 2.4.3 and 3.1.1, also covered in detail in Bingham and Fry (2010). *Time-varying regression* models are regression models whose parameters or coefficients are time-varying and typically slowly evolving. Such models are used routinely in economics, where regression coefficients correspond to dynamic covariates and are expected to change over time, and in other social science fields such as politics, see e.g. Beck (1983). A number of theoretical developments with applications are discussed in Kalaba and Tesfatsion (1988), Tesfatsion and Kalaba (1989), Pankratz (1991), Rao (2000), Commandeur and Koopman (2007) and in Montana et al. (2009). In this section we briefly describe the basic formulation of time-varying regression using state space models.

The time-varying regression is introduced in this book as a special form of the state space model in Sect. 3.1.3. Indeed, the model is defined by

$$y_t = x_{1t}\beta_{1t} + x_{2t}\beta_{2t} + \cdots + x_{pt}\beta_{pt} + \epsilon_t = x_t^\top \beta_t + \epsilon_t, \qquad (4.17)$$

where y_t is a response variable, x_{it} is the ith time-varying covariate, β_{it} is the ith time-varying regression coefficient and ϵ_t is the usual white noise innovation series. Here, $x_t = [x_{1t}, \ldots, x_{pt}]^\top$ is the design vector, comprising all time-varying covariates and $\beta_t = [\beta_{1t}, \ldots, \beta_{pt}]^\top$ is the vector of time-varying coefficients. Each of β_{it} is assumed to vary or evolve according to a random walk, i.e. $\beta_{it} = \beta_{i,t-1} + \zeta_{it}$, or equivalently $\beta_t = \beta_{t-1} + \zeta_t$, where $\zeta_t \sim N(0, \mathbf{Z}_t)$ and $\mathbf{Z}_t$ is a diagonal covariance matrix, whose diagonal elements are the respective variances of $\zeta_{1t}, \ldots, \zeta_{pt}$, i.e.

$$\mathbf{Z}_t = \begin{bmatrix} Z_{1t} & 0 & \cdots & 0 \\ 0 & Z_{2t} & \cdots & 0 \\ \vdots & \vdots & \ddots & \vdots \\ 0 & 0 & \cdots & Z_{pt} \end{bmatrix}.$$

Other evolutions of β_{it} may be considered, but the random walk effectively captures the slow evolution of the coefficients, expressed as $\beta_{it} \approx \beta_{i,t-1}$; see also the relevant discussion in Sect. 3.1.3. Model (4.17) does not include an intercept; if an intercept is required, this can be embedded into the above formulation by setting the first covariate equal to one, i.e. $x_{1t} = 1$. We can observe that a *static* or *time-invariant* regression model can be obtained by the above model, just by setting $\beta_{it} = \beta_{i,t-1}$, for all i and for all t, i.e. by setting $Z_t = 0$. Model (4.17) is very flexible, in the sense that some parameters may be time-varying and some may be static; this can be achieved by setting $Z_{it} = 0$, for the static coefficients, and $Z_{it} > 0$, for the time-varying coefficients.

Given a working data set $y_{1:t} = \{y_1, \ldots, y_t\}$ and a set of covariates $x_{1t}, \ldots, x_{pt}$, estimation and forecasting of model (4.17) follows immediately by the filtering, smoothing and forecasting algorithms of Chap. 3, after setting the transition matrix $\mathbf{F}_t$ equal to the identity $\mathbf{I}$.

Example 4.7 (Pollution Data)   In this example we consider data consisting of a response variable y_t with values of the pollutant nitric oxide NO (in milligrams per square meter), together with measurements of three covariates temperature (in ^{0}C), humidity (%) and wind speed (in meters per second); these covariates are denoted by x_{1t}, x_{2t}, x_{3t}, respectively. This data set is collected on a daily frequency over a period of one year, from 1 January 2001 until 31 December 2001, and is part of a larger data set, which will be discussed later. The data, obtained by one of 16 pollution monitoring stations in the city of Athens, is plotted in Fig. 4.8.

It is postulated that y_t is related to x_{it}; e.g. in the summer months when the temperature is high, the value of NO is also expected to be high. There are several interesting questions we may ask: can we use the values of the covariates, in order to forecast future values of NO? Is NO affected by all three covariates, and if it is, in what extent? Can we issue warning signals by projecting when the values of NO will exceed health thresholds, based on forecasts of the three covariates? We can answer these questions (or part of them) by building a regression model. Our starting point

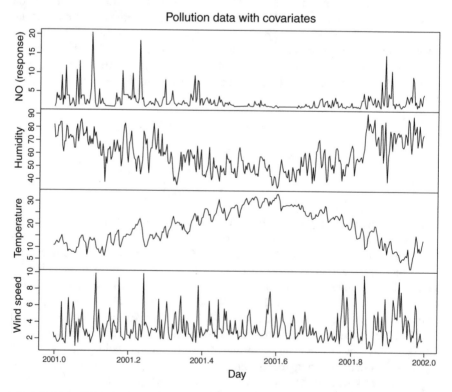

Fig. 4.8 Air-pollution levels of NO_2 and three covariates (humidity, temperature and wind speed)

is to consider the static regression model

$$y_t = \beta_0 + \beta_1 x_{1t} + \beta_2 x_{2t} + \beta_3 x_{3t} + \epsilon_t, \quad \epsilon_t \sim N(0, 1).$$

This model can be trivially obtained by the time-varying regression model (4.17), if we set $Z_t = 0$. Here we use the analysis produced by R, using the lm function:

```
> # read data
> poll <- read.table("Pollution.txt")
> # variables 1-3 indicate time,
> # variables 4-6 are the 3 covariates,
> # variables 7-9 are 3 responses
> # define response NO
> y <- poll[,9]
> # define covariates
> x <- poll[,4:6]
> # fit static linear regression
> lm(y ~ x[,1] + x[,2] + x[,3] )
```

This returns the estimates of β_i as $\hat{\beta}_0 = 3.07625$, $\hat{\beta}_1 = 0.03142$, $\hat{\beta}_2 = -0.04442$ and $\hat{\beta}_3 = -0.43174$ ($\hat{\beta}_i$ are the traditional least squares estimates produced by the static regression model). A Bayesian analysis of the above statistic model can also be produced by using the commands

```
> # define the model
> x0 <- cbind(rep(1,365),x[,1],x[,2],x[,3])
> x1 <- list(); for (i in 1:365) x1 <- updt (x1,   x0[i,])
> F0 <- diag(4)
> Z0 <- matrix(0, 4, 4)
> beta0 <- rep(0,4)
> P0 <- 1000*diag(4)
> # fit the model
> fit0 <- bts.filter(y, x0=x1, F0=F0, beta0=beta0, P0=P0,
+ Z0=matrix(0,nrow=4,ncol=4), obsvar=1, ONEX=F, DISO=FALSE,
+ VAREST=FALSE)
```

The first two commands define the design vector x_t including the intercept and the three covariates, the next command defines the transition matrix $\mathbf{F}_t = \mathbf{I}$ and the following command defines a zero covariance transition matrix $\mathbf{Z}_t =)$, so that $\beta_t = \beta_{t-1}$, for all t, or that $\beta_t = \beta$ (time-invariant regression). The remaining commands define the prior mean $\beta_0 = [2, 0, 0, 0]^\top$ and the prior covariance matrix $\mathbf{P}_{0|0} = 1000\mathbf{I}$. The values of the elements of $\hat{\beta}_{0|0}$ are chosen somewhat arbitrarily; the rationale here is to set all prior coefficient means to zero, except for the intercept; prior to observing the data we have $E(y_t) = 2$ (this value is known from historical data or past studies). Applying the above static regression model produces identical posterior means of $\beta_0, \beta_1, \beta_2, \beta_3$ with the respective least squares estimates of the standard regression model mentioned above.

However, since the data are collected over time, it is natural to consider that the coefficients β_i change over time; Fig. 4.8 indicates clearly that the response and the covariates are time-dependent. Thus, a second model is the time-varying regression

$$y_t = \beta_{0t} + \beta_{1t} x_{1t} + \beta_{2t} x_{2t} + \beta_{3t} x_{3t} + \epsilon_t, \quad \epsilon_t \sim N(0, 1),$$

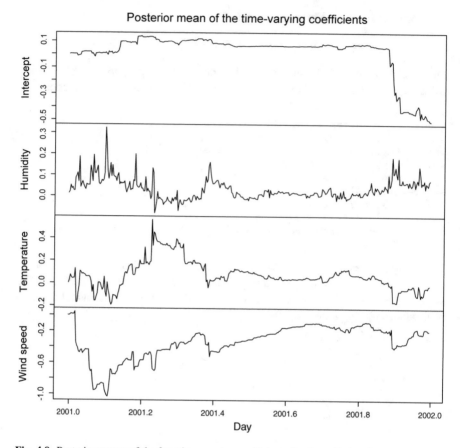

Fig. 4.9 Posterior means of the four time-varying coefficients for the pollution data

where now each β_{it} follows a random walk, i.e. $\beta_{it} = \beta_{i,t-1} + \zeta_{it}$ and we set $\zeta_{it} \sim N(0, 10)$, to allow β_{it} to exhibit some variation around its zero mean. The prior distribution of $\beta_0 = [\beta_{0,0}, \beta_{1,0}, \beta_{2,0}, \beta_{3,0}]^\top$ is set to $\beta_0 \sim N\{[2, 0, 0, 0]^\top, 100\mathbf{I}\}$ as before.

Figure 4.9 shows the posterior means of the four coefficients β_{it}. We observe that these estimates indicate a clear time-varying effect, and it would be wrong to assume them to be static. For fitting the time-varying model and for Fig. 4.9 the following R code was used:

```
> # define and fit the model
> # x1, F0, beta0, P0 as defined above
> Z0 <- 10*diag(4)
> fit <- bts.filter(y, x0=x1, F0=F0, Z0=Z0, beta0=beta0, P0=P0,
+ obsvar=1, ONEX=F, DISO=FALSE, VAREST=FALSE)

> # time series plot
```

```
> m <- ts(fit$m, names=c("Intercept","Humidity","Temperature",
+   "Wind speed"), start=c(2001,0), frequency=365)
> plot.ts(m, main=expression("Posterior mean of the
+   time-varying coefficients"),xlab="Day")
```

4.1.6   Time-Varying Autoregressions

Autoregressive models are popular time series models and are covered in most time series textbooks, see e.g. Box et al. (2008) or Shumway and Stoffer (2017). They form the basic model blocks of time-dependent data and are closely associated with the development of time series analysis. In this section we describe their basic structure and discuss how they can be analysed using state space models. Section 7.2 further discusses autoregressive models and their relation to stationarity. The Kalman filter can be used for inference of time-varying moving average models, such as discussed in Triantafyllopoulos and Nason (2007, 2009), but these models are not discussed further in this book. It is worthwhile to note that a moving average model can be seen as an infinite-order (large-order for practical modelling reasons) autoregressive model; see e.g. Box et al. (2008) for more discussion on this topic.

Suppose that $\{y_t\}$ is a time series which possesses some time-dependent structure. This means that the current value of y_t is likely to depend on previous values $y_{t-1}, y_{t-2}, \ldots$. Perhaps the simplest way to describe this dependence is via a linear regression model, in which the covariates are lagged values of y_t, hence the name *autoregressive model*—literally regressing y_t against itself. Formally, an autoregressive model of order d, abbreviated as AR(d), is defined by

$$y_t = \phi_1 y_{t-1} + \phi_2 y_{t-2} + \cdots + \phi_d y_{t-d} + \varepsilon_t, \quad \varepsilon_t \sim N(0, \sigma_\varepsilon^2), \qquad (4.18)$$

where $\phi_1, \phi_2, \ldots, \phi_d$ are the static or time-invariant AR parameters or coefficients, ε_t is a white noise process with some variance σ_ε^2 and d is a positive integer. This model is referred to as the *static* AR model, or just as AR model, for simplicity.

We can put model (4.18) in state space form by defining the observation model as

$$y_t = [1, 0, \ldots, 0]\beta_t = x_t^\top \beta_t \qquad (4.19)$$

and the transition model as

$$\beta_t = \begin{bmatrix} y_t \\ y_{t-1} \\ \vdots \\ y_{t-d+1} \end{bmatrix} = \begin{bmatrix} \phi_1 & \phi_2 & \cdots & \phi_{d-1} & \phi_d \\ 1 & 0 & \cdots & 0 & 0 \\ \vdots & \vdots & \ddots & \vdots & \vdots \\ 0 & 0 & \cdots & 1 & 0 \end{bmatrix} \begin{bmatrix} y_{t-1} \\ y_{t-2} \\ \vdots \\ y_{t-d} \end{bmatrix} + \begin{bmatrix} \varepsilon_t \\ 0 \\ \vdots \\ 0 \end{bmatrix}$$

$$= \mathbf{F}\beta_{t-1} + \zeta_t,$$

with

$$\mathbf{F}_t = \mathbf{F} = \begin{bmatrix} \phi_1 & \phi_2 & \cdots & \phi_{d-1} & \phi_d \\ 1 & 0 & \cdots & 0 & 0 \\ \vdots & \vdots & \ddots & \vdots & \vdots \\ 0 & 0 & \cdots & 1 & 0 \end{bmatrix} \quad \text{and} \quad \zeta_t = \begin{bmatrix} \varepsilon_t \\ 0 \\ \vdots \\ 0 \end{bmatrix}.$$

We note that ζ_t is random only in one variable (in ε_t), and so the related distribution of ζ_t is a singular normal distribution, i.e. $\zeta_t \sim N(0, \mathbf{Z}_t)$, with

$$\mathbf{Z}_t = \mathbf{Z} = \begin{bmatrix} \sigma_\varepsilon^2 & 0 & \cdots & 0 \\ 0 & 0 & \cdots & 0 \\ \vdots & \vdots & \ddots & \vdots \\ 0 & 0 & \cdots & 0 \end{bmatrix}.$$

The above state space representation of model (4.18) is the classical one and is discussed in many textbooks, see e.g. Brockwell and Davis (1991), Kitagawa and Gersch (1996) and De Jong and Penzer (2004) among others. It is important to note that after model (4.18) is cast in state space form, the Kalman filter provides a powerful tool for estimation and forecasting. In particular, R uses this representation, in order to compute the exact likelihood function of model (4.18) (see the help pages of the function `arima`). A second point of interest is that a general and popular class of time series models, namely autoregressive integrated moving average (ARIMA) models, can be described in state space form, see Brockwell and Davis (1991) for details. As a result state space provides a very general time series model and the Kalman filter (with its extensions) provides an extremely useful modelling framework.

It should be appreciated that there are several state space representations of model (4.18). A second one is

$$y_t = x_t^\top \beta + \epsilon_t, \tag{4.20}$$

where x_t now includes all lagged values y_{t-i} and β is the vector with coefficients ϕ_i, i.e.

$$x_t = \begin{bmatrix} y_{t-1} \\ y_{t-2} \\ \vdots \\ y_{t-d} \end{bmatrix} \quad \text{and} \quad \beta = \begin{bmatrix} \phi_1 \\ \phi_2 \\ \vdots \\ \phi_d \end{bmatrix}$$

and $\epsilon_t = \varepsilon_t$, with $\sigma^2 = \sigma_\varepsilon^2$.

This model is basically a static regression model. It has the advantage that estimation of β (via Kalman filtering) results in estimation of ϕ_i, while in the state space representation (4.19) the ϕ_i's are included in matrix **F**, hence the Kalman filter does not provide estimation of the ϕ_i's. In such a case the ϕ_i's can be estimated by maximum likelihood methods, as described in Sect. 4.3.

The model formulation (4.20) motivates the situation in which the coefficients ϕ_i are allowed to vary with time, following similar thinking to that of Sect. 4.1.5 in linear regression. According to this, we adopt (4.20), but replace the static vector of coefficients β by a time-varying $\beta_t = [\phi_{1t}, \phi_{2t}, \ldots, \phi_{dt}]^\top$, so that each of ϕ_{it} follows a random walk evolution, or some other suitable Markovian evolution. This model, which is known as *time-varying autoregressive model* of order d, is defined by

$$y_t = \sum_{i=1}^{d} \phi_{it} y_{t-i} + \epsilon_t = x_t^\top \beta_t + \epsilon_t \quad \text{and} \quad \beta_t = \beta_{t-1} + \zeta_t, \tag{4.21}$$

where

$$x_t = \begin{bmatrix} y_{t-1} \\ y_{t-2} \\ \vdots \\ y_{t-d} \end{bmatrix}, \quad \beta = \begin{bmatrix} \phi_{1t} \\ \phi_{2t} \\ \vdots \\ \phi_{dt} \end{bmatrix}$$

and $\mathbf{Z}_t$, the covariance matrix of ζ_t is a diagonal matrix, with each element in the main diagonal of $\mathbf{Z}_t$ being the variance of ζ_{it}, with $\zeta_t = [\zeta_{1t}, \zeta_{2t}, \ldots, \zeta_{dt}]^\top$. Similar models are discussed in West et al. (1999), Lundbergh et al. (2003), Triantafyllopoulos (2007b), Triantafyllopoulos (2011b), Triantafyllopoulos and Montana (2011) and Prado and West (2010) among others. A time-varying autoregressive model applied in the context of pairs trading in finance is discussed in Sect. 7.5.3.

4.2 Decomposition of State Space Models

4.2.1 Historical Note and Motivation

The problem of representing time series as sum of simpler (component) time series has been studied in the literature at least since the work of Yule (1927) and Davis (1941). The classical decomposition of a time series y_t considers an additive structure,

$$y_t = T_t + S_t + r_t,$$

where T_t denotes the trend, S_t denotes the seasonal or periodic variation and r_t denotes the random or irregular component. In the classical decomposition, starting with some observations $y_1, \ldots, y_n$, estimated values $\hat{T}_t, \hat{S}_t$ and $\hat{r}_t$ of T_t, S_t and r_t are obtained, for $t = 1, \ldots, n$, so that we can write $y_t \approx \hat{T}_t + \hat{S}_t + \hat{r}_t$. The procedure is based on trend estimation using the technique of moving averages and then estimating the seasonal component. Finally the random component is effectively what is left if the trend and seasonal components are accounted for. Such a breakdown of time series into simpler components is helpful in understanding the data and proposing suitable models, also known as components time series models. More information is provided in most time series textbooks, see e.g. Whittle (1984, Chapter 8). The same approach can work with a multiplicative decomposition, if the time series and its three components (trend, seasonal and random) are positive, $y_t = T_t S_t r_t$ so that the logarithm of y_t admits an additive decomposition. In R the classical decomposition (additive or multiplicative) can be applied by using the command decompose.

Hilmer and Tiao (1982) propose a model-based approach to decomposition assuming an autoregressive integrated moving average (ARIMA) model for the generating process of the data. West (1997) proposes a methodology for decomposing a time series signal which is cast in state space form, into component state space models. This methodology is based on the similarity of the transition matrix **F**, using Jordan forms. In that article an autoregressive model of order 20 is decomposed into several quasi cyclical components. Godolphin and Johnson (2003) and Godolphin and Triantafyllopoulos (2006) propose a decomposition based on the rational canonical decomposition of **F**. The proposed decomposition of component state space models can be regarded as the inverse process of superposition discussed in Sect. 4.1.2. This approach, which is described in the next sections, uses as building block the companion matrix.

4.2.2 Rational Canonical Form

This section provides the matrix algebra background necessary for the development of the theory of decomposition of Sect. 4.2.3.

Let $p(x)$ be a monic polynomial of order n

$$p(x) = a_0 + a_1 x + \cdots + a_{n-1} x^{n-1} + x^n, \tag{4.22}$$

where $a_0, a_1, \ldots, a_{n-1}$ are known coefficients of the real field. Associated to $p(x)$ is a matrix, known as *companion matrix*, defined as

$$\mathbf{C}_p = \mathbf{C}(p) = \begin{bmatrix} 0 & 1 & 0 & \cdots & 0 \\ 0 & 0 & 1 & \cdots & 0 \\ \vdots & \vdots & \vdots & \ddots & \vdots \\ 0 & 0 & 0 & \cdots & 1 \\ -a_0 & -a_1 & -a_2 & \cdots & -a_{n-1} \end{bmatrix}. \tag{4.23}$$

For $n = 1$, by convention, the companion matrix $\mathbf{C}_p = -a_0$, with the associated polynomial $p(x) = a_0 + x$. The companion matrix has the following interesting properties.

1. **Characteristic polynomial.** The characteristic polynomial of $\mathbf{C}_p$ is equal to $p(x)$, i.e. $|x\mathbf{I} - \mathbf{C}_p| = p(x)$, where $|\cdot|$ denotes determinant.
   To prove this use induction for n. For $n = 1$ the characteristic polynomial is $|x - \mathbf{C}_p| = x + a_0$ and the statement is true. For $n = 2$ we have

$$|x\mathbf{I} - \mathbf{C}_p| = \begin{vmatrix} x & -1 \\ a_0 & x + a_1 \end{vmatrix} = a_0 + a_1 x + x^2 = p(x).$$

Suppose that the statement that the characteristic polynomial of $\mathbf{C}_p$ is equal to (4.22) for order $n - 1$. We shall prove the statement for n.

$$|x\mathbf{I} - \mathbf{C}_p| = \begin{vmatrix} x & -1 & 0 & \cdots & 0 \\ 0 & x & -1 & \cdots & 0 \\ \vdots & \vdots & \vdots & \ddots & \vdots \\ 0 & 0 & 0 & \cdots & -1 \\ a_0 & a_1 & a_2 & \cdots & x + a_{n-1} \end{vmatrix}.$$

We shall expand this determinant using the cofactors of the first column. We have

$$|x\mathbf{I} - \mathbf{C}_p| = x \begin{vmatrix} x & -1 & \cdots & 0 \\ 0 & x & \cdots & 0 \\ \vdots & \vdots & \ddots & \vdots \\ 0 & 0 & \cdots & -1 \\ a_1 & a_2 & \cdots & x + a_{n-1} \end{vmatrix}$$

$$+ (-1)^{n+1} a_0 \begin{vmatrix} -1 & 0 & \cdots & 0 & 0 \\ x & -1 & \cdots & 0 & 0 \\ \vdots & \vdots & \ddots & \vdots & \vdots \\ 0 & 0 & \cdots & -1 & 0 \\ 0 & 0 & \cdots & 0 & -1 \end{vmatrix}$$

$$= x(a_1 + a_2 x + \cdots + a_{n-1} x^{n-2} + x^{n-1}) + (-1)^{n+1} a_0 (-1)^{n-1}$$

$$= a_0 + a_1 x + a_2 x^2 + \cdots + a_{n-1} x^{n-1} + x^n.$$

2. **Similarity condition.** Two $n \times n$ matrixes $\mathbf{A}$ and $\mathbf{B}$ are said to be similar if there exists a non-singular $n \times n$ matrix $\mathbf{S}$, so that $\mathbf{SAS}^{-1} = \mathbf{B}$. The following theorem provides conditions to ensure that an $n \times n$ matrix is similar to $\mathbf{C}_p$.

Theorem 4.1 *Let* **A** *be an* $n \times n$ *matrix and* $\mathbf{C}_p$ *be an* $n \times n$ *companion matrix as defined above. Then the following two statements are equivalent.*

a. *There exists a vector* $v \in \mathbb{R}^n$ *such that the row vectors* $v^\top$, $v^\top \mathbf{A}$, $v^\top \mathbf{A}^2$, ..., $v^\top \mathbf{A}^{n-1}$ *form a basis of* $\mathbb{R}^n$.

b. *There exists an* $n \times n$ *non-singular matrix* **S** *so that* $\mathbf{SAS}^{-1} = \mathbf{C}_p$.

Proof First suppose (a) holds true and prove (b). Define

$$
\mathbf{S} = \begin{bmatrix} v^\top \\ v^\top \mathbf{A} \\ \vdots \\ v^\top \mathbf{A}^{n-1} \end{bmatrix}.
$$

From (a) we know that the rows of **S** are independent, hence **S** is invertible. We will prove $\mathbf{SA} = \mathbf{SC}_p$. The i-th row of the matrix **SA** is $v^\top \mathbf{A}^{i-1}$, for $i = 1, \ldots, n$. Let $a = [a_0, a_1, \ldots, a_{n-1}]^\top$ and let $e_i = [0, \ldots, 0, 1, 0, \ldots, 0]^\top$ be the vector with a unit at the i-th element and all other elements equal to 0. We can write

$$
\mathbf{C}_p\mathbf{S} = \begin{bmatrix} e_2^\top \\ e_3^\top \\ \vdots \\ e_n^\top \\ -a \end{bmatrix} \mathbf{S} = \begin{bmatrix} v^\top \mathbf{A} \\ v^\top \mathbf{A}^2 \\ \vdots \\ v^\top \mathbf{A}^{n-1} \\ -a^\top \mathbf{S} \end{bmatrix},
$$

because the row vector $e_i^\top \mathbf{S}$ is equal to the i-th row of **S**, which is $v^\top \mathbf{A}^{i-1}$. Hence to show $\mathbf{SA} = \mathbf{C}_p\mathbf{S}$ it suffices to show

$$
v^\top \mathbf{A}^n = -a^\top \mathbf{S}. \tag{4.24}
$$

Write a as $a = a_0 e_1 + a_1 e_2 + \cdots + a_{n-1} e_n$. Then

$$
\begin{aligned}
a^\top \mathbf{S} &= (a_0 e_1 + a_1 e_2 + \cdots + a_{n-1} e_n)^\top \mathbf{S} \\
&= a_0 e_1^\top \mathbf{S} + a_1 e_2^\top \mathbf{S} + \cdots + a_{n-1} e_n^\top \mathbf{S} \\
&= a_0 v^\top + a_1 v^\top \mathbf{A} + \cdots + a_{n-1} v^\top \mathbf{A}^{n-1} \\
&= v^\top (a_0 \mathbf{I} + a_1 \mathbf{A} + \cdots + a_{n-1} \mathbf{A}^{n-1}) \\
&= -v^\top \mathbf{A}^n,
\end{aligned}
$$

where the last equation follows from an application of the Cayley-Hamilton theorem, or

$$
\mathbf{O} = p(\mathbf{A}) = a_0 \mathbf{I} + a_1 \mathbf{A} + \cdots + a_{n-1} \mathbf{A}^{n-1} + \mathbf{A}^n.
$$

A proof of the Cayley-Hamilton theorem can be found in Jacobson (1953).
This establishes (4.24) and so $\mathbf{SA} = \mathbf{C}_p\mathbf{S}$ is true. From this and the invertibility
of $\mathbf{S}$, it follows $\mathbf{SAS}^{-1} = \mathbf{C}_p$.

Now suppose (b) holds true and prove (a). Define $v^\top = e_1^\top\mathbf{S}$ to be the
first row vector of $\mathbf{S}$. We prove that with this choice of v the row vectors
$v^\top, v^\top\mathbf{A}, \ldots, v^\top\mathbf{A}^{n-1}$ form a basis of $\mathbb{R}^n$. We have

$$\mathbf{SA} = \begin{bmatrix} e_1^\top\mathbf{SA} \\ e_2^\top\mathbf{SA} \\ \vdots \\ e_{n-1}^\top\mathbf{SA} \\ e_n^\top\mathbf{SA} \end{bmatrix} \quad \text{and} \quad \mathbf{C}_p\mathbf{S} = \begin{bmatrix} e_2^\top\mathbf{S} \\ e_3^\top\mathbf{S} \\ \vdots \\ e_n^\top\mathbf{S} \\ -a^\top\mathbf{S} \end{bmatrix}.$$

From (b) we have $\mathbf{SA} = \mathbf{C}_p\mathbf{S}$ and by equating terms we get $e_1^\top\mathbf{SA} = e_2^\top\mathbf{S}$,
$e_2^\top\mathbf{SA} = e_3^\top\mathbf{S}, \ldots, e_{n-1}^\top\mathbf{SA} = e_n^\top\mathbf{S}, e_n^\top\mathbf{SA} = -a^\top\mathbf{S}$. This implies that the row
vectors $v^\top, v^\top\mathbf{A}, \ldots, v^\top\mathbf{A}^{n-1}$ can be expressed as

$$v^\top = e_1^\top\mathbf{S} \quad \text{(by construction)}$$

$$v^\top\mathbf{A} = e_1^\top\mathbf{SA} = e_2^\top\mathbf{S}$$

$$v^\top\mathbf{A}^2 = (v^\top\mathbf{A})\mathbf{A} = e_2^\top\mathbf{SA} = e_3^\top\mathbf{S}$$

$$\vdots$$

$$v^\top\mathbf{A}^{n-1} = (v^\top\mathbf{A}^{n-2})\mathbf{A} = e_{n-1}^\top\mathbf{SA} = e_n^\top\mathbf{S},$$

or in short

$$v^\top\mathbf{A}^i = e_{i+1}^\top\mathbf{S}, \quad \text{for} \quad i = 0, 1, \ldots, n-1. \tag{4.25}$$

In order to prove $v^\top, v^\top\mathbf{A}, \ldots, v^\top\mathbf{A}^{n-1}$ are independent row vectors, we
consider the linear combination

$$c_1 v^\top + c_2 v^\top\mathbf{A} + \cdots + c_n v^\top\mathbf{A}^{n-1} = 0,$$

for some $c_1, \ldots, c_n \in \mathbb{R}$. From Eq. (4.25) and the invertibility assumption of $\mathbf{S}$
we have

$$c_1 e^\top\mathbf{S} + c_2 e_2^\top\mathbf{S} + \cdots + c_n e_n^\top\mathbf{S} = 0$$

$$(c_1 e_1^\top + c_2 e_2 + \cdots + c_n e_n^\top)\mathbf{S} = 0$$

$$c_1 e_1^\top + c_2 e_2 + \cdots + c_n e_n^\top = 0,$$

which is satisfied only for $c_1 = c_2 = \cdots = c_n = 0$, since $e_1, e_2, \ldots, e_n$ form a basis of $\mathbf{R}^n$. Hence $v^\top, v^\top \mathbf{A}, \ldots, v^\top \mathbf{A}^{n-1}$ are independent.

To show that $v^\top, v^\top \mathbf{A}, \ldots, v^\top \mathbf{A}^{n-1}$ form a basis of $\mathbb{R}^n$ we need to show that any row vector u of $\mathbb{R}^n$ can be written as a linear combination of these vectors. Since $e_1, e_2, \ldots, e_n$ is a basis of $\mathbb{R}^n$ the row vector, there exists some non-zero scalars $\lambda_1, \lambda_2, \ldots, \lambda_n$ so that

$$
\begin{aligned}
u^\top \mathbf{S}^{-1} &= \lambda_1 e_1^\top + \lambda_2 e_2^\top + \cdots + \lambda_n e_n^\top \\
&= (\lambda_1 e_1^\top \mathbf{S} + \lambda_2 e_2^\top \mathbf{S} + \cdots + \lambda_n e_n^\top \mathbf{S}) \mathbf{S}^{-1} \\
&= (\lambda v^\top + \lambda_2 v^\top \mathbf{A} + \cdots + \lambda_n v^\top \mathbf{A}^{n-1}) \mathbf{S}^{-1}.
\end{aligned}
$$

Hence $u^\top = \lambda v^\top + \lambda_2 v^\top \mathbf{A} + \cdots + \lambda_n v^\top \mathbf{A}^{n-1}$.

Since the vectors $v^\top, v^\top \mathbf{A}, \ldots, v^\top \mathbf{A}^{n-1}$ are independent and generate any vector of $\mathbb{R}^n$, they form a basis of $\mathbb{R}^n$ and the proof of (a) is completed.

□

3. The minimal polynomial $m(x)$ of a matrix $\mathbf{A}$ is defined as the polynomial of least degree satisfying $m(\mathbf{A}) = \mathbf{O}$; see e.g. (Jacobson, 1953, p. 98) and (Horn and Johnson, 2013, Chapter 3). The minimal polynomial of $\mathbf{C}(p)$ is equal to $p(x)$. It follows that $\mathbf{C}(p)$ is a non-derogatory matrix (the minimal and characteristic polynomials are identical). The minimal polynomial can be factorised as follows

$$
m(x) = p_1^{m_1}(x) p_2^{m_2}(x) \cdots p_k^{m_k}(x), \tag{4.26}
$$

where $p_i(x)$ are irreducible polynomials of degree d_i and $m_1 d_1 + \cdots + m_k d_k = n$. It follows that $p_i(x)$ are the elementary divisors of $m(x)$.

Consider now a matrix $\mathbf{A}$, with characteristic polynomial $p(x) = |x\mathbf{I} - \mathbf{A}$, given by (4.22), where $a_0, a_1, \ldots, a_{n-1}$ are given as functions of the elements of $\mathbf{A}$. If condition (a) of Theorem 4.1 is satisfied, then $\mathbf{A}$ will be similar to a companion matrix. The minimal polynomial $m(x)$ of this matrix will have the form of (4.26). Each of the elementary divisors $p_i(x)$ of $m(x)$ will correspond to a companion matrix $\mathbf{C}(p_i)$, for $i = 1, 2, \ldots, k$. It follows that the matrix $\mathbf{A}$ is similar to $\mathbf{C}$ the direct sum of companion matrixes $\mathbf{C}(p_1), \ldots, \mathbf{C}(p_k)$, hence $\mathbf{C}$ is a $n \times n$ block diagonal matrix

$$
\mathbf{C} = \begin{bmatrix}
\mathbf{C}(p_1) & \mathbf{O} & \cdots & \mathbf{O} \\
\mathbf{O} & \mathbf{C}(p_2) & \cdots & \mathbf{O} \\
\vdots & \vdots & \ddots & \vdots \\
\mathbf{O} & \mathbf{O} & \cdots & \mathbf{C}(p_k)
\end{bmatrix}. \tag{4.27}
$$

This is known as the *rational canonical decomposition* of $\mathbf{A}$; for more details see Jacobson (1953, Chapter 3) and Horn and Johnson (2013, Chapter 3).

4.2.3 Decomposition of Linear State Space Models

Consider the state space model (3.10a)–(3.10b) as defined in Sect. 3.1.3, or

$$y_t = \mu_t + \epsilon_t = x^\top \beta_t + \epsilon_t, \tag{4.28}$$

$$\beta_t = \mathbf{F}\beta_{t-1} + \zeta_t, \tag{4.29}$$

where μ_t is the mean response and $\zeta_t \sim N(0, \mathbf{Z})$. It is assumed that the components x, $\mathbf{F}$ and $\mathbf{Z}$ are time-invariant.

Assumption 1 To what follows we will assume that the model is observable or that the $p \times p$ observability matrix O in (3.32) has full rank p. Observability in state space models is discussed in some detail in Sect. 3.5.1. This implies that the row vectors $x^\top, x^\top \mathbf{F}, \ldots, x^\top \mathbf{F}^{p-1}$ are linearly independent and hence they form a basis of the column vector space $C(O)$. It also follows that $C(O)$ is a cyclic space, with x being its generator. Hence $\mathbf{F}$ is non-derogatory. As a result the minimal polynomial of $\mathbf{F}$ is equal to its characteristic polynomial

$$\Phi(\lambda) = |\lambda \mathbf{I} - \mathbf{F}| = \phi_0 + \phi_1 \lambda + \cdot + \phi_{p-1}\lambda^{p-1} + \lambda^p,$$

for some real coefficients $\phi_0, \phi_1, \ldots, \phi_{p-1}$.

Suppose that $\Phi(\lambda)$ can be expressed as a product of ℓ relatively prime factors

$$\Phi(\lambda) = \Phi_1(\lambda)\Phi_2(\lambda) \cdots \Phi_\ell(\lambda),$$

for $\ell \leq p$. The factors $\Phi_j(\lambda)$ are powers of factors from the system of the elementary divisors of $\Phi(\lambda)$.

With the assumption of observability (Assumption 1) we have that condition (1) of Theorem 4.1 is satisfied and so $\mathbf{F}$ is similar to a companion matrix. More specifically, $\mathbf{F}$ is similar to a direct sum of companion matrices, each one corresponding to the factor $\Phi_j(\lambda)$. So

$$\mathbf{SFS}^{-1} = \mathbf{C} = \mathbf{C}(\Phi_1) \oplus \mathbf{C}(\Phi_2) \oplus \cdots \oplus \mathbf{C}(\Phi_\ell),$$

for some non-singular similarity matrix $\mathbf{S}$.

Our aim is to define a linear transformation of the state vector β_t to a new state vector γ_t, so that the mean response μ_t is decomposed to a sum of ℓ components $\chi_t^{(1)}, \ldots, \chi_t^{(\ell)}$, each of which following a state space model, or

$$\mu_t = \chi_t^{(1)} + \chi_t^{(2)} + \cdots + \chi_t^{(\ell)}, \tag{4.30}$$

$$\chi_t^{(j)} = e_1^\top \gamma_{jt}, \tag{4.31}$$

$$\gamma_{jt} = \mathbf{C}(\Phi_j)\gamma_{j,t-1} + \xi_{jt}, \tag{4.32}$$

for some ξ_{jt} to be defined, where $e_1 = [1, 0, \ldots, 0]^\top$ is the $p_j \times 1$ column vector with a unit as its first element and elsewhere zero, and the dimension of the vector is implicit. Here γ_{jt} is a state vector and should not be confused with the natural parameter γ_t of the dynamic generalised linear model (see Eq. (6.4) of Sect. 6.2).

In order to form the transformation, for $j = 1, \ldots, \ell$, we define the $p_j \times p_j$ matrix

$$\Omega_j = \begin{bmatrix} x_j^\top \\ x_j^\top C(\Phi_j) \\ \vdots \\ x_j^\top C(\Phi_j)^{p_j-1} \end{bmatrix}, \tag{4.33}$$

where $p_1 + \cdots + p_\ell = p$ and $x_j^\top$ is the $1 \times p_j$ row component vector of the matrix

$$x^\top S^{-1} = [x_1^\top, x_2^\top, \ldots, x_\ell^\top].$$

Subsequently we define the $p \times p$ matrix Ω to be the direct sum of $\Omega_1, \ldots, \Omega_\ell$, or

$$\Omega = \Omega_1 \oplus \Omega_2 \oplus \cdots \oplus \Omega_\ell. \tag{4.34}$$

With these definitions in place we have the following lemma, which is instrumental in the decomposition (4.32) of the state vector β_t.

Lemma 4.1 *With the definitions of C and Ω as in (4.27) and (4.34), C and Ω commute.*

Proof Since both C and Ω are block diagonal matrices it suffices to prove that the $p_j \times p_j$ matrices Ω_j and $C(\Phi_j)$ commute, for $j = 1, 2, \ldots, \ell$. First we will show that Ω_j can be expressed as a polynomial of $C(\Phi_j)$ as

$$\Omega_j = x_{j,1}I + x_{j,2}C(\Phi_j) + \cdots + x_{j,p_j}C(\Phi_j)^{p_j-1}, \tag{4.35}$$

where $x_j^\top = [x_{j,1}, x_{j,2}, \ldots, x_{j,p_j}]$. To show this first we consider the matrix polynomial

$$p[C(\Phi_j)] = x_{j,1}I + x_{j,2}C(\Phi_j) + \cdots + x_{j,p_j}C(\Phi_j)^{p_j-1}.$$

Deploying induction for i we shall prove the following statement $P(i)$

$$P(i): \quad e_i^\top p[C(\Phi_j)] = x_j^\top C(\Phi_j)^{i-1},$$

for any $1 \le i \le p_j$.

For $i = 1$ and the definition of the companion matrix $\mathbf{C}(\Phi_j)$ (see Eq. (4.23)) we have

$$e_1^\top \mathbf{C}(\Phi_j)^{k-1} = e_1^\top \mathbf{C}(\Phi_j)\mathbf{C}(\Phi_j)^{k-2} = e_2^\top \mathbf{C}(\Phi_j)^{k-2} = \cdots = e_k^\top,$$

for $k = 1, \ldots, p_j$. Hence

$$e_1^\top p[\mathbf{C}(\Phi_j)] = x_{j,1}e_1^\top + x_{j,2}e_2^\top + \cdots + x_{j,p_j}e_{p_j}^\top = x_j^\top,$$

which proves $P(1)$.

Suppose now that statement $P(i)$ is true for some $1 \le i \le p_j - 1$. We shall prove the statement holds true for $i + 1$. We have

$$\begin{aligned}
e_{i+1}^\top p[\mathbf{C}(\Phi_j)] &= e_i^\top \mathbf{C}(\Phi_j)p[\mathbf{C}(\Phi_j)] \\
&= e_i^\top p[\mathbf{C}(\Phi_j)]\mathbf{C}(\Phi_j) \\
&= x_j^\top \mathbf{C}(\Phi_j)^{i-1}\mathbf{C}(\Phi_j) \\
&= x_j^\top \mathbf{C}(\Phi_j)^{i+1-1},
\end{aligned}$$

since $p[\mathbf{C}(\Phi_j)]$ and $\mathbf{C}(\Phi_j)$ commute. Hence $P(i+1)$ holds true and the statement is true for any i. This implies that

$$\mathbf{\Omega}_j = \begin{bmatrix} x_j^\top \\ x_j^\top \mathbf{C}(\Phi_j) \\ \vdots \\ x_j^\top \mathbf{C}(\Phi_j)^{p_j-1} \end{bmatrix} = \begin{bmatrix} e_1^\top \\ e_2^\top \\ \vdots \\ e_{p_j}^\top \end{bmatrix} p[\mathbf{C}(\Phi_j)] = p[\mathbf{C}(\Phi_j)],$$

which proves (4.35).

With (4.35) established we see

$$\mathbf{\Omega}_j\mathbf{C}(\Phi_j) = \mathbf{C}(\Phi_j)\mathbf{\Omega}_j$$

and so $\mathbf{\Omega}_j$ and $\mathbf{C}(\Phi_j)$ commute.   □

Coming back to the original state space model (4.28)–(4.29) we define the $p \times 1$ state vector γ_t as $\gamma_t = \mathbf{\Omega}\mathbf{S}\beta_t$. This definition implies

$$\begin{aligned}
\mathbf{\Omega}\mathbf{S}\mathbf{F}\beta_{t-1} &= \mathbf{\Omega}\mathbf{S}\mathbf{F}\mathbf{S}^{-1}\mathbf{S}\beta_{t-1} \\
&= \mathbf{\Omega}\mathbf{C}\mathbf{S}\beta_{t-1} \\
&= \mathbf{C}\mathbf{\Omega}\mathbf{S}\beta_{t-1} \\
&= \mathbf{C}\gamma_{t-1},
\end{aligned}$$

where Lemma 4.1 is used.

By left multiplying the state equation (4.29) we have

$$\gamma_t = \boldsymbol{\Omega} \mathbf{S} \beta_t = \boldsymbol{\Omega} \mathbf{S} \mathbf{F} \beta_{t-1} + \boldsymbol{\Omega} \mathbf{S} \zeta_t = \mathbf{C} \gamma_{t-1} + \xi_t,$$

where $\xi_t = \boldsymbol{\Omega} \mathbf{S} \zeta_t \sim N(0, \boldsymbol{\Omega} \mathbf{S} \mathbf{Z} \mathbf{S}^\top \boldsymbol{\Omega}^\top)$.

If we now write $\gamma_t = [\gamma_{1t}, \gamma_{2t}, \ldots, \gamma_{\ell,t}]^\top$ and $\xi_t = [\xi_{1t}, \xi_{2t}, \ldots, \xi_{\ell,t}]^\top$ and taking into account that $\mathbf{C}$ is the direct sum of $\mathbf{C}(\Phi_1), \ldots, \mathbf{C}(\Phi_\ell)$, it follows that

$$\gamma_{jt} = \mathbf{C}(\Phi_j) \gamma_{j,t-1} + \xi_{jt},$$

which is Eq. (4.32).

Considering now the linear predictor η_t we have

$$
\begin{aligned}
\mu_t &= x_t^\top \beta_t = x^\top \mathbf{S}^{-1} \mathbf{S} \beta_t \\
&= [x_1^\top, x_2^\top, \ldots, x_\ell^\top] \mathbf{S} \beta_t \\
&= [e_1^\top \boldsymbol{\Omega}_1, e_1^\top \boldsymbol{\Omega}_2, \ldots, e_1^\top \boldsymbol{\Omega}_\ell] \mathbf{S} \beta_t \\
&= [e_1^\top, e_1^\top, \ldots, e_1^\top] \boldsymbol{\Omega} \mathbf{S} \beta_t \\
&= [e_1^\top, e_1^\top, \ldots, e_1^\top]
\begin{bmatrix} \gamma_{1t} \\ \gamma_{2t} \\ \vdots \\ \gamma_{\ell,t} \end{bmatrix} \\
&= \sum_{j=1}^{\ell} e_1^\top \gamma_{jt} = \chi_t^{(1)} + \chi_t^{(2)} + \cdots + \chi_t^{(\ell)},
\end{aligned}
$$

which give Eqs. (4.30)–(4.31). Hence we have established that the linear predictor η_t has the decomposition of Eqs. (4.30)–(4.32).

Some comments are in order.

1. If the polynomials $\Phi_j(\lambda)$ are relative prime factors, the decomposition is known as the *irreducible decomposition*, see Godolphin and Johnson (2003). Otherwise, $\Phi_j(\lambda)$ will be a polynomial obtained by elementary operations from the system of the elementary divisors of $\Phi(\lambda)$.

2. There is some interest to establish whether $\chi_t^{(i)}$ is independent of $\chi_t^{(j)}$, for all t and for all $i \neq j$. If this is the case, the resulting decomposition is referred to as *independent decomposition*. This property is useful if we wish to compute the variance of $\eta_t = \sum_{j=1}^{\ell} \chi_t^{(j)}$, so that $\mathrm{Cov}(\chi_t^{(i)}, \chi_t^{(j)}) = 0$.
From (4.31), $\chi_t^{(i)}$ is independent of $\chi_t^{(j)}$, if γ_{it} is independent of γ_{jt}. From (4.32) and the direct sum $\mathbf{C} = \mathbf{C}(\Phi_1) \oplus \cdots \oplus \mathbf{C}(\Phi_\ell)$, it follows that γ_{it} is independent of γ_{jt}, if the innovations ξ_{it} and ξ_{jt} are independent, for $i \neq j$. The definition $\xi_t = \boldsymbol{\Omega} \mathbf{S} \zeta_t$ implies that ξ_{it} and ξ_{jt} are independent when $\mathbf{S} \mathbf{Z} \mathbf{S}^\top$ has the same

block diagonal structure as $\boldsymbol{\Omega}$. This is satisfied if $\mathbf{Z}$ and $\mathbf{S}$ are direct sums

$$\mathbf{Z} = \mathbf{Z}_1 \oplus \cdots \oplus \mathbf{Z}_\ell \quad \text{and} \quad \mathbf{S} = \mathbf{S}_1 \oplus \cdots \oplus \mathbf{S}_\ell,$$

for some $p_j \times p_j$ matrices $\mathbf{Z}_j$ and $\mathbf{S}_j$, $j = 1, \ldots, \ell$.

3. An application of this model is in forecasting. Consider the linear state space model (3.10a)–(3.10b), where the design vector $x_t = x$, the transition matrix $\mathbf{F}_t = \mathbf{F}$ and the innovation variances are all time-invariant. Suppose that at time t the posterior distribution of β_t, given data $y_{1:t}$ is obtained from the Kalman filter (Theorem 3.2 in Sect. 3.2) and hence the posterior mean $\hat{\beta}_{t|t}$ is computed. From decomposition (4.30)–(4.32) and the transformation of β_t we can compute $\hat{\gamma}_{t|t} = \boldsymbol{\Omega S} \hat{\beta}_{t|t}$. This can then decomposed to obtain $\hat{\gamma}_{j,t|t}$, where $\hat{\gamma}_{t|t}^\top = [\hat{\gamma}_{1,t|t}^\top, \ldots, \hat{\gamma}_{\ell,t|t}^\top]$. As a result the k-step ahead forecast mean of the observation y_{t+k} is given by

$$\hat{y}_{t+k|t} = \hat{\chi}_{t+k|t}^{(1)} + \hat{\chi}_{t+k|t}^{(2)} + \cdots + \hat{\chi}_{t+k|t}^{(\ell)},$$

$$\hat{\chi}_{t+k|t}^{(j)} = e_1^\top \hat{\gamma}_{j,t+k|t} \quad \text{and} \quad \hat{\gamma}_{j,t+k|t} = \mathbf{C}(\boldsymbol{\Phi}_j)^k \hat{\gamma}_{j,t|t},$$

after using the standard result of the k-step ahead forecast mean of state space models (see Theorem 3.6 of Sect. 3.4). This has the interesting application that if a state space model is decomposed as above, the forecasts are also decomposed into a sum of ℓ forecasts. Therefore, if the forecast performance is poor, perhaps evidenced by some outliers, the modeller may concentrate to fix a particular model component (trend, seasonal) rather than dealing with the overall model which will be more complex.

4.2.4 Turkey Data Revisited

We revisit the Turkey data, discussed in Example 4.6 of Sect. 4.1.4. There a linear Gaussian state space model was fitted with design vector and transition matrix

$$x = \begin{bmatrix} 1 \\ 0 \\ 1 \\ 0 \\ 1 \end{bmatrix} \quad \text{and} \quad \mathbf{F} = \begin{bmatrix} 1 & 1 & 0 & 0 & 0 \\ 0 & 1 & 0 & 0 & 0 \\ 0 & 0 & 0 & -1 & 0 \\ 0 & 0 & 1 & 0 & 0 \\ 0 & 0 & 0 & 0 & -1 \end{bmatrix} \tag{4.36}$$

This model is essentially the superposition of two independent state space models, a linear growth and a seasonal component model, with period 4. This model is fitted in Sect. 4.6 and Fig. 4.7 illustrates its forecast performance.

Below we obtain the rational canonical decomposition of this model. The characteristic polynomial of $\mathbf{F}$ is

$$\Phi(\lambda) = |\lambda\mathbf{I} - \mathbf{F}| = 1 - \lambda - \lambda^4 + \lambda^5 \tag{4.37}$$

and the eigenvalues of $\mathbf{F}$ are 1 (multiplicity 2) and $-1, \pm i$ (multiplicity 1).
The irreducible factorisation of (4.37) is

$$\Phi(\lambda) = (\lambda - 1)^2(\lambda + 1)(\lambda^2 + 1), \tag{4.38}$$

with the elementary divisors of $\Phi(\lambda)$ being $\lambda - 1$, $\lambda + 1$ and $\lambda^2 + 1$.

Each of the polynomials $\Phi_1(\lambda) = (\lambda - 1)^2$, $\Phi_2(\lambda) = (\lambda + 1)$ and $\Phi_3(\lambda) = (\lambda^2 + 1)$ corresponds to companion matrices $\mathbf{C}(\Phi_1)$, $\mathbf{C}(\Phi_2)$, $\mathbf{C}(\Phi_3)$, hence $\mathbf{F}$ is similar to the direct sum of companion matrices

$$\mathbf{C} = \mathbf{C}(\Phi_1) \oplus \mathbf{C}(\Phi_2) \oplus \mathbf{C}(\Phi_3) = \begin{bmatrix} 0 & 1 & 0 & 0 & 0 \\ -1 & 2 & 0 & 0 & 0 \\ 0 & 0 & 0 & 1 & 0 \\ 0 & 0 & 1 & -1 & 0 \\ 0 & 0 & 0 & 0 & -1 \end{bmatrix}. \tag{4.39}$$

We can see that the y_t can be decomposed to two models

$$y_t = \chi_t^{(1)} + \chi_t^{(2)} + \epsilon_t,$$

where $\chi_t^{(1)}$ is a trend component

$$\chi_t^{(1)} = [1, 0]\gamma_t^{(1)} \quad \text{and} \quad \gamma_t^{(1)} = \begin{bmatrix} 0 & 1 \\ -1 & 2 \end{bmatrix}\gamma_{t-1}^{(1)} + \xi_{1t} \tag{4.40}$$

and $\chi_t^{(2)}$ is a seasonal component

$$\chi_t^{(2)} = [1, 0, 0]\gamma_t^{(2)} \quad \text{and} \quad \gamma_t^{(2)} = \begin{bmatrix} 0 & 1 & 0 \\ 1 & -1 & 0 \\ 0 & 0 & -1 \end{bmatrix}\gamma_{t-1}^{(2)} + \xi_{2t}.$$

It can be shown (see Exercise 5) that the trend component can be written as

$$\chi_t^{(1)} = [1, 0]\gamma_t^{(3)} \quad \text{and} \quad \gamma_t^{(3)} = \begin{bmatrix} 1 & 1 \\ 0 & 1 \end{bmatrix}\gamma_{t-1}^{(3)} + \xi_{3t}, \tag{4.41}$$

for some innovations ξ_{3t}, which are given as linear functions of the innovations $\xi_{11}, \ldots, \xi_{1t}$.

We can see that models (4.40) and (4.41) are identical and as a result the transition matrix $\mathbf{F}$ in (4.36) is identical to the matrix $\mathbf{C}$, if the block $\mathbf{C}(\Phi_1)$ is replaced by the Jordan block

$$\mathbf{J} = \begin{bmatrix} 1 & 1 \\ 0 & 1 \end{bmatrix}.$$

Notice that both matrices $\mathbf{C}(\Phi_1)$ and $\mathbf{J}$ have a unit eigenvalue of multiplicity two and they have the same characteristic polynomial $|\lambda \mathbf{I} - \mathbf{C}(\Phi_1)| = |\lambda \mathbf{I} - \mathbf{J}| = (\lambda - 1)^2$. It follows that the state space model of Sect. 4.6, which was the superposition of a trend and a seasonal component, is obtained by the irreducible decomposition proposed in the previous section.

Consider now the (non-irreducible) factorisation

$$\Phi(\lambda) = (\lambda - 1)^2(\lambda^3 + \lambda^2 + \lambda + 1),$$

which follows directly from (4.38) by expanding the seasonal factor corresponding to the term $(\lambda + 1)(\lambda^2 + 1)$.

This time y_t is decomposed to two components $y_t = \chi_t^{(1)} + \chi_t^{(2)} + \epsilon_t$, the trend with is given by (4.40) and the seasonal given by

$$\chi_t^{(2)} = [1, 0, 0]\gamma_t^{(2)} \quad \text{and} \quad \gamma_t^{(2)} = \begin{bmatrix} 0 & 1 & 0 \\ 0 & 0 & 1 \\ -1 & -1 & -1 \end{bmatrix} \gamma_{t-1}^{(2)} + \xi_{2t}.$$

Hence, $\mathbf{F}$ is similar to the matrix $\mathbf{C}$ given by

$$\mathbf{C} = \begin{bmatrix} 0 & 1 & 0 & 0 & 0 \\ -1 & 2 & 0 & 0 & 0 \\ 0 & 0 & 0 & 1 & 0 \\ 0 & 0 & 0 & 0 & 1 \\ 0 & 0 & -1 & -1 & -1 \end{bmatrix}.$$

The similarity matrix of $\mathbf{F}$ and $\mathbf{C}$ is

$$\mathbf{S} = \begin{bmatrix} 1 & -1 & 0 & 0 & 0 \\ 1 & 0 & 0 & 0 & 0 \\ 0 & 0 & -1 & -1 & -1 \\ 0 & 0 & 1 & -1 & 1 \\ 0 & 0 & 1 & 1 & -1 \end{bmatrix}$$

as it can be verified that $\mathbf{SFS}^{-1} = \mathbf{C}$.

In order to establish whether this is an independent decomposition we need to calculate matrix $\boldsymbol{\Omega}$ (see point (3) of p. 150; see also Eq. (4.33) for the definition of $\boldsymbol{\Omega}$). Recall that $x^\top \mathbf{S}^{-1} = [x_1^\top, x_2^\top]$, which is needed for the computation of $\boldsymbol{\Omega}_j$, $j = 1, 2$. We have

$$
\mathbf{S}^{-1} = \begin{bmatrix} 0 & 1 & 0 & 0 & 0 \\ -1 & 1 & 0 & 0 & 0 \\ 0 & 0 & 0 & \frac{1}{2} & \frac{1}{2} \\ 0 & 0 & -\frac{1}{2} & -\frac{1}{2} & 0 \\ 0 & 0 & -\frac{1}{2} & 0 & -\frac{1}{2} \end{bmatrix}
$$

and so $x^\top \mathbf{S}^{-1} = [0, 1, -1/2, 1/2, 0]^\top$.

Hence

$$
\boldsymbol{\Omega}_1 = \begin{bmatrix} x_1^\top \\ x_1^\top \mathbf{C}(\Phi_1) \end{bmatrix} = \begin{bmatrix} 0 & 1 \\ -1 & 2 \end{bmatrix}
$$

and

$$
\boldsymbol{\Omega}_2 = \begin{bmatrix} x_2^\top \\ x_2^\top \mathbf{C}(\Phi_2') \\ x_2^\top \mathbf{C}(\Phi_2')^2 \end{bmatrix} = \begin{bmatrix} -\frac{1}{2} & \frac{1}{2} & 0 \\ 0 & -\frac{1}{2} & \frac{1}{2} \\ -\frac{1}{2} & -\frac{1}{2} & -1 \end{bmatrix},
$$

where $\Phi_2'(\lambda) = \lambda^3 + \lambda^2 + \lambda + 1$. Therefore

$$
\boldsymbol{\Omega} = \boldsymbol{\Omega}_1 \oplus \boldsymbol{\Omega}_2 = \begin{bmatrix} 0 & 1 & 0 & 0 & 0 \\ -1 & 2 & 0 & 0 & 0 \\ 0 & 0 & -\frac{1}{2} & \frac{1}{2} & 0 \\ 0 & 0 & 0 & -\frac{1}{2} & \frac{1}{2} \\ 0 & 0 & -\frac{1}{2} & -\frac{1}{2} & -1 \end{bmatrix}.
$$

With the choice of $\mathbf{Z} = 100\mathbf{I}$, the covariance matrix of ζ_t (see the analysis of the turkey data in Sect. 4.1.4), the condition of independence (point (3), p. 150) is met. Alternatively we can see that

$$
\boldsymbol{\Omega} \mathbf{S} \mathbf{Z} \mathbf{S}^\top \boldsymbol{\Omega}^\top = \begin{bmatrix} 100 & 100 & 0 & 0 & 0 \\ 100 & 200 & 0 & 0 & 0 \\ 0 & 0 & 200 & -100 & 0 \\ 0 & 0 & -100 & 200 & -100 \\ 0 & 0 & 0 & -100 & 200 \end{bmatrix},
$$

which clearly shows that the two components $\gamma_t^{(1)}$ and $\gamma_t^{(2)}$ are uncorrelated and hence independent as the vector $\gamma_t^\top = [\gamma_t^{(1)\top}, \gamma_t^{(2)\top}]$ follows a normal distribution.

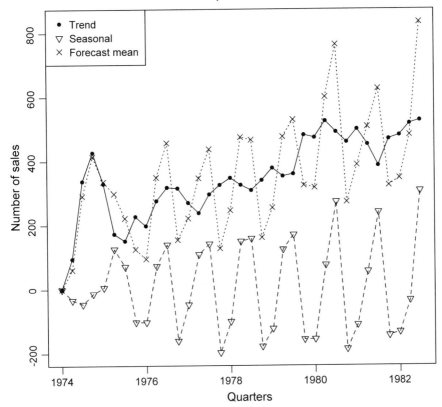

Fig. 4.10 Decomposition for the turkey data. Shown are one-step ahead forecast mean of the trend component $\hat{\chi}_t^{(1)}$ (solid line and solid points), one-step forecast mean of the seasonal component $\hat{\chi}_t^{(2)}$ (dashed line and triangle points) and the overall forecast mean of y_t, which is equal to $\hat{\chi}_t^{(1)} + \hat{\chi}_t^{(2)}$ (dotted line and cross points)

This means that the k-step forecast variance of y_{t+k} is decomposed into the respective forecast variances of the trend and seasonal components.

Figure 4.10 gives an illustration of the decomposition as far as forecasting is concerned. Shown is the one-step ahead forecast mean of y_t, $\hat{y}_{t|t-1} = \mathrm{E}(y_t \mid y_{1:t-1})$ decomposed as the sum of the forecast mean of the trend component $\hat{\chi}_t^{(1)} = \mathrm{E}(\chi_t^{(1)} \mid y_{1:t-1})$ (solid lines and solid points) and the one-step ahead forecast mean of the seasonal component $\hat{\chi}_t^{(2)} = \mathrm{E}(\chi_t^{(2)} \mid y_{1:t-1})$ (dashed lines and triangle points). The forecast means $\hat{\chi}_t^{(1)} + \hat{\chi}_t^{(2)}$ match exactly those of $\hat{y}_{t|t-1}$ provided by the application of the Kalman filter (see Fig. 4.7). Figure 4.10 suggests that the amplitude of the seasonal cycle is stable in the quarters between 1976 and 1980, while increasing after 1980.

4.3   Estimation of Hyperparameters

In the previous sections we dealt with the specification of x_t and $\mathbf{F}_t$, for some popular state space models. In general, even if we know x_t and $\mathbf{F}_t$, these may depend on some unknown hyperparameters. In addition to these σ^2 and $\mathbf{Z}_t$ (the observation variance and the transition covariance matrix) may be either unknown or depend also on some unknown hyperparameters. The successful application of the Kalman filter and related algorithms invokes the choice or specification of these hyperparameters. Perhaps, the most popular approach for their specification is to use the maximum likelihood principle, i.e. to choose the hyperparameters that maximise the log-likelihood function. Below we describe this approach, together with two other popular approaches for specifying the transition covariance matrix $\mathbf{Z}_t$ and estimating the observation variance σ^2.

4.3.1   Maximum Likelihood Estimation

Denote by θ the vector of hyperparameters, so that x_t, $\mathbf{F}_t$, σ^2 and $\mathbf{Z}_t$ (or a subset of these components) may depend on θ. We will further assume that θ includes time-invariant hyperparameters. One simple example is to consider $\theta = \sigma^2$, where σ^2 is the observation variance (see the definition of the state space model in Sect. 3.1.3). Another example is to consider that the design vector is $x_t = [\phi, 0]^T$ and the transition matrix is

$$\mathbf{F}_t = \begin{bmatrix} 1 & \psi \\ 0 & 1 \end{bmatrix},$$

for some hyperparameters ϕ and ψ, subject to estimation or specification. This may be suitable for a linear trend model with smooth trend (closer to the local level) if $0 \le \psi < 1$ and abrupt trend if $\psi > 1$. We observe that if we define $\theta = [\phi, \psi]^T$, then x_t and $\mathbf{F}_t$ depend on θ. In this set-up another desirable option would be to include σ^2 (the observational variance) in θ, i.e. $\theta = [\phi, \psi, \sigma^2]^T$. A third example is the static AR model (4.19) of Sect. 4.1.6, in which the AR coefficients $\phi_1, \ldots, \phi_d$ and the variance σ_ν^2 may be considered as hyperparameters; so we can write $\theta = [\phi_1, \ldots, \phi_d, \sigma_\varepsilon^2]^T$.

Returning to the general case, after θ is defined to include any hyperparameters subject to estimation, the likelihood function of θ based on a collection of data $y_{1:n} = \{y_1, \ldots, y_n\}$ is

$$L(\theta; y_{1:n}) = p(y_1, \ldots, y_n \mid \theta)$$
$$= p(y_n \mid y_{1:n-1}, \theta) p(y_1, \ldots, y_{n-1} \mid \theta)$$

$$= \cdots$$

$$= \prod_{t=1}^{n} p(y_t \mid y_{1:t-1}, \theta),$$

where $p(y_1 \mid y_{1:0}, \theta) \equiv p(y_1 \mid \theta)$ by definition. Now we note that given θ, the density $p(y_t \mid y_{1:t-1}, \theta)$ is just the one-step forecast distribution at time t, which is given by the Kalman filter as $y_t \mid y_{1:t-1}, \theta \sim N(\hat{y}_{t|t-1}, q_{t|t-1})$, where $\hat{y}_{t|t-1}$ and $q_{t|t-1}$ are implicitly conditional on θ. Substituting this Gaussian density to the above likelihood we obtain

$$L(\theta; y_{1:n}) = \prod_{t=1}^{n} \frac{1}{\sqrt{2\pi q_{t|t-1}}} \exp\left[-\frac{(y_t - \hat{y}_{t|t-1})^2}{2q_{t|t-1}} \right]$$

$$= \frac{1}{(2\pi)^{n/2}} \left(\prod_{t=1}^{n} q_{t|t-1}^{-1/2} \right) \exp\left[-\frac{1}{2} \sum_{t=1}^{n} \frac{(y_t - \hat{y}_{t|t-1})^2}{q_{t|t-1}} \right]$$

and the log-likelihood is

$$\ell(\theta; y_{1:n}) = -\frac{n}{2} \log(2\pi) - \frac{1}{2} \sum_{t=1}^{n} \log q_{t|t-1} - \frac{1}{2} \sum_{t=1}^{n} \frac{(y_t - \hat{y}_{t|t-1})^2}{q_{t|t-1}}.$$

Maximisation of $\ell(\cdot)$ usually cannot be performed analytically, as $\ell(\cdot)$ is a non-linear function on θ, even in the simple case of $\theta = \sigma^2$. Such a maximisation may be performed using iterative numerical methods, such as the Newton-Raphson method, or the more sophisticated expectation maximisation (EM) algorithm, both discussed in Shumway and Stoffer (2017). Direct maximisation may be appropriate if the dimension of θ is small, but when θ includes many parameters, the maximisation algorithm is likely to be stack to some local maxima of the likelihood. Compared to the direct maximisation the EM algorithm provides a more sophisticated efficient maximisation procedure and is discussed next; for a background discussion of the EM algorithm see Sect. 2.4.2.

The EM-algorithm The EM algorithm for state space models was first discussed in Shumway and Stoffer (1982); for a general description of the EM algorithm see Sect. 2.4.2. Consider the state space model (3.10a)–(3.10b) with information $y_{1:n} = \{y_1, \ldots, y_n\}$, where $\mathbf{Z}_t = \mathbf{Z}$ is time-invariant and the hyperparameter vector θ includes the distinct elements of $\mathbf{Z}$, σ^2, the prior mean of β_0, $\hat{\beta}_{0|0}$, and the distinct elements of the prior covariance matrix of β_0, $\mathbf{P}_{0|0}$. We write

$$\theta = [\sigma^2, \text{vech}(\mathbf{Z})^\top, \hat{\beta}_{0|0}^\top, \text{vech}(\mathbf{P}_{0|0})^\top]^\top,$$

where $\text{vech}(\mathbf{Z})$ denotes the column stacking vector of the distinct elements of $\mathbf{Z}$ (see Sect. 2.1).

The likelihood function of θ based on the observed data $y_{1:n}$ and the unobserved data (consisting of the states) $\beta_{0:n} = (\beta_0^\top, \beta_1^\top, \ldots, \beta_n^\top)^\top$ is

$$
\begin{aligned}
L(\theta; y_{1:n}, \beta_{0:n}) &= p(y_1, \ldots, y_n, \beta_0, \beta_1, \ldots, \beta_n \mid \theta) \\
&= p(y_1, \ldots, y_n \mid \beta_0, \beta_1, \ldots, \beta_n, \theta) p(\beta_1, \ldots, \beta_n \mid \theta) \\
&= \prod_{t=1}^{n} p(y_t \mid \beta_t, \theta) \prod_{t=1}^{n} p(\beta_t \mid \beta_{t-1}, \theta) p(\beta_0 \mid \theta)
\end{aligned}
$$

and so the log-likelihood of θ is

$$
\begin{aligned}
\ell(\theta; y, \beta) = &-\frac{n}{2} \log \sigma^2 - \frac{1}{2\sigma^2} \sum_{t=1}^{n} (y_t - x_t^\top \beta_t)^2 \\
&-\frac{n}{2} \log |\mathbf{Z}| - \frac{1}{2} \sum_{t=1}^{n} (\beta_t - \mathbf{F}_t \beta_{t-1})^\top \mathbf{Z}^{-1} (\beta_t - \mathbf{F}_t \beta_t) \\
&-\frac{1}{2} \log |\mathbf{P}_{0|0}| - \frac{1}{2} (\beta_0 - \hat{\beta}_{0|0})^\top \mathbf{P}_{0|0}^{-1} (\beta_0 - \hat{\beta}_{0|0}), \qquad (4.42)
\end{aligned}
$$

where from the observation equation (3.10a) we have $y_t \mid \beta_t, \theta \sim N(x_t^\top \beta_t, \sigma^2)$, from the transition equation (3.10b) we have $\beta_t \mid \beta_{t-1}, \theta \sim N(\mathbf{F}_t \beta_{t-1}, \mathbf{Z})$ and from the initial distribution (3.12) we have $\beta_0 \mid \theta \sim N(\hat{\beta}_{0|0}, \mathbf{P}_{0|0})$.

In the **E-step** we calculate the quantity

$$
Q(\theta \mid \theta^{(i)}) = E[\ell(\theta; y_{1:n}, \beta_{0:n}) \mid y_{1:n}, \theta^{(i)}],
$$

the conditional expectation of the log-likelihood function of θ, conditioned upon the observed data $y_{1:n}$ and the current estimate $\theta^{(i)}$.

Define

$$
S_1 = E\left[-\frac{n}{2} \log \sigma^2 - \frac{1}{2\sigma^2} \sum_{t=1}^{n} (y_t - x_t^\top \beta_t)^2 \mid y_{1:n}, \theta^{(i)} \right],
$$

$$
S_2 = E\left[-\frac{n}{2} \log |\mathbf{Z}| - \frac{1}{2} \sum_{t=1}^{n} (\beta_t - \mathbf{F}_t \beta_{t-1})^\top \mathbf{Z}^{-1} (\beta_t - \mathbf{F}_t \beta_t) a_t)^2 \mid y_{1:n}, \theta^{(i)} \right],
$$

$$
S_3 = E\left[-\frac{1}{2} \log |\mathbf{P}_{0|0}| - \frac{1}{2} (\beta_0 - \hat{\beta}_{0|0})^\top \mathbf{P}_{0|0}^{-1} (\beta_0 - \hat{\beta}_{0|0}) a_t)^2 \mid y_{1:n}, \theta^{(i)} \right],
$$

so that $Q(\theta \mid \theta^{(i)}) = S_1 + S_2 + S_3$. In the sequel we calculate S_1, S_2, S_3.

From the definition of S_1 above we have

$$S_1 = -\frac{n}{2}\log\sigma^2 - \frac{1}{2\sigma^2}\sum_{t=1}^{n}\mathrm{E}\left[(y_t - x_t^{\top}\beta_t)^2 \mid y_{1:n}, \theta^{(i)}\right]. \tag{4.43}$$

Now by expanding the square in the expectation term and taking expectations, we obtain

$$\mathrm{E}\left[(y_t - x_t^{\top}\beta_t)^2 \mid y_{1:n}, \theta^{(i)}\right] = \mathrm{E}\left[y_t^2 + (x_t^{\top}\beta_t)^2 - 2y_t x_t^{\top}\beta_t \mid y_{1:n}, \theta^{(i)}\right]$$

$$= y_t^2 - 2y_t x_t^{\top}\mathrm{E}(\beta_t \mid y_{1:n}, \theta^{(i)}) + x_t^{\top}x_t\mathrm{E}(\beta_t^{\top}\beta_t \mid y_{1:n}, \theta^{(i)})$$

$$= y_t^2 - 2y_t x_t^{\top}\hat{\beta}_{t|n} + x_t^{\top}x_t\sum_{j=1}^{n}(p_{jj,t|n} + \hat{\beta}_{j,t|n}^2)$$

$$= y_t^2 - 2y_t x_t^{\top}\hat{\beta}_{t|n} + x_t^{\top}\hat{\beta}_{t|n}\hat{\beta}_{t|n}^{\top}x_t + x_t^{\top}x_t\mathrm{trace}(\mathbf{P}_{t|n})$$

$$= (y_t - x_t^{\top}\hat{\beta}_{t|n})^2 + x_t^{\top}\mathbf{P}_{t|n}x_t, \tag{4.44}$$

where

$$\mathrm{E}(\beta_t^{\top}\beta_t \mid y_{1:n}, \theta^{(i)}) = \mathrm{E}\left(\sum_{j=1}^{n}\beta_{jt}^2 \mid y_{1:n}, \theta^{(i)}\right) = \sum_{j=1}^{n}\mathrm{E}(\beta_{jt}^2 \mid y_{1:n}, \theta^{(i)})$$

$$= \sum_{j=1}^{n}\left[\mathrm{Var}(\beta_{jt} \mid y_{1:n}, \theta^{(i)}) + \mathrm{E}(\beta_{jt} \mid y_{1:n}, \theta^{(i)})^2\right]$$

$$= \sum_{j=1}^{n}(p_{jj,t|n} + \hat{\beta}_{j,t|n}^2)$$

$$= \hat{\beta}_{t|n}^{\top}\hat{\beta}_{t|n} + \mathrm{trace}(\mathbf{P}_{t|n})$$

and we have used $\hat{\beta}_{t|n} = [\hat{\beta}_{1,t|n}, \ldots, \hat{\beta}_{p,t|n}]^{\top} = \mathrm{E}(\beta_t \mid y_{1:n}, \theta^{(i)})$ and $\mathbf{P}_{t|n} = \{p_{jk,t|n}, j, k = 1, \ldots, p\} = \mathrm{Var}(\beta_t \mid y_{1:n}, \theta^{(i)})$, provided by the fixed-interval smoothing algorithm of Sect. 3.3 (Theorem 3.4).

Substituting (4.44) into (4.43) we obtain

$$S_1 = -\frac{n}{2}\log\sigma^2 - \frac{1}{2\sigma^2}\sum_{t=1}^{n}(y_t - x_t^{\top}\hat{\beta}_{t|n})^2 - \frac{1}{2\sigma^2}x_t^{\top}\mathbf{P}_{t|n}x_t. \tag{4.45}$$

Moving on to S_2, from its definition above, we have

$$S_2 = -\frac{n}{2}\log|\mathbf{Z}| - \frac{1}{2}\sum_{t=1}^{n}\mathrm{E}\left[(\beta_t - \mathbf{F}_t\beta_{t-1})^{\top}\mathbf{Z}^{-1}(\beta_t - \mathbf{F}_t\beta_{t-1}) \mid y_{1:n}, \theta^{(i)}\right].$$

(4.46)

In order to evaluate the expectation, as before, we expand the terms and taking expectations

$$\mathrm{E}\left[(\beta_t - \mathbf{F}_t\beta_{t-1})^{\top}\mathbf{Z}^{-1}(\beta_t - \mathbf{F}_t\beta_{t-1}) \mid y_{1:n}, \theta^{(i)}\right] = \mathrm{E}(\beta_t^{\top}\mathbf{Z}^{-1}\beta_t \mid y_{1:n}, \theta^{(i)})$$

$$-2\mathrm{E}(\beta_{t-1}^{\top}\mathbf{F}_t^{\top}\mathbf{Z}^{-1}\beta_t \mid y_{1:n}, \theta^{(i)}) + \mathrm{E}(\beta_{t-1}^{\top}\mathbf{F}_t^{\top}\mathbf{Z}^{-1}\mathbf{F}_t\beta_t \mid y_{1:n}, \theta^{(i)})$$

(4.47)

These three expectations are

$$\mathrm{E}(\beta_t^{\top}\mathbf{Z}^{-1}\beta_t \mid y_{1:n}, \theta^{(i)}) = \mathrm{E}\left[\mathrm{trace}(\beta_t^{\top}\mathbf{Z}^{-1}\beta_t) \mid y_{1:n}, \theta^{(i)}\right]$$

$$= \mathrm{E}\left[\mathrm{trace}(\beta_t\beta_t^{\top}\mathbf{Z}^{-1}) \mid y_{1:n}, \theta^{(i)}\right]$$

$$= \mathrm{trace}\left[\mathrm{E}(\beta_t\beta_t^{\top} \mid y_{1:n}, \theta^{(i)})\mathbf{Z}^{-1}\right]$$

$$= \mathrm{trace}\left[(\mathbf{P}_{t|n} + \hat{\beta}_{t|n}\hat{\beta}_{t|n}^{\top})\mathbf{Z}^{-1}\right],$$

$$\mathrm{E}(\beta_{t-1}^{\top}\mathbf{F}_t^{\top}\mathbf{Z}^{-1}\beta_t \mid y_{1:n}, \theta^{(i)}) = \mathrm{trace}\left[\mathrm{E}(\beta_t\beta_{t-1}^{\top} \mid y_{1:n}, \theta^{(i)})\mathbf{F}_t^{\top}\mathbf{Z}^{-1}\right]$$

$$= \mathrm{trace}\left[[\mathrm{E}[(\beta_t - \hat{\beta}_{t|n})(\beta_{t-1} - \hat{\beta}_{t-1|n})^{\top} \mid y_{1:n}, \theta^{(i)}]\right.$$

$$\left. + \mathrm{E}(\beta_t \mid y_{1:n}, \theta^{(i)})\mathrm{E}(\beta_{t-1}^{\top} \mid y_{1:n}, \theta^{(i)})]\mathbf{F}_t^{\top}\mathbf{Z}^{-1}\right]$$

$$= \mathrm{trace}\left[(\mathbf{P}_{t,t-1|n} + \hat{\beta}_{t|n}\hat{\beta}_{t-1|n}^{\top})\mathbf{F}_t^{\top}\mathbf{Z}^{-1}\right]$$

and

$$\mathrm{E}(\beta_{t-1}^{\top}\mathbf{F}_t^{\top}\mathbf{Z}^{-1}\mathbf{F}_t\beta_{t-1} \mid y_{1:n}, \theta^{(i)}) = \mathrm{trace}\left[\mathrm{E}(\beta_{t-1}\beta_{t-1}^{\top} \mid y_{1:n}, \theta^{(i)})\mathbf{F}_t^{\top}\mathbf{Z}^{-1}\mathbf{F}_t\right]$$

$$= \mathrm{trace}\left[(\mathbf{P}_{t-1|n} + \hat{\beta}_{t-1|n}\hat{\beta}_{t-1|n}^{\top})\mathbf{F}_t^{\top}\mathbf{Z}^{-1}\mathbf{F}_t\right].$$

By substituting the three expectations above to (4.47) and (4.46) we obtain

$$S_2 = -\frac{n}{2}\log|\mathbf{Z}| - \frac{1}{2}\sum_{t=1}^{n}\mathrm{trace}\left[(\mathbf{P}_{t|n} + \hat{\beta}_{t|n}\hat{\beta}_{t|n}^{\top} - 2(\mathbf{P}_{t,t-1|n} + \hat{\beta}_{t|n}\hat{\beta}_{t-1|n}^{\top})\mathbf{F}_t^{\top}\right.$$

$$\left. + \mathbf{F}_t(\mathbf{P}_{t-1|n} + \hat{\beta}_{t-1|n}\hat{\beta}_{t-1|n}^{\top})\mathbf{F}_t^{\top})\mathbf{Z}^{-1}\right]$$

$$= -\frac{n}{2}\log|\mathbf{Z}| - \frac{1}{2}\sum_{t=1}^{n}\text{trace}\left[\{(\hat{\beta}_{t|n} - \mathbf{F}_t\hat{\beta}_{t-1|n})(\hat{\beta}_{t|n} - \mathbf{F}_t\hat{\beta}_{t-1|n})^\top + \right.$$

$$\left. \mathbf{P}_{t|n} - \mathbf{P}_{t,t-1|n}\mathbf{F}_t^\top - \mathbf{F}_t\mathbf{P}_{t,t-1|n}^\top + \mathbf{F}_t\mathbf{P}_{t-1|n}\mathbf{F}_t^\top\}\mathbf{Z}^{-1}\right]. \tag{4.48}$$

Finally, from the definition of S_3 we have

$$S_3 = -\frac{1}{2}\log|\mathbf{P}_{0|0}| - \frac{1}{2}\text{E}\left[(\beta_0 - \hat{\beta}_{0|0})^\top\mathbf{P}_{0|0}^{-1}(\beta_0 - \hat{\beta}_{0|0}) \mid y_{1:n}, \theta^{(i)}\right]. \tag{4.49}$$

The expectation term can be written as

$$\text{E}\left[(\beta_0 - \hat{\beta}_{0|0})^\top\mathbf{P}_{0|0}^{-1}(\beta_0 - \hat{\beta}_{0|0}) \mid y_{1:n}, \theta^{(i)}\right]$$

$$= \text{E}(\beta_0^\top\mathbf{P}_{0|0}^{-1}\beta_0 \mid y_{1:n}, \theta^{(i)}) - 2\hat{\beta}_{0|0}^\top\mathbf{P}_{0|0}^{-1}\text{E}(\beta_0 \mid y_{1:n}, \theta^{(i)}) + \hat{\beta}_{0|0}^\top\mathbf{P}_{0|0}^{-1}\hat{\beta}_{0|0}.$$

Now

$$\text{E}(\beta_0^\top\mathbf{P}_{0|0}^{-1}\beta_0 \mid y_{1:n}, \theta^{(i)}) = \text{trace}\left[\text{E}(\beta_0\beta_0^\top \mid y_{1:n}, \theta^{(i)})\mathbf{P}_{0|0}^{-1}\right]$$

$$= \text{trace}\left[(\mathbf{P}_{0|n} + \hat{\beta}_{0|n}\hat{\beta}_{0|n}^\top)\mathbf{P}_{0|0}^{-1}\right]$$

and so replacing these expectations in (4.49) we obtain

$$S_3 = -\frac{1}{2}\log|\mathbf{P}_{0|0}| - \frac{1}{2}\text{trace}\left[((\hat{\beta}_{0|n} - \hat{\beta}_{0|0})(\hat{\beta}_{0|n} - \hat{\beta}_{0|0})^\top + \mathbf{P}_{0|n})\mathbf{P}_{0|0}^{-1}\right]. \tag{4.50}$$

The E-step is completed by noting that

$$Q(\theta \mid \theta^{(i)}) = S_1 + S_2 + S_3,$$

where S_1, S_2, S_3 are computed by (4.45), (4.48), (4.50).

Moving on to the **M-step** we first note that $S_1 = S_1(\sigma^2)$ is a function of σ^2 only, not depending on $\mathbf{Z}$, $\hat{\beta}_{0|0}$ and $\mathcal{P}_{0|0}$. Likewise S_2 is a function of $\mathbf{Z}$ only and S_3 is a function of $\hat{\beta}_{0|0}$ and $\mathbf{P}_{0|0}$ only. Thus in the M-step we have

$$\frac{\partial S_1}{\partial \theta} = \frac{\partial S_1}{\partial \sigma^2} = -\frac{n}{2\sigma^2} + \frac{1}{2\sigma^4}\sum_{t=1}^{n}\left[(y_t - x_t^\top\hat{\beta}_{t|n})^2 + x_t^\top\mathbf{P}_{t|n}x_t\right]$$

and equating this to zero we find

$$\hat{\sigma}^2 = \frac{1}{n}\sum_{t=1}^{n}\left[(y_t - x_t^\top\hat{\beta}_{t|n})^2 + x_t^\top\mathbf{P}_{t|n}x_t\right]. \tag{4.51}$$

For S_2 we have

$$\frac{\partial S_2}{\partial \theta} = \frac{\partial S_2}{\partial \text{vech}(\mathbf{Z})} = -\frac{n}{2}\frac{\partial \log |\mathbf{Z}|}{\partial \text{vech}(\mathbf{Z})} - \frac{1}{2}\sum_{t=1}^{n}\frac{\text{trace}(\mathbf{A}_t\mathbf{Z}^{-1})}{\partial \text{vech}(\mathbf{Z})}$$

$$= -\frac{n}{2}\left[2\mathbf{Z}^{-1} - \text{diag}(\mathbf{Z}^{-1})\right] - \frac{1}{2}\sum_{t=1}^{n}[-2\mathbf{Z}^{-1}\mathbf{A}_t\mathbf{Z}^{-1} + \text{diag}(\mathbf{Z}^{-1}\mathbf{A}_t\mathbf{Z}^{-1})],$$

where

$$\mathbf{A}_t = (\hat{\beta}_{t|n} - \mathbf{F}_t\hat{\beta}_{t-1|n})(\hat{\beta}_{t|n} - \mathbf{F}_t\hat{\beta}_{t-1|n})^\top + \mathbf{P}_{t|n} - \mathbf{P}_{t,t-1|n}\mathbf{F}_t^\top - \mathbf{F}_t\mathbf{P}_{t,t-1|n}^\top$$

$$+\mathbf{F}_t\mathbf{P}_{t-1|n}\mathbf{F}_t^\top.$$

Now, equating $\partial S_2/\partial \theta$ to zero we obtain

$$\frac{n}{2}\left[2\hat{\mathbf{Z}}^{-1} - \text{diag}(\hat{\mathbf{Z}}^{-1})\right] = \frac{1}{2}\sum_{t=1}^{n}[-2\hat{\mathbf{Z}}^{-1}\mathbf{A}_t\hat{\mathbf{Z}}^{-1} + \text{diag}(\hat{\mathbf{Z}}^{-1}\mathbf{A}_t\hat{\mathbf{Z}}^{-1})]$$

or

$$\sum_{t=1}^{n}\hat{\mathbf{Z}}^{-1}\mathbf{A}_t\hat{\mathbf{Z}}^{-1} = n\hat{\mathbf{Z}}^{-1}$$

or

$$\hat{\mathbf{Z}} = \frac{1}{n}\sum_{t=1}^{n}\mathbf{A}_t = \frac{1}{n}\sum_{t=1}^{n}\left[(\hat{\beta}_{t|n} - \mathbf{F}_t\hat{\beta}_{t-1|n})(\hat{\beta}_{t|n} - \mathbf{F}_t\hat{\beta}_{t-1|n})^\top\right.$$

$$\left. +\mathbf{P}_{t|n} - \mathbf{P}_{t,t-1|n}\mathbf{F}_t^\top - \mathbf{F}_t\mathbf{P}_{t,t-1|n}^\top + \mathbf{F}_t\mathbf{P}_{t-1|n}\mathbf{F}_t^\top\right].$$

Moving on to S_3, simultaneous estimation of $\hat{\beta}_{0|0}$ and $\mathbf{P}_{0|0}$ is not available. Specification of these quantities outside of the EM algorithm is discussed in some detail in Sect. 4.5. Following a similar treatment as that adopted in Shumway and Stoffer (2017), $\hat{\beta}_{0|0}$ can be set equal to $\hat{\beta}_{0|n}$ and $\mathbf{P}_{0|0}$ can be optimised so as to maximise S_3. Following a similar maximisation approach as that in S_2, we conclude that

$$\hat{\mathbf{P}}_{0|0} = \frac{1}{n}\left[(\hat{\beta}_{0|n} - \hat{\beta}_{0|0})(\hat{\beta}_{0|n} - \hat{\beta}_{0|0})^\top + \mathbf{P}_{0|n}\right].$$

Summing up, after the M-step we update the iteration vector

$$\hat{\theta}^{(i+1)} = [\hat{\sigma}^2, \text{vech}(\hat{\mathbf{Z}})^\top, \hat{\beta}_{0|0}^\top, \text{vech}(\hat{\mathbf{P}}_{0|0})^\top]^\top,$$

and the algorithm proceeds with the E-step and M-step conditioned upon $\hat{\theta}^{(i+1)}$. The EM-algorithm is summarised below.

EM Algorithm for State Space Models

In the state space model (3.10a)–(3.10b) with information $y_{1:n} = \{y_1, \ldots, y_n\}$ the maximum likelihood estimate of the parameter vector

$$\theta = [\sigma^2, \text{vech}(\mathbf{Z})^\top, \hat{\beta}_{0|0}^\top, \text{vech}(\mathbf{P}_{0|0})^\top]^\top,$$

is approximately $\hat{\theta}^{(N)}$, where $\hat{\theta}^{(i)}$ is iteratively computed as follows:

1. Initial estimate $\theta^{(0)} = \left[(\hat{\sigma}^2)^{(0)}, \text{vech}(\hat{\mathbf{Z}}^{(0)})^\top, (\hat{\beta}_{0|0}^{(0)})^\top, \text{vech}(\hat{\mathbf{P}}_{0|0}^{(0)})^\top\right]^\top$.
2. For each $i = 0, 1, 2, \ldots, N - 1$:

    a. For each $t = 0, 1, \ldots, n$ Compute $\hat{\beta}_{t|n}$, $P_{t|n}$, conditioned upon $\theta^{(i)}$;
    b. Compute

$$(\hat{\sigma}^2)^{(i+1)} = \frac{1}{n} \sum_{t=1}^{n} \left[(y_t - x_t^\top \hat{\beta}_{t|n})^2 + x_t^\top \mathbf{P}_{t|n} x_t \right];$$

$$\hat{\mathbf{Z}}^{(i+1)} = \frac{1}{n} \sum_{t=1}^{n} \left[(\hat{\beta}_{t|n} - \mathbf{F}_t \hat{\beta}_{t-1|n})(\hat{\beta}_{t|n} - \mathbf{F}_t \hat{\beta}_{t-1|n})^\top \right.$$

$$\left. + \mathbf{P}_{t|n} - \mathbf{P}_{t,t-1|n} \mathbf{F}_t^\top - \mathbf{F}_t \mathbf{P}_{t,t-1|n}^\top + \mathbf{F}_t \mathbf{P}_{t-1|n} \mathbf{F}_t^\top \right];$$

$$\hat{\beta}_{0|0}^{(i+1)} = \hat{\beta}_{0|n};$$

$$\hat{\mathbf{P}}_{0|0}^{(i+1)} = \frac{1}{n} \left[(\hat{\beta}_{0|n} - \hat{\beta}_{0|0})(\hat{\beta}_{0|n} - \hat{\beta}_{0|0})^\top + \mathbf{P}_{0|n} \right].$$

    c. Set $\hat{\theta}^{(i+1)} = \left[(\hat{\sigma}^2)^{(i+1)}, \text{vech}(\hat{\mathbf{Z}}^{(i+1)})^\top, (\hat{\beta}_{0|0}^{(i+1)})^\top, \text{vech}(\hat{\mathbf{P}}_{0|0}^{(i+1)})^\top \right]^\top$.

Some comments are in order.

- In step 2(a) the quantities $\hat{\beta}_{t|n}$ and $\mathbf{P}_{t|n}$ are the respective smoothed mean vector and covariance matrix of β_t, given $\theta = \hat{\theta}^{(i)}$. These are provided routinely by the fixed-interval smoothing theorem (Theorem 3.4).
- If estimation of the initial mean vector $\hat{\beta}_{0|0}$ and covariance matrix $\mathbf{P}_{0|0}$ are not required, then θ contains only the observation variance σ^2 and the transition covariance matrix $\mathbf{Z}$. As noted earlier, $\hat{\beta}_{0|0}$, $\mathbf{P}_{0|0}$ can be selected a priori by the modeller, as discussed in Sect. 4.5.

- The EM algorithm can be used to estimate the design and transition components $x_t = x$ and $\mathbf{F}_t = \mathbf{F}$, if those are time-invariant. However, for simplicity we have not included this consideration in the above version of the EM algorithm. In this book we propose that these components be specified according to each model adoption.
- The EM-algorithm terminates at N iterations when $\hat{\theta}^{(i+1)}$ is very close to $\hat{\theta}^{(i)}$, for any $i \geq N$. Basically, this indicates that $\hat{\theta}^{(i)}$ has converged to θ. The above closeness is usually checked empirically by ensuring that the Eucledian distance of $\theta^{(i+1)}$ and $\theta^{(i)}$ is smaller than some tolerance limit (for several iterations i), i.e.

$$\| \hat{\theta}^{(i+1)} - \hat{\theta}^{(i)} \| = \left[(\hat{\theta}^{(i+1)} - \hat{\theta}^{(i)})^\top (\hat{\theta}^{(i+1)} - \hat{\theta}^{(i)}) \right]^{1/2} < \text{Tol},$$

where Tol is the tolerance limit, usually 0.001 or smaller. Discussion of the termination of the algorithm and of further results on the asymptotic properties of the maximum likelihood can be found in Shumway and Stoffer (2017) and in Hannan and Deistler (1988).

For implementation in R, we use the function `bts.EM` of the package `bts`. The command below provides maximum likelihood estimates of the observation and transition variances σ^2 and Z of the local level model used for the annual central temperatures of England example.

```
# needs data temp
> fit <- bts.EM( temp, x0=1, F0=1, Sigma0=1, Z0=1,
+ beta0=1, P0=1000, k=10 )
```

This function requires the argument `temp` (the data) and a list of components (here $x_t = 1$ and $F_t = 1$) are specified for the local level, as well as the prior mean $\beta_{0|0} = 1$ and prior variance $P_{0|0} = 1000$ of β_0. The values of `Sigma0=1` and `Z0=1` are the initial values $\sigma^{2(0)}$ and $Z^{(0)}$ of σ^2 and Z. The function returns the estimated values $\sigma^{2(i)}$ and $Z^{(i)}$ of σ^2 and Z as well as the log-likelihood value of σ^2, Z at each iteration i and they are shown in Table 4.1, for the first five iterations.

We observe that the log-likelihood stabilises to the value of -309.9480 after just two iterations and the estimated values of σ^2 and Z are 0.2923737 and 0.006257308; the algorithm has converged after just three iterations.

To explore more the likelihood estimation Fig. 4.11 shows the contour plot of the log=likelihood (left panel) and the log-likelihood of just σ^2 when $Z = 0.006$

Table 4.1 EM algorithm for the estimation of σ^2 and Z for the annual England temperatures

Iteration (i)	$\sigma^{(i)}$	$Z^{(i)}$	Log-likelihood
0	1	1	–
1	0.2923771	0.02082655	−311.5735
2	0.2923737	0.006257376	−309.9480
3	0.2923737	0.006257308	−309.9480
5	0.2923737	0.006257308	−309.9480

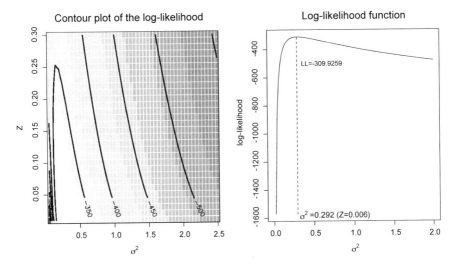

Fig. 4.11 Contour plot and marginal likelihood of the annual England temperature data)

(right panel). The contour plot shows that the maximum is obtained on the white-shaded area, agreeing with the results provided from the EM algorithm. Because it is difficult to visualise the exact maximum of the log-likelihood in that white-shaded area, the right panel of Fig. 4.11 shows the log-likelihood of σ^2 when Z is estimated as 0.006. We observe that the log-likelihood is maximised for $\hat{\sigma}^2 = 0.292$ with a corresponding maximum likelihood at -309.9259, which again agree with the results of the EM algorithm shown in Table 4.1. In order to obtain the contour plot we had to run the Kalman filter many times and this results in a very slow approach to finding the maximum; indeed this approach is used here only for visualisation purposes, while we recommend the application of the EM algorithm, which is remarkably faster and more accurate.

4.3.2 Specification of Z_t Using Discount Factors

Out of the possible hyperparameters that require specification, the transition covariance matrix Z_t is of particular interest. If Z_t is time-invariant and if its dimensions are small (e.g. as in the examples above), then maximum likelihood may be applied. However, one may observe that as Z_t is a $p \times p$ covariance matrix, there are $p(p + 1)/2$ distinct elements to be specified, for each t. Indeed, note that such a matrix may be written as

$$
Z_t = \begin{bmatrix}
Z_{11,t} & Z_{12,t} & \cdots & Z_{1p,t} \\
Z_{12,t} & Z_{22,t} & \cdots & Z_{2p,t} \\
\vdots & \vdots & \ddots & \vdots \\
Z_{1p,t} & Z_{2p,t} & \cdots & Z_{pp,t}
\end{bmatrix},
$$

so that $Z_{ij,t} = Z_{ji,t}$ and the distinct elements are the p elements of the main diagonal plus half of $p^2 - p$ remaining elements, or $p + (p^2 - p)/2 = p(p+1)/2$. We observe that the problem of the specification of $Z_{ij,t}$ is inflated by the dimension p. Even for relatively low values of p, e.g. $p = 5$, there are too many elements to specify (for $p = 5$ there are 15 distinct elements to specify). Furthermore, it is likely that over time $\mathbf{Z}_t$ will fail to be time-invariant; the inclusion of the time-index in $\mathbf{Z}_t$ allows for different stochastic evolutions of the unobserved states β_t, since $\mathbf{Z}_t$ is the covariance matrix of $\zeta_t = \beta_t - \mathbf{F}_t\beta_{t-1}$ (see the transition equation (3.10b)). In other words $\mathbf{Z}_t$ controls the stochastic magnitude of change of β_t from $\mathbf{F}_t\beta_{t-1}$ and as such, $\mathbf{Z}_t$ is likely to change with the course of time. The assumption of a time-invariant $Z_t = Z$ (adopted in the previous section) is usually done for convenience and it may be valid only for short periods of time.

For these reasons a practical approach of the specification of $\mathbf{Z}_t$ is required. Consider first the dynamic (or time-varying) regression state space model (3.9a)–(3.9b) and see that the Kalman filter recursions for $\hat{\beta}_{t|t}$ and $\mathbf{P}_{t|t}$ are exactly the same as those of the recursive least squares (RLS) algorithm, described in Sect. 3.1.2, if we set $\sigma^2 = 1$ and

$$\mathbf{Z}_t = \delta^{-1}(1 - \delta)\mathbf{P}_{t-1|t-1}, \tag{4.52}$$

where δ is the discount factor of the RLS and $\mathbf{P}_{t-1|t-1}$ is the covariance matrix of β_{t-1}, given information $y_{1:t-1}$. To see this just write down the Kalman filter recursions for the above state space model, i.e.

$$\hat{\beta}_{t|t} = \hat{\beta}_{t|t-1} + k_t e_t, \quad \mathbf{P}_{t|t} = \mathbf{P}_{t|t-1} - K_t K_t^\top / q_{t|t-1}$$

$$K_t = \mathbf{P}_{t|t-1}x_t / q_{t|t-1}, \quad q_{t|t-1} = x_t^\top \mathbf{P}_{t|t-1}x_t + 1, \quad \mathbf{P}_{t|t-1} = \mathbf{P}_{t-1|t-1} + \mathbf{Z}_t$$

and e_t is the residual at time t. Then replacing $\mathbf{Z}_t$ as in (4.52) and $\sigma^2 = 1$ we obtain the recursions of the RLS, with $\hat{\beta}_{t|t} = \hat{\beta}$ and $\mathbf{P}_{t|t} = \mathbf{P}_t$.

It is important to explain the contribution of the discount factor δ in (4.52). We first note from the discussion of Sect. 3.1.2 that $0 < \delta \leq 1$. If $\delta = 1$, this reduces $\mathbf{Z}_t$ to zero and basically this implies $\beta_t = \beta_{t-1}$, with probability 1, for all t, i.e. $\beta_t = \beta$ is time-invariant and model (3.9a)–(3.9b) reduces to an ordinary regression model, as its parameter vector β is time-invariant (the transition equation of the state space model vanishes). In this case the Kalman filter recursions (with $\mathbf{Z}_t = 0$) and the RLS recursions (with $\lambda = 1$) reduce to the well-known ordinary least squares (OLS), presented in Sect. 3.1.1.

On the other extreme when δ approaches zero, the factor $(1 - \delta)/\delta$ tends to infinity, which means that $\mathbf{Z}_t$ introduces erratic shocks on the evolution of β_t. In any case values close, but less than one should considered, typically in the range [0.7, 0.999], as argued by West and Harrison (1997, Chapter 6) and their co-authors.

Motivated by the above discussion, and considering the general state space model (3.10a)–(3.10b), $\mathbf{Z}_t$ may be specified using a forgetting factor λ as

$$\mathbf{Z}_t = \delta^{-1}(1-\delta)\mathbf{F}_t\mathbf{P}_{t-1|t-1}\mathbf{F}_t^\top. \tag{4.53}$$

We can observe that $\mathbf{Z}_t$ of (4.53) is reduced to $\mathbf{Z}_t$ of (4.52), if we set $\mathbf{F}_t = \mathbf{I}$ (i.e. in the above time-varying regression model). Similar comments as above apply for the behaviour of $\mathbf{Z}_t$ for extreme values of δ (at zero and at one). By specifying $\mathbf{Z}_t$ using a discount factor, we reduce a problem of specification of $p(p+1)/2$ elements to just a single element (the discount factor). Furthermore, we automatically make $\mathbf{Z}_t$ time-varying, since it depends on the time-varying component $\mathbf{F}_t\mathbf{P}_{t-1|t-1}\mathbf{F}_t^\top$, which is the covariance matrix of $\mathbf{F}_t\beta_{t-1|t-1}$, given information $y_{1:t-1}$. Unlike the maximum likelihood approach of the previous section (which requires a full set of data $y_{1:n}$), the approach using discount factors makes the specification of $\mathbf{Z}_t$ suitable for on-line application (if δ is chosen). In the implementation of complex state space models, more than one discount factor may be used (for an example of this see Sect. 4.1.4); for more information about such models and the concept of discounting in the specification of $\mathbf{Z}_t$ the reader is referred to West and Harrison (1997) and to references therein.

4.3.3 Estimation of σ^2: Conjugate Bayesian Estimation

In the previous sections the hyperparameter vector θ is assumed implicitly to be a vector of constants, subject to estimation. Within a Bayesian framework, one may assume that θ is a random vector and a priori distribution may be chosen for θ, with density function say $p(\theta)$. Then, given a set of observed data $y_{1:n} = \{y_1, \ldots, y_n\}$, the posterior $p(\theta \mid y_{1:n})$ is provided by the formula

$$p(\theta \mid y_{1:n}) \propto p(y_{1:n} \mid \theta)p(\theta) = p(y_1, \ldots, y_n \mid \theta)p(\theta)$$
$$= p(y_n \mid \theta, y_{1:n-1})p(y_1, \ldots, y_{n-1} \mid \theta)p(\theta)$$
$$= p(y_n \mid \theta, y_{1:n-1})p(\theta \mid y_{1:n-1}).$$

This formula may be applied sequentially. At time $t = 1$ starting at a prior $p(\theta)$, the posterior density $p(\theta \mid y_1)$ is proportional to $p(y_1 \mid \theta)$ (the likelihood using the single observation y_1) times the prior $p(\theta)$. Then at $t = 2$, the posterior density $p(\theta \mid y_{1:2})$ is proportional to $p(y_2 \mid \theta, y_1)$ (the one-step forecast density which, conditional on θ, is available from the Kalman filter) times the posterior at time $t = 1$, $p(\theta \mid y_1)$, which is equal to the prior distribution of θ at time $t = 2$ (i.e. before observing y_2). This process is repeated over time, giving the sequential updating

$$p(\theta \mid y_{1:t}) \propto p(y_t \mid \theta, y_{1:t-1})p(\theta \mid y_{1:t-1}), \tag{4.54}$$

for any $t = 1, \ldots, n$. In Eq. (4.54) $p(y_t \mid \theta, y_{1:t-1})$ is the one-step forecast distribution of y_t, given θ, which is provided by the Kalman filter and is a Gaussian distribution, and $p(\theta \mid y_{1:t-1})$ is the posterior distribution of θ at time $t - 1$, which is provided by Eq. (4.54), if we start from an initial prior $p(\theta)$.

The above framework can only provide the posterior distribution of θ up to a proportionality constant. This proportionality constant involves integration over complex non-linear functions and usually in high dimensions. Typically, for most cases, this means no closed calculations and thus for the evaluation of $p(\theta \mid y_{1:t})$, one may have to resort to simulation-based methods, such as Markov chain Monte Carlo or particle filtering, or approximations. Such approaches are discussed in many textbooks, see e.g. Gamerman and Lopes (2006) and Petris et al. (2009, Section 4.4).

In this section, in the very special, but important, case of $\theta = 1/\sigma^2$, we discuss conjugate Bayesian estimation, which does not need to rely on simulation or numerical approximations.

We start by considering state space model (3.10a)–(3.10b), where $\sigma^2 = \mathrm{Var}(\epsilon_t)$ is subject to estimation and the transition covariance matrix is scaled by σ^2, so that $\mathrm{Var}(\zeta_t) = \mathbf{Z}_t = \sigma^2 \mathbf{Z}_t^*$ for known covariance matrix $\mathbf{Z}_t^*$. Thus, the current working model can be written as

$$y_t = x_t^\top \beta_t + \epsilon_t, \qquad \epsilon_t \sim N(0, \sigma^2), \tag{4.55a}$$

$$\beta_t = \mathbf{F}_t \beta_{t-1} + \zeta_t, \quad \zeta_t \sim N(0, \sigma^2 \mathbf{Z}_t^*). \tag{4.55b}$$

Furthermore, the prior (initial) covariance matrix of β_0 is also scaled by σ^2, i.e. $\mathrm{Var}(\beta_0) = \sigma^2 \mathbf{P}_{0|0}^*$, where $\mathbf{P}_{0|0}^*$ is assumed known. This model, which is known as *scaled observational model*, is described in some detail in West and Harrison (1997, Section 4.5); some generalisations of this model are given in Triantafyllopoulos and Harrison (2008).

Define $\theta = 1/\sigma^2$ and assume that initially θ follows a gamma distribution with parameters $n_0/2$ and $d_0/2$, written as

$$\theta = \frac{1}{\sigma^2} \sim G\left(\frac{n_0}{2}, \frac{d_0}{2}\right). \tag{4.56}$$

Suppose that at time $t - 1$, the posterior distribution of θ is

$$\theta \mid y_{1:t-1} \sim G\left(\frac{n_{t-1}}{2}, \frac{d_{t-1}}{2}\right), \tag{4.57}$$

for some known values of n_{t-1} and d_{t-1}.

Note that conditionally on θ (or on σ^2), from the Kalman filter (see Theorem 3.2 in Sect. 3.2) it is $y_t \mid \theta, y_{1:t-1} \sim N(\hat{y}_{t|t-1}, \sigma^2 q_{t|t-1}^*)$, with $\hat{y}_{t|t-1}$ as in the above theorem and $q_{t|t-1}^* = x_t^\top \mathbf{P}_{t|t-1}^* x_t + 1$. Thus, conditionally on σ^2 all recursions of $\sigma^2 \mathbf{P}^*_{t|t}$ and $\sigma^2 \mathbf{P}^*_{t|t-1}$ follow from an application of the Kalman filter, where the recursions of $\mathbf{P}^*_{t|t}$ and $\mathbf{P}^*_{t|t-1}$ are provided by Theorem 3.2.

Then, applying formula (4.54) we obtain

$$p(\theta \mid y_{1:t}) \propto p(y_t \mid \theta, y_{1:t-1}) p(\theta \mid y_{1:t-1})$$

$$\propto \sqrt{\theta} \exp \left[-\frac{(y_t - \hat{y}_{t|t-1})^2 \theta}{2q^*_{t|t-1}} \right] \theta^{n_{t-1}/2-1} \exp \left(-\frac{d_{t-1}\theta}{2} \right)$$

$$= \theta^{(n_{t-1}+1)/2-1} \exp \left\{ -\frac{1}{2} \left[\frac{(y_t - \hat{y}_{t|t-1})^2}{q^*_{t|t-1}} + d_{t-1} \right] \theta \right\},$$

so that $p(\theta \mid y_{1:t})$ is proportional to a gamma distribution with parameters $n_t/2$ and $d_t/2$, or

$$\theta \mid y_{1:t} \sim G \left(\frac{n_t}{2}, \frac{d_t}{2} \right), \tag{4.58}$$

where

$$n_t = n_{t-1} + 1 \quad \text{and} \quad d_t = d_{t-1} + \frac{e_t^2}{q^*_{t|t-1}}, \tag{4.59}$$

and as usual $e_t = y_t - \hat{y}_{t|t-1}$ is the residual at time t.

Equations (4.56), (4.57) and (4.58) prove by induction the gamma posterior distribution (4.58) of $\theta \mid y_{1:t}$. Indeed, for $t = 1$, (4.57) is just the assumed prior (4.56).

Considering estimation of β_t, we first observe that conditionally on θ, the posterior distribution of β_t is given by $\beta_t \mid \theta, y_{1:t} \sim N(\hat{\beta}_{t|t}, \sigma^2 \mathbf{P}^*_{t|t})$, where $\hat{\beta}_{t|t}$ and $\mathbf{P}^*_{t|t}$ are provided by the Kalman filter (see Theorem 3.2 in Section 3.2). Consequently, the posterior distribution of β_t is found by integrating out θ from the joint distribution of β_t and θ, i.e.

$$p(\beta_t \mid y_{1:t}) = \int_{\mathbb{R}^p} p(\beta_t, \theta \mid y_{1:t}) \, d\theta$$

$$= \int_{\mathbb{R}^p} p(\beta_t \mid \theta, y_{1:t}) p(\theta \mid y_{1:t}) \, d\theta$$

$$\propto \int_{\mathbb{R}^p} \theta^{p/2} \exp \left[-\frac{1}{2} (\beta_t - \hat{\beta}_{t|t})^\top \mathbf{P}^{*-1}_{t|t} (\beta - \hat{\beta}_{t|t}) \theta \right]$$

$$\times \theta^{n_t/2-1} \exp \left(-\frac{d_t \theta}{2} \right) d\theta$$

$$= \int_{\mathbb{R}^p} \theta^{(n_t+p)/2-1}$$

$$\times \exp\left\{-\frac{1}{2}\left[(\beta_t - \hat{\beta}_{t|t})^\top \mathbf{P}_{t|t}^{*-1}(\beta_t - \hat{\beta}_{t|t}) + d_t\right]\theta\right\} d\theta.$$

Recall that if a random variable X follows a gamma distribution, $X \sim G(\alpha, \beta)$, then

$$\int_0^\infty x^{\alpha-1} \exp(-\beta x)\, dx = \frac{\Gamma(\alpha)}{\beta^\alpha}. \tag{4.60}$$

By applying this formula for $\alpha = (n_t + p)/2$ and $\beta = 2^{-1}(\beta_t - \hat{\beta}_{t|t})^\top \mathbf{P}_{t|t}^{*-1}(\beta_t - \hat{\beta}_{t|t}) + d_t$, we have that

$$p(\beta_t \mid y_{1:t}) \propto \left[(\beta_t - \hat{\beta}_{t|t})^\top \mathbf{P}_{t|t}^{*-1}(\beta_t - \hat{\beta}_{t|t}) + d_t\right]^{-(n_t+p)/2}.$$

By defining $S_t = d_t/n_t$, the above density is proportional to a multivariate t distribution $\beta_t \mid y_{1:t} \sim t(n_t, \hat{\beta}_{t|t}, \mathbf{P}_{t|t})$, with n_t degrees of freedom, location vector $\hat{\beta}_{t|t}$ and scale matrix $\mathbf{P}_{t|t}$, where $\mathbf{P}_{t|t} = S_t \mathbf{P}_{t|t}^*$. Similarly, it can be established that $\beta_t \mid y_{1:t-1} \sim t(n_{t-1}, \hat{\beta}_{t|t-1}, \mathbf{P}_{t|t-1})$ and $y_t \mid y_{1:t-1} \sim t(n_{t-1}, \hat{y}_{t|t-1}, q_{t|t-1})$, with $\mathbf{P}_{t|t-1} = S_{t-1}\mathbf{P}_{t|t-1}^*$, $q_{t|t-1} = S_{t-1}q_{t|t-1}^* = x_t^\top \mathbf{P}_{t|t-1}x_t + S_{t-1}$ and as mentioned earlier the recursions of $\mathbf{P}_{t|t}^*$, $\mathbf{P}_{t|t-1}^*$ and $q_{t|t-1}^*$ are provided by the Kalman filter. This shows that

$$\mathbf{P}_{t|t} = S_t \mathbf{P}_{t|t}^* = \frac{S_t}{S_{t-1}}(\mathbf{P}_{t|t-1} - K_t K_t^\top q_{t|t-1}).$$

Furthermore, we can write $d_t = n_t S_t$ and from (4.59) we obtain

$$n_t S_t = n_{t-1}S_{t-1} + \frac{S_{t-1}e_t^2}{q_{t|t-1}},$$

which can be simplified further to give

$$n_t S_t = n_{t-1}S_{t-1} + r_t e_t, \tag{4.61}$$

where $r_t = y_t - x_t^\top \hat{\beta}_{t|t}$. To establish this recursion, write

$$r_t = y_t - x_t^\top \hat{\beta}_{t|t} = y_t - x_t^\top \hat{\beta}_{t|t-1} - x_t^\top K_t e_t = (1 - x_t^\top K_t)e_t$$

and observe that

$$1 - x_t^\top K_t = 1 - x_t^\top \mathbf{P}_{t|t-1}x_t/q_{t|t-1} = \frac{q_{t|t-1} - x_t^\top \mathbf{P}_{t|t-1}x_t}{q_{t|t-1}} = \frac{S_{t-1}}{q_{t|t-1}}.$$

Some comments are in order.

- First note that since $1/\sigma^2 \mid y_{1:t} \sim G(n_t/2, d_t/2)$, it follows that, given $y_{1:t}$, σ^2 follows an inverse gamma distribution with scale and shape parameters $n_t/2$ and $d_t/2$, i.e. $\sigma^2 \mid y_{1:t} \sim IG(n_t/2, d_t/2)$. A point estimate of σ^2 can be taken as the mode of $IG(n_t/2, d_t/2)$, which from the properties of the inverse gamma distribution (see Chap. 2) is

$$\hat{\sigma}_t^2 = \text{mode}(\sigma^2 \mid y_{1:t}) = \frac{d_t/2}{n_t/2 + 1} = \frac{n_t S_t}{n_t + 2}. \tag{4.62}$$

  We note that for large t, the ratio $n_t/(n_t + 2)$ is close to one, hence $\hat{\sigma}_t^2 \approx S_t$; the posterior distribution of β_t and the forecast distribution of y_t are approximately Gaussian.
- For the estimation of σ^2 formula (4.61) coincides with the formula providing maximum likelihood estimation in recursive least squares (see Theorem 3.1 in Sect. 3.1.2). Obviously, (4.61) refers to a more general model (e.g. when $\mathbf{F}_t$ is not equal to the identity matrix). Moreover, the above approach of Bayesian estimation for σ^2 provides much more information than maximum likelihood estimation, e.g. it offers the availability of credible bounds based on the gamma distribution.
- It is worthwhile pointing out that using Eq. (4.61) the point estimate $\hat{\sigma}_t^2$ can be calculated recursively.
- We can see that the evaluation of the Kalman gain is unaffected by the estimation, i.e. $K_t^* = K_t$. Indeed

$$K_t^* = \frac{\mathbf{P}_{t|t-1}^* x_t}{q_{t|t-1}^*} = \frac{S_{t-1}\mathbf{P}_{t|t-1}^* x_t}{S_{t-1}q_{t|t-1}^*} = \frac{\mathbf{P}_{t|t-1} x_t}{q_{t|t-1}} = K_t.$$

Below is a summary of the algorithm.

Scaled Observational Precision (SOP)

1. Prior distributions at $t = 0$:
   $\beta_0 \sim t(n_0, \hat{\beta}_{0|0}, \mathbf{P}_{0|0})$ and $\sigma^{-2} \sim G(n_0/2, n_0 S_0/2)$;
2. Posterior distribution of β_{t-1} at time $t - 1$:
   $\beta_{t-1} \mid y_{1:t-1} \sim t(n_{t-1}, \hat{\beta}_{t-1|t-1}, \mathbf{P}_{t-1|t-1})$;
3. Prior distribution of β_t at time t:
   $\beta_t \mid y_{1:t-1} \sim t(n_{t-1}, \hat{\beta}_{t|t-1}, \mathbf{P}_{t|t-1})$,
   where $\hat{\beta}_{t|t-1} = \mathbf{F}_t \hat{\beta}_{t-1|t-1}$ and $\mathbf{P}_{t|t-1} = \mathbf{F}_t \mathbf{P}_{t-1|t-1} \mathbf{F}_t^\top + \mathbf{Z}_t$;

(continued)

4. Posterior distributions at time t:
   $\beta_t \mid y_{1:t} \sim t(n_t, \hat{\beta}_{t|t}, \mathbf{P}_{t|t})$ and $\sigma^{-2} \mid y_{1:t} \sim G(n_t/2, n_t S_t/2)$, where

$$\hat{\beta}_{t|t} = \hat{\beta}_{t|t-1} + K_t e_t, \quad \mathbf{P}_{t|t} = \frac{S_t}{S_{t-1}}\left(\mathbf{P}_{t|t-1} - q_{t|t-1} K_t K_t^{\top}\right),$$

$$n_t = n_{t-1} + 1 \quad \text{and} \quad n_t S_t = n_{t-1} S_{t-1} + r_t e_t,$$

$$\hat{y}_{t|t-1} = x_t^{\top} \hat{\beta}_{t|t-1}, \quad e_t = y_t - \hat{y}_{t|t-1}, \quad r_t = y_t - x_t^{\top} \hat{\beta}_{t|t},$$

$$K_t = \mathbf{P}_{t|t-1} x_t / q_{t|t-1} \quad q_{t|t-1} = x_t^{\top} \mathbf{P}_{t|t-1} x_t + S_{t-1}.$$

We note that in the application of the above algorithm $\mathbf{P}_{0|0}$ and $\mathbf{Z}_t$ are specified without any reference to $\mathbf{P}_{0|0}^*$, $\mathbf{Z}_t^*$ and to S_0. Without loss of generality, $\mathbf{P}_{0|0}$ is set, as in the Kalman filter algorithm (see also Sect. 4.5) and $\mathbf{Z}_t$ is set via maximum likelihood or via discount factors. It turns out that the above approach to the estimation of σ^2 combined with specifying $\mathbf{Z}_t$ by forgetting factors as in Sect. 4.3.2 provides an attractive methodology for estimating/specifying the hyperparameters σ^2 and $\mathbf{Z}_t$. This is illustrated in the example below.

Example 4.8 (Annual Temperature Example Continued)   We consider again the annual temperatures in England, discussed in Example 3.1. In that example a local level model, given by Eqs. (3.13a) and (3.13b), was fitted with a somewhat arbitrary choice of $\sigma^2 = 1$ and $Z = 10$. In this section we refit the model using the observational scaled model, for the estimation of σ^2, and employing forgetting factors to specify Z_t.

To implement the model in R, we used the function sop.ss available from the book website.

```
> # read data
> # need temp
> # fit SOP model with delta=0.95
> fit1 <- bts.filter(temp, x0=1, F0=1, delta=0.95,
+ beta0=9, P0=1000, n0=1/100, S0=1)
> # fit SOP model with delta=0.8
> fit2 <- bts.filter(temp, x0=1, F0=1, delta=0.8,
+ beta0=9, P0=1000, n0=1/100, S0=1)

> # time series plot
> pred1 <- ts(fit1$FittedMean, start=1659, frequency=1)
> pred2 <- ts(fit2$FittedMean, start=1659, frequency=1)
> tempts<-ts(temp, start=1659, frequency=1)
> par(mfrow=c(2,1))
> ts.plot(tempts,main=expression("SOP model
+ with discount factor 0.95"), xlab="Year",
+ ylab="Degrees in Celsius")
> lines(pred1, lty=2,lwd=2, col=2)
> legend("bottomright",c("Observations","Forecast mean"),
```

```
+ pch=c(20, 1), col=c(1,2))
> ts.plot(tempts,main=expression("SOP model
+ with discount factor 0.8"), xlab="Year",
+ ylab="Degrees in Celsius")
> lines(pred2, lty=2,lwd=2, col=2)
> legend("bottomright",c("Observations","Forecast mean"),
+ pch=c(20, 1), col=c(1,2))
```

Figure 4.12 shows the one-step forecast means $\hat{y}_{t|t-1}$ (for values of the discount factor $\delta = 0.95$ and $\delta = 0.8$) against the actual data. We observe that when $\delta = 0.95$, the forecasts are smooth, not being able to follow closely the dynamics of the time series; for $\delta = 0.8$, the forecasts are more adaptive following better the local fluctuations of the time series. Figure 4.13 shows the estimate S_t of σ^2, plotted against the actual time series data of the temperatures. We observe that at the start S_t fluctuates much more than in later years, for which it is remarkably small. We also observe that the evolution of S_t follows fluctuations of the observed data (as shown in the top panel of Fig. 4.13), e.g. near the year 1750 there seems to be an outlier in the data and this is reflected in S_t by a high estimated observation variance. This pinpoints one of the advantages of the SOP model: unlike the maximum likelihood

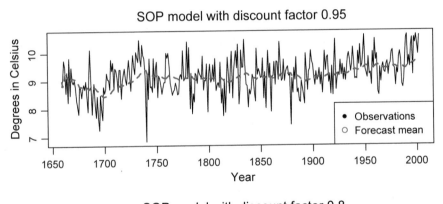

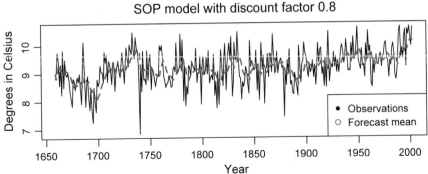

Fig. 4.12 One-step forecast means (dashed lines) of temperature values of central England (solid lines)

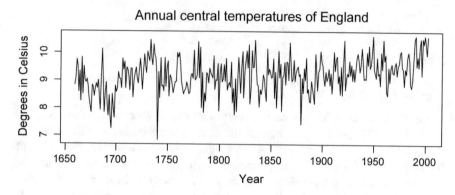

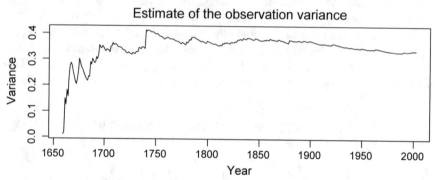

Fig. 4.13 One-step forecasts of the observation variance of central England temperatures

estimation, it can capture local effects in the variance as well as in the mean of
the data, while the likelihood approach (similarly as in least squares estimation)
only captures an overall optimum model performance. The R code for producing
Fig. 4.13 is

```
> par(mfrow=c(2,1))
> ts.plot(tempts, xlab="Year", ylab="Degrees in Celsius",
+ main=expression("Annual central temperatures of England"))
> var1 <- ts(fit1$ObsVar, start=1659, frequency=1)
> ts.plot(var1, xlab="Year", ylab="Variance",
+ main=expression("Estimate of the observation variance"))
```

4.4   Error Analysis

In the above sections we have seen how a state space model can be built, how the
design vector x_t and the transition matrix $\mathbf{F}_t$ can be chosen and how the observation
variance σ^2 and the transition covariance matrix $\mathbf{Z}_t$ may be estimated or specified.
Furthermore, we have seen how hyperparameters included in these models may be

estimated. In this section we deal with the goodness of fit of a chosen model and with the closely related area of model choice.

We start by exploring the goodness of fit problem, known variously as *error analysis* or as *residual analysis*. We will suppose we have specified a state space model (3.10a)–(3.10b) and that the above model components $(x_t, \mathbf{F}_t, \sigma^2, \mathbf{Z}_t)$ have been chosen or estimated, based on an observed data set $y_{1:t} = \{y_1, \dots, y_{1:n}\}$, for some positive integer n. We will also assume that the prior mean vector $\hat{\beta}_{0|0}$ and covariance matrix $\mathbf{P}_{0|0}$ (and perhaps the prior of $1/\sigma^2$) have been chosen or estimated; for a related discussion see Sect. 4.5. Error analysis attempts to answer the question of how well the model explains the data. The model performance is usually understood in terms of forecast accuracy. If the model is a good fit to the data, then the forecasts should be close to the observed data. Consequently, measures of goodness of fit are usually based on the residuals or one-step forecast errors, defined as the difference of the one-step forecasts $\hat{y}_{t|t-1}$ from the observations y_t, i.e.

$$e_t = y_t - \hat{y}_{t|t-1}, \tag{4.63}$$

Hence their analysis (residual or error analysis) is used to assess the goodness of model fit.

The following theorem provides some properties of the residuals.

Theorem 4.2 *In the state space model* (3.10a)–(3.10b) *and with the definitions of the Kalman filter recursions (Theorem 3.2) for some data* $y_{1:n} = \{y_1, \dots, y_n\}$, *the following apply for the residuals.*

1. *Given* $y_{1:t-1}$, *the distribution of the residuals is* $e_t \mid y_{1:t-1} \sim N(0, q_{t|t-1})$, *where* $q_{t|t-1}$ *is the forecast variance of* y_t, *provided by the Kalman filter.*
2. $e_1, \dots, e_n$ *are independent.*
3. *With the definitions of* e_t *and* $q_{|t-1}$ *as above,*

$$\sum_{t=1}^{n} \frac{e_t^2}{q_{t|t-1}} \sim \chi_n^2,$$

where χ_n^2 *denotes the chi-square distribution with n degrees of freedom.*

Proof (1) follows by noting that the one-step forecast distribution of y_t is $y_t \mid y_{1:t-1} \sim N(\hat{y}_{t|t-1}, q_{t|t-1})$, see Theorem 3.2.

For (2), first we prove that e_t is uncorrelated with any function $h_t = h(y_1, \dots, y_{t-1})$ of $y_1, \dots, y_{t-1}$. From (1) we have $E(e_t \mid y_{1:t-1}) = 0$, and so using the tower property (5a) we have $E(e_t) = 0$. Then using again the tower property

$$\text{Cov}(e_t, h_t) = E(e_t h_t) = E[E(e_t h_t \mid y_{1:t-1})] = E[h_t E(e_t \mid y_{1:t-1})] = 0.$$

Note that from (4.63), e_t is a function of $y_t, y_{t-1}, \ldots, y_1$, since $\hat{y}_{t|t-1}$ is a function of $y_{t-1}, \ldots, y_1$. Thus, by replacing $h_t = e_{t-s}$, for any $s = 1, \ldots, t-1$, we have that e_t and e_{t-s} are uncorrelated, and since from (1) they are normally distributed, they are also independent. By repeating this result, for each t, the independence of $e_1, \ldots, e_t$ follows.

From (1) we have that $e_t/\sqrt{q_{t|t-1}} \sim N(0,1)$ and so $e_t^2/q_{t|t-1} \sim \chi_1^2$. Since $e_1, \ldots, e_n$ are independent from (2), the sum $\sum_{t=1}^{n} e_t^2/q_{t|t-1}$ follows the stated chi-square distribution.   □

The above theorem provides important information that can be used to form an error analysis. The following gives a summary: From (1), the mean of e_t is zero. So an informal criterion of goodness of fit is to plot the residuals over time. If the fit is good, the residuals should fluctuate around 0. But because the variance of e_t is $q_{t|t-1}$, which can vary over time, it is better to work with the *standardised residuals*, defined as

$$e_t^* = \frac{e_t}{\sqrt{q_{t|t-1}}},$$

which, from the proof of (3) in Theorem 4.2, follow a $N(0,1)$ standard normal distribution. Again a plot of e_t^* should not reveal any structure, e_t^* should fluctuate around zero having variance 1. So, in order to assess the goodness of fit, we have the following:

- Plot e_t^* against 95% credible bounds (or more generally, against $(1-\alpha)\%$ credible bounds). For a good fit, approximately 5% or $\alpha\%$ of e_t should lie outside the bounds.
- Compute the mean of squared residuals or errors (MSE), the mean of squared standardised residuals or errors (MSSE), and the mean absolute deviation (MAD), defined by

$$\text{MSE} = \frac{1}{n}\sum_{t=1}^{n} e_t^2, \quad \text{MSSE} = \frac{1}{n}\sum_{i=1}^{n} e_t^{*2}, \quad \text{MAD} = \frac{1}{n}\sum_{t=1}^{n} |e_t|,$$

respectively. For a good fit, MSE and MAD should be close to 0, while MSSE should be close to 1. The MSE and MAD have been used extensively in signal processing (Haykin, 2001) and in econometrics (Harvey, 1989), but they do not take into account the variance of the forecasts $q_{t|t-1}$. MSSE does take this variance into account, but it is usually harder to interpret it, in particular it is hard to distinguish the effects of the residual mean from the residual variance. For example, if the value of MSSE is 1.5 (higher than 1), we do not know whether this is due to high average residuals or too low variance $q_{t|t-1}$. Likewise, it is hard to interpret differences between say values of MSSE equal to 0.8 and 1.2. As a result, usually MSSE is presented together with MSE or MAD. Sometimes MSSE is avoided, because it is very hard to forecast accurately the variance and its mis-specification may lead to several types of bias.

- Since the residuals $e_1, \ldots, e_n$ are independent, see (2) of Theorem 4.2, another goodness of fit tool is to assess whether they are not correlated or autocorrelated. This can be done by plotting the autocorrelation function (ACF), for several lags (this plot is known as a *correlogram*), for a discussion of the ACF and its theoretical properties the reader is referred to Box et al. (2008) or to Shumway and Stoffer (2017). In short, the ACF is a sequence, each value of which comprises the sample correlations of e_{t+k} and e_t, for a lag $k = 1, 2, \ldots$. If the residuals are uncorrelated, all values of the ACF should lie within $\pm 1.96/\sqrt{n}$. More formal tests exist and they can be found in the above references. In R, the ACF and the correlogram plot can be obtained by using the function $\texttt{acf}$.
- Finally, based on part (3) of Theorem 4.2, we can assess whether overall the residuals are low to see whether the sum of $(e_t^{*2}$ follows a chi-square distribution with n degrees of freedom, for example one may compare the observed sum with the 95% quantile of χ_n^2. This test is relevant to the popular portmanteau test of Box and Pierce (see Box et al. (2008)), aimed at ARMA models.

Example 4.9 (Turkey Example Continued)   We revisit Example 4.1.4 of the turkey sales data. We are assuming that the trend-seasonal model described in that example is fitted in R and the objects $\texttt{fit\$f}$ and q (referring to $\hat{y}_{t|t-1}$ and $q_{t|t-1}$, respectively) are obtained (see p. 134 for the R code). Here, we examine the quality of the fit, using the error analysis described above. The residuals and standardised residuals are computed in R using the commands

```
> # need turkey (for the data) and object fit (fitted Kalman filter)
> pred <- ts(fit$FittedMean,start=c(1974,1), frequency=4)
> predVar <- ts(fit$FittedVar, start=c(1974,1), frequency=4)
> # define residuals
> e <- q <- rep(NA,35)
> for(t in 1:35){
+ e[t]  <- turkey[t]-fit$FittedMean[[t]]
+ q[t]  <- fit$FittedVar[[t]]
+ }
> estar <- e/sqrt(q)
```

The left panel of Fig. 4.14 plots the standardised residuals e_t^* together with the 95% credible bound at ± 1.96 of the standard normal distribution. Moving on to the independence check for the residuals, the left Fig. 4.14 is the correlogram of the standardised residuals. We observe that all values of the ACF (the value of the ACF at lag $k = 0$ is always equal to 1) are within the 95% credible bounds, and so we conclude that the standardised residuals appear to be uncorrelated.

The R code used for Fig. 4.14 is

```
> # plot of the stand residuals
> par(mfrow=c(1,2))
> ts.plot(estar, main=expression("Standardized residuals
+ for the turkey data"),xlab="Quarters",ylab="Residual")
> points(estar, pch=20)
> abline(h=1.96, lty=3)
> abline(h=-1.96, lty=3)
> acf(estar,main=expression("ACF of the standardized residuals"))
```

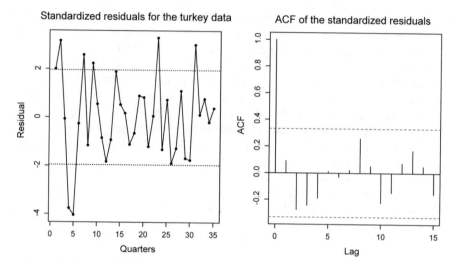

Fig. 4.14 Residuals (left panel) and correlogram of the residuals (right panel) for the turkey data

We can see that the mean of the residuals is close to 0, but the variance is above 3, e.g. in R

```
> mean(estar)
[1] 0.01101288
> var(estar)
[1] 3.143806
```

The proportion of outliers (exceeding the 95% credible bounds of $N(0, 1)$) is 22.86%, which is too high (normally we would expect only a 5% of e_t^* to exceed the bounds). This can be checked easily in R by using the command

```
> # proportion of outliers of residuals
> length(estar[abs(estar)>1.96])/length(estar)
[1] 0.2285714
```

As we observe in Fig. 4.14 most of the outliers appear in the start of the data. One option would be to discard the first say 10 observations, accounting them to the learning stage of the model. If we apply this, the proportion of outliers falls to 8%, still quite high. This may be an indicator that the forecast variance $q_{t|t-1}$ is underestimated (providing a smaller divider in e_t^* than it should be). In turn, this could be the cause of a mis-specification of the observation variance σ^2. If, for example σ^2 is underestimated, this will result in underestimation of $q_{t|t-1}$.

The MSE, MSSE and MAD are computed as

```
> # MSE
> mean(e^2)
[1] 11196.55
> # MSSE
> mean(estar^2)
```

```
[1] 3.054104
> # MAD
> mean(abs(e))
[1] 76.38005
```

We note the MSE and MAD appear to be too large and it is hard to make any interpretation, based only on this. This is because they both depend on the scale of the observations $\{y_t\}$. The MSSE does not suffer from this effect, but here it appears to be very high, reflecting the miss-specification or underestimation of $q_{t|t-1}$.

Finally, checking overall the residuals we have that $\sum_{t=1}^{35}(e_t^*)^2 = 106.111$, while the one-sided 95% quantile of the chi-square distribution with 35 degrees of freedom is only 49.802 (provided by the command qchisq(0.95, 35)). This implies that the overall quality of fit is poor, producing some large absolute residuals. When referenced against χ_{25}^2 (excluding the first 10 residuals as suggested earlier), the result is clearly significant at conventional levels—the p-value is 0.00336.

The above error analysis and model fit assessment are based on the assumptions of Theorem 4.2, which again adopts the assumptions of the Kalman filter algorithm (see Theorem 3.2 in Sect. 3.2). In particular, the normality assumption is used in all parts (1)–(3) in Theorem 4.2. If model hyperparameters are estimated, e.g. via maximum likelihood or via conjugate Bayesian estimation, then the normality of the residuals e_t is either approximate, or it is lost. When we estimate the hyperparameters using the maximum likelihood principle, we can consider the residuals conditional on the maximum likelihood estimates. A similar approach can be followed when we specify the transition matrix Z_t using forgetting factors. When we use the Bayesian conjugate inference for the estimation of the observation variance σ^2, we know that, unconditionally on σ^2, the forecast distribution of y_t is a Student t distribution, i.e. $y_t \mid y_{1:t-1} \sim t(n_{t-1}, \hat{y}_{t|t-1}, q_{t|t-1})$, where n_{t-1} are the degrees of freedom, $\hat{y}_{t|t-1}$ is the mode and $q_{t|t-1}$ is the scale of the above t distribution, see also Sect. 4.3.3. We note that in this case $\hat{y}_{t|t-1}$ is the forecast mean (since the mode is the same as the mean of the t distribution), but the forecast variance is $n_{t-1}(n_{t-1} - 2)^{-1}q_{t|t-1}$ and not $q_{t|t-1}$ (as it is in the Kalman filter).

Based on the above, related to parts (1)–(2) of Theorem 4.2, we have

1. The distribution of the residual e_t is $e_t \mid y_{1:t-1} \sim t(n_{t-1}, 0, q_{t|t-1})$.
2. The residuals $e_1, \ldots, e_n$ are uncorrelated.

(1) follows immediately from the definition of y_t and the t distribution of $y_t \mid y_{1:t-1}$. Thus, the standardised residual $e_t^* = e_t/\sqrt{q_{t|t-1}}$ follows a standard Student t distribution with n_{t-1} degrees of freedom, i.e. $e_t^* \mid y_{1:t-1} \sim t(n_{t-1}, 0, 1)$. This implies that $E(e_t^* \mid y_{1:t-1}) = 0$ and $Var(e_t^* \mid y_{1:t-1}) = n_{t-1}(n_{t-1} - 2)^{-1}$, for $n_{t-1} > 2$, or infinity otherwise. Thus, the three performance measures MSE, MAD and MSSE are computed as above, but now the value of MSSE should be compared to $n_{t-1}(n_{t-1} - 2)^{-1}$, instead of 1, which was the case for the normal distribution discussed above. From the distribution of e_t^* we can calculate $(1 - \alpha)\%$ credible bounds of e_t^*, given by $\pm t_{n_{t-1}, 1-\alpha/2}$, where $t_{n_{t-1}, 1-\alpha/2}$ is the $(1 - \alpha/2)$-quantile of

the t distribution for n_{t-1} degrees of freedom. In R we can use the function qt for the quantile of the t distribution, e.g. for $n_{t-1} = 10$ and $\alpha = 0.05$, we have

```
> qt(df=10, 1-0.05/2)
[1] 2.228139
```

so that a 95% credible bound for e_t^* is $[-2.228, 2.228]$. One point to note is that in this case the credible bounds for the standardised residuals are time-varying, starting at wider intervals and converging to the time-invariant bounds of the normal distribution, e.g. we can check that $t_{100000,1-0.05/2} = 1.959988$, which is very close to 1.959964, the respective quantile using the normal distribution, provided in R by qnorm(1-0.05/2). We observe that for early points of time the credible bounds are much wider using the Student t distribution, and this reflects the additional uncertainty in the early period in the estimation of σ^2. With the course of time, more data become available, and the estimation of σ^2 becomes more accurate, and thus the credible bounds approach the limiting credible bounds of the normal distribution. This is theoretically justified as the probability density function of the Student t distribution converges to that of the normal distribution as $t \to \infty$ or $v_t \to \infty$. The proof of (2) is the same as that in Theorem 4.2.

4.5 Prior Specification

The estimation procedures described in Chaps. 3 and 4, namely the Kalman filter, smoothing, forecasting and the algorithm of the SOP model, require us to specify values of the elements of $\hat{\beta}_0$ and $\mathbf{P}_{0|0}$ in the prior state distribution

$$\beta_0 \sim N(\hat{\beta}_{0|0}, \mathbf{P}_{0|0}), \tag{4.64}$$

as well as values of n_0 and d_0 in the prior distribution of $\theta = 1/\sigma^2$

$$\theta \sim G\left(\frac{n_0}{2}, \frac{d_0}{2}\right). \tag{4.65}$$

The success of the above mentioned estimation algorithms depend in some degree on the choice of the above quantities. Their choice is known as *prior specification*. The problem of prior specification is also known as initialisation, a detailed account of which can be found in Durbin and Koopman (2012, Chapter 5). The main idea is that the initial state β_0 is composed of some stochastic components with known joint distribution and by some other elements that may be deterministic and not of interest. Then the Kalman filter and smoothing filters, presented in the previous chapter, can be re-formulated as functions of the above structure of the initial state β_0; for more details the reader is referred to Koopman (1997) and Durbin and Koopman (2012, Chapter 5). In our experience the influence of the priors above is small, in particular when there is a "reasonable" amount of data available. As we have shown in Theorem 3.7 (Sect. 3.5.3) for a wide class of models, for which the design vector $x_t = x$ and the transition matrix $\mathbf{F}_t = \mathbf{F}$ are time-invariant, the posterior covariance matrix $\mathbf{P}_{t|t}$ converges very rapidly (as $t \to \infty$) and the resulting limiting matrix is independent of the prior covariance matrix of β_0. In the sequel we describe practical specifications for β_0 and σ^2, and we illustrate their sensitivity.

4.5.1 Prior Specification of β_0

We begin with the specification of $\hat{\beta}_{0|0}$ and $\mathbf{P}_{0|0}$. $\hat{\beta}_{0|0}$ reflects the prior "beliefs" we may have on β_0, $\mathbf{P}_{0|0}$ on the associated uncertainty around this belief. For example, in the local level model of Sect. 3.1.3, β_t represents the (conditional) level of the time series y_t, i.e. $\beta_t = E(y_t \mid \beta_t)$, and so β_0 represents the level of data prior to $t = 1$, should this have been observed. Thus, it is reasonable to set $\hat{\beta}_{0|0}$ what we would expect the level of such "prior data" would be. This may be done using historical data, or simply $\hat{\beta}_{0|0}$ may represent our rough belief. In this case $P_{0|0}$ is a variance and setting $P_{0|0}$ close to 0 a strong statement about $\hat{\beta}_{0|0}$ is made, i.e. the uncertainty of β_0 around its mean $\hat{\beta}_{0|0}$ is small. This is sometimes expressed by defining the precision of β_0 as the inverse of $P_{0|0}$; a large precision $P_{0|0}^{-1}$ implies small uncertainty around $\hat{\beta}_{0|0}$. On the other hand, if we are not certain (if for example our prior beliefs are based on little prior knowledge or information), then $P_{0|0}$ should be large (or equivalently the precision $P_{0|0}^{-1}$ should be small). This is illustrated in Fig. 4.15, where the density functions of three normal distributions are plotted $\beta_0 \sim N(0, 0.025)$, $\beta_0 \sim N(0, 1)$ and $\beta_0 \sim N(0, 4)$. We notice that as

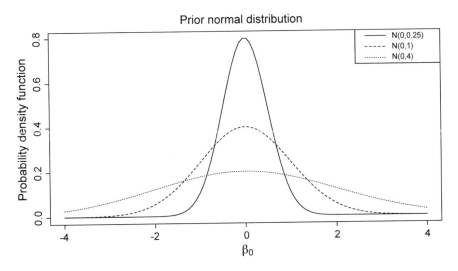

Fig. 4.15 Prior distribution of β_0

the variance gets smaller the density concentrates around the zero mean, but as the variance gets larger the uncertainty around the zero mean belief is increasing. In most examples considered in so far we have set the variance equal to 1000, to reflect a vague prior variance specification with increased uncertainty around $\hat{\beta}_{0|0}$ (it will be rare that we have precise prior information).

Moving on to other state space models, a similar approach is followed, i.e. $\hat{\beta}_{0|0}$ is a prior belief of what β_0 is expected or postulated to be (perhaps by using historical or past data) and $\mathbf{P}_{0|0}$ is set to be proportional to the identity, a popular choice being $\mathbf{P}_{0|0} = 1000\mathbf{I}$. This implies that the diagonal elements of $\mathbf{P}_{0|0}$, which are the prior variances of $\beta_{i,0}$ ($\beta_0 = [\beta_{1,0}, \ldots, \beta_{p,0}]^{\top}$), are very large (reflecting on high uncertainty on the specification of $\hat{\beta}_{i,0}$), while the off-diagonal elements of $\mathbf{P}_{0|0}$ are zero (reflecting a lack of information on the correlation of $\beta_{i,0}$ and $\beta_{j,0}$, for $i \neq j$). For example, in Example 4.2 of aluminium prices in Sect. 4.1.1, we used

$$\beta_{0|0} = \begin{bmatrix} 1800 \\ 1 \end{bmatrix} \quad \text{and} \quad \mathbf{P}_{0|0} = \begin{bmatrix} 1000 & 0 \\ 0 & 1000 \end{bmatrix}.$$

In this case, the level is $\beta_{1t} = \mathrm{E}(y_t \mid \beta_t)$, which a priori (at $t = 0$) is expected to be around \$1800, while $\beta_{2,0}$, which takes part in the evolution of the trend, see e.g. Eq. (4.1c) of the linear growth model, is set to 1. The uncertainty around this prior belief is set as high, by specifying that the variances of β_{1t} and β_{2t} are equal to 1000, while their covariance is zero (this is supported in the absence of information that would suggest $\beta_{1,0}$ and $\beta_{2,0}$ to be correlated). The value of 1000 is customary; any large number would suffice, such that $\mathbf{P}_{0|0}^{-1} \approx \mathbf{0}$ (precision matrix is near zero).

This approach of prior specification for $\hat{\beta}_{0|0}$ and $\mathbf{P}_{0|0}$ is known as *weakly informative* prior specification, as there is little information we feed in to the system, basically coming from $\hat{\beta}_{0|0}$. However, sometimes even this information may not be available. For example the practitioner may not have historical data or may be reluctant to specify such a prior. In this case, just setting $\hat{\beta}_{0|0} = 0$ will work well. The explanation is as follows. Suppose that the "true" value of $\hat{\beta}_{0|0} = c \neq 0$ is different from the zero vector and that we miss-specify it by "wrongly" setting $\hat{\beta}_{0|0} = 0$. At $t = 1$, after y_1 is observed, the "correct" posterior mean $\hat{\beta}_{1|1}$ is $\hat{\beta}_{1|1} = \mathbf{F}_1 c + K_1 e_1$. Now, $x_1^{\top} \mathbf{F}_1 c$ must be close to y_1 (since c is the true prior mean) and so $e_1 = y_1 - x_1^{\top} \mathbf{F}_1 c \approx 0$. Thus, $\hat{\beta}_{1|1} = \mathbf{F}_1 c$, and the Kalman gain K_t here does not play a crucial role as $e_1 \approx 0$. Now, if we set $\hat{\beta}_{0|0} = 0$, we have $\hat{\beta}_{1|1} = K_1 e_1 = K_1 y_1$, since $c = 0$ here. Thus, in this case our posterior mean is a linear function of y_1, and it will adapt to y_1 using the Kalman gain K_1. Projecting this to a few more observations, it follows that the posterior mean using the "wrong" prior mean will converge to that using the "correct" prior mean. In other words, the posterior mean $\hat{\beta}_{t|t}$ will quickly adapt to observed data and its dependence upon the prior will quickly diminish.

This effect is illustrated in Fig. 4.16, which plots the one-step forecast mean $\hat{y}_{t|t-1}$ of the data y_t (solid points), using the prior mean $\hat{\beta}_{0|0} = [1800, 1]^{\top}$ (solid line

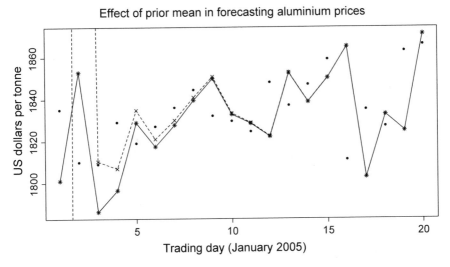

Fig. 4.16 Effect of the prior mean $\hat{\beta}_{0|0}$ in the forecasting of the aluminium prices

and stars) and the zero prior mean $\hat{\beta}_{0|0} = [0, 0]^\top$ (dashed line and ticks). We observe that initially the forecasts (using the zero prior) are very far from the respective forecasts using the "more informative prior" $\hat{\beta}_{0|0} = [1800, 1]^\top$. But, after only a few points, the two forecasts converge to each other (after time $t = 8$ the two forecasts are nearly inseparable). This is more clearly seen in the following table, which tabulates the forecasts of each prior together with the original data (Table 4.2).

Moving on to the specification of the prior covariance matrix $\mathbf{P}_{0|0}$, Fig. 4.17 shows the one-step forecast variance $q_{t|t-1}$ produced using $\mathbf{P}_{0|0} = 1000\mathbf{I}$ (solid line and points), $\mathbf{P}_{0|0} = \mathbf{I}$ (dashed line and ticks) and $\mathbf{P}_{0|0} = 0.001\mathbf{I}$ (dotted line and stars). This is a numerical justification of the convergence of $\mathbf{P}_{t|t}$ and its independence of $\mathbf{P}_{0|0}$ established in Theorem 3.7. We observe that the three variances converge very quickly: after $t = 8$, the three lines coincide. The conclusion is that in this example, convergence is remarkably fast, and effectively time points 1–9 could be considered as a training mode for the model to adapt. If this is considered so, the prior mean vector $\hat{\beta}_{0|0}$ and the prior covariance matrix $\mathbf{P}_{0|0}$ are not critical for the performance of the model. However, we note that the above convergence, which is a phenomenon that applies in all state space models with time-invariant components $x_t = x$, $\mathbf{F}_t = \mathbf{F}$, $\mathbf{Z}_t = \mathbf{Z}$ and σ^2 (see Theorem 3.7), will not apply generally, i.e. when these components are time-varying, as in the regression models of Sect. 4.1.5. In such situations, the priors $\hat{\beta}_{0|0}$ and $\mathbf{P}_{0|0}$ will still adapt quickly to the data but the forecast variance or the posterior covariance matrix will not converge to stable values.

The study of the convergence of the posterior covariance matrix $\mathbf{P}_{t|t}$, which is considered in Exercise 3.7 (Chap. 3) for the local level model, has been the topic of much of theoretical research in the past decades. The topic is interesting, because

Table 4.2 Observed aluminium prices (left column), one-step forecast mean using the relatively "informative" prior (middle column) and one-step forecast mean using the zero prior (right column)

t	y_t	$\hat{y}_{t\mid t-1}$ with $\hat{\beta}_{0\mid 0} = [1800, 1]^\top$	$\hat{y}_{t\mid t-1}$ with $\hat{\beta}_{0\mid 0} = [0, 0]^\top$
1	1835.000	1801.000	0.000
2	1810.000	1852.890	2746.569
3	1809.000	1786.135	1810.328
4	1829.000	1796.361	1806.862
5	1819.000	1828.788	1834.682
6	1827.000	1817.344	1820.929
7	1836.000	1827.619	1829.864
8	1844.500	1839.489	1840.912
9	1832.000	1849.834	1850.741
10	1829.500	1832.674	1833.252
11	1824.500	1828.326	1828.695
12	1848.000	1822.097	1822.333
13	1837.000	1852.533	1852.684
14	1847.000	1838.662	1838.759
15	1859.000	1850.114	1850.176
16	1811.000	1865.019	1865.059
17	1835.000	1802.608	1802.633
18	1827.000	1832.593	1832.609
19	1863.000	1824.818	1824.828
20	1866.000	1871.042	1871.048

it implies that for a very general class of state space models, posterior and forecast variances will converge to stable forms. Thus in the Kalman filter recursion of $\hat{\beta}_{t\mid t}$, $\mathbf{P}_{t\mid t}$ can be replaced by its limit $\mathbf{P}$, leading to what is known the *steady state*. What is important in prior specification is that this limit does not depend on the priors $\hat{\beta}_{0\mid 0}$ and $\mathbf{P}_{0\mid 0}$ (see also Exercise 3.7). More details on the steady state of state space models can be found in Jazwinski (1970, Chapter 7), Anderson and Moore (1979, p. 77), Whittle (1984, p. 113), Chan et al. (1984) and in Harvey (1989, p. 119). Triantafyllopoulos (2007a) proves the convergence of $\mathbf{P}_{t\mid t}$ when $\mathbf{Z}_t$ is time-varying and is specified using multiple forgetting factors (see Sect. 4.3.2).

4.5.2 Prior Specification of σ^2

In this section suppose that σ^2 (the observation variance) is unknown and subject to the conjugate Bayesian estimation of Sect. 4.3.3. We discuss the specification of the prior degrees of freedom n_0 and the prior d_0 in (4.65). Starting with the degrees of freedom, it is desirable to set $n_0 \approx 0$, because for a transition matrix $\mathbf{F}_t = \mathbf{I}$ the estimator $S_t = d_t/n_t$ (see also Eq. (4.61)) is the maximum likelihood estimator

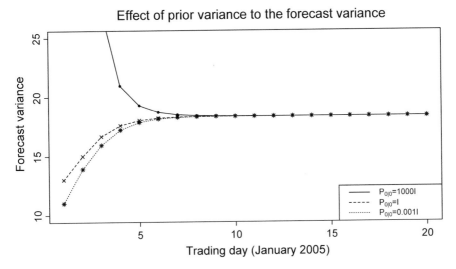

Fig. 4.17 Effect of the prior covariance matrix $\mathbf{P}_{0|0}$ in the one-step forecast variance of the aluminium prices

of σ^2 (for more details on this see the relevant discussion on p. 171). Obviously, for the prior gamma distribution (4.65) of $\theta = 1/\sigma^2$ to be defined one must have $n_0 > 0$. In any case the degrees of freedom n_0 are not crucial for the performance of the estimator S_t, as $n_t = n_{t-1} + 1 = n_0 + t$ (see Sect. 4.3.3) and n_t converges to infinity as $t \to \infty$ (not depending on n_0). However, for early time points it is recommended to use the setting $n_0 \approx 0$ and indeed this is followed in Sect. 4.3.3 by setting $n_0 = 1/1000$.

Once n_0 has been set, the specification of d_0 commences by noticing that $d_0 = n_0 S_0$. Thus, by setting a prior for S_0, this implies a prior value of d_0. However, it may be difficult to estimate σ^2 prior to observing the data. As before, it turns out that the choice of S_0 is not important, as the estimator S_t of σ^2 converges to a stable value as $t \to \infty$ independently of the choice of S_0. Next we prove this argument.

From Eq. (4.58) in Sect. 4.3.3, the posterior distribution of $1/\sigma^2$ is the gamma distribution $G(n_t/2, d_t/2)$. Thus the posterior distribution of σ^2 is an inverted gamma distribution, i.e. $\sigma^2 \mid y_{1:t} \sim IG(n_t/2, d_t/2)$, from which we have that the posterior variance of σ^2 is

$$\text{Var}(\sigma^2 \mid y_{1:t}) = \frac{2d_t^2}{(n_t - 1)^2(n_t - 2)} = \frac{2(n_0 + t)^2 S_t^2}{(n_0 + t - 1)^2(n_0 + t - 2)}, \tag{4.66}$$

for $n_t > 2$, where it is used $d_t = \nu_t S_t$ and $n_t = n_0 + t$. With the proposed prior degrees of freedom $n_0 = 1/1000$, the above variance is finite for $t \geq 2$, while for $t = 1$ it is infinite.

We can establish that as $t \to \infty$, the variance $\mathrm{Var}(\sigma^2 \mid y_{1:t})$ tends to zero. To arrive to this result, notice that the first part of the numerator in (4.66) is a polynomial in t of order 2, while the denominator is a polynomial of order 3 and the sequence $\{S_t\}$ is bounded. To see this, we prove that if S_{t-1} is bounded, then S_t is bounded too (the boundedness of $\{S_t\}$ follows by indication since S_0 is bounded). First observe that with the definitions of r_t and e_t (see p. 170), we have

$$
r_t e_t = (1 - x_t^\top K_t) e_t^2 = \left(1 - \frac{x_t^\top \mathbf{P}_{t|t-1} x_t}{q_{t|t-1}}\right) e_t^2 = \frac{S_{t-1} e_t^2}{q_{t|t-1}},
$$

which is bounded, since e_t^2, $q_{t|t-1}$ are bounded and by our hypothesis S_{t-1} is bounded too. Then, re-writing recursion (4.61) as

$$
S_t = \left(1 - \frac{1}{n_0 + t}\right) S_{t-1} + \frac{r_t e_t}{n_0 + t},
$$

we have that S_t is bounded, because it is a sum of two bounded sequences. Therefore, $\mathrm{Var}(\sigma^2 \mid y_{1:t})$ tends to zero as $t \to \infty$ and this means that as t increases, the density function of $\sigma^2 \mid y_{1:t}$ concentrates about its mode (4.62) asymptotically degenerating.

4.6   Automatic Sequential Monitoring

4.6.1   Model Monitoring

So far we have described error analysis and hyperparameter estimation or specification, for a single state space model (see Sects. 4.3 and 4.4). In other words, given observed data $y_{1:n} = \{y_1, \ldots, y_n\}$, for some n, we may propose a state space model and assess the goodness of fit as described in Sect. 4.4. However, in many real-life situations a chosen state space model is subject to continuous assessment, i.e. at each time t a decision needs to be made on whether the current model, denoted by $\mathcal{M}_0$, is adequate against an alternative model, denoted by $\mathcal{M}_A$. There may be reasons to doubt $\mathcal{M}_0$ in favour of $\mathcal{M}_A$, for example in the presence of outliers, or systemic deterioration of the model performance of $\mathcal{M}_0$ as compared to $\mathcal{M}_A$. Therefore a quantitative toolkit is required to compare the two models at each time t and to propose, if required, possible modes of corrective action. The model comparison part of such an analysis is known as *model monitoring* and the corrective action is known as *intervention analysis*. This relates to diagnostics and outlier detection, topics which have received some considerable attention in the literature, see e.g. Harvey and Koopman (1992), McCulloch and Tsay (1993), Shephard (1994b), Chib and Tiwari (1994) and Atkinson et al. (1997). Below we describe monitoring based on Bayes factors, introduced by West (1986) and subsequently developed in West and Harrison (1986), West and Harrison (1997, Chapter 11) , Salvador and Gargallo (2003) and Salvador and Gargallo (2004); an alternative approach to model diagnostics and intervention analysis is given in De Jong and Penzer (1998). A similar development as that we follow is presented in the master's thesis of Molinari (2009). In the context of statistical quality control Bersimis et al. (2007) discuss process monitoring using control charts.

We start by discussing the monitoring procedure in general terms, assuming that the current model $\mathcal{M}_0$ is the state space model (3.10a)–(3.10b), where the components x_t, $\mathbf{F}_t$, σ^2 and $\mathbf{Z}_t$ together with the priors $\hat{\beta}_{0|0}$ and $\mathbf{P}_{0|0}$ have been estimated, specified or selected, as discussed in the sections above. It is also assumed that the alternative model $\mathcal{M}_A$ is in the form of the state space model (3.10a)–(3.10b), with model components specified. Thus, there is no uncertainty on the parameters within each of the two models; there is, however, between-model uncertainty. At each time t, a decision of the most adequate between two models is required. This task is typically utilising the cumulative Bayes factor, which is defined as the ratio of the joint forecast distribution of the two models. In short, the model that produces largest forecast distribution for a given collection of data up to time t is thought to perform better with respect to its forecast ability.

More formally, define $B_t(k)$ the cumulative Bayes factor as

$$B_t(k) = \frac{p(y_t, y_{t-1}, \ldots, y_{t-k+1} \mid y_{1:t-k}, \mathcal{M}_0)}{p(y_t, y_{t-1}, \ldots, y_{t-k+1} \mid y_{1:t-k}, \mathcal{M}_A)}, \quad \text{for } k = 1, 2, \ldots, t.$$

The value of k indicates how many past observations are included in the joint distributions above. For $k = 1$ only the forecast distribution of the current observation y_t is used, while for $k = t$, the forecast distribution of the entire past $y_1, y_2, \ldots, y_t$ is used. For $k = 1$, $B_t(1)$ is just reduced to the ratio of the one-step forecast distributions of y_t, i.e.

$$B_t(1) = \frac{p(y_t \mid y_{1:t-1}, \mathcal{M}_0)}{p(y_t \mid y_{1:t-1}, \mathcal{M}_A)},$$

which sometimes is referred to as the Bayes factor of $\mathcal{M}_0$ against $\mathcal{M}_A$. Here we make the convention that $y_{1:0}$ does not include any past observations, so that $B_t(k)$ is defined for $k = 1$ and $k = t$. Note for example that under $\mathcal{M}_0$, $y_1 \mid y_{1:0} \equiv y_1 \sim N(\hat{y}_{1|0}, q_{1|0})$, where $\hat{y}_{1|0}$ and $q_{1|0}$ are the one-step forecast mean and variance of y_1, computed using only the prior mean vector and covariance matrix $\hat{\beta}_{0|0}$ and $\mathbf{P}_{0|0}$ of β_0 (provided by the Kalman filter).

Bayes factor $B_t(1)$ enables us to compare the one-step forecast distribution of $\mathcal{M}_0$ and $\mathcal{M}_A$ evaluated at the observed value of y_t. When $B_t(1) > 1$, then $\mathcal{M}_0$ is thought to have a better forecast performance at time t, as its forecast density at the observed value y_t is larger than that of $\mathcal{M}_A$. This, will be discussed in some detail later, but first we give a simple example to illustrate this point.

Example 4.10 Consider two state space models $\mathcal{M}_0$ and $\mathcal{M}_A$ that have produced the respective forecast distributions at time t:

$$y_t \mid y_{1:t-1}, \mathcal{M}_0 \sim N(10, 9) \quad \text{and} \quad y_t \mid y_{1:t-1}, \mathcal{M}_A \sim N(10 + \lambda, 9),$$

where $\lambda > 0$ measures positive deviations from the mean 10 of the forecast distribution of $\mathcal{M}_0$. In fact the two forecast distributions differ only via the value of λ. It is of interest to examine for which values of λ, $\mathcal{M}_0$ is preferred to $\mathcal{M}_0$ in terms of forecast performance.

The Bayes factor of $\mathcal{M}_0$ against $\mathcal{M}_A$ is

$$
\begin{aligned}
B_t(1) &= \frac{p(y_t \mid y_{1:t-1}, \mathcal{M}_0)}{p(y_t \mid y_{1:t-1}, \mathcal{M}_A)} \\
&= \frac{(2\pi)^{-1/2} 3^{-1} \exp[-18^{-1}(y_t - 10)^2]}{(2\pi)^{-1/2} 3^{-1} \exp[-18^{-1}(y_t - 10 - \lambda)^2]} \\
&= \exp[-18^{-1}\{2\lambda(y_t - 10) - \lambda^2\}].
\end{aligned}
$$

Thus, λ, $\mathcal{M}_0$ is preferred at time t, if $B_t(1) = \exp[-18^{-1}\{2\lambda(y_t - 10) - \lambda^2\}] > 1$ or by taking logarithms, if $-18^{-1}\{2\lambda(y_t - 10) - \lambda^2\} > 0$, which implies $\lambda > 2(y_t - 10)$. For example, if $\lambda = 2$ and y_t was observed to be $y_t = 10.5$, then $\lambda = 2 > 2 \times (10.5 - 10)$, hence model λ, $\mathcal{M}_0$ is thought to be a better model than λ, $\mathcal{M}_A$, as it produces a larger forecast distribution evaluated at the observation $y_t = 10.5$.

Returning at the cumulative Bayes factor, $B_t(k)$ may be computed by using the following recursion:

$$B_t(k) = B_t(1) B_{t-1}(k-1), \tag{4.67}$$

for $k = 2, 3, \ldots, t$.

Indeed, from the definition of $B_t(k)$ we obtain

$$
\begin{aligned}
B_t(1) B_{t-1}(k-1) &= \frac{p(y_t \mid y_{1:t-1}, \mathcal{M}_0) p(y_{t-1}, \ldots, y_{t-k+1} \mid y_{1:t-k}, \mathcal{M}_0)}{p(y_t \mid y_{1:t-1}, \mathcal{M}_A) p(y_{t-1}, \ldots, y_{t-k+1} \mid y_{1:t-k}, \mathcal{M}_A)} \\
&= \frac{p(y_t, y_{t-1}, \ldots, y_{t-k+1} \mid y_{1:t-k}, \mathcal{M}_0)}{p(y_t, y_{t-1}, \ldots, y_{t-k+1} \mid y_{1:t-k}, \mathcal{M}_A)} \\
&= B_t(k).
\end{aligned}
$$

Iterating recursion (4.67) we obtain

$$B_t(k) = B_t(1) B_{t-1}(1) B_{t-2}(k-2) \cdots B_{t-k+1}(1) = \prod_{i=t-k+1}^{t} B_i(1),$$

for $k = 1, 2, \ldots, t$. We can see that $B_t(k)$ is the product of k Bayes factors $B_i(1)$, hence the name *cumulative Bayes factor*.

If, for some time t, $B_t(1) < 1$, we have some evidence of y_t being an outlier (as the forecast distribution evaluated at y_t under $\mathcal{M}_0$ is smaller compared to the respective distribution under $\mathcal{M}_0$). If, on the other hand, $B_t(1) > 1$, this would imply that y_t is not an outlier, while $B_t(1) = 1$ would give no evidence for or against y_t being an outlying observation. As we will see later, monitoring is not only concerned with the detection of outliers; a model may well deteriorate because of systemic poor performance. If $B_t(t) > 1$, this does not mean that model $\mathcal{M}_0$ is acceptable. To see this, suppose that $t = 10$ and $B_i(1) = 2$, for $i = 1, 2, 3, 4, 5, 6$, while $B_i(1) = 1/2$, for $i = 7, 8, 9, 10$. Then the cumulative Bayesian factor is $B_{10}(10) = 2^6 0.5^4 = 4 > 1$, but clearly for $i = 7, 8, 9, 10$ the model is not acceptable with $B_i(1) < 1$.

In order to detect the most likely point of change, we need to identify the most recent group of incompatible consecutive observations by minimising $B_t(k)$, with respect to k. The next theorem, due to West (1986), settles this minimisation.

Theorem 4.3 *Let* $L_t = \min\limits_{1 \leq k \leq t} B_t(k)$, *with* $L_1 = B_1(1)$. *Then* L_t *can be computed by the recursion*

$$L_t = B_t(1) \min\{1, L_{t-1}\}, \quad \text{for } t \geq 2.$$

The minimum at time t is taken at $k = k_t$, with $L_t = B_t(k_t)$, where the positive integers k_t are updated by

$$k_t = \begin{cases} 1 + k_{t-1}, & \text{if } L_{t-1} < 1 \\ 1, & \text{if } L_{t-1} \geq 1 \end{cases}, \quad t \geq 2,$$

with $k_1 = 1$.

Proof From the definition of L_t and from Eq. (4.67), we have

$$L_t = \min\left\{B_t(1), \min_{2 \leq k \leq t} B_t(k)\right\}$$

$$= \min\left\{B_t(1), \min_{2 \leq k \leq t} [B_t(1)B_{t-1}(k-1)]\right\}$$

$$= B_t(1) \min\left\{1, \min_{2 \leq k \leq t} B_{t-1}(k-1)\right\}$$

$$= B_t(1) \min\left\{1, \min_{1 \leq j \leq t-1} B_{t-1}(j)\right\}$$

$$= B_t(1) \min\{1, L_{t-1}\}.$$

To establish the recursion for k_t, we examine separately the cases $L_t \geq 1$ and $L_t < 1$. If $L_{t-1} \geq 1$, then $\min\{1, L_{t-1}\} = 1$ and so $L_t = B_t(1)$, with $k_t = 1$. If $L_{t-1} <$

1, then $\min\{1, L_{t-1}\} = L_{t-1}$, hence $L_t = B_t(1)L_{t-1}$. From (4.67) this implies $B_t(k_t) = B_t(1)B_{t-1}(k_{t-1}) = B_t(1 + k_{t-1})$, from which the recursion $k_t = 1 + k_{t-1}$ follows.   □

Theorem 4.3 provides key results. The sequence $\{L_t\}$ is used to perform continual monitoring of model $\mathcal{M}_0$.

- If at some $t > 0$, $L_{t-1} < 1$, there is possible evidence for deterioration in $\mathcal{M}_0$, which started k_{t-1} steps backwards in time.
- If at some $t > 0$, $L_{t-1} \geq 1$, then evidence exists in favour of $\mathcal{M}_0$ and any possible evidence against $\mathcal{M}_0$ should be based on $L_t = Bt(1)$, if such value is small. Then one needs to look at L_t:

$$\text{If } \tau \leq L_t < 1, \quad \mathcal{M}_0 \quad \text{is accepted;}$$

$$\text{If } L_t < \tau, \quad \mathcal{M}_0 \quad \text{is rejected.}$$

Usually the threshold τ is set around 0.1 or 0.2. If $L_t < \tau$, we need to further consider the value of k_t, in particular:

- If $k_t = 1$, then a single observation y_t has activated the monitoring signal. If y_t is thought to be an outlier, then no further action should be considered; perhaps one may remove this observation in calculating filtered estimates and forecasts, for $t + 1, t + 2, \ldots$.
- If $k_t > 1$, this is evidence suggesting deterioration of $\mathcal{M}_0$, which started k_t steps backward in time.

This procedure can be applied for any time t, providing an assessment of $\mathcal{M}_0$ against the alternative $\mathcal{M}_A$.

4.6.2 Specification of Alternative Models

We turn our attention to the specification of the alternative model. In the application of the monitoring procedure described above, the models $\mathcal{M}_0$ and $\mathcal{M}_A$ need to be specified. $\mathcal{M}_0$ is the current model and is determined according to the analysis described in Sects. 4.1–4.5. Once the current model is fitted, the standardised residuals e_t^* are formed and if the model is good, this should follow the standard Gaussian distribution $N(0, 1)$ (or the standard Student t distribution, if the observation variance σ^2 is estimated by the data); in Sect. 4.4 there is a detailed discussion of the standardised residuals. The alternative model $\mathcal{M}_A$ measures deviations from $\mathcal{M}_0$ and its purpose is to quantify how far from $N(0, 1)$ the standardised residuals may be. Thus, it is natural to consider

$$\mathcal{M}_A: \quad e_t^* \mid y_{1:t-1} \sim N(\mu, \delta^2),$$

where μ is a mean-shift (measuring deviations from the zero mean of $\mathcal{M}_0$) and δ^2 a variance-shift (measuring deviations from the unit variance of $\mathcal{M}_0$).

Based on this specification, the Bayes factor $B_t(1)$ of $\mathcal{M}_0$ against $\mathcal{M}_A$ is

$$
\begin{aligned}
B_t(1) &= \frac{p(e_t^* \mid y_{1:t-1}, \mathcal{M}_0)}{p(e_t^* \mid y_{1:t-1}, \mathcal{M}_0)} \\
&= \frac{(2\pi)^{-1/2} \exp(-2^{-1} e_t^{*2})}{(2\pi \delta^2)^{-1/2} \exp[-(2\delta^2)^{-1}(y_t^* - \mu)^2]} \\
&= \delta \exp\left[\frac{(e_t^* - \mu)^2 - \delta^2 e_t^{*2}}{2\delta^2} \right].
\end{aligned}
$$

Given the observed value of y_t, we have the observed standardised residual e_t^*, hence for the evaluation of $B_t(1)$, we need to know the values of μ and δ. Obviously, if μ and δ are too close to 0 and 1, respectively, then the two models $\mathcal{M}_0$ and $\mathcal{M}_0$ may be indistinguishable. We note that in general, small values of e_t^* in modulus are expected to result in validating model $\mathcal{M}_0$, while large (in modulus) values of e_t^* are associated with rejecting $\mathcal{M}_0$ in favour of $\mathcal{M}_A$. As a result, we can set up μ and δ in such a way that they be consistent with the smallest values that would suggest rejection of $\mathcal{M}_0$.

For example, suppose that $e_t^* = 1.65$ is observed to be the 95% quantile of the $N(0,1)$ distribution, so as the probability that $|e_t^*| \le 1.65$ is equal to 0.90. The value 1.65 for the standardised residual is considered to be small enough, so that the values of μ and δ of the Gaussian distribution $N(\mu, \delta^2)$ of the alternative model $\mathcal{M}_A$ should be chosen so that the Bayes factor is close to 1 (i.e. since $|e_t^*| = 1.65$ is considered to be small enough both models $\mathcal{M}_0$ and $\mathcal{M}_A$ should have similar forecast performance, hence $B_t(1) \approx 1$). Likewise, fixing μ and δ as above, if we consider $e_t^* = 2.33$, the 99% quantile of the $N(0,1)$ distribution, as being an extreme value for $N(0,1)$ (or an unlikely value under model $\mathcal{M}_0$), then we can suggest the value of the threshold $\tau = B_t(1)$, so that any value of the Bayes factor below τ would classify y_t (which is used to compute e_t^*) as a potential outlier.

Table 4.3 shows values of the Bayes factor $B_t(1) \equiv B_t(1)[\mu, \delta]$, for five possible alternative models $\mathcal{M}_A^{(j)}$, for $j = 1, 2, 3, 4, 5$. Each of these models assumes a Gaussian distribution $N(\mu, \delta^2)$, for some values of μ and δ. The proportions in the brackets indicate the % quantile of the $N(0,1)$ distribution, i.e. 1.65 is the 95.1% quantile of $N(0,1)$ (rounded in two decimal places). Model 1 considers a variance 2.58 while the mean is equal to 0 and so in comparison with the $N(0,1)$ of the current model $\mathcal{M}_0$, it assesses departures from the variance.

Models 2 and 3 assess positive departures from the mean of $\mathcal{M}_0$ and Models 4 and 5 assess negative departures from the mean of $\mathcal{M}_0$. From Table 4.3, we see that for $e_t^* = 1.65$, all models return a Bayes factor close to 1 (Model 1 slightly less than 1, Models 2 and 4 exactly equal to 1 and Models 3 and 5 over 1); the values of μ and δ for each model are chosen to be consistent with those Bayes factor and to yield an acceptable forecast performance of model $\mathcal{M}_0$. Looking over the rows of

Table 4.3 Bayes factors $B_t(1) = B_t(1)[\mu, \delta]$ of the current model $\mathcal{M}_0$ against five alternative models $\mathcal{M}_A^{(j)}$, for $j = 1, 2, 3, 4, 5$

Model $\mathcal{M}_A^{(j)}$	μ	δ	$B_t(1)[\mu, \delta]$			
			$e_t^* = 1.65$ (95.1%)	$e_t^* = 2.33$ (99%)	$e_t^* = 2.40$ (99.2%)	$e_t^* = 2.58$ (99.5%)
1	0	2.58	0.81	0.26	0.22	0.15
2	3.61	1	1.75	0.15	0.12	0.06
3	3.70	1.46	1.00	0.15	0.12	0.07
Model $\mathcal{M}_A^{(j)}$	μ	δ	$B_t(1)[\mu, \delta]$			
			$e_t^* = -1.65$ (5%)	$e_t^* = -2.33$ (1%)	$e_t^* = -2.40$ (0.8%)	$e_t^* = -2.58$ (0.5%)
4	−3.61	1	1.75	0.15	0.12	0.06
5	−3.70	1.46	1.00	0.15	0.12	0.07

the table, we observe that the Bayes factor lowers as e_t^* gets large, indicating lack of support for $\mathcal{M}_0$. These values of the Bayes factor are used to define the threshold τ for the outlier detection step when $B_t(1) < \tau$. In the following we use $\tau = 0.12$ and so with the alternative models 2–4, any value y_t with $|e_t^*| > 2.4$ would be classified as an outlier. However, we note that the signal detection rule is not just a mere consideration of extreme standardised residuals e_t^*; for example under model 1, an observation y_t with respective $e_t^* = 2.58$ has a Bayes factor equal to 0.15 and thus y_t is not classified as an outlier (as $0.15 > 0.12$), but considering models 2 and 3 y_t is classified as an outlier, as the respective Bayes factor is equal to 0.07 (lower than 0.12).

After the alternative models $\mathcal{M}_A^{(j)}$ are set up as above and the threshold $\tau = 0.12$ is been fixed, we need to consider the signal rules. As mentioned in the previous section, the monitor signals when $L_t < \tau$ (if $k_t = 1$ this is due to y_t as a potential outlier and is equivalent to $B_t(1) < \tau$; while if $k_t > 1$, then deterioration of model $\mathcal{M}_0$ started k_t steps backward in time). In addition to those two signals, we consider two more signalling rules: (a) when there are too many successive outliers (two such outliers are considered as the threshold to warrant lack of support for $\mathcal{M}_0$) and (b) when $\tau \leq L_t < 1$ without being lower than the threshold τ, but the corresponding value of k_t is too large (here this is set to $k_t = 4$). The rationale of (a) is that an occasional outlier (responsible for triggering $B_t(1) < \tau$) may not warrant a problem with, but two or more successive outliers are likely to be the result of a poor forecast performance of $\mathcal{M}_0$. The rationale of (b) is that even if $\tau \leq L_t < 1$, the relatively large value of k_t suggests a tendency for deterioration, which may later develop further to a more clear signal (e.g. an outlier); given $\tau \leq L_t < 1$, a threshold of $k_t = 4$ is applied in what follows, i.e. four observations backward in time drive a relatively low value of L_t. Table 4.4 summarises the signal rules that may be triggered by any of the alternative models $\mathcal{M}_A^{(j)}$, $j = 1.2.3.4.5$, described earlier.

Table 4.4 Signal rules of the monitoring procedure

Signal	Meaning	Condition
1	Potential outlier	$B_t(1) \leq \tau$ or $L_t < \tau$ and $k_t = 1$
2	Tendency of deterioration	$\tau \leq L_t < 1$ and $k_t \geq 4$
3	Model deterioration	$L_t < \tau$ and $k_t \geq 2$
4	At least two consecutive outliers	$B_{t-1}(1) < \tau$ and $B_t(1) < \tau$

According to Table 4.4 more than one signals may be issued at a particular time. For example, if Signal 4 is issued, then clearly Signal 1 is issued too. Considering this and the five alternative models, at any time t, we might have several simultaneous signals from the monitor. The intervention policy of the next section, will be determined by the maximum number of recent observations giving rise to a signalling event, throughout all the alternative models considered. To summarise the monitoring procedure, we have the current model $\mathcal{M}_0$, a state space model, with sequential fitting, produces the standardised residuals e_t^*. Setting up the five alternative models as in Table 4.3 and considering the signalling rules of Table 4.4, at each time t, either we assess model $\mathcal{M}_0$ as acceptable, or we issue a signal. In the latter case corrective action may be required, or we may decide to take no action (e.g. in Signal 1 of 4.4). If corrective action is decided, this is known as intervention analysis, and is discussed in the next section. We close this section with an example illustrating the above monitoring procedure.

4.6.3 Monitoring for the Tobacco total sales data—CP6

We illustrate the monitoring procedure described above, by considering a real data set, consisting of monthly total sales (in some standard scale) of UK tobacco and related products in the period 1955 to 1959. The data, described in more detail in West and Harrison (1997, Chapter 11), are depicted in Fig. 4.18. We observe that the data follow an upward linear growth trend, but there are some potential outliers together with several structural changes. For example, the 12th observation, corresponding to December of 1955, is a clear outlying observation. There seem to be two main structural changes. The first is on January 1957 (25th observation), as we can see that the observations following this observation have a much increased level from the observations up to December 1956. Likewise, the observations following January 1958 exhibit a level change compared to those observations up to December 1957. These three points of time (December 1955, January 1957 and January 1958) are depicted in Fig. 4.18 by the circles.

Therefore, it is expected that a linear growth model without a monitoring scheme will not perform well over those structural changes and outliers. The linear growth state space model was formally discussed in Sect. 4.1.1; according to this if y_t is the value of the CP6 time series at time t, then the model is given by Eqs. (4.1a)–

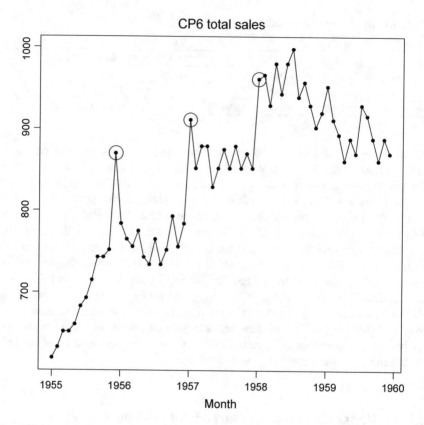

Fig. 4.18 Tobacco CP6 sales data. The circles indicate possible outliers

(4.1c). We have set up the model as the scaled observation precision (SOP) model, so as to be able to estimate the observation variance σ^2, according to the Bayesian estimation described in Sect. 4.3.3 and the covariance matrix $\mathbf{Z}_t$ specified using the discount factor $\delta = 0.8$ (see Sect. 4.3.2 for the full specification of this covariance matrix using discount factors). Finally, we use vague or weakly informative priors for the mean vector and the covariance matrix of the initial state vector β_0, i.e. $\hat{\beta}_{0|0} = [0, 0]^\top$, $\mathbf{P}_{0|0} = 1000\mathbf{I}$, $S_0 = 1$, $n_0 = 1$, where, following standard notation, $\hat{\beta}_{|0} = E(\beta_0)$, $\mathbf{P}_{0|0} = \text{Var}(\beta_0)$, n_0 are the prior degrees of freedom and S_0 is the prior estimate of σ^2. For more details on the prior specification the reader is referred to Sect. 4.5.

Figure 4.19 plots one-step forecast means (dashed line) with 95% predictive intervals together with the observed data. We see that in the start of the time series the model seems to forecast well the data (the predictive intervals are very tight and most of the observations lie inside them). But at time point $t = 12$ there seems to be an outlier (this is indicated in Fig. 4.18 by a circle), after which the model deteriorates as we see that the forecast mean is consistently higher than

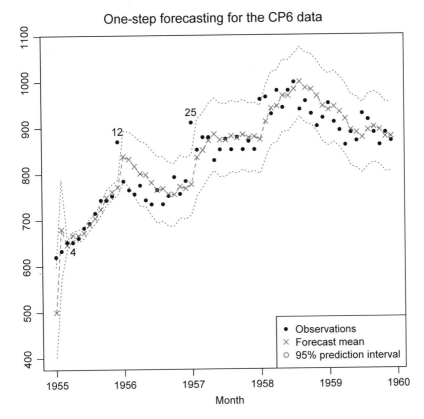

Fig. 4.19 One-step forecast mean with corresponding 95% predictive intervals, for the tobacco CP6 data

the observations and the forecast intervals are a lot wider, which suggests that the forecast variance has considerably increased.

We apply the monitoring procedure of the above section in order to detect which observations cause the poor forecast performance. We use the five alternative models of Table 4.3 and the four signals of Table 4.4. The analysis below is performed in R by first obtaining the standardised residuals using the SOP model

```
# Fit data using the SOP model
> fit <- bts.filter(y, x0=c(1,0), F0=matrix(c(1,0,1,1),2,2),
+ delta=0.8, beta0=c(500,0),  P0=1000*diag(2))

# standardised residuals
> f <- as.numeric(fit$FittedMean)
> Q <- as.numeric(fit$FittedVar)
> res.std <- (y-f) / sqrt(Q)
```

For the monitoring part we use the function `monitor`, which returns the values of L_t, k_t and the signals, for Models 1–5 (see Table 4.3).

```
# Monitoring using alternative Model 1
> mod1 <- monitor(res.std, 0, 2.58)
# Monitoring using alternative Model 2
> mod2 <- monitor(res.std, 3.61, 1)
# Monitoring using alternative Model 3
> mod3 <- monitor(res.std, 3.70, 1.46)
# Monitoring using alternative Model 4
> mod4 <- monitor(res.std, -3.61, 1)
# Monitoring using alternative Model 5
> mod5 <- monitor(res.std, -3.70, 1.46)
```

The summary of the signals are presented in Table 4.5. Some comments are in order. First of all we notice that no model issues Signal 4 (at least two consecutive outliers). This is expected, as it appears that the model adjusts reasonably to the data after an outlier. Signal 2 and 3 are relevant (2 signals tendency for deterioration and 3 model deterioration). Signal 3 is issued for a lot of times indicating that the model seems to deteriorate for long periods; for such times signal 2 is not issued as much, because it is suppressed by Signal 3. The model deterioration is not necessarily too negative for this model, but it is suggestive that the model can be improved. Signal 1 is issued at points of time 4, 12 and 25. Indeed the most prominent outlier of the data is at point of time 12, with a standardised residual $e_{12}^* = 12.176$ (this is issued by Models 2, 3 and 5). At point of time $t = 25$ is another outlier, which is picked by Model 2 (the respective standardised residual is $e_{25}^* = 4.017$). These two outliers can be guessed from Fig. 4.19. The monitoring algorithm picks another outlier at time $t = 4$ (Models 4 and 5 issue Signal 1, with corresponding standardised residual equal to $e_4^* = -3.920$). In Fig. 4.19 the time-points $t = 4, 12, 25$, which correspond to outliers, are indicated on these observations. We remark that the third circle of point $t = 37$ (January 1958) of Fig. 4.18 is not an outlier; we see from Fig. 4.19 that this observation is marginally falling out of the predictive interval and has standardised residual $e_t^* = 2.322$.

Table 4.5 Monitoring for the CP6 data set; shown are time-points, which Model j (see Table 4.3) has issued a signal i, from the range of signals of Table 4.4

Signal	Model 1	Model 2	Model 3	Model 4	Model 5
1	–	12, 25	12	4	4, 12
2	–	–	28	–	22
3	$4 - 60$	9, 13, 14, 26	$9, 13 - 27$	5	$13 - 21$
4	–	–	–	–	–

4.7 Exercises

1. Find the state space form of the autoregressive moving average model of orders 2 and 1 ARMA(2,1):

$$y_t - \mu = \phi_1(y_{t-1} - \mu) + \phi_2(y_{t-2} - \mu) + \varepsilon_t + \psi_1\varepsilon_{t-1},$$

   where μ is the mean of y_t, ϕ_i are the autoregressive (AR) coefficients, ψ_1 is the moving average (MA) coefficient and ε_t is white noise with variance σ_ε^2. For extensive coverage of this model the reader is referred to Brockwell and Davis (1991) or Box et al. (2008).

2. Motivated from Exercise 4.1 find a state space representation of the autoregressive moving average model of orders m, q ARMA(m, q):

$$y_t - \mu = \phi_1(y_{t-1} - \mu) + \cdots + \phi_m(y_{t-m} - \mu) + \varepsilon_t + \psi_1\varepsilon_{t-1} + \cdots + \psi_q\varepsilon_{t-q},$$

   where μ is the mean of y_t, ϕ_i are the autoregressive (AR) coefficients, ψ_1 is the moving average (MA) coefficient and ε_t is white noise with variance σ_ε^2. For extensive coverage of this model the reader is referred to Brockwell and Davis (1991) or Box et al. (2008). Thus, suggest a possible estimation procedure using state space methods; explain how the parameters ϕ_i and ψ_j may be estimated.

3. A company sets up a time series model for the yield y_t of an investment at time t as

$$y_t = \psi_t\theta_t + \alpha y_{t-1} + \varepsilon_t,$$

   where $\theta_t = \theta_{t-1} + \eta_t$ and $\varepsilon_t = 0.9\varepsilon_{t-1} + v_t$. Here, ψ_t is a time-varying covariate, θ_t and α are regression and autoregression coefficients and the innovations η_t and v_t are assumed to be independent, each following a white noise process. It is further assumed that η_t is independent of α, for all t.

   a. Define the state vector

$$\beta_t = \begin{bmatrix} \theta_t \\ \alpha \\ \varepsilon_t \end{bmatrix}$$

   and use it to express y_t in state space form, i.e.

$$y_t = x_t\beta_t + \epsilon_t \quad \text{and} \quad \beta_t = \mathbf{F}\beta_{t-1} + \zeta_t,$$

   hence determine x_t, $\mathbf{F}$ and define the innovations of the model ϵ_t and ζ_t.

b. Suppose this model was fitted to data $y_1, \ldots, y_n$. With information $y_{1:n} = \{y_1, \ldots, y_n\}$, suppose that the posterior distributions of θ_n and α were

$$\theta_n \mid y_{1:n} \sim N(1, 10) \quad \text{and} \quad \alpha \mid y_{1:n} \sim N(1, 2).$$

With information $y_{1:n}$,
show that the one-step ahead forecast mean $\hat{y}_{n+1|n}$ of y_{n+1} is

$$\hat{y}_{n+1|n} = y_n + \psi_{n+1};$$

and that the two-step ahead forecast mean $\hat{y}_{n+2|n}$ of y_{n+2} is

$$\hat{y}_{n+2|n} = y_n + \psi_{n+2} - \psi_{n+1}.$$

4. Consider that a time series $\{y_t\}$ is generated from an ARIMA(1,1,1) model, so that

$$y_t - y_{t-1} = \alpha(y_{t-1} - y_{t-2}) + \epsilon_t + \gamma \epsilon_{t-1},$$

where α is the AR parameter, γ is the MA parameter and $\{\epsilon_t\}$ is a Gaussian white noise sequence with variance equal to 1.
Define the state vector

$$\beta_t = \begin{bmatrix} y_t \\ y_{t-1} \\ \epsilon_t \end{bmatrix}.$$

a. Write down a state space representation for y_t, i.e. express y_t as a state space model:

$$y_t = x^\top \beta_t + \delta_t,$$
$$\beta_t = \mathbf{F}\beta_{t-1} + \zeta_t.$$

and hence determine the components x, $\mathbf{F}$, δ_t and ζ_t, and write down the distributions of δ_t and ζ_t.

b. Suppose that at time $t = 2$, the posterior distribution of β_2 is

$$\beta_2 \mid y_{1:2} \sim N \left\{ \begin{bmatrix} 0 \\ 1 \\ 0 \end{bmatrix}, \begin{bmatrix} 0 & 0 & 0 \\ 0 & 0 & 0 \\ 0 & 0 & 1 \end{bmatrix} \right\},$$

where $y_{1:2} = \{y_1 = 1, y_2 = 0\}$ denotes the information available at time $t = 2$.

c. If $y_3 = 1$, perform a step of the Kalman filter and hence derive the posterior distribution of

$$\beta_3 \mid y_{1:3},$$

where $y_{1:3} = \{y_1 = 1, y_2 = 0, y_3 = 1\}$.

Given information $y_{1:3}$, find the posterior distribution of ϵ_3.

5. In the context of Sect. 4.2.4 show that the trend component model (Eq. (4.40))

$$\chi_t^{(1)} = [1, 0]\gamma_t^{(1)} \quad \text{and} \quad \gamma_t^{(1)} = \begin{bmatrix} 0 & 1 \\ -1 & 2 \end{bmatrix} \gamma_{t-1}^{(1)} + \xi_{1t}$$

can be written as (see Eq. (4.41))

$$\chi_t^{(1)} = [1, 0]\gamma_t^{(3)} \quad \text{and} \quad \gamma_t^{(3)} = \begin{bmatrix} 1 & 1 \\ 0 & 1 \end{bmatrix} \gamma_{t-1}^{(3)} + \xi_{3t},$$

for some innovations ξ_{3t}, which are given as linear functions of the innovations $\xi_{11}, \ldots, \xi_{1t}$.

6. Consider the local level model

$$y_t = \beta_t + \epsilon_t \quad \text{and} \quad \beta_t = \beta_{t-1} + \zeta_t,$$

with the usual assumptions of ϵ_t and ζ_t. For observed data $y_{1:t} = \{y_1, \ldots, y_t\}$:

a. Using the Kalman filter (Theorem 3.2), show that the posterior distribution of ϵ_t is $\epsilon_t \mid y_{1:t} \sim N[(1 - K_t)e_t, K_t\sigma^2]$, where K_t is the Kalman gain and σ^2 is the variance of ϵ_t.

b. A time series z_t follows an integrated autoregressive moving average model of orders $m, 1, q$ (ARIMA$(m, 1, q)$), if the first order difference $z_t - z_{t-1}$ follows the ARMA(m, q) of Exercise 4.2. For the time series y_t of Exercise 4.3 prove

$$y_t - y_{t-1} = e_t - (1 - K_{t-1})e_{t-1},$$

so that y_t is an ARIMA$(0,1,1)$ type model. Is it exactly an ARIMA$(0,1,1)$ and if not why? How can we justify the argument that y_t basically is an ARIMA$(0,1,1)$?

c. Prove

$$\hat{\beta}_{t|t} = K_t y_t + (1 - K_t)\hat{\beta}_{t-1|t-1},$$

which is an EWMA (exponentially weighted moving average). Then, give an interpretation of K_t.

7. Suppose that for a time series y_t a time-varying polynomial regression model is first considered:

$$y_t = \beta_{0t} + \beta_{1t}t + \beta_{2t}t^2 + \beta_{3t}t^3 + \epsilon_t,$$

where $\beta_{0t}, \beta_{1t}, \beta_{2t}, \beta_{3t}$ are dynamic regression coefficients and ϵ_t is a white noise process with variance. If β_{it} are evolving according to a random walk, then write down y_t in state space form and determine the components of the model (i.e. x_t, $\mathbf{F}_t$, σ^2, $\mathbf{Z}_t$).

8. For some data y_t with covariate x_t, we consider a simple static regression model

$$y_t = \beta_0 + \beta_1 x_t + \varepsilon_t, \quad \text{(Model 1)}$$

where ε_t is white noise with variance 1.

An alternative model is considered, with autocorrelated errors, i.e. ε_t is not a white noise:

$$y_t = \beta_0 + \beta_1 x_t + \varepsilon_t \quad \text{and} \quad \varepsilon_t = \phi \varepsilon_{t-1} + v_t, \quad \text{(Model 2)}$$

where ϕ is an autoregressive coefficient (assumed to lie in the unit circle) and v_t is white noise with variance 2.

Based on some data $y_{1:t} = \{y_1, \ldots, y_t\}$, find the form of the forecast function $\hat{y}_{t+k|t} = E(y_{t+k} \mid y_{1:t})$, for each model and prove that when $\phi \approx 0$ or when $k \to \infty$ the two models are equivalent. Thus, suggest in which case model 2 is expected to perform better than model 1.

9. For the data of Exercise 3.7 (Chap. 3) carry out an error analysis and comment on the quality of the model fit.

10. The following table (source: Cooper and Harrison (1997)) shows the number of five-year-old (y_t) and four-year-old (x_t) cattle that are confirmed as having bovine spongiform encephalopathy (BSE) at a specific quarter. It is postulated that a simple regression model will help to forecast y_t (numbers of five-year-old cattle suffering from BSE) using x_t (number of four-year-old confirmed as BSE at the same quarter). It will also be useful in identifying whether the relationship has changed over the years.

	Qtr1		Qtr2		Qtr3		Qtr4	
Year	y_t	x_t	y_t	x_t	y_t	x_t	y_t	x_t
1989	559	139	486	209	590	346	904	536
1990	1172	570	837	586	904	694	1185	831
1991	1338	1178	1090	1037	1532	1192	2394	1497
1992	2821	2027	2172	1861	2989	2543	4620	3634
1993	4831	4654	3907	4035	3618	3934	2809	2558
1994	2099	1992	1304	1338	1204	1342	1256	1220
1995	1075	1048	849	923	777	846	641	558
1996	484	472	310	430	252	417	–	439
1997	–	475	–	331	–	182	–	–

Fit an appropriate state space model up to the 4th Quarter of 1996 and assess the evidence of time-varying regression coefficients (or change of the relationship over time). Provide the one-step ahead forecasts of y_t, for the 4th Quarter of 1996 and the first three quarters of 1997. Can you forecast y_t for the 4th Quarter of 1997?

11. The following table shows the turkey sales data described in Sect. 4.1.4; they consist of number of sales (in thousands) of quarterly turkey chick sales from autumn 1974 up to the summer of 1983.

Year	Qtr1	Qtr2	Qtr3	Qtr4
1974				131.7
1975	322.6	285.6	105.7	80.4
1976	285.1	347.8	68.9	203.3
1977	375.9	415.9	65.8	177.0
1978	438.3	463.2	136.0	192.2
1979	442.8	509.6	201.2	196.0
1980	478.6	688.6	259.8	352.5
1981	508.1	701.5	325.6	305.9
1982	422.2	771.0	329.3	384.0
1983	472.0	852.0		

a. In the state space model that is proposed in p. 134, suppose that the observation variance σ^2 and the transition covariance matrix $\mathbf{Z}_t = \mathbf{Z}$ are unknown. Use the EM-algorithm to estimate σ^2 and $\mathbf{Z}$ and then plot the one-step forecast mean of the data, together with the respective 95% forecast bounds, using the estimated values of σ^2 and $\mathbf{Z}$. Comment on the quality of forecasting compared to that of the state space model used in p. 134.

b. Fit the scaled observational model (SOP) of Sect. 4.3.3, where the transition covariance matrix $\mathbf{Z}_t$ is specified with a discount factor δ. Experiment with values of $\delta = 0.99, 0.8, 0.5, 0.1$. Perform an error analysis and choose the optimal value of δ. Compare the estimated values of σ^2 and $\mathbf{Z}_t$ with those in (a) estimated by the EM-algorithm and comment.

12. The air passenger data consist of numbers of passengers (in thousands) carried by international airlines each month, from January 1949 to December 1960. The data are described in Sect. 4.1.4 and are plotted in Fig. 4.6. The data can be found in R by uploading the library MASS and looking at the data set AirPassengers; in the R console type

```
> library(MASS)
> data(AirPassengers)
> y <- AirPassengers
```

Find a suitable state space model and fit it to the air passenger data. Perform an error analysis and give an assessment of the fit of the chosen model.

13. Quarterly earnings (US dollars) per Johnson and Johnson share in the period 1960–1980 are available in R by executing the following commands

```
> library(MASS)
> data(JohnsonJohnson)
> y <- JohnsonJohnson
```

The data are reported in Shumway and Stoffer (2017). Suggest a suitable state space model and analyse this data using this model. Assess the goodness of fit and provide 95% forecast intervals for the quarters of 1981.

14. Yearly averages above 570ft of the levels of lake Huron in the period 1875–1972 are available in R by executing the following commands

```
> library(MASS)
> data(LakeHuron)
> y <- LakeHuron
```

The question here is whether there is any regularity in the fluctuations, and if so how to describe it, and forecast levels. Fit an appropriate state space model for this data set. Using both smoothed and filtered estimates of the mean level, assess the evidence that the mean level of the lake has changed over the years.

15. Twenty-four observations consisting of annual number of telephone calls made (in millions of calls) in Belgium in the period 1950–1973 are available in R by executing the following commands

```
> library(MASS)
> data(phones)
> y <- phones
```

The data set is in the object y$calls, while y$year includes the years. Suggest a state space model for this data set. Provide a plot with the one-step forecasts against the actual data and perform an error analysis in order to evaluate the goodness of the fit.

16. Hundred observations consisting of measurements of the annual flow of the river Nile at Aswan in the period 1871–1970 are available in R by executing the following commands

```
> library(MASS)
> data(Nile)
> y <- Nile
```

The data are first reported in Cobb (1978)[Table 1, p. 249]. There appears to be a drop in the level of the annual flow near year 1898. Fit a suitable state space model. See if your model can detect the change-point of the level mentioned above.

17. Seventy-two observations consisting of monthly totals of accidental deaths in the USA in the period of 1973–1978 available in R by executing the following commands

```
> library(MASS)
> data(USAccDeaths)
> y <- USAccDeaths
```

The data are reported in Brockwell and Davis (1991). Suggest a state space model for this data set. Provide a plot with the one-step forecasts against the actual data and perform an error analysis in order to evaluate the goodness of the fit.

18. Monthly average air temperatures at Nottingham Castle (in degrees Fahrenheit) in the period 1920–1939 are available in R by executing the following commands

```
> library(MASS)
> data(nottem)
> y <- nottem
```

The data are reported in Anderson (1976). Suggest a state space model for this data set. Fit the model to the data and perform an error analysis in order to evaluate the goodness of the fit.

19. One hundred and fourteen observations consisting of annual numbers of lynx trappings for 1821–1934 in Canada are available in R by executing the following commands

```
> library(MASS)
> data(lynx)
> y <- lynx
```

The data are reported in Brockwell and Davis (1991).

   a. Suggest a state space model for this data set. Fit the model to the data and perform an error analysis in order to evaluate the goodness of the fit.
   b. A closer look at the data reveals that the histogram of the observation is right-skewed. Suggest a state space model on $\log y_t$, where y_t denotes the value of the annual lynx trappings at time t. Fit this model and make see if it fits better than the model in part (a).

20. Two hundred and eighty-nine observations on annual numbers of sunspots from 1700 to 1988 are available in R by executing the following commands

```
> library(MASS)
> data(sunspot.year)
> y <- sunspot.year
```

The data is reported in Tong (1996). Suggest a state space model for this data. Fit the model to the data and assess the goodness of fit.

21. Monthly time series data (source: Harvey (1989)) consisting of monthly totals of car drivers killed or seriously injured in the period January 1969 to December 1984 are available in R by executing the following commands

```
> library(MASS)
> data("drivers")
> y <- drivers
```

It is interesting to describe and quantify the form of variation here. Under-standing/quantification could enable the value of road safety measures to be assessed. From the plot of the data there is an evident seasonal element, but how consistent is it? Various road safety measures have been introduced during the period (for example, the compulsory wearing of helmets by motor-cyclists in 1973–1974 and of seat belts for car drivers and front-seat passengers on 31 January 1983). How large an effect did they have?

Analyse this data set by building a state space model and assess the forecast performance of this model.

22. The following data contains the number of hectolitres of whisky produced each month in UK between 1980 and 1987.

	1980	1981	1982	1983	1984	1985	1986	1987
Jan	34.6	53.5	31.5	10.4	10.7	17.1	12.3	11.7
Feb	59.1	67.9	57.4	50.6	54.7	33.4	35.7	32.5
Mar	82.5	50.5	61.9	73.6	78.3	90.7	91.1	52.1
Apr	9.3	6.6	7.1	8.7	7.5	7.6	7.6	14.0
May	12.2	12.3	12.4	16.2	12.9	16.8	14.4	21.5
Jun	19.5	18.5	21.9	23.4	17.6	21.7	22.3	30.6
Jul	28.5	24.0	19.7	20.9	23.5	25.7	24.3	32.3
Aug	29.3	29.6	26.8	27.2	26.9	29.6	28.1	29.0
Sep	35.1	34.6	30.9	31.4	26.9	32.2	34.5	35.2
Oct	58.9	53.8	49.7	50.2	56.2	56.3	53.8	53.6
Nov	79.8	73.3	72.4	82.8	73.8	78.2	77.9	80.6
Dec	52.8	52.6	55.9	49.3	44.8	51.6	54.2	53.0

a. Find an appropriate state space model to describe the data;
b. Provide the one-step forecasts at each time and plot them against the data;
c. Carry out an error analysis and comment on the performance of the chosen model.

23. Consider the state space model (3.10a)–(3.10b), with

$$
x_t = \begin{bmatrix} z_t \\ 1 \\ 0 \\ 0 \end{bmatrix} \quad \text{and} \quad F = \begin{bmatrix} 1 & 0 & 0 & 0 \\ 0 & 1 & 1 & 0 \\ 0 & 0 & 1 & 1 \\ 0 & 0 & 0 & 1 \end{bmatrix},
$$

where z_t is a time-varying covariate.
Give a name of this model and suggest what type of time series data this model may be useful for.

24. Consider the state space model (3.10a)–(3.10b), with

$$
x = \begin{bmatrix} 1 \\ 0 \end{bmatrix} \quad \text{and} \quad F = \begin{bmatrix} 1 & \lambda \\ 0 & 1 \end{bmatrix},
$$

for some constant λ.

a. Give an expression of the forecast function of this model and suggest what type of time series data this model may be useful for. Give an interpretation of λ (e.g. by examining the effect several values of λ have in the forecast function) and suggest how λ can be estimated given some observed data $y_1, \ldots, y_n$.
b. (Overparameterisation). A state space model is overparameterised if its forecast function is equal to the forecast function of another state space model with fewer parameters. Show that the above state space model is overparameterised when $\lambda = 0$ and that it can be reduced to the local level model.

25. At each time t, I_t denotes the lead indicator for a company's sales S_t. The following time-varying regression model is suggested to relate y_t the quarterly change in S_t to x_t the quarterly change in I_{t-2}:

$$
y_t = x_t \beta_t + \epsilon_t \quad \text{and} \quad \beta_t = \beta_{t-1} + \zeta_t,
$$

where ϵ_t is white noise with variance 1, ζ_t is a white noise with variance 10, and both ϵ_t, ζ_t are normally distributed.

a. Show that $P_{t|t}$ the posterior variance of β_t satisfies

$$
\frac{1}{P_{t|t}} = \frac{1}{P_{t-1|t-1} + 10} + x_t^2.
$$

b. If $x_1 = 4$, $x_2 = 4$, $y_1 = 12$, $y_2 = 11$ and the prior of β_0 is $\beta_0 \sim N(2, 0.81)$, then use the result above to calculate the posterior means $\hat{\beta}_{1|1}$, $\hat{\beta}_{2|2}$ and the posterior variances $P_{1|1}$, $P_{2|2}$.

c. Using (b) obtain the one-step forecast mean of $y_3 = 9$ and the associated residual. Comment on the quality of this forecast.

d. If instead of $\beta_0 \sim N(2, 0.81)$ the prior distribution of β_0 is set to either of the following:

    i. $\beta_0 \sim N(10, 0.81)$ or

    ii. $\beta_0 \sim N(2, 100)$,

comment on whether you expect an improvement on forecasting and the general model performance for all y_t.

e. Suggest how the model performance can be improved.

26. In the scaled observational precision (SOP) model (4.55a)–(4.55b) of Sect. 4.3.3 prove the following fixed-interval smoothing algorithm:

a. Initial state distributions at $t = n$: $\beta_n \mid y_{1:n} \sim t(n_n, \hat{\beta}_{n|n}, \mathbf{P}_{n|n})$ and $\sigma^{-2} \mid y_{1:n} \sim G(n_n/2, n_n S_n/2)$, where $\hat{\beta}_{n|n}$, $\mathbf{P}_{n|n}$, v_n and S_n are computed by the SOP algorithm of Sect. 4.3.3.

b. Smoothed state distribution, for $t = 1, 2, \ldots, n-1$:
$\beta_t \mid y_{1:n} \sim t(n_n, \hat{\beta}_{t|n}, \mathbf{P}_{t|n})$, where

$$\hat{\beta}_{t|n} = \hat{\beta}_{t|t} + \mathbf{L}_t(\hat{\beta}_{t+1|n} - \hat{\beta}_{t+1|t}),$$

$$\mathbf{P}_{t|n} = \frac{S_n}{S_t}\left[\mathbf{P}_{t|t} + \mathbf{L}_t(S_t S_n^{-1}\mathbf{P}_{t+1|n} - \mathbf{P}_{t+1|t})\mathbf{L}_t^\top\right],$$

with $\mathbf{L}_t = \mathbf{P}_{t|t}\mathbf{F}_{t+1}^\top\mathbf{P}_{t+1|t}^{-1}$ and $\hat{\beta}_{t|t}$, $\hat{\beta}_{t+1|t}$, $\mathbf{P}_{t|t}$, $\mathbf{P}_{t+1|t}$ being calculated by the SOP algorithm.

c. Smoothed observation distribution, for $t = 1, 2, \ldots, n-1$:
$y_t \mid y_{1:n} \sim t(n_n, \hat{y}_{t|n}, q_{t|n})$, where $\hat{y}_{t|n} = x_t^\top\hat{\beta}_{t|n}$ and $q_{t|n} = S_n(x_t^\top\mathbf{P}_{t|n}x_t/S_t + 1)$.

27. Show that the polynomial trend model with design vector and transition matrix defined as in (4.2) is observable.

28. Consider the state space model (3.10a)–(3.10b), with

$$x_t = x = \begin{bmatrix} a_1 \\ a_2 \\ \vdots \\ a_p \end{bmatrix} \quad \text{and} \quad \mathbf{F}_t = \mathbf{F} = \begin{bmatrix} \lambda & 1 & 0 & \cdots & 0 \\ 0 & \lambda & 1 & \cdots & 0 \\ \vdots & \vdots & \vdots & \ddots & \vdots \\ 0 & 0 & 0 & \cdots & \lambda \end{bmatrix},$$

where $a_1, \ldots, a_p, \lambda$ are constants, satisfying $a_1^2 + \cdots + a_p^2 \neq 0$ (at least one of a_j is non-zero); notice that $\mathbf{F}$ is a Jordan block matrix. Suppose that $\mathbf{Z}_t$, the transition covariance matrix of the model, is specified using a discount factor δ.

a. If $\sigma^2 > 0$ and the prior covariance matrix $\mathbf{P}_{0|0}$ is non-singular, then show that the posterior covariance matrix $\mathbf{P}_{t|t}$ is non-singular, for all $t = 1, 2, \ldots$.

b. If $a_1 \neq 0$ and $\lambda^2 > \delta$ (together with the conditions in (a)) show that the limit of $\mathbf{P}_{t|t}^{-1}$ exists and is not depending on $\mathbf{P}_0$.

c. With the conditions of (a) and (b), show that the kl-th element $p_{kl}^{(-1)}$ of the precision limiting matrix $\mathbf{P}^{-1} = \lim_{t\to\infty} \mathbf{P}_{t|t}^{-1}$ satisfies the following equation

$$p_{kl}^{(-1)} = \delta \sum_{i=1}^{k} \sum_{j=1}^{l} (-1)^{k+l-i-j} \lambda^{i+j-k-l-2} p_{ij}^{(-1)} + a_k a_l / \sigma^2,$$

for $k, l = 1, \ldots, p$.

29. a. In the context of Exercise 4.21 show that the precision covariance matrix $\mathbf{P}_{t|t}^{-1}$ of the $(p-1)$-th order polynomial trend model of Sect. 4.1.1 when the transition covariance matrix is specified with a discount factor converges to a matrix.

b. In the context of Exercise 4.21 consider a state space model (3.10a)–(3.10b), with

$$x = \begin{bmatrix} 1 \\ 1 \\ 1 \end{bmatrix}, \quad F = \begin{bmatrix} 0.8 & 1 & 0 \\ 0 & 0.8 & 1 \\ 0 & 0 & 0.8 \end{bmatrix}$$

and $\sigma^2 = 5$.

   i. Find the range of δ so that $\lim_{t\to\infty} \mathbf{P}_{t|t}^{-1}$ exists.
   ii. If $\delta = 0.5$, then calculate the limiting posterior covariance matrix $\mathbf{P} = \lim_{t\to\infty} \mathbf{P}_{t|t}$.

30. Show that, for $\lambda \neq 0$, the forecast function of the state space model of Exercise 4.20 is

$$\hat{y}_{t+k|t} = \sum_{i=1}^{p} \sum_{j=1}^{i} \binom{k}{i-j} \lambda^{k-i+j} a_j \hat{\beta}_{t|t,i},$$

where $\hat{\beta}_{t|t} = [\hat{\beta}_{t|t,1}, \ldots, \hat{\beta}_{t|t,p}]^{\mathsf{T}}$.
If $a_1 = 1$ and $a_j = 0$, for all $j > 1$, show that the above forecast function reduces to

$$\hat{y}_{t+k|t} = \sum_{i=1}^{p} \binom{k}{i-1} \lambda^{k-i+1} \hat{\beta}_{t|t,i},$$

If, in addition $\lambda = 1$ show that the above forecast function reduces to the forecast function of the $(p-1)$-th polynomial trend model of Sect. 4.1.1. Thus, we have generalised the polynomial trend models to allow for any non-zero λ.

31. Consider the local level model

$$y_t = \beta_t + \epsilon_t \quad \text{and} \quad \beta_t = \beta_{t-1} + \zeta_t,$$

where $\epsilon_t \sim N(0, \sigma^2)$ and $\zeta_t \sim N(0, Z_t)$, and the innovations are as usual individually and mutually independent and independent of the assumed prior $\beta_0 \sim N(\hat{\beta}_{0|0}, P_{0|0})$. Suppose that information $y_{1:n} = (y_1, \ldots, y_n)$ is available, for observed time series $y_1, \ldots, y_n$ and for some positive integer n.
If Z_t is modelled with a discount factor δ, show that, for $t = 1, \ldots, n-1$, the recursions of the smoothed mean $\hat{\beta}_{t|n}$ and the smoothed variance $P_{t|n}$ can be simplified as

$$\hat{\beta}_{t|n} = (1 - \delta)\hat{\beta}_{t|t} + \delta\hat{\beta}_{t+1|n}$$

and

$$P_{t|n} = (1 - \delta)P_{t|t} + \delta^2 P_{t+1|n},$$

where $\hat{\beta}_{t|t}$ and $P_{t|t}$ are the posterior mean and variance of β_t, provided by the Kalman filter. Thus, the smoothed mean $\hat{\beta}_{t|n}$ is an EWMA (exponentially weighted moving average) of the sequence of posterior means $\hat{\beta}_{1|1}, \ldots, \hat{\beta}_{n|n}$. Use this fact to interpret the role of δ in the dynamical behaviour of $\hat{\beta}_{t|n}$.

Chapter 5
Multivariate State Space Models

This chapter studies multivariate Gaussian state space models, aimed at situations where several time series are observed at each time t. The foundation of the chapter is an extension of the univariate state space models of Chaps. 3 and 4 to the multivariate case. This is achieved in a relatively painless way by replacing scalar components with vectors and vector components with matrices in the main univariate state space model of Chap. 3. Particular emphasis in this chapter is placed on the estimation of the covariance matrices of the innovation terms of the model, which can be exploited in order to estimate the cross-correlation of the several time series observed at each time.

Section 5.1 introduces the multivariate state space model and discusses the Kalman filter for this extended class of state space models. Model design and error analysis follow in Sects. 5.2 and 5.4 and they are developed in parallel with the univariate case (Chap. 4). Section 5.5 considers in detail covariance estimation and in particular provides two generalisations of the scaled observational precision (SOP) univariate model of Sect. 4.3.3. The next section provides an illustrative example of a trivariate time series consisting of pollution variables in the wider area of Athens. Section 5.7 develops Markov chain Monte Carlo estimation methods for the multivariate state space model.

5.1 The Kalman Filter

So far we have assumed that the response time series y_t is scalar, i.e. we are interested on a single time series. There are many situations that instead we are interested in modelling several time series jointly. Modelling each series separately is not desirable, because it is usually of interest to estimate the interdependence or correlation structure between the component time series. To set up notation we consider d time series $y_{1t}, y_{2t}, \ldots, y_{dt}$, for some integer $d \geq 1$ and $t = 1, 2, \ldots$.

© The Author(s), under exclusive license to Springer Nature Switzerland AG 2021
K. Triantafyllopoulos, *Bayesian Inference of State Space Models*, Springer Texts in Statistics, https://doi.org/10.1007/978-3-030-76124-0_5

We form the vector time series

$$y_t = \begin{bmatrix} y_{1t} \\ y_{2t} \\ \vdots \\ y_{dt} \end{bmatrix}$$

and we are interested in setting up a state space model that can describe the stochastic dynamics and variation of y_t. Sometimes the scalar time series $\{y_{it}\}$ is referred to as *component time series*, because the element y_{it} is a component of the vector y_t at each time t.

A multivariate state space model for y_t generalises the univariate state space model (3.10a)–(3.10b) by setting

$$y_t = \mathbf{x}_t^\top \beta_t + \epsilon_t \qquad \text{(observation model)}, \tag{5.1a}$$

$$\beta_t = \mathbf{F}_t \beta_{t-1} + \zeta_t \qquad \text{(transition model)}, \tag{5.1b}$$

where $\mathbf{x}_t$ is a $p \times d$ *design* matrix, β_t is a $p \times 1$ state vector and $\mathbf{F}_t$ is a $p \times p$ *transition* matrix. The *innovation* vectors ϵ_t and ζ_t are assumed to follow multivariate Gaussian distributions, i.e.

$$\epsilon_t \sim N(0, \mathbf{\Sigma}) \quad \text{and} \quad \zeta_t \sim N(0, \mathbf{Z}_t), \tag{5.2}$$

where $\mathbf{\Sigma}$ is a $d \times d$ covariance matrix and $\mathbf{Z}_t$ is a $p \times p$ covariance matrix.

The innovation sequences $\{\epsilon_t\}$ and $\{\zeta_t\}$ are each assumed to be independent as well as mutually independent, i.e. $\mathrm{E}(\epsilon_t \epsilon_s^\top) = 0$, $\mathrm{E}(\zeta_t \zeta_s^\top) = 0$, for any $t \neq s$, and $\mathrm{E}(\epsilon_t \zeta_t^\top) = 0$, for any t, s. The model is complete by specifying the initial distribution of β_0, which is assumed to be independent of $\{\epsilon_t\}$ and $\{\zeta_t\}$, as

$$\beta_0 \sim N(\hat{\beta}_{0|0}, \mathbf{P}_{0|0}), \tag{5.3}$$

for some $p \times 1$ prior vector $\hat{\beta}_{0|0}$ and $p \times p$ prior covariance matrix $\mathbf{P}_{0|0}$.

Some comments are in order.

- Model (5.1a)–(5.1b) provides a complete generalisation of model (3.10a)–(3.10b); indeed we can observe that for $d = 1$, $y_t = y_{1t}$ is a scalar time series, $\mathbf{x}_t = x_t$ is a $p \times 1$ vector, $\mathbf{\Sigma} = \sigma^2$ is a variance and (5.1a)–(5.1b) are reduced to (3.10a)–(3.10b).
- If $\mathbf{x}_t$ is a matrix of time-varying covariates and $\mathbf{F}_t = \mathbf{I}$, then (5.1a)–(5.1b) describes a time-varying regression model for a vector of time series $\mathbf{y}_t$. This in turn is a direct generalisation of the time-varying regression model for univariate time series (3.9a)–(3.9b) described in Sect. 3.1.3.
- If $p = d$, $\mathbf{x}_t = \mathbf{F}_t = \mathbf{I}$, then the state space model (5.1a)–(5.1b) describes y_t to exhibit local variation around the level vector β_t, which generalises the univariate

local level model of Sect. 3.1.3; this model may be referred to as *multivariate local level* model.
- From the above points it becomes clear that, within the state space model (5.1a)–(5.1b), all models of Sect. 4.1 can be extended to accommodate a vector of observations y_t.
- If we denote by x_{it} the i-th column of $\mathbf{x}_t$ $(i = 1, 2, \ldots, d)$ so that

$$\mathbf{x}_t = [x_{1t}, x_{2t}, \ldots, x_{dt}],$$

then we can write the observation equation (5.1a) as

$$
\begin{bmatrix} y_{1t} \\ y_{2t} \\ \vdots \\ y_{dt} \end{bmatrix}
=
\begin{bmatrix} x_{1t}^\top \\ x_{2t}^\top \\ \vdots \\ x_{dt}^\top \end{bmatrix}
\beta_t +
\begin{bmatrix} \epsilon_{1t} \\ \epsilon_{2t} \\ \vdots \\ \epsilon_{dt} \end{bmatrix}
$$

where $\epsilon_t = [\epsilon_{1t}, \epsilon_{2t}, \ldots, \epsilon_{dt}]^\top$. Thus, from the state model (5.1a)–(5.1b) to d we can write

$$y_{it} = x_{it}^\top \beta_t + \epsilon_t, \qquad \epsilon_{it} \sim N(0, \sigma_{ii}), \tag{5.4a}$$

$$\beta_t = \mathbf{F}_t \beta_{t-1} + \zeta_t, \qquad \zeta_t \sim N(0, \mathbf{Z}_t), \tag{5.4b}$$

where σ_{ii} is the ii-th element of Σ. First notice that for all y_{it} $(i = 1, 2, \ldots, d)$, the transition equation (5.4b), hence the above d state space models are not independent, because estimation of one model (fixing y_{it}) will influence estimation of another model (fixing y_{jt}, for $i \neq j$). Secondly, notice that the system of the d models (5.4a)–(5.4b) is not able to take into account the effects of the covariances of ϵ_{it} and ϵ_{jt} (the off-diagonal elements of Σ), which imply an effect of the correlation of y_{it} and y_{jt}. Therefore, when we observe time series that are correlated, it is then incorrect to model each of them using a state space model of the form (5.4a)–(5.4b); in such situations we need to apply the multivariate state space model (5.1a)–(5.1b), which explicitly define the inter-dependence of $y_{1t}, y_{2t}, \ldots, y_{dt}$.

With the definition of the state space model (5.1a)–(5.1b) the Kalman filter (used in Sect. 3.2 for filtering of univariate time series) is updated for multivariate time series as follows.

Theorem 5.1 (Kalman Filter) *Consider the state space model (5.1a)–(5.1b) together with the error or innovations distribution (5.2) and the initial distribution (5.3). Then, for each time $t = 1, \ldots, n$, the following apply:*

1. *The forecast distribution of β_t at time $t - 1$ is $\beta_t \mid y_{1:t-1} \sim N(\hat{\beta}_{t|t-1}, \mathbf{P}_{t|t-1})$, where $\hat{\beta}_{t|t-1} = \mathbf{F}_t \hat{\beta}_{t-1|t-1}$ and $\mathbf{P}_{t|t-1} = \mathbf{F}_t \mathbf{P}_{t-1|t-1} \mathbf{F}_t^\top + \mathbf{Z}_t$.*

2. *The posterior distribution of β_t at time t is $\beta_t \mid y_{1:t} \sim N(\hat{\beta}_{t|t}, \mathbf{P}_{t|t})$, where $\hat{\beta}_{t|t} = \hat{\beta}_{t|t-1} + \mathbf{K}_t e_t$, $\hat{y}_{t|t-1} = \mathbf{x}_t^\top \hat{\beta}_{t|t-1}$, $e_t = y_t - \hat{y}_{t|t-1}$, $Q_{t|t-1} = \mathbf{x}_t^\top \mathbf{P}_{t|t-1}\mathbf{x}_t + \Sigma$, $\mathbf{K}_t = \mathbf{P}_{t|t-1}\mathbf{x}_t Q_{t|t-1}^{-1}$ and $\mathbf{P}_{t|t} = \mathbf{P}_{t|t-1} - \mathbf{K}_t Q_{t|t-1}\mathbf{K}_t^\top$.*

The proof of this result is very similar to that of Theorem 3.2 and is left to the reader as an exercise.

Some comments are in order. Theorem 3.2 for a univariate state space model is obtained as a special case of Theorem 5.1, for $d = 1$. If $d \geq 2$, the *Kalman gain* $\mathbf{K}_t$ is a $p \times d$ matrix and the forecast variance $Q_{t|t-1}$ is now a $d \times d$ covariance matrix. Many of the aspects of univariate state space modelling discussed in Chap. 4 can be extended to the multivariate case. Below we point out several of these extensions.

- **Fixed-interval smoothing.**
  For each $t = 1, \ldots, n$, the smoothed state distribution is: $\beta_t \mid y_{1:n} \sim N(\hat{\beta}_{t|n}, \mathbf{P}_{t|n})$, where $\hat{\beta}_{t|n} = \hat{\beta}_{t|t} + \mathbf{L}_t(\hat{\beta}_{t+1|n} - \hat{\beta}_{t+1|t})$ and $\mathbf{P}_{t|n} = \mathbf{P}_{t|t} + \mathbf{L}_t(\mathbf{P}_{t+1|n} - \mathbf{P}_{t+1|t})\mathbf{L}_t^\top$, with $\mathbf{L}_t = \mathbf{P}_{t|t}\mathbf{F}_{t+1}^\top \mathbf{P}_{t+1|t}^{-1}$ and $\hat{\beta}_{t|t}, \hat{\beta}_{t+1|t}, \mathbf{P}_{t|t}, \mathbf{P}_{t+1|t}$ being calculated via the Kalman filter (Theorem 5.1). Similarly, the smoothed observation distribution is: $y_t \mid y_{1:n} \sim N(\hat{y}_{t|n}, Q_{t|n})$, where $\hat{y}_{t|n} = \mathbf{x}_t^\top \hat{\beta}_{t|n}$ and $Q_{t|n} = \mathbf{x}_t^\top \mathbf{P}_{t|n}\mathbf{x}_t + \Sigma$. These results reduce to Theorem 3.4 when $d = 1$ (univariate case).

- **Forecasting.** Assuming that the transition matrix $\mathbf{F}_t = \mathbf{F}$ is time-invariant and with information $y_{1:t} = \{y_1, \ldots, y_t\}$, the k-step ahead forecast state distribution is given by $\beta_{t+k} \mid y_{1:t} \sim N(\hat{\beta}_{t+k|t}, \mathbf{P}_{t+k|t})$, with $\hat{\beta}_{t+k|t} = \mathbf{F}^k \hat{\beta}_{t|t}$ and

$$\mathbf{P}_{t+k|t} = \mathbf{F}^k \mathbf{P}_{t|t}(\mathbf{F}^k)^\top + \sum_{j=0}^{k-1} \mathbf{F}^j \mathbf{Z}_{t+k-j}(\mathbf{F}^j)^\top,$$

where $\hat{\beta}_{t|t}, \mathbf{P}_{t|t}$ are computed by the Kalman filter (Theorem 5.1).
Similarly, the k-step ahead forecast observation distribution is given by $y_{t+k} \mid y_{1:t} \sim N(\hat{y}_{t+k|t}, Q_{t+k|t})$, where $\hat{y}_{t+k|t} = \mathbf{x}_{t+k}^\top \hat{\beta}_{t+k|t}$ and $Q_{t+k|t} = \mathbf{x}_{t+k}^\top \mathbf{P}_{t+k|t}\mathbf{x}_{t+k} + \Sigma$. As in Chap. 3 the forecast mean vector $\hat{y}_{t+k|t} = E(y_{t+k} \mid y_{1:t})$ (if viewed as a function of $k = 1, 2, 3, \ldots$) is known as the *forecast function* and can play an important role in the model design.

Below is a summary of the Kalman filter.

Kalman Filter

1. Initial distribution at $t = 0$: $\beta_0 \sim N(\hat{\beta}_{0|0}, \mathbf{P}_{0|0})$;
2. Posterior distribution of β_{t-1} at time $t - 1$:
   $\beta_{t-1} \mid y_{1:t-1} \sim N(\hat{\beta}_{t-1|t-1}, \mathbf{P}_{t-1|t-1})$;

(continued)

3. Prior distribution of β_t at time $t-1$:
$$\beta_t \mid y_{1:t-1} \sim N(\hat{\beta}_{t|t-1}, \mathbf{P}_{t|t-1}),$$
where $\hat{\beta}_{t|t-1} = \mathbf{F}_t \hat{\beta}_{t-1|t-1}$ and $\mathbf{P}_{t|t-1} = \mathbf{F}_t \mathbf{P}_{t-1|t-1} \mathbf{F}_t^\top + \mathbf{Z}_t$;
4. Posterior distribution at time t:
$$\beta_t \mid y_{1:t} \sim N(\hat{\beta}_{t|t}, \mathbf{P}_{t|t}), \text{ where}$$

$$\hat{\beta}_{t|t} = \hat{\beta}_{t|t-1} + \mathbf{K}_t e_t, \quad \mathbf{P}_{t|t} = \mathbf{P}_{t|t-1} - \mathbf{K}_t \mathbf{Q}_{t|t-1} \mathbf{K}_t^\top,$$

$$\hat{y}_{t|t-1} = \mathbf{x}_t^\top \hat{\beta}_{t|t-1}, \quad e_t = y_t - \hat{y}_{t|t-1}, \quad \mathbf{Q}_{t|t-1} = \mathbf{x}_t^\top \mathbf{P}_{t|t-1} \mathbf{x}_t + \Sigma,$$

$$\mathbf{K}_t = \mathbf{P}_{t|t-1} \mathbf{x}_t \mathbf{Q}_{t|t-1}^{-1}.$$

5.2 Model Specification and Design

This section discusses specification of the design matrix $\mathbf{x}_t$, the transition matrix $\mathbf{F}_t$ as well as specification or estimation of the observation covariance matrix Σ and the transition covariance matrix $\mathbf{Z}_t$.

Specification of $\mathbf{x}_t$ and $\mathbf{F}_t$. In the core of model specification and design is the specification or selection of the design matrix $\mathbf{x}_t$, the transition matrix $\mathbf{F}_t$, the observation covariance matrix Σ and the transition covariance matrix $\mathbf{Z}_t$. Leaving aside these two covariance matrices (their specification is discussed later), the components $\mathbf{x}_t$ and $\mathbf{F}_t$ characterise the forecast function $y_{t+k|t}$ (assuming $\hat{\beta}_{t|t}$ being calculated), hence their specification (or estimation) plays a critical role for filtering, forecasting and smoothing. Noting the state space breakdown of y_t into d univariate state space models $y_{1t}, \ldots, y_{dt}$ as in (5.4a)–(5.4b), we can think of choosing each row $x_{it}^\top$ of $\mathbf{x}_t^\top$ to describe each of the scalar time series y_{it}. This can be facilitated as in Chap. 4 (see e.g. Sect. 4.1), but here we note that the state vector β_t is common for all d state space models, hence the dynamics of β driven by the transition matrix $\mathbf{F}_t$ must be the same for all d models). For example, consider a bivariate time series $y_t = [y_{1t}, y_{2t}]^\top$, for which we consider a bivariate state space model (5.1a)–(5.1b). Suppose that both y_{1t} and y_{2t} represent a linear trend (linear growth), but they may differ in their variability, i.e. the components of the observation and transition variances. Then a possible specification for $\mathbf{x}_t$ and $\mathbf{F}_t$ is

$$\mathbf{x}_t = \begin{bmatrix} 1 & 1 \\ 0 & 0 \end{bmatrix} \quad \text{and} \quad \mathbf{F}_t = \begin{bmatrix} 1 & 1 \\ 0 & 1 \end{bmatrix}$$

so that both

$$y_{1t} = [1, 0]\beta_t + \epsilon_{1t} \quad \text{and} \quad y_{2t} = [1, 0]\beta_t + \epsilon_{2t},$$

represent linear growth state space models, with common transition equation

$$\beta_t = \begin{bmatrix} 1 & 1 \\ 0 & 1 \end{bmatrix} \beta_{t-1} + \zeta_t,$$

where $\epsilon_t = [\epsilon_{1t}, \epsilon_{2t}]^\top$.

This approach will not work if y_{1t} and y_{2t} represent structurally different state space models, e.g. if y_{1t} represents linear growth, while y_{2t} represents a seasonal time series. This is because their respective state space vectors will not be the same; they will be driven by different transition matrices and hence will exhibit different dynamics (see e.g. the different state representations in Sect. 4.1). In such a case it may be worth modelling each y_{it} separately using a univariate state space model.

Specification of Σ and Z_t. Apart from the specification of the design and transition matrices x_t and F_t, the estimation and forecasting provided by the Kalman filter are subject to the specification of the observation covariance matrix Σ (the covariance matrix of the observation innovation vector ϵ_t) and the transition covariance matrix Z_t (the covariance matrix of the state innovation vector ζ_t). Z_t may be specified using discount factors, as in Sect. 4.3.2, i.e. setting up Z_t as in Eq. (4.53). Alternatively, if $Z_t = Z$ is time-invariant, it can be estimated from the data, together with Σ by using the EM algorithm. The EM algorithm is very similar to that of the univariate case (see p. 158). Indeed, with the multivariate state space model (5.1a)–(5.1b), the log-likelihood function (4.42) is upgraded to

$$\ell(\theta; y, \beta) = -\frac{n}{2} \log |\Sigma| - \frac{1}{2} \sum_{t=1}^{n} (y_t - x_t^\top \beta_t)^\top \Sigma^{-1} (y_t - x_t^\top \beta_t)$$

$$-\frac{n}{2} \log |Z| - \frac{1}{2} \sum_{t=1}^{n} (\beta_t - F_t \beta_{t-1})^\top Z^{-1} (\beta_t - F_t \beta_t)$$

$$-\frac{1}{2} \log |P_{0|0}| - \frac{1}{2} (\beta_0 - \hat{\beta}_{0|0})^\top P_{0|0}^{-1} (\beta_0 - \hat{\beta}_{0|0}), \qquad (5.5)$$

where now θ is the vector including the unknown parameters of Σ, Z, $\hat{\beta}_{0|0}$, $P_{0|0}$. A similar procedure to the EM algorithm for univariate state space models involves the expectation(E-step) and differentiation (M-step) leading to the following recursions of the EM algorithm.

EM Algorithm for Multivariate State Space Models

In the state space model (5.1a)–(5.1b) with information $y_{1:n} = \{y_1, \ldots, y_n\}$ the maximum likelihood estimate of the parameter vector

$$\theta = [\text{vech}(\boldsymbol{\Sigma}), \text{vech}(\mathbf{Z})^\top, \hat{\beta}_{0|0}^\top, \text{vech}(\mathbf{P}_{0|0})^\top]^\top,$$

is approximately $\hat{\theta}^{(N)}$, where $\hat{\theta}^{(i)}$ is iteratively computed as follows:

1. Initial estimate

$$\theta^{(0)} = \left[\text{vech}(\hat{\boldsymbol{\Sigma}}^{(0)})^\top, \text{vech}(\hat{\mathbf{Z}}^{(0)})^\top, (\hat{\beta}_{0|0}^{(0)})^\top, \text{vech}(\hat{\mathbf{P}}_{0|0}^{(0)})^\top\right]^\top.$$

2. For each $i = 0, 1, 2, \ldots, N - 1$:

   a. For each $t = 0, 1, \ldots, n$ Compute $\hat{\beta}_{t|n}, P_{t|n}$, conditioned upon $\theta^{(i)}$;
   b. Compute

$$\hat{\boldsymbol{\Sigma}}^{(i+1)} = \frac{1}{n} \sum_{t=1}^{n} \left[(y_t - \mathbf{x}_t^\top \hat{\beta}_{t|n})(y_t - \mathbf{x}_t^\top \hat{\beta}_{t|n})^\top + \mathbf{x}_t^\top P_{t|n} \mathbf{x}_t\right];$$

$$\hat{\mathbf{Z}}^{(i+1)} = \frac{1}{n} \sum_{t=1}^{n} \left[(\hat{\beta}_{t|n} - \mathbf{F}_t \hat{\beta}_{t-1|n})(\hat{\beta}_{t|n} - \mathbf{F}_t \hat{\beta}_{t-1|n})^\top\right.$$

$$\left. + \mathbf{P}_{t|n} - \mathbf{P}_{t,t-1|n}\mathbf{F}_t^\top - \mathbf{F}_t \mathbf{P}_{t,t-1|n}^\top + \mathbf{F}_t \mathbf{P}_{t-1|n}\mathbf{F}_t^\top\right];$$

$$\hat{\beta}_{0|0}^{(i+1)} = \hat{\beta}_{0|n};$$

$$\hat{\mathbf{P}}_{0|0}^{(i+1)} = \frac{1}{n} \left[(\hat{\beta}_{0|n} - \hat{\beta}_{0|0})(\hat{\beta}_{0|n} - \hat{\beta}_{0|0})^\top + \mathbf{P}_{0|n}\right];$$

   c. Set

$$\hat{\theta}^{(i+1)} = \left[\text{vech}(\hat{\boldsymbol{\Sigma}}^{(i+1)})^\top, \text{vech}(\hat{\mathbf{Z}}^{(i+1)})^\top, (\hat{\beta}_{0|0}^{(i+1)})^\top, \text{vech}(\hat{\mathbf{P}}_{0|0}^{(i+1)})^\top\right]^\top.$$

The proof of this version of the EM algorithm is very similar to that in the univariate case (see Sect. 4.3.1 in Chap. 4); we note that the proof of $\hat{\boldsymbol{\Sigma}}$ is obtained in a similar way as that of $\hat{\mathbf{Z}}$ in the univariate case (Sect. 4.3.1). The same stopping rules can be applied as in the EM algorithm for the univariate case.

Finally, we briefly discuss about the steady state of multivariate state space models.

Steady State of the Kalman Filter Considering the state space model (5.1a)–(5.1b) with time-invariant components $\mathbf{x}_t = \mathbf{x}$, $\mathbf{F}_t = \mathbf{F}$ and $\mathbf{Z}_t = \mathbf{Z}$, then Theorem 3.7 is easily extended to the multivariate case. Thus, the limit of the posterior covariance matrix $\mathbf{P}_{t|t}$ exists and does not depend on the prior covariance matrix $\mathbf{P}_{0|0}$. The proof is almost identical to that of Theorem 3.7, the modifications taking care that y_t is known a vector and $\mathbf{Q}_{t|t-1}$ is a covariance matrix (replacing the variance $q_{t|t-1}$). The steady state of the Kalman filter replaces $\mathbf{P}_{t|t}$ in $\hat{\beta}_{t|t}$ by its limit $\mathbf{P}$, hence enabling significant computational savings. The next section considers the multivariate local level model, for which an explicit expression of the limit of the posterior covariance matrix is available.

5.3   Steady State of the Multivariate Local Level Model

In Sect. 3.5.2 the steady state of the Kalman filter for univariate local level models was studied; in effect the limit of the posterior variance of the states, obtained from the Kalman filter, was derived. The concept of the steady state for multivariate state space models provides a complete analogue of that for univariate models, discussed in Sect. 3.5. In essence under certain conditions the posterior covariance matrix of the state vector converges to a stable matrix, which can replace the posterior covariance matrix in the recursions of the Kalman filter providing computational savings in the application of the Kalman filter. In the following we consider a slight generalisation of the local level state space model and we derive in close form the limit of the covariance matrix of the state vector.

Before we proceed with the description of the local level model, we briefly discuss square root of symmetric matrices. We only discuss here the square root based on the spectral decomposition of a symmetric matrix; other approaches are also possible, e.g. based on the Choleski decomposition. Suppose that $\mathbf{X}$ is a symmetric matrix with elements from the real field. Then we can write

$$\mathbf{X} = \mathbf{\Gamma}\mathbf{\Lambda}\mathbf{\Gamma}^{\top}, \tag{5.6}$$

where $\mathbf{\Lambda}$ is the diagonal matrix with its diagonal including the eigenvalues of $\mathbf{X}$ and $\mathbf{\Gamma}$ is the matrix consisting of the respective normalised eigenvectors; the normalisation concerns restricting the eigenvectors to have a unit Eucledean norm. This implies that the matrix $\mathbf{\Gamma}$ is orthogonal, or $\mathbf{\Gamma}^{\top}\mathbf{\Gamma} = \mathbf{I}$ For the proof of (5.6) the reader is referred to Horn and Johnson (2013) and Harville (1997).

Suppose now that $\mathbf{X}$ is a covariance matrix, so that it is symmetric and positive definite. It follows that the eigenvalues of $\mathbf{X}$ are real and strictly positive and so we can define the *symmetric square root matrix* of $\mathbf{X}$ as the matrix

$$\mathbf{X}^{1/2} = \mathbf{\Gamma}\mathbf{\Lambda}^{1/2}\mathbf{\Gamma}^{\top},$$

or in words, the matrix whose eigenvalues are the square root of the eigenvalues of $\mathbf{X}$. From this definition and Eq. (5.6) it is easy to verify that

$$\mathbf{X}^{1/2}\mathbf{X}^{1/2} = \mathbf{\Gamma}\mathbf{\Lambda}^{1/2}\mathbf{\Gamma}^{\top}\mathbf{\Gamma}\mathbf{\Lambda}^{1/2}\mathbf{\Gamma}^{\top} = \mathbf{\Gamma}\mathbf{\Lambda}\mathbf{\Gamma}^{\top} = \mathbf{X},$$

since the matrix $\mathbf{\Gamma}$ is orthogonal. The inverse of $\mathbf{X}^{1/2}$ will be denoted by $\mathbf{X}^{-1/2}$ and is the covariance matrix with eigenvalues equal to the inverse of the square root of the eigenvalues of $\mathbf{X}$.

Returning to the local level model, consider that d-dimensional observation vector time series $\{y_t\}$ is generated by the state space model

$$y_t = \beta_t + \epsilon_t \quad \text{and} \quad \beta_t = \phi\beta_{t-1} + \zeta_t, \tag{5.7}$$

where ϕ is a known scalar, ϵ_t is a white noise process, ζ_t is a white noise process, ϵ_t is independent of ζ_s, for any t, s. Moreover, the distribution of the observation vector ϵ_t is a d-dimensional Gaussian distribution with zero mean vector and some covariance matrix $\mathbf{\Sigma}$, written as $\epsilon_t \sim N(0, \mathbf{\Sigma})$. Likewise, the distribution of ζ_t is $\zeta_t \sim N(0, \mathbf{\Sigma}^{1/2}\mathbf{Z}^*\mathbf{\Sigma}^{1/2})$, where here $\mathbf{\Sigma}^{1/2}$ denotes the symmetric square root matrix of $\mathbf{\Sigma}$, based on the spectral decomposition of symmetric matrices (see above). The above factorisation of the covariance matrix of ζ_t, i.e. $\text{Var}(\zeta_t) = \mathbf{Z} = \mathbf{\Sigma}^{1/2}\mathbf{Z}^*\mathbf{\Sigma}^{1/2}$ proposes a suitable proportionality between $\mathbf{\Sigma}$ and $\mathbf{Z}$ and is considered because it facilitates estimation of $\mathbf{\Sigma}$, which is developed in Triantafyllopoulos (2011a). The prior covariance matrix of β_0 is also scaled in the same way, hence $\text{Var}(\beta_0) = \mathbf{P}_{0|0} = \mathbf{\Sigma}^{1/2}\mathbf{P}^*_{0|0}\mathbf{\Sigma}^{1/2}$, so that the prior distribution of β_0 is

$$\beta_0 \sim N(\hat{\beta}_{0|0}, \mathbf{\Sigma}^{1/2}\mathbf{P}^*_{0|0}\mathbf{\Sigma}^{1/2}),$$

for some known $d \times d$ covariance matrix $\mathbf{P}^*_{0|0}$. In the discussion of this section we assume that $\mathbf{\Sigma}$ is known; its estimation is studied in Triantafyllopoulos (2011a). Strictly speaking model (5.7) is a local level when $\phi = 1$, but here we expand the definition by considering a slightly wider class of state space models, for any value of ϕ. In particular, we remark the following:

1. If $\phi = 0$, then model (5.7) is reduced to $y_t = \epsilon_t + \zeta_t$, which is essentially a white noise with covariance matrix $\mathbf{\Sigma} + \mathbf{Z}$.
2. If $\phi = 1$, then model (5.7) is the traditional local level model; the level β_t follows is a random walk process.
3. If $|\phi| < 1$, then the level β_t follows a multivariate autoregressive process of order one, with AR parameter matrix $\phi\mathbf{I}$.
4. If $|\phi| > 1$, the model behaves similarly as a local level, but the level evolution expanding erratically, depending on the value of ϕ.

The following theorem gives the main result, the convergence of the posterior covariance matrix $\mathbf{P}_{t|t}$.

Theorem 5.2 *In the state space model (5.7) with the prior* $\mathbf{P}^*_{0|0} = p_0 I_p$, *for a known constant* $p_0 > 0$, *the limit of the sequence of posterior covariance matrices* $\{\mathbf{P}_{t|t}\}$ *exists and it is given by*

$$\mathbf{P} = \lim_{t \to \infty} \mathbf{P}_{t|t} = \mathbf{\Sigma}^{1/2} \mathbf{P}^* \mathbf{\Sigma}^{1/2},$$

where

$$\mathbf{P}^* = \frac{1}{2\phi^2} \left[\left\{ (\mathbf{Z}^* + (1 - \phi^2)\mathbf{I})^2 + 4\mathbf{Z}^* \right\}^{1/2} - \mathbf{Z}^* - (1 - \phi^2)\mathbf{I} \right],$$

for $\phi \neq 0$ *and* $\mathbf{P}^* = \mathbf{Z}^*(\mathbf{Z}^* + \mathbf{I})^{-1}$, *for* $\phi = 0$.

The principles of boundedness, monotonicity and convergence of a sequence of matrices (in particular covariance matrices) were discussed in Sect. 3.5.3. Here we show that if $\mathbf{A} > 0$ and $\mathbf{B} > 0$ are two positive definite matrices, such that $\mathbf{A} - \mathbf{B} > 0$ is a positive definite matrix (this is written as $\mathbf{A} > \mathbf{B}$, as discussed in Sect. 3.5.3), then $\mathbf{A}^{-1} < \mathbf{B}^{-1}$. Indeed, note that $\mathbf{A} - \mathbf{B} > 0$ implies that there is a positive definite matrix $\mathbf{C}$ such that $\mathbf{A} = \mathbf{B} + \mathbf{C}$. Now from the property $\mathbf{B}^{-1} - \mathbf{A}^{-1} = \mathbf{A}^{-1}(\mathbf{A} - \mathbf{B})\mathbf{B}^{-1}$ we have that $\mathbf{B}^{-1} - \mathbf{A}^{-1} > 0$, since $\mathbf{A}^{-1} > 0$, $\mathbf{A} - \mathbf{B} > 0$ and $\mathbf{B}^{-1} > 0$. Hence $\mathbf{B}^{-1} > \mathbf{A}^{-1}$. A book length discussion on matrix analysis with particular emphasis to covariance matrices can be found in the classic textbook Horn and Johnson (2013).

With this result in place, the existence of the limit of $\mathbf{P}_{t|t}$ is proven in the next lemma.

Lemma 5.1 *With the assumptions of Theorem 5.2, the sequence of* $d \times d$ *positive definite matrices* $\{\mathbf{P}_t\}$ *is convergent.*

Proof With the prior of β_0 and the Kalman filter, we will show that $\mathbf{P}_{t|t} = \mathbf{\Sigma}^{1/2}\mathbf{P}^*_{t|t}\mathbf{\Sigma}^{1/2}$, hence the limit of $\mathbf{P}_{t|t}$ exists if and only if the limit of $\mathbf{P}^*_{t|t}$ exists, where $\mathbf{P}^*_{t|t}$ is updated by the Kalman filter as follows. First we note that $\mathbf{P}_{t|t-1} = \mathbf{\Sigma}^{1/2}(\mathbf{P}^*_{t-1|t-1} + \mathbf{Z}^*)\mathbf{\Sigma}^{1/2} = \mathbf{\Sigma}^{1/2}\mathbf{P}^*_{t|t-1}\mathbf{\Sigma}^{1/2}$ and the Kalman gain is $\mathbf{K}_t = \mathbf{P}_{t|t-1}\mathbf{Q}^{-1}_{t|t-1} = \mathbf{\Sigma}^{1/2}\mathbf{K}^*_t\mathbf{\Sigma}^{-1/2}$, where $\mathbf{Q}_{t|t-1} = \mathbf{\Sigma}^{1/2}(\mathbf{P}^*_{t|t-1} + \mathbf{I})\mathbf{\Sigma}^{1/2} = \mathbf{\Sigma}^{1/2}\mathbf{Q}^*_{t|t-1}\mathbf{\Sigma}^{1/2}$ is used. Thus, the posterior covariance matrix of β_t is $\mathbf{P}_{t|t} = \mathbf{\Sigma}^{1/2}\mathbf{P}^*_{t|t}\mathbf{\Sigma}^{1/2}$, where

$$\begin{aligned}
\mathbf{P}^*_{t|t} &= \mathbf{P}^*_{t|t-1} - \mathbf{K}^*\mathbf{Q}^*_{t|t-1}\mathbf{K}^*_t \\
&= \mathbf{P}^*_{t|t-1}(\mathbf{I} - \mathbf{Q}^{*-1}_{t|t-1}\mathbf{P}^*_{t|t-1}) \\
&= \mathbf{P}^*_{t|t-1}\mathbf{Q}^*_{t|t-1}(\mathbf{Q}^*_{t|t-1} - \mathbf{P}^*_{t|t-1}) \\
&= \mathbf{P}^*_{t|t-1}\mathbf{Q}^*_{t|t-1}.
\end{aligned}$$

First suppose that $\phi = 0$. Then $\mathbf{P}^*_{t|t-1} = \mathbf{Z}^*$, for all t, and so $\mathbf{P}^*_{t|t} = \mathbf{Z}^*(\mathbf{Z}^*+\mathbf{I})^{-1}$, which of course is convergent.

Suppose now that $\phi \neq 0$. To prove the existence of the limit of $\{\mathbf{P}^*_{t|t}\}$ It suffices to prove that $\{\mathbf{P}^*_{t|t}\}$ is bounded and monotonic. Clearly, $0 \leq \mathbf{P}^*_{t|t}$ and since $\phi^2 > 0$ and $\mathbf{Z}^*$ is positive definite $0 < \mathbf{P}^*_{t|t}$, for all $t > 0$. Since $(\mathbf{P}^*_{t|t-1} + \mathbf{I})^{-1} > 0$, $(\mathbf{P}^*_{t|t-1} + \mathbf{I} - \mathbf{P}^*_{t|t-1})(\mathbf{P}^*_{t|t-1} + \mathbf{I})^{-1} > 0 \Rightarrow \mathbf{P}^*_{t|t} = \mathbf{P}^*_{t|t}(\mathbf{P}^*_{t|t-1} + \mathbf{I})^{-1} < \mathbf{I}$ and so $0 < \mathbf{P}^*_{t|t} < \mathbf{I}$.

For the monotonicity it suffices to prove that, if $\mathbf{P}^{*-1}_{t-1|t-1} > \mathbf{P}^{*-1}_{t-2|t-2}$ (equivalent $\mathbf{P}^{*-1}_{t-1|t-1} < \mathbf{P}^{*-1}_{t-2|t-2}$), then $\mathbf{P}^{*-1}_{t|t} > \mathbf{P}^{*-1}_{t-1|t-1}$ (equivalent $\mathbf{P}^{*-1}_{t|t} < \mathbf{P}^{*-1}_{t-1}$). From $\mathbf{P}^{*-1}_{t-1|t-1} > \mathbf{P}^{*-1}_{t-2|t-2}$ we have $\mathbf{P}^*_{t-1|t-1} < \mathbf{P}^*_{t-2|t-2} \Rightarrow \mathbf{P}^*_{t|t-1} < \mathbf{P}^*_{t-1|t-2} \Rightarrow \mathbf{P}^{*-1}_{t|t} > \mathbf{P}^{*-1}_{t-1|t-2} \Rightarrow \mathbf{P}^{*-1}_{t|t} - \mathbf{P}^{*-1}_{t-1|t-1} = \mathbf{P}^{*-1}_{t|t-1} - \mathbf{P}^{*-1}_{t-1|t-2} > 0$, since $\mathbf{P}^{*-1}_{t|t} = (\mathbf{P}^*_{t|t-1} + \mathbf{I})\mathbf{P}^{*-1}_{t|t-1} = \mathbf{I} + \mathbf{P}^{*-1}_{t|t-1}$. With an analogous argument we have that if $\mathbf{P}^{*-1}_{t-1|t-1} < \mathbf{P}^{*-1}_{t-2|t-2}$, then $\mathbf{P}^{*-1}_{t|t} - \mathbf{P}^{*-1}_{t-1|t-1} < 0$, from which the monotonicity follows. □

The next lemma proves a property, which is used to derive a closed form expression of the limit of $\{\mathbf{P}^*_{t|t}\}$ below.

Lemma 5.2 Let $\{\mathbf{P}^*_{t|t}\}$ be the sequence of Lemma 5.1. Then, with $\mathbf{Z}^*$ as in Lemma 5.1, the limiting matrix $\mathbf{P}^* = \lim_{t\to\infty} \mathbf{P}^*_{t|t}$ commutes with $\mathbf{Z}^*$.

Proof First we prove that if $\mathbf{P}^*_{t-1|t-1}$ commutes with $\mathbf{Z}^*$, then $\mathbf{P}^*_{t|t}$ also commutes with $\mathbf{Z}^*$. Indeed from $\mathbf{P}^*_{t|t} = (\phi^2\mathbf{P}^*_{t-1|t-1} + \mathbf{Z}^*)(\phi^2\mathbf{P}^*_{t-1|t-1} + \mathbf{Z}^* + \mathbf{I})^{-1}$ we have that $\mathbf{P}^{*-1}_{t|t} = \mathbf{I} + (\phi^2\mathbf{P}^*_{t-1|t-1} + \mathbf{Z}^*)^{-1}$ and then

$$\mathbf{P}^{*-1}_{t|t}\mathbf{Z}^{*-1} = \mathbf{Z}^{*-1} + (\phi^2\mathbf{Z}^*\mathbf{P}^*_{t-1|t-1} + \mathbf{Z}^{*2})^{-1}$$
$$= \mathbf{Z}^{*-1} + (\phi^2\mathbf{P}^*_{t-1|t-1}\mathbf{Z}^* + \mathbf{Z}^{*2})^{-1} = \mathbf{Z}^{*-1}\mathbf{P}^{*-1}_{t|t}$$

which implies that $\mathbf{Z}^*\mathbf{P}^*_{t|t} = (\mathbf{P}^{*-1}_{t|t}\mathbf{Z}^{*-1})^{-1} = (\mathbf{Z}^{*-1}\mathbf{P}^{*-1}_{t|t})^{-1} = \mathbf{P}^*_{t|t}\mathbf{Z}^*$ and so $\mathbf{P}^*_{t|t}$ and $\mathbf{Z}^*$ commute. Because $\mathbf{P}^*_{0|0} = p_0\mathbf{I}$, $\mathbf{P}_{0|0}$ commutes with $\mathbf{Z}^*$ and so by induction it follows that the sequence of matrices $\{\mathbf{P}_{t|t}, t \geq 0\}$ commutes with $\mathbf{Z}^*$. Since $\mathbf{P}^* = \lim_{t\to\infty} \mathbf{P}^*_{t|t}$ exists (Lemma 5.1) we have

$$\mathbf{P}^*\mathbf{Z}^* = \lim_{t\to\infty}(\mathbf{P}^*_{t|t}\mathbf{Z}^*) = \lim_{t\to\infty}(\mathbf{Z}^*\mathbf{P}^*_{t|t}) = \mathbf{Z}^*\mathbf{P}^*$$

and so $\mathbf{P}^*$ commutes with $\mathbf{Z}^*$. □

Finally, we can give the proof of Theorem 5.2.

Proof (Proof of Theorem 5.2) From Lemma 5.1 we know that the limit of $\mathbf{P}_{t|t} = \mathbf{\Sigma}^{1/2}\mathbf{P}^*_{t|t}\mathbf{\Sigma}^{1/2}$ exists, hence it is

$$\mathbf{P} = \lim_{t\to\infty} \mathbf{P}_{t|t} = \mathbf{\Sigma}^{1/2}\mathbf{P}^*\mathbf{\Sigma}^{1/2},$$

where $\mathbf{P}^* = \lim_{t \to \infty} \mathbf{P}^*_{t|t}$, which exists by Lemma 5.1. The rest of the proof concerns the derivation of $\mathbf{P}^*$.

If $\phi = 0$, then from Lemma 5.1 we have $\mathbf{P}^*_{t|t} = \mathbf{P}^* = \mathbf{Z}^*(\mathbf{Z}^* + \mathbf{I})^{-1}$. Let $\phi \neq 0$; from Lemma 5.1 we have that $\mathbf{P}^*$ exists and from Lemma 5.2 we have that $\mathbf{P}^*$ and $\mathbf{Z}^*$ commute. From $\mathbf{P}^*_{t|t} = (\phi^2 \mathbf{P}^*_{t-1|t-1} + \mathbf{Z}^*)(\mathbf{P}^*_{t-1|t-1} + \mathbf{Z}^* + \mathbf{I})^{-1}$ we have $\mathbf{P}^* = (\phi^2 \mathbf{P}^* + \mathbf{Z}^*)(\phi^2 \mathbf{P}^* + \mathbf{Z}^* + \mathbf{I})^{-1}$ from which we get the equation $\mathbf{P}^{*2} + \phi^{-2} \mathbf{P}^*(\mathbf{Z}^* + \mathbf{I} - \phi^2 \mathbf{I}) - \phi^{-2} \mathbf{Z}^* = 0$. In order to solve this matrix equation for $\mathbf{P}^*$, we complete the square, by using the result that $\mathbf{P}^*$ and $\mathbf{Z}^*$ commute (Lemma 5.2). Thus

$$\mathbf{P}^{*2} + \frac{1}{2\phi^2} \mathbf{P}^*(\mathbf{Z}^* + (1 - \phi^2)\mathbf{I}) + \frac{1}{2\phi^2}(\mathbf{Z}^* + (1 - \phi^2)\mathbf{I})\mathbf{P}^*$$

$$+ \frac{1}{4\phi^4}(\mathbf{Z}^* + (1 - \phi^2)\mathbf{I})^2 - \frac{1}{4\phi^4}(\mathbf{Z}^* + (1 - \phi^2)\mathbf{I})^2 - \mathbf{Z}^* = 0$$

$$\text{or} \quad \left(\mathbf{P}^* + \frac{1}{2\phi^2}(\mathbf{Z}^* + (1 - \phi^2)\mathbf{I}) \right)^2 = \frac{1}{4\phi^4}(\mathbf{Z}^* + (1 - \phi^2)\mathbf{I})^2 + \mathbf{Z}^*$$

$$\text{or} \quad \mathbf{P}^* = \frac{1}{2\phi^2} \left[\left\{ (\mathbf{Z}^* + (1 - \phi^2)\mathbf{I})^2 + 4\mathbf{Z}^* \right\}^{1/2} - \mathbf{Z}^* - (1 - \phi^2)I_p \right],$$

after rejecting the negative definite root.   □

Theorem 5.2 generalises results for the univariate local level, discussed in Sect. 3.5.2. Indeed, we observe that for $d = 1$ and if we set $\phi = 1$ (local level model), $\sigma^2 = \Sigma$ and $Z = \sigma^2 Z^*$ we obtain that the limiting covariance matrix P is

$$P = \lim_{t \to \infty} P_{t|t} = \sigma^2 P^* = \frac{\sigma^2}{2} \left(\sqrt{Z^{*2} + 4Z^*} - Z^* \right) = \frac{Z}{2} \left(\sqrt{1 + \frac{4\sigma^2}{Z}} - 1 \right),$$

which is the same as in Sect. 3.5.2. Indeed, it can be observed that the proof for the multivariate case parallels the proof for the univariate case in Sect. 3.5.2. However, if the proposed scaling of $\mathbf{Z} = \Sigma^2 \mathbf{Z}^* \Sigma^{1/2}$ (the proportionality of Σ and $\mathbf{Z}$) is not considered, then the expression of $\mathbf{P}$ in closed form in Theorem 5.2 is no more available.

Based on the above discussion and on the Kalman filter (Theorem 5.1) the steady state of the multivariate local level considered above replaces $\mathbf{P}_{t|t}$ by its limit $\Sigma^{1/2} \mathbf{P}^* \Sigma^{1/2}$ in $\hat{\beta}_{t|t}$ or

$$\hat{\beta}_{t|t} = \phi \hat{\beta}_{t-1|t-1} + \Sigma^{1/2} \mathbf{K}^* \Sigma^{-1/2} e_t,$$

where $e_t = y_t - \phi \hat{\beta}_{t-1|t-1}$ and

$$\mathbf{K}^* = \lim_{t \to \infty} \mathbf{K}^*_t = (\mathbf{P}^* + \mathbf{Z}^*)(\mathbf{P}^* + \mathbf{Z}^* + \mathbf{I})^{-1}.$$

We note that $\mathbf{P}^*$ as well as $\mathbf{K}^*$ depend on ϕ, $\mathbf{Z}^*$ and $\boldsymbol{\Sigma}$, hence $\mathbf{P}^*$ and $\mathbf{K}^*$ can be calculated before any data is collected (as long as the covariance matrices $\mathbf{Z}^*$ and $\boldsymbol{\Sigma}$, and ϕ are known a priori). This introduces important computational savings, especially for large values of d of the dimension of y_t. Indeed, we observe that there is only one matrix inversion required that of the matrix $\mathbf{P}^* + \mathbf{Z}^* + \mathbf{I}$ involved in the calculation of $\mathbf{K}^*$; instead the application of the Kalman filter (Theorem 5.1) requires a matrix inverse for each time t.

5.4 Error Analysis

In the above section we have described how the design matrix $\mathbf{x}_t$, the transition matrix $\mathbf{F}_t$ as well as the covariances $\boldsymbol{\Sigma}$ and $\mathbf{Z}_t$ may be specified or estimated, considering the multivariate state space model (5.1a)–(5.1b); more detailed discussion on sequential estimation of $\boldsymbol{\Sigma}$ follows in Sect. 5.5 below. Here, assuming that the matrices $\mathbf{x}_t$, the transition matrix $\mathbf{F}_t$ as well as the covariances $\boldsymbol{\Sigma}$ and $\mathbf{Z}_t$ are estimated or specified, we consider model assessment in the lines of residual analysis of Sect. 4.4.

The residual vector (or one-step ahead forecast errors) e_t are defined by

$$e_t = y_t - \hat{y}_{t|t-1} \tag{5.8}$$

where $\hat{y}_{t|t-1}$ is the one-step ahead forecast mean vector of y_t, which is provided recursively by the Kalman filter (Theorem 5.1). The goodness of fit is assessing whether the forecasts $\hat{y}_{t|t-1}$ are close enough to the observation vectors, or whether the residuals e_t are close to zero. The following result provides a generalisation of Theorem 4.2 and provides the main properties of the residuals.

Theorem 5.3 *In the state space model (5.1a)–(5.1b) and with the definitions of the Kalman filter recursions (Theorem 5.1) for some data* $y_{1:n} = \{y_1, \ldots, y_n\}$, *the following apply for the residuals.*

1. *Given* $y_{1:t-1}$, *the distribution of the residuals is* $e_t \mid y_{1:t-1} \sim N(0, \mathbf{Q}_{t|t-1})$ *(a d-variate Gaussian distribution), where* $\mathbf{Q}_{t|t-1}$ *is the forecast covariance matrix of* y_t, *provided by the Kalman filter.*
2. $e_1, \ldots, e_n$ *are independent.*
3. *With the definitions of* e_t *and* $\mathbf{Q}_{|t-1}$ *as above,*

$$\sum_{t=1}^{n} e_t^\top \mathbf{Q}_{t|t-1}^{-1} e_t \sim \chi_{nd}^2,$$

where χ_{nd}^2 *denotes the chi-square distribution with nd degrees of freedom.*

Proof (1) follows immediately from the definition of e_t by noting that the one-step forecast distribution of y_t is $y_t \mid y_{1:t-1} \sim N(\hat{y}_{t|t-1}, \mathbf{Q}_{t|t-1})$, see Theorem 5.1. The proof of (2) is identical to that of Theorem 4.2 except that now e_t is a residual vector, while in Theorem 4.2 e_t was a scalar. Moving on to the proof of (3) we note that from part (1) we have $e_t^* = \mathbf{Q}_{t|t-1}^{-1/2} e_t \sim N(0, \mathbf{I})$, so that $e_t^{*\top} e_t^* = e_t^\top \mathbf{Q}_{t|t-1}^{-1} e_t$ follows a chi-square distribution with d degrees of freedom, where $\mathbf{Q}_{t|t-1}^{-1/2}$ denotes the inverse of square root matrix, based on spectral decomposition theorem (see Sect. 5.3 for the definition of square root matrix). Now, since $e_1, \ldots, e_n$ are independent (from part (2)), this implies that

$$\sum_{t=1}^{n} e_t^{*\top} e_t^* = \sum_{t=1}^{n} e_t^\top \mathbf{Q}_{t|t-1}^{-1} e_t \sim \chi_{nd}^2$$

and the proof is completed.   □

From the proof of the above theorem we see that if we define by

$$e_t^* = \mathbf{Q}_{t|t-1}^{-1/2} e_t$$

the standardised residuals, the following properties apply.

1. $e_t^* \sim N(0, \mathbf{I})$;
2. $e_1^*, \ldots, e_n^*$ are independent;
3. For each t e_{it}^* and e_{jt}^* are independent and $e_{it} \sim N(0, 1)$, where $e_t^* = [e_{1t}^*, e_{2t}^*, \ldots, e_{dt}^*]^\top$.

Property (1) says that e_t^* follows a Gaussian distribution with zero mean vector and identity covariance matrix. This means that the elements of e_t^* are independent (property (3)) and e_t^* is independent of e_s^*, for $t \neq s$ (property (2)). These properties of e_t^* may be used to construct formal or informal diagnostic checks to assess the goodness of fit. Informal analysis may include:

- Histograms or normal probability plots to check whether e_{it}^* follows a Gaussian distribution.
- A plot of e_{it} against time t together with ± 1.96 credible intervals of the $N(0, 1)$ distribution. For a good fit we would expect about 5% of residuals to lie outside the intervals.
- Plots of the autocorrelation function (ACF) on $e_{i1}^*, e_{i2}^*, \ldots, e_{i,n}^*$ to check whether $e_{i1}^*, e_{i2}^*, \ldots, e_{in}^*$ are independent, for a fixed $i = 1, 2, \ldots, d$. For independence we would expect the autocorrelations in all lags of the residuals to lie inside the credible intervals $\pm 1.96/\sqrt{n}$.

Furthermore, more formal tests may be developed by using part (3) of Theorem 5.3, for example for a good model fit the value $X^2 = \sum_{t=1}^{n} e_t^{*\top} e_t^*$ should be smaller or equal to the 95% quantile of the chi-square distribution χ_{nd}^2.

Other than the above model diagnostics, measures of goodness of fit may be considered. These include the mean square error (MSE), mean of squared standardised residuals (MSSE) and mean absolute deviation (MAD), all being in line with the respective measures discussed in Sect. 4.4 for univariate state space models. These three measures are defined as

$$\text{MSE} = \frac{1}{n} \sum_{t=1}^{n} \left[e_{1t}^2, e_{2t}^2, \dots, e_{dt}^2 \right]^\top, \quad \text{MSSE} = \frac{1}{n} \sum_{t=1}^{n} \left[e_{1t}^{*2}, e_{2t}^{*2}, \dots, e_{dt}^{*2} \right]^\top,$$

$$\text{MAD} = \frac{1}{n} \sum_{t=1}^{n} \left[|e_{1t}|, |e_{2t}|, \dots, |e_{dt}| \right]^\top,$$

where it should be noted that, in contrast to the univariate case, the MSE, MSSE and MAD are now d-dimensional vectors. For a good model fit MSE and MAD should be close to the zero vector $[0, 0, \dots, 0]^\top$, while the MSSE should be close to the unit vector $[1, 1, \dots, 1]^\top$.

5.5 Covariance Estimation in State Space Models

5.5.1 Variance Estimation

The problem of estimating $\boldsymbol{\Sigma}$ and $\mathbf{Z}_t$ in the state space model (5.1a)–(5.1b) is known as *covariance estimation* (of state space models). As discussed above, one solution is to use the EM algorithm to estimate $\boldsymbol{\Sigma}$ and $\mathbf{Z}_t = \mathbf{Z}$, if $\mathbf{Z}_t = \mathbf{Z}$ is assumed to be time-invariant. In this section and in the next section we will assume that $\mathbf{Z}_t$ is known or is specified using discount factors and we discuss inference for $\boldsymbol{\Sigma}$. Unfortunately, a similar analysis as in the univariate case for the scaled observational model (SOP), discussed in Sect. 4.3.3, for the state space model (5.1a)–(5.1b) is not available. In the univariate case the variance of the observation innovation ϵ_t is σ^2 and the transition covariance matrix $\mathbf{Z}_t = \sigma^2 \mathbf{Z}_t^*$ is scaled by σ^2, for some covariance matrix $\mathbf{Z}_t^*$.

In this section we consider a multivariate generalisation of the SOP model of Sect. 4.3.3, whereby the observation and transition covariance matrices $\boldsymbol{\Sigma}$ and $\mathbf{Z}_t$ are known up to a scaling factor (or variance) σ^2, i.e.

$$\boldsymbol{\Sigma} = \sigma^2 \mathbf{V} \quad \text{and} \quad \mathbf{Z}_t = \sigma^2 \mathbf{Z}_t^*, \tag{5.9}$$

where $\mathbf{V}$ is a known $d \times d$ covariance matrix and $\mathbf{Z}_t^*$ is a known $p \times p$ covariance matrix. As stated above $\mathbf{Z}_t^*$ may be specified using a discount factor. A first consideration for $\mathbf{V}$ is to be proportional to the identity matrix; more advanced settings may include specifying $\mathbf{V}$ as a diagonal matrix or including in $\mathbf{V}$ some common correlation structure for all its off-diagonal elements. Any of these settings may be done using historical data or considering a priori beliefs of the component time series.

Consider the state space model (5.1a)–(5.1b), with Σ and $\mathbf{Z}_t$ as in (5.9), where the prior covariance matrix of β_0 is also scaled by σ^2, i.e.

$$\beta_0 \sim N(\hat{\beta}_{0|0}, \sigma^2 \mathbf{P}^*_{0|0}),$$

for some known covariance matrix $\mathbf{P}^*_{0|0}$. Furthermore, we assume that the prior distribution of $1/\sigma^2 = \theta$ is a gamma distribution

$$\theta = \frac{1}{\sigma^2} \sim G\left(\frac{n_0}{2}, \frac{d_0}{2}\right),$$

where n_0 and d_0 are known positive values. This model provides a first generalisation of the univariate SOP model of Sect. 4.3.3, which is obtained from the above model formulation for $d = 1$ and $\mathbf{V} = 1$. The derivation of the posterior distribution of θ and β_t, given $y_{1:t}$, follows that of the univariate SOP model in Sect. 4.3.3.

First of all we observe that conditionally on θ (or on σ^2), the prior distribution of $\beta_t \mid \theta, y_{1:t-1}$, the forecast distribution of $y_t \mid \theta, y_{1:t-1}$ and the posterior distribution of $\beta_t \mid \theta, y_{1:t}$ are given by the Kalman filter (Theorem 5.1) as

$$\beta_{t-1} \mid \theta, y_{1:t-1} \sim N(\hat{\beta}_{t|t-1}, \sigma^2 \mathbf{P}^*_{t|t-1}),$$

$$y_t \mid \theta, y_{1:t-1} \sim N(\hat{y}_{t|t-1}, \sigma^2 \mathbf{Q}^*_{t|t-1}), \tag{5.10}$$

$$\beta_t \mid \theta, y_{1:t} \sim N(\hat{\beta}_{t|t}, \sigma^2 \mathbf{P}^*_{t|t}), \tag{5.11}$$

where $\hat{\beta}_{t|t-1} = \mathbf{F}_t \hat{\beta}_{t-1|t-1}$, $\mathbf{P}^*_{t|t-1} = \mathbf{F}_t \mathbf{P}^*_{t-1|t-1} \mathbf{F}_t^\top$, $\hat{y}_{t|t-1} = \mathbf{x}_t^\top \hat{\beta}_{t|t-1}$, $\mathbf{Q}^*_{t|t-1} = \mathbf{x}_t^\top \mathbf{P}^*_{t|t-1} \mathbf{x}_t + \mathbf{V}$, $\hat{\beta}_{t|t} = \hat{\beta}_{t|t-1} + \mathbf{K}_t e_t$, $\mathbf{K}_t = \mathbf{P}_{t|t-1} \mathbf{x}_t \mathbf{Q}^{-1}_{t|t-1}$ and $e_t = y_t - \hat{y}_{t|t-1}$. Suppose that at time $t - 1$, the posterior distribution of θ is

$$\theta \mid y_{1:t-1} \sim G\left(\frac{n_{t-1}}{2}, \frac{d_{t-1}}{2}\right),$$

for some n_{t-1} and d_{t-1}. This prior is combined with (5.10) to give the posterior distribution of θ at t

$$p(\theta \mid y_{1:t}) \propto p(y_t \mid \theta, y_{1:t-1}) p(\theta \mid y_{1:t-1})$$

$$\propto \theta^{d/2} \exp\left(-\frac{1}{2} e_t^\top \mathbf{Q}^{*-1}_{t|t-1} e_t \theta\right) \theta^{n_{t-1}/2-1} \exp\left(-\frac{d_{t-1}\theta}{2}\right)$$

$$= \theta^{(n_{t-1}+d)/2-1} \exp\left[-\frac{1}{2}(e_t^\top \mathbf{Q}^{*-1}_{t|t-1} e_t + d_{t-1})\theta\right],$$

so that $\theta \mid y_{1:t}$ is proportional to the gamma distribution

$$\theta \mid y_{1:t} \sim G\left(\frac{n_t}{2}, \frac{d_t}{2}\right),$$

where $n_t = n_{t-1} + d$ and $d_t = d_{t-1} + e_t^\top \mathbf{Q}_{t|t-1}^{*-1} e_t$.

Making use of the conditional on θ posterior distribution of β_t (5.11), by integrating out θ, we obtain the posterior distribution of β_t as

$$p(\beta_t \mid y_{1:t}) = \int_{\mathbb{R}^p} p(\beta_t, \theta \mid y_{1:t}) \, d\theta$$

$$= \int_{\mathbb{R}^p} p(\beta_t \mid \theta, y_{1:t}) p(\theta \mid y_{1:t}) \, d\theta$$

$$\propto \int_{\mathbb{R}^p} \theta^{p/2} \exp\left[-\frac{1}{2}(\beta_t - \hat\beta_{t|t})^\top \mathbf{P}_{t|t}^{*-1}(\beta - \hat\beta_{t|t})\theta\right]$$

$$\times \theta^{n_t/2-1} \exp\left(-\frac{d_t \theta}{2}\right) d\theta$$

$$= \int_{\mathbb{R}^p} \theta^{(n_t+p)/2-1}$$

$$\times \exp\left\{-\frac{1}{2}\left[(\beta_t - \hat\beta_{t|t})^\top \mathbf{P}_{t|t}^{*-1}(\beta_t - \hat\beta_{t|t}) + d_t\right]\theta\right\} d\theta$$

$$\propto \left[(\beta_t - \hat\beta_{t|t})^\top \mathbf{P}_{t|t}^{*-1}(\beta_t - \hat\beta_{t|t}) + d_t\right]^{-(n_t+p)/2}.$$

which is obtained from the gamma integral (4.60).

Thus, the posterior distribution of β_t at time t is proportional to the multivariate Student t distribution

$$\beta_t \mid y_{1:t} \sim t(n_t, \hat\beta_{t|t}, \mathbf{P}_{t|t}),$$

where $\mathbf{P}_{t|t} = S_t \mathbf{P}_{t|t}^*$ and $S_t = d_t/n_t$.

Similarly we see that $\beta_t \mid y_{1:t-1} \sim t(n_{t-1}, \hat\beta_{t|t-1}, \mathbf{P}_{t|t-1})$ and $y_t \mid y_{1:t-1} \sim t(n_{t-1}, \hat{y}_{t|t-1}, \mathbf{Q}_{t|t-1})$, where $\mathbf{P}_{t|t-1} = S_{t-1} \mathbf{P}_{t|t-1}^*$ and $\mathbf{Q}_{t|t-1} = S_{t-1} \mathbf{Q}_{t|t-1}^* = \mathbf{x}_t^\top \mathbf{P}_{t|t-1} \mathbf{x}_t + S_{t-1} \mathbf{V}$. From these equations and the recursion of $\mathbf{P}_{t|t}^*$ given above we obtain the recursion of $\mathbf{P}_{t|t}$ as

$$\mathbf{P}_{t|t} = S_t \mathbf{P}_{t|t}^* = \frac{S_t}{S_{t-1}}(\mathbf{P}_{t|t-1} - \mathbf{K}_t \mathbf{Q}_{t|t-1} \mathbf{K}_t^\top).$$

If we define $r_t = y_t - \mathbf{x}_t^\top \hat{\beta}_{t|t-1}$ and we substitute the Kalman filter recursion $\hat{\beta}_{t|t} = \hat{\beta}_{t|t-1} + \mathbf{K}_t e_t$ in r_t, we observe that

$$r_t = y_t - \mathbf{x}_t^\top \hat{\beta}_{t|t-1} - \mathbf{x}_t^\top \mathbf{K}_t e_t = (\mathbf{I} - \mathbf{x}_t^\top \mathbf{K}_t) e_t = (\mathbf{I} - \mathbf{x}_t^\top \mathbf{P}_{t|t-1} \mathbf{x}_t \mathbf{Q}_{t|t-1}^{-1}) e_t$$

$$= (\mathbf{Q}_{t|t-1} - \mathbf{x}_t^\top \mathbf{P}_{t|t-1} \mathbf{x}_t) \mathbf{Q}_{t|t-1}^{-1} e_t = S_{t-1} \mathbf{V} \mathbf{Q}_{t|t-1}^{-1} e_t.$$

Hence, by using $S_t = d_t / n_t$, the recursion of d_t above can be written as

$$n_t S_t = n_{t-1} S_{t-1} + r_t^\top \mathbf{V}^{-1} e_t.$$

Below is a summary of the algorithm.

Multivariate Scaled Observational Precision I (MSOP-I)
In the state space model (5.1a)–(5.1b), with $\boldsymbol{\Sigma}$ and $\mathbf{Z}_t$ as in (5.9), the following apply:

1. Prior distributions at $t = 0$:
   $\beta_0 \sim t(n_0, \hat{\beta}_{0|0}, \mathbf{P}_{0|0})$ and $\sigma^{-2} \sim G(n_0/2, n_0 S_0/2)$;
2. Posterior distribution of β_{t-1} at time $t-1$:
   $\beta_{t-1} \mid y_{1:t-1} \sim t(n_{t-1}, \hat{\beta}_{t-1|t-1}, \mathbf{P}_{t-1|t-1})$;
3. Prior distribution of β_t at time t:
   $\beta_t \mid y_{1:t-1} \sim t(n_{t-1}, \hat{\beta}_{t|t-1}, \mathbf{P}_{t|t-1})$,
   where $\hat{\beta}_{t|t-1} = \mathbf{F}_t \hat{\beta}_{t-1|t-1}$ and $\mathbf{P}_{t|t-1} = \mathbf{F}_t \mathbf{P}_{t-1|t-1} \mathbf{F}_t^\top + \mathbf{Z}_t$;
4. Posterior distributions at time t:
   $\beta_t \mid y_{1:t} \sim t(n_t, \hat{\beta}_{t|t}, \mathbf{P}_{t|t})$ and $\sigma^{-2} \mid y_{1:t} \sim G(n_t/2, n_t S_t/2)$, where

$$\hat{\beta}_{t|t} = \hat{\beta}_{t|t-1} + \mathbf{K}_t e_t, \quad \mathbf{P}_{t|t} = \frac{S_t}{S_{t-1}} \left(\mathbf{P}_{t|t-1} - \mathbf{K}_t \mathbf{Q}_{t|t-1} \mathbf{K}_t^\top \right),$$

$$n_t = n_{t-1} + d \quad \text{and} \quad n_t S_t = n_{t-1} S_{t-1} + r_t^\top \mathbf{V}^{-1} e_t,$$

$$\hat{y}_{t|t-1} = \mathbf{x}_t^\top \hat{\beta}_{t|t-1}, \quad e_t = y_t - \hat{y}_{t|t-1}, \quad r_t = y_t - \mathbf{x}_t^\top \hat{\beta}_{t|t},$$

$$\mathbf{K}_t = \mathbf{P}_{t|t-1} \mathbf{x}_t \mathbf{Q}_{t|t-1}^{-1} \quad \mathbf{Q}_{t|t-1} = \mathbf{x}_t^\top \mathbf{P}_{t|t-1} \mathbf{x}_t + S_{t-1} \mathbf{V}.$$

In this application of the above model, apart from the specification of the model components $\mathbf{x}_t$, $\mathbf{F}_t$ and $\mathbf{Z}_t$ discussed in Sect. 5.2, the covariance matrix $\mathbf{V}$ must be specified. The obvious option is $\mathbf{V} = \mathbf{I}$, but this is limited as it postulates that ϵ_{it} is independent of ϵ_{jt}, for $i \neq j$ and $\mathrm{Var}(\epsilon_{1t}) = \mathrm{Var}(\epsilon_{2t}) = \cdots = \mathrm{Var}(\epsilon_{dt}) = \sigma^2$, where $\epsilon_t = [\epsilon_{1t}, \epsilon_{2t}, \ldots, \epsilon_{dt}]^\top$ is the observation innovation vector. In other words, given the states β_t, the component time series y_{it} and y_{jt} are conditionally independent and also they are homoscedastic (they have the same variance). This is usually not supported in practice, and although specifying $\mathbf{V}$ as a diagonal matrix with different elements in its diagonal would resolve the problem of homoscedasticity, the problem of independence of y_{it} and y_{jt} still remains. Usually, we resort to multivariate modelling, specifically with the aim to estimate cross-dependencies among the component time series of y_t. One possibility to address this point is to estimate $\mathbf{V}$ by employing the EM algorithm (described in Sect. 5.2). For this to commence there is no need to consider scaling $\boldsymbol{\Sigma}$, $\mathbf{Z}$ and $\mathbf{P}_{0|0}$ by σ^2, since $\boldsymbol{\Sigma}$ and $\mathbf{Z}_t = \mathbf{Z}$ (if the transition matrix is assumed to be time-invariant) can be estimated directly by the EM algorithm; in this case all distributions involved are Gaussian. However, the disadvantage is that the EM algorithm relies upon smoothing estimation and hence real-time estimation and forecasting, in the sense of Kalman filter, is not available. Given the generalisation of the SOP model discussed above, it would be desirable to have a similar algorithm where we could scale the matrices $\mathbf{Z}_t$ and $\mathbf{P}_{0|0}$ by the observation matrix $\boldsymbol{\Sigma}$. If that were possible, then we should be able to replace the gamma prior distribution of σ^{-2} with a matrix-variate distribution appropriate to describe covariance matrices. This is discussed in the next section.

5.5.2 Covariance Structure and Matrix-Variate Probability Distributions

In the previous section, matrices $\mathbf{Z}_t$ and $\mathbf{P}_{0|0}$ were scaled by the variance σ^2 in order to facilitate closed form estimation within the Gaussian, gamma and Student t family of distributions. If $\boldsymbol{\Sigma}$ is a $d \times d$ covariance matrix, this scaling is generally not any more available, because $\mathbf{Z}_t$ is a $p \times p$ covariance matrix, e.g. $\boldsymbol{\Sigma}$ and $\mathbf{Z}_t$ cannot be multiplied, if $p \neq d$. Even in the special case of $d = p$, it is not possible to set $\mathbf{Z}_t = \boldsymbol{\Sigma}\mathbf{Z}_t^*$, because in this specification $\mathbf{Z}_t$ is not a symmetric matrix. This problem has attracted considerable interest; early efforts include the formulation of a new multivariate state space model, developed initially by Harvey (1986) and Quintana and West (1987) and further discussed in Harvey (1989), West and Harrison (1997) and in Triantafyllopoulos (2008b). This new state space model, which generalises the SOP model of Sects. 4.3.3 and 5.5.1, allows for a particular scaling making use of matrix-variate Gaussian distributions. The formulation of Quintana and West (1987) proposes Bayesian estimation of $\boldsymbol{\Sigma}$ based on a Wishart prior for the precision matrix (in an analogous way as in the SOP univariate model), while Harvey (1986) discusses maximum likelihood estimation of $\boldsymbol{\Sigma}$. Other efforts of covariance estimation of the multivariate state space model include the exponentially weighted regression models of Triantafyllopoulos and Pikoulas (2002); Triantafyllopoulos (2006b), covariance estimation for the general state space model (5.1a)–(5.1b) Triantafyllopoulos and Harrison (2008); Triantafyllopoulos (2007b) and covariance estimation for the multivariate local level model when $\mathbf{x}_t = \mathbf{F}_t = \mathbf{I}$ (Triantafyllopoulos, 2011a).

In the sequel we describe the model formulation of Harvey (1986) and we discuss Bayesian and maximum likelihood estimation for Σ; this discussion is based on Triantafyllopoulos (2008a,b); related models are discussed in West and Harrison (1997) and in Prado and West (2010).

Before we start we give a discussion on matrix-variate normal and inverse Wishart distributions, necessary for the definition and development of the related state space models. We have already seen the definition of a multivariate normal distribution in Sect. 2.3.3 and used throughout in Chaps. 3 and 4. The matrix-variate normal distribution is a generalisation of the multivariate normal distribution to a random matrix. Let $X_1, \ldots, X_d$ be p-dimensional column random vectors and form the $p \times d$ random matrix $\mathbf{X} = [X_1, X_2, \ldots, X_d]$. The random matrix $\mathbf{X}$ is said to follow the *matrix-variate normal* distribution if its density is

$$p(\mathbf{X}) = c_1 \exp\left\{-\frac{1}{2}\text{trace}[(\mathbf{X} - \mathbf{M})^\top \mathbf{P}^{-1}(\mathbf{X} - \mathbf{M})\Sigma^{-1}]\right\},$$

where $c_1 = (2\pi)^{-dp/2}|\mathbf{P}|^{-d/2}|\Sigma|^{-p/2}$, $\mathbf{M}$ is a $p \times d$ mean matrix and $\mathbf{P}$ and Σ are $p \times p$ and $d \times d$ covariance matrices respectively. In terms of notation we will write $\mathbf{X} \sim N(\mathbf{M}, \mathbf{P}, \Sigma)$; sometimes $\mathbf{P}$ is referred to as *left covariance matrix* and Σ as *right covariance matrix*. We observe that for $d = 1$ we obtain the multivariate distribution of Sect. 2.3.3 and for $p = d = 1$ we obtain the univariate normal distribution. A basic property of the above distribution is that if $\mathbf{X} \sim N(\mathbf{M}, \mathbf{P}, \Sigma)$, then the random vector $Y = \text{vec}(\mathbf{X})$ follows a multivariate normal distribution $Y \sim N[\text{vec}(\mathbf{M}), \Sigma \otimes \mathbf{P}]$, where $\text{vec}(\cdot)$ is the column stacking vector operation and $\otimes$ denotes the Kronecker product, both of which are discussed in Sect. 2.1. Based on this property and the properties of the multivariate normal distribution it follows that $\mathbf{M}$ is the mean matrix of $\mathbf{X}$ and that $\Sigma \otimes \mathbf{P}$ is the covariance matrix of Y.

Related to the above normal distribution is the inverse Wishart distribution. A $d \times d$ random covariance matrix (symmetric and positive definite) Σ is said to follow the inverse Wishart distribution with n degrees of freedom and scale covariance matrix $\mathbf{S}$, if its density is

$$p(\Sigma) = c_2|\Sigma|^{-(n+d+1)/2}\exp\left[-\frac{1}{2}\text{trace}(\mathbf{S}\Sigma^{-1})\right],$$

where $c_2 = 2^{-nd/2}\Gamma_d(n/2)^{-1}|\mathbf{S}|^{n/2}$ is the proportionality constant, $n > d - 1$ and $\Gamma_d(\cdot)$ is the multivariate gamma function. By way of notation we write $\Sigma \sim IW(n, \mathbf{S})$. Since the density function $p(\mathbf{X})$ integrates to 1, it follows that

$$\int_{\Sigma>0} |\Sigma|^{-(n+d+1)/2}\exp\left[-\frac{1}{2}\text{trace}(\mathbf{S}\Sigma^{-1})\right] d\Sigma = 2^{nd/2}\Gamma_d(n/2)|\mathbf{S}|^{-n/2}.$$

(5.12)

The distribution of the inverse of $\boldsymbol{\Sigma}$ is the well-known Wishart distribution, written as $\boldsymbol{\Sigma}^{-1} \sim W(n, \mathbf{S}^{-1})$, with density

$$p(\boldsymbol{\Sigma}^{-1}) = c_3 |\boldsymbol{\Sigma}|^{(n-d-1)/2} \exp\left[-\frac{1}{2}\operatorname{trace}(\mathbf{S}\boldsymbol{\Sigma}^{-1})\right],$$

where $c_3 = 2^{-nd/2}\Gamma_d(n/2)^{-1}|\boldsymbol{\Sigma}|^{n/2}$ and $n > d - 1$. The Wishart distribution can be seen as a generalisation of the gamma distribution and the inverse Wishart as a generalisation of the inverse gamma distribution; see Sect. 2.3.3 for a discussion of the gamma and inverse gamma distributions. Finally, we give the definition of the matrix-variate Student t distribution. A $p \times d$ random matrix $\mathbf{X}$ is said to follow the matrix-variate t distribution with mean $\mathbf{M}$, spread covariances matrices $\mathbf{P}$ and $\mathbf{S}$ and degrees of freedom v, if its density is given by

$$p(\mathbf{X}) = c_4 |\mathbf{I} + (\mathbf{X} - \mathbf{M})\mathbf{P}^{-1}(\mathbf{X} - \mathbf{M})^{\top}\mathbf{S}^{-1}|^{-(n+p+d-1)/2},$$

where $c_4 = \Gamma_d[(n + p + d - 1)/2]\pi^{-dp/2}\Gamma_d[(n + d - 1)/2]^{-1}|\mathbf{P}|^{-d/2}|\mathbf{S}|^{-p/2}$. By way of notation we write $\mathbf{X} \sim t(n, \mathbf{M}, \mathbf{P}, \mathbf{S})$. We observe that for $d = 1$ the above is reduced to the multivariate t distribution of Sect. 2.3.3. A detailed discussion of matrix-variate distributions such as the normal, the t and the inverse Wishart can be found in Gupta and Nagar (1999).

5.5.3 The Multivariate Scaled Observational Model

Returning to the state space models, suppose that y_t is a d-dimensional observation vector and consider the multivariate state space model

$$y_t^{\top} = x_t^{\top}\boldsymbol{\beta}_t + \epsilon_t^{\top} \quad \text{(observation model)}, \tag{5.13a}$$

$$\boldsymbol{\beta}_t = \mathbf{F}_t\boldsymbol{\beta}_{t-1} + \boldsymbol{\zeta}_t \quad \text{(transition model)}, \tag{5.13b}$$

where x_t is a p-dimensional vector, $\boldsymbol{\beta}_t$ is a $p \times d$ state matrix and $\mathbf{F}_t$ is a $p \times p$ transition matrix. The d-dimensional observation innovation ϵ_t and the $p \times d$ state innovation $\boldsymbol{\zeta}_t$ are assumed independent for any t, ϵ_t is independent of ϵ_s, $\boldsymbol{\zeta}_t$ is independent of $\boldsymbol{\zeta}_s$, for any $t \neq s$ and ϵ_t follows a d-dimensional multivariate Gaussian distribution and $\boldsymbol{\zeta}_t$ follows a $p \times d$ matrix Gaussian distribution, written

$$\epsilon_t \sim N(0, \boldsymbol{\Sigma}) \quad \text{and} \quad \boldsymbol{\zeta}_t \sim N(0, \mathbf{Z}_t^*, \boldsymbol{\Sigma}),$$

so that the vector $\operatorname{vec}(\boldsymbol{\zeta}_t)$ follows the pd-dimensional Gaussian distribution $\operatorname{vec}(\boldsymbol{\zeta}_t) \sim N(0, \mathbf{Z}_t) \equiv N(0, \boldsymbol{\Sigma} \otimes \mathbf{Z}_t^*)$.

The state space model (5.13a)–(5.13b) is known as *seemingly unrelated equations model* (Harvey, 1986) or *common components model* (Harvey, 1986), but here we name it as *multivariate scaled observational precision model* (MSOP), because the model is scaled by the observation precision covariance matrix Σ^{-1} and for $d = 1$ (univariate case) the model is reduced to the univariate SOP model discussed in Section 4.3.3.

For the state space model (5.13a)–(5.13b) the following prior specification is adopted

$$\boldsymbol{\beta}_0 \mid \Sigma \sim N(\hat{\boldsymbol{\beta}}_{0|0}, \mathbf{P}_{0|0}, \Sigma) \quad \text{and} \quad \Sigma \sim IW(n_0, n_0 S_0), \tag{5.14}$$

for some known $p \times d$ mean matrix $\hat{\boldsymbol{\beta}}_{0|0}$, a $p \times p$ left covariance matrix $\mathbf{P}_{0|0}$, a $d \times d$ scale covariance matrix $n_0 S_0$ and degrees of freedom $n_0 > d - 1$. The following theorem provides inference for the MSOP model (5.13a)–(5.13b).

Theorem 5.4 *In the state space model (5.13a)–(5.13b), with the priors (5.14), for any $t = 1, 2, \ldots, n$ the following apply.*

1. *Conditionally on Σ, the one-step ahead forecast distribution of y_t and the posterior distribution of $\boldsymbol{\beta}_t$ are*

$$y_t \mid \Sigma, y_{1:t-1} \sim N(\hat{y}_{t|t-1}, q_{t|t-1}\Sigma) \quad \text{and} \quad \boldsymbol{\beta}_t \mid \Sigma, y_{1:t} \sim N(\hat{\boldsymbol{\beta}}_{t|t}, \mathbf{P}_{t|t}, \Sigma),$$

where $\hat{y}_{t|t-1} = \hat{\boldsymbol{\beta}}_{t|t-1}^\top x_t$, $q_{t|t-1} = x_t^\top \mathbf{P}_{t|t-1} x_t + 1$, $\hat{\boldsymbol{\beta}}_{t|t} = \hat{\boldsymbol{\beta}}_{t|t-1} + K_t e_t^\top$ and $\mathbf{P}_{t|t} = \mathbf{P}_{t|t-1} - q_{t|t-1} K_t K_t^\top$, with $\hat{\boldsymbol{\beta}}_{t|t-1} = \mathbf{F}_t \hat{\boldsymbol{\beta}}_{t-1|t-1}$, $\mathbf{P}_{t|t-1} = \mathbf{F}_t \mathbf{P}_{t-1|t-1} \mathbf{F}_t^\top + \mathbf{Z}_t$, $K_t = q_{t|t-1}^{-1} \mathbf{P}_{t|t-1} x_t$ and $e_t = y_t - \hat{y}_{t|t-1}$.
2. *The posterior distribution of Σ is $\Sigma \mid y_{1:t} \sim IW(n_t, n_t S_t)$, with $n_t = n_{t-1} + 1$ and $n_t S_t = n_{t-1} S_{t-1} + r_t e_t^\top$, where $r_t = y_t - \hat{\boldsymbol{\beta}}_{t|t}^\top x_t$.*
3. *Unconditionally of Σ, the one-step forecast distribution of y_t and the posterior distribution of $\boldsymbol{\beta}_t$ are $y_t \mid y_{1:t-1} \sim t(n_{t-1} - d + 1, \hat{y}_{t|t-1}, q_{t|t-1} n_{t-1} S_t)$ and $\boldsymbol{\beta}_t \mid y_{1:t} \sim t(n_t - d + 1, \hat{\boldsymbol{\beta}}_{t|t}, \mathbf{P}_{t|t}, n_t S_t)$.*

Proof First we prove (1). Suppose that at time $t - 1$ the posterior distribution of $\boldsymbol{\beta}_{t-1}$ is $\boldsymbol{\beta}_{t-1} \mid \Sigma, y_{1:t-1} \sim N(\hat{\boldsymbol{\beta}}_{t-1|t-1}, \mathbf{P}_{t-1|t-1}, \Sigma)$, for some known $\hat{\boldsymbol{\beta}}_{t-1|t-1}$ and $\mathbf{P}_{t-1|t-1}$. This combined to the transition and observation equations (5.13a)–(5.13b) give the distributions $\boldsymbol{\beta}_t \mid \Sigma, y_{1:t-1} \sim N(\hat{\boldsymbol{\beta}}_{t|t-1}, \mathbf{P}_{t|t-1}, \Sigma)$ and $y_t \mid \Sigma, y_{1:t-1} \sim N(\hat{y}_{t|t-1}, q_{t|t-1}\Sigma)$, with $\hat{\boldsymbol{\beta}}_{t|t-1}, \mathbf{P}_{t|t-1}, \hat{y}_{t|t-1}$ and $q_{t|t-1}$ as stated in the theorem.

Given Σ, model (5.13a)–(5.13b) can be written as a multivariate state space model (5.1a)–(5.1b). Indeed, apply the vec operation

$$y_t = \text{vec}(y_t^\top) = (\mathbf{I} \otimes x_t^\top)\text{vec}(\boldsymbol{\beta}_t) + \epsilon_t, \quad \epsilon_t \sim N(0, \Sigma),$$

with transition

$$\text{vec}(\boldsymbol{\beta}_t) = (\mathbf{I} \otimes \mathbf{F}_t)\text{vec}(\boldsymbol{\beta}_{t-1}) + \text{vec}(\boldsymbol{\zeta}_t), \quad \text{vec}(\boldsymbol{\zeta}_t) \sim N(0, \Sigma \otimes \mathbf{Z}_t),$$

which is in the form of (5.1a)–(5.1b). From the prior distribution $\text{vec}(\boldsymbol{\beta}_0) \sim N(0, \boldsymbol{\Sigma} \otimes \mathbf{P}_{0|0})$ and the Kalman filter (Theorem 5.1) we have that the posterior distribution of $\text{vec}(\boldsymbol{\beta}_t)$ is $\text{vec}(\boldsymbol{\beta}_t) \mid \boldsymbol{\Sigma}, y_{1:t} \sim N[\text{vec}(\hat{\boldsymbol{\beta}}_{t|t}), \boldsymbol{\Sigma} \otimes \mathbf{P}_{t|t}]$, so that $\boldsymbol{\beta}_t \mid y_{1:t} \sim N(\hat{\boldsymbol{\beta}}_{t|t}, \mathbf{P}_{t|t}, \boldsymbol{\Sigma})$, with $\hat{\boldsymbol{\beta}}_{t|t}$ and $\mathbf{P}_{t|t}$ as stated in the theorem.

Proceeding now to (2) by applying the Bayes theorem we have

$$p(\boldsymbol{\Sigma} \mid y_{1:t}) \propto p(y_t \mid \boldsymbol{\Sigma}, y_{1:t-1})p(\boldsymbol{\Sigma} \mid y_{1:t-1})$$

$$\propto |\boldsymbol{\Sigma}|^{-1/2} \exp\left\{-\frac{1}{2q_{t|t-1}}\left[(y_t - \hat{y}_{t|t-1})^\top \boldsymbol{\Sigma}^{-1}(y_t - \hat{y}_{t|t-1})\right]\right\}$$

$$\times |\boldsymbol{\Sigma}|^{-(n_{t-1}+d+1)/2}\exp\left[-\frac{1}{2}\text{trace}(n_{t-1}\mathbf{S}_{t-1}\boldsymbol{\Sigma}^{-1})\right]$$

$$\propto |\boldsymbol{\Sigma}|^{-(n_{t-1}+1+d+1)/2}\exp\left\{-\frac{1}{2}\text{trace}[(q_{t|t-1}^{-1}e_t e_t^\top + n_{t-1}\mathbf{S}_{t-1})\boldsymbol{\Sigma}^{-1}]\right\}$$

$$= |\boldsymbol{\Sigma}|^{-(n_{t-1}+1+d+1)/2}\exp\left\{-\frac{1}{2}\text{trace}[(r_t e_t^\top + n_{t-1}\mathbf{S}_{t-1})\boldsymbol{\Sigma}^{-1}]\right\},$$

since

$$r_t = y_t - \hat{\boldsymbol{\beta}}_{t|t-1}^\top x_t - e_t K_t^\top x_t = (1 - K_t^\top x_t)e_t = \frac{e_t}{q_{t|t-1}}.$$

Hence $\boldsymbol{\Sigma} \mid y_{1:t}$ follows an inverse Wishart distribution, $\boldsymbol{\Sigma} \mid y_{1:t} \sim IW(n_t, n_t\mathbf{S}_t)$, with $n_t = n_{t-1} + 1$ and $n_t\mathbf{S}_t = n_{t-1}\mathbf{S}_{t-1} + r_t e_t^\top$.

By combining the results in (1) and (2) we can integrate out $\boldsymbol{\Sigma}$ in order to find the posterior and forecast distributions unconditionally of $\boldsymbol{\Sigma}$. For the posterior distribution of $\boldsymbol{\beta}_t$ we have

$$p(\boldsymbol{\beta}_t \mid y_{1:t}) = \int_{\boldsymbol{\Sigma}>0} p(\boldsymbol{\beta}_t, \boldsymbol{\Sigma} \mid y_{1:t})\,d\boldsymbol{\Sigma}$$

$$= \int_{\boldsymbol{\Sigma}>0} p(\boldsymbol{\beta}_t \mid \boldsymbol{\Sigma}, y_{1:t})p(\boldsymbol{\Sigma} \mid y_{1:t})\,d\boldsymbol{\Sigma}$$

$$\propto \int_{\boldsymbol{\Sigma}>0} |\boldsymbol{\Sigma}|^{-p/2}\exp\left\{-\frac{1}{2}\text{trace}[(\boldsymbol{\beta}_t - \hat{\boldsymbol{\beta}}_t)\mathbf{P}_{t|t}^{-1}(\boldsymbol{\beta}_t - \hat{\boldsymbol{\beta}}_t)^\top \boldsymbol{\Sigma}^{-1}]\right\}$$

$$\times |\boldsymbol{\Sigma}|^{-(n_t+d+1)/2}\exp\left[-\frac{1}{2}\text{trace}(n_t\mathbf{S}_t\boldsymbol{\Sigma}^{-1})\right]\,d\boldsymbol{\Sigma}$$

$$= \int_{\boldsymbol{\Sigma}>0} |\boldsymbol{\Sigma}|^{-(n_t+p+d+1)/2}$$

$$\times \exp\left\{-\frac{1}{2}\text{trace}[(\boldsymbol{\beta}_t - \hat{\boldsymbol{\beta}}_t)\mathbf{P}_{t|t}^{-1}(\boldsymbol{\beta}_t - \hat{\boldsymbol{\beta}}_t)^\top + n_t\mathbf{S}_t]\boldsymbol{\Sigma}^{-1}]\right\}\,d\boldsymbol{\Sigma}$$

$$\propto |(\boldsymbol{\beta}_t - \hat{\boldsymbol{\beta}}_t)\mathbf{P}_{t|t}^{-1}(\boldsymbol{\beta}_t - \hat{\boldsymbol{\beta}}_t)^\top + n_t\mathbf{S}_t|^{-(n_t+p)/2}, \tag{5.15}$$

which is computed by (5.12), if we set $n = n_t + p$ and $S = (\boldsymbol{\beta}_t - \hat{\boldsymbol{\beta}}_t)\mathbf{P}_{t|t}^{-1}(\boldsymbol{\beta}_t - \hat{\boldsymbol{\beta}}_t)^\top + n_t\mathbf{S}_t$. We can see that (5.15) is proportional to

$$p(\boldsymbol{\beta}_t \mid y_{1:t}) \propto |\mathbf{I} + n_t^{-1}(\boldsymbol{\beta}_t - \hat{\boldsymbol{\beta}}_t)\mathbf{P}_{t|t}^{-1}(\boldsymbol{\beta}_t - \hat{\boldsymbol{\beta}}_t)^\top \mathbf{S}_t^{-1}|^{-(n_t-d+1+p+d-1)/2},$$

which is proportional to a matrix Student t density, hence $\boldsymbol{\beta}_t \mid y_{1:t} \sim t(n_t - d + 1, \hat{\boldsymbol{\beta}}_{t|t}, \mathbf{P}_{t|t}, n_t\mathbf{S}_t)$, as required. The proof of the forecast distribution of y_t unconditional of $\boldsymbol{\Sigma}$ is very similar to that of the posterior distribution of $\boldsymbol{\beta}_t$, given above, and it is omitted.

$\square$

Some comments are in order.

- For $d = 1$ (y_t is scalar time series and $\boldsymbol{\Sigma} = \sigma^2$ is reduced to a scalar variance) Theorem 5.4 provides similar results to those of Sect. 4.3.3 for the univariate SOP model; the differences being on the parameterisation of the gamma and inverse gamma distributions.
- From the inverse Wishart posterior distribution of $\boldsymbol{\Sigma}$ we can extract estimators such as the posterior mean or the posterior mode, i.e.

$$\mathrm{E}(\boldsymbol{\Sigma} \mid y_{1:t}) = \frac{n_t\mathbf{S}_t}{n_t - d - 1} \quad (n_t > d+1) \quad \text{and} \quad \mathrm{mode}(\boldsymbol{\Sigma} \mid y_{1:t}) = \frac{n_t\mathbf{S}_t}{n_t + d + 1}.$$

- From the updating of $\mathbf{S}_t$ it follows that for large t the posterior mean of $\boldsymbol{\Sigma}$ is approximately

$$\mathrm{E}(\boldsymbol{\Sigma} \mid y_{1:t}) \approx \frac{1}{t}\sum_{i=1}^{t} r_i e_i^\top,$$

or in words that the estimator of $\boldsymbol{\Sigma}$ is the average of $\{r_1 e_1^\top, \ldots, r_t e_t^\top\}$.
- Given information $y_{1:n} = \{y_1, y_2, \ldots, y_n\}$, for some positive integer n, the maximum likelihood estimator of $\boldsymbol{\Sigma}$ is

$$\hat{\boldsymbol{\Sigma}} = \frac{1}{n}\sum_{t=1}^{n} r_t e_t^\top.$$

Indeed, the log likelihood function of $\boldsymbol{\Sigma}$ is

$$\ell(\boldsymbol{\Sigma}; y_{1:n}) = \log p(y_1, y_2, \ldots, y_n \mid \boldsymbol{\Sigma})$$

$$= \log \prod_{t=1}^{n} p(y_t \mid \boldsymbol{\Sigma}, y_{1:t-1})$$

$$= -\frac{dn}{2}\log(2\pi) - \frac{d}{2}\sum_{t=1}^{n}\log q_{t|t-1} - \frac{n}{2}\log|\mathbf{\Sigma}|$$

$$-\frac{1}{2}\sum_{t=1}^{n}q_{t|t-1}^{-1}\mathrm{trace}(e_t e_t^\top \mathbf{\Sigma}^{-1}).$$

From Eqs. (2.9) and (2.14) we have

$$\frac{\partial \ell(\mathbf{\Sigma}; y_{1:n})}{\partial \mathbf{\Sigma}} = -n\mathbf{\Sigma}^{-1} + \frac{n}{2}\mathrm{diag}(\mathbf{\Sigma}^{-1})$$

$$-\frac{1}{2}\sum_{t=1}^{n}q_{t|t-1}^{-1}\left[-2\mathbf{\Sigma}^{-1}e_t e_t^\top \mathbf{\Sigma}^{-1} + \mathrm{diag}(\mathbf{\Sigma}^{-1}e_t e_t^\top \mathbf{\Sigma}^{-1})\right].$$

Equating this to zero, we obtain the matrix equation

$$n\hat{\mathbf{\Sigma}}^{-1} = \sum_{t=1}^{n}q_{t|t-1}^{-1}\hat{\mathbf{\Sigma}}^{-1}e_t e_t^\top \hat{\mathbf{\Sigma}}^{-1},$$

which solution is

$$\hat{\mathbf{\Sigma}} = \frac{1}{n}\sum_{t=1}^{n}q_{t|t-1}^{-1}e_t e_t^\top = \frac{1}{n}\sum_{t=1}^{n}r_t e_t^\top,$$

with r_t as defined in Theorem 5.4 above.

It can be shown that the second partial derivative of $\ell(\mathbf{\Sigma}; y_{1:n})$ with respect to $\mathbf{\Sigma}$ is a negative definite matrix, hence $\hat{\mathbf{\Sigma}}$ maximises the log-likelihood function; for a proof of this result the reader is referred to Triantafyllopoulos (2008b).

Below a summary of the MSOP algorithm is given.

Multivariate Scaled Observation Model II (MSOP-II)
In the state space model (5.13a)–(5.13b), for each $t = 1, 2, \ldots, n$ the following apply:

1. Prior distributions at $t = 0$: $\mathbf{\Sigma} \sim IW(n_0, n_0 S_0)$ and $\beta_0 \sim t(n_0 - d + 1, \hat{\beta}_{0|0}, P_{0|0}, n_0 S_0)$, for some $n_0, S_0, \hat{\beta}_{0|0}$ and $P_{0|0}$.
2. The one-step ahead distribution of y_t is $y_t \mid y_{1:t-1} \sim t(n_{t-1} - d + 1, \hat{y}_{t|t-1}, n_{t-1}q_{t|t-1}S_{t-1})$ and the posterior distribution of β_t is $\beta_t \mid y_{1:t} \sim t(n_t - d + 1, \hat{\beta}_{t|t}, P_{t|t}, n_t S_t)$,
where $\hat{y}_{t|t-1} = \hat{\beta}_{t|t-1}^\top x_t$, $q_{t|t-1} = x_t^\top P_{t|t-1}x_t + 1$, $\hat{\beta}_{t|t} = \hat{\beta}_{t|t-1} + K_t e_t^\top$

(continued)

and $\mathbf{P}_{t|t} = \mathbf{P}_{t|t-1} - q_{t|t-1} K_t K_t^\top$, with $\hat{\beta}_{t|t-1} = \mathbf{F}_t \hat{\beta}_{t-1|t-1}$, $\mathbf{P}_{t|t-1} = \mathbf{F}_t \mathbf{P}_{t-1|t-1} \mathbf{F}_t^\top + \mathbf{Z}_t$, $K_t = q_{t|t-1}^{-1} \mathbf{P}_{t|t-1} x_t$ and $e_t = y_t - \hat{y}_{t|t-1}$.

3. The posterior distribution of Σ is $\Sigma \mid y_{1:t} \sim IW(n_t, n_t S_t)$, with $n_t = n_{t-1} + 1$ and $n_t S_t = n_{t-1} S_{t-1} + r_t e_t^\top$, where $r_t = y_t - \hat{\beta}_{t|t}^\top x_t$.

5.6   Forecasting Pollution Time Series

In this section we consider a larger data set than that of Example 4.7, which considers values of the pollutant nitric oxide NO over a course of 1 year. Here we look at three pollutants, ozone (O_3), nitrogen dioxide (NO_2) and nitric oxide (NO) all measured in milligrams per square metre. The data are daily observations for 2 years covering the period 1 January 2001 to 31 December 2002 (732 observations in total). We denote with y_{1t} the observation of O_3 at time t, with y_{2t} the observation of NO_2 at time t, with y_{3t} the value of NO at time t and we form the observation vector $y_t = [y_{1t}, y_{2t}, y_{3t}]^\top$. The data, obtained by one of 16 pollution monitoring stations in the city of Athens, is plotted in Fig. 5.1; missing data were imputed by regression and moving average methods and are indicated in Fig. 5.1 by the straight lines. In addition to the observations $\{y_t\}$, the measurements of the covariates: temperature (in °C), humidity (%) and wind speed (in metres per second) are available; these covariates are denoted by x_{1t}, x_{2t}, x_{3t}, respectively. The aim of this example is to propose a model for the vector time series y_t, which will be capable of forecasting future values of y_t as well as estimating the correlations of ozone, nitrogen dioxide and nitric oxide over time.

By comparing this data set with that of Example 4.7 we see that the NO observations considered in Chap. 4 covered only a year, hence it was not possible to explore the possibility of the values being affected by annual seasonality. Indeed, it is clear from Fig. 5.1 that the observations of y_{it} ($i = 1, 2, 3$) exhibit annual seasonality, which is most notable for the ozone. Therefore, a plausible state space model should include three time-varying covariates (temperature, humidity and wind speed), linear trend and seasonal components. Thus, the proposed model for y_t combines time-varying regression (see e.g. Sect. 4.1.5) and trend-seasonal components (see e.g. Sect. 4.1.4). Below we discuss the model in more detail.

Let β_t be a time-varying 15×3 unobserved state matrix for some positive integer d which drives the dynamics of y_t in the following state space model

$$y_t = R_{1t} + R_{2t} + R_{3t} + T_t + s_t + \epsilon_t = \beta_t^\top x_t + \epsilon_t, \tag{5.16}$$

so that y_t comprises three dynamic regression components $R_{it} = \beta_{it}^\top x_{it}$, a trend component $T_t = \beta_{4t}^\top [1, 0]^\top$, a seasonal component $s_t = \beta_{5t}^\top [1, 0, 1, 0, 1, 0, 1, 0,$

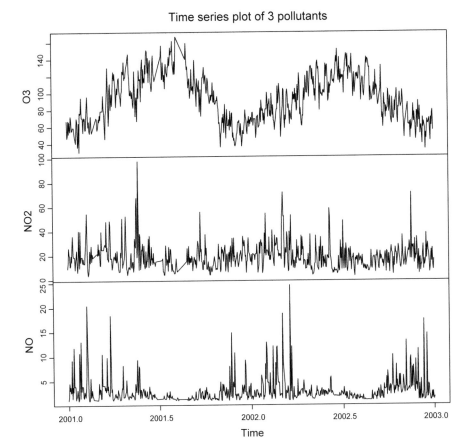

Fig. 5.1 Daily observations of ozone O_3, dyadic oxide NO_2 and monoxide NO over a period of 2 years

$1, 0]^\top$ and a random innovation term ϵ_t, where $\boldsymbol{\beta}_t^\top = [\boldsymbol{\beta}_{1t}^\top, \boldsymbol{\beta}_{2t}^\top, \boldsymbol{\beta}_{3t}^\top, \boldsymbol{\beta}_{4t}^\top, \boldsymbol{\beta}_{5t}^\top]$. The model (5.16) can be written as

$$y_t^\top = x_t^\top \boldsymbol{\beta}_t + \epsilon_t^\top \quad \text{and} \quad \boldsymbol{\beta}_t = \mathbf{F}\boldsymbol{\beta}_{t-1} + \boldsymbol{\zeta}_t, \tag{5.17}$$

with design vector x_t and the transition matrix $\mathbf{F}$

$$x_t = \left[\underbrace{x_{1t}^\top, x_{2t}^\top, x_{3t}^\top}_{\text{covariates}}, \underbrace{1, 0}_{\text{trend}}, \underbrace{1, 0, 1, 0, 1, 0, 1, 0, 1, 0}_{\text{seasonal}} \right]^\top ,$$

$$F = \begin{bmatrix} \mathbf{I} & 0 & 0 & 0 & 0 & 0 & 0 \\ 0 & \mathbf{J} & 0 & 0 & 0 & 0 & 0 \\ 0 & 0 & \mathbf{F}_1 & 0 & 0 & 0 & 0 \\ 0 & 0 & 0 & \mathbf{F}_2 & 0 & 0 & 0 \\ 0 & 0 & 0 & 0 & \mathbf{F}_3 & 0 & 0 \\ 0 & 0 & 0 & 0 & 0 & \mathbf{F}_4 & 0 \\ 0 & 0 & 0 & 0 & 0 & 0 & \mathbf{F}_5 \end{bmatrix},$$

where $\mathbf{F}$ is a block diagonal matrix with components $\mathbf{I}$ for the identity matrix, $\mathbf{J}$ for the transition matrix of the linear growth model, responsible for the trend variation, and $\mathbf{F}_i$ ($i = 1, 2, 3, 4, 5$) for the harmonic component, responsible for the seasonal variation, i.e.

$$\mathbf{J} = \begin{bmatrix} 1 & 1 \\ 0 & 1 \end{bmatrix} \quad \text{and} \quad \mathbf{F}_i = \begin{bmatrix} \cos(i\omega) & \sin(i\omega) \\ -\sin(i\omega) & \cos(i\omega) \end{bmatrix}.$$

The above state space model is a multivariate time-varying regression model with trend-seasonal components, for the seasonal variation of which, a reduced form state space representation of the full Fourier expansion is used. For this data the cycle is $c = 365$ (daily data with annual seasonality) with frequency $2\pi/c = \omega$; here for computational efficiency, we use only the first 5 harmonics out of a total of $(c - 1)/2 = 182$, hence the reduced form.

The distributions of ϵ_t and ζ_t are multivariate and matrix-variate Gaussian, i.e. $\epsilon_t \sim N(0, \boldsymbol{\Sigma})$ and $\zeta_t \sim N(0, \mathbf{Z}_t, \boldsymbol{\Sigma})$, where $\boldsymbol{\Sigma}$ is a 3×3 observation covariance matrix subject to estimation and the 15×15 transition covariance matrix $\mathbf{Z}_t$ is specified by using a discount factor δ. The prior distribution of $\boldsymbol{\beta}_0$ is also a matrix-variate Gaussian distribution, i.e. $\boldsymbol{\beta}_0 \sim N(\hat{\boldsymbol{\beta}}_{0|0}, \mathbf{P}_{0|0}, \boldsymbol{\Sigma})$, for some $\mathbf{P}_{0|0}$. The prior distribution of $\boldsymbol{\Sigma}$ is an inverse Wishart distribution with n_0 degrees of freedom and scale matrix $n_0 \mathbf{S}_0$, written as $\boldsymbol{\Sigma} \sim IW(n_0, n_0 \mathbf{S}_0)$.

In R we can specify the above model components by using the commands

```
> # read data
> data <- read.table("dataPollution.txt")
> # use the first 732 time points and the first 3 variables
> y <- data[1:732,4:6]

> # create design vector
> h<-5
> des <- matrix(0, 2922, 3+2+2*h)
> des[,1] <- y[,1]
> des[,2] <- y[,2]
> des[,3] <- y[,3]
> des[,seq(4,2*h+4,2)] <- 1
> x1 <- list(); for(i in 1:732) x1 <- updt(x1, des[i,])

> # define transition matrix
> w <- 2*pi/365
```

```
> BD <- blockDiagMat(diag(3),Jtrend(1))
>    for(k in 1:h){
>      BD <- blockDiagMat(BD, Jseasonal(k,w))
>    }
```

We have applied the MSOP-II algorithm described above, with values of the hyperparameters: $\delta = 0.98$, $\hat{\beta}_{0|0} = \mathbf{0}$, $P_{0|0} = 1000\mathbf{I}$, $n_0 = 10$, $S_0 = 0.1\mathbf{I}$. Some uncertainty about the specification of these parameters may arise; the following comments may be useful for general application. The reliance of the discount factor is suggested because $\mathbf{Z}_t$ is a 15×15 covariance matrix and estimating it by the EM algorithm could be very inefficient; in addition to that it is desirable that $\mathbf{Z}_t$ is time-varying (as is the case when we specify it with a discount factor) because this allows for the variance in the evolution of the states to change over time. Overall we achieve to specify a 15×15 covariance matrix just by specifying a scalar, the discount factor δ. Several values of this discount factor may be considered and tuned using measures of goodness of fit (e.g. MSE, MSSE, MAD or the likelihood function). Typically, values of δ close, but less than, one should be considered and here we have found that $\delta = 0.98$ works well. For more details on the specification of transition covariance matrices using discount factors see the discussion in Sect. 4.3.2 and West and Harrison (1997). The specification of $\hat{\beta}_{0|0} = \mathbf{0}$ is motivated by convenience, since $\hat{\beta}_{0|0}$ is a 15×3 matrix. As we have seen in Sect. 4.5 for univariate state space models, the specification of $\hat{\beta}_{0|0}$ does not play a crucial role in estimation and forecasting; instead the model is adaptive and very quickly corrects its ability to forecast accurately, as evidenced by Table 4.2 of Sect. 4.5. Following the discussion in Sect. 4.5 and throughout the book, a weakly informative prior covariance matrix $P_{0|0} = 1000\mathbf{I}$ is chosen, reflecting high uncertainty around the zero mean matrix of β_0. The values of the degrees of freedom n_0 and the scale matrix S_0 are picked so that $n_0 S_0 = \mathbf{I}$ and n_0 (the degrees of freedom) must be larger than $p - 1 = 2$ for the inverse Wishart distribution to exist; here we have picked $n_0 = 10$ and $S_0 = 10^{-1}\mathbf{I}$. This implies that a prior point estimate of Σ (the prior mode of Σ) is $n_0 S_0/(n_0 + p + 1) = 10^{-1}\mathbf{I}/14 \approx 0.7\mathbf{I}$.

After the above prior settings are in place, we apply the MSOP-II algorithm to the data. The model is fitted in R using the command

```
> fit <- bts.msop(y, x1, F0=BD, beta0=matrix(0, 15, 3),
+ n0=10,S0=0.1*diag(3), delta=0.99)
```

The standardised residuals (not shown here) fluctuate around zero, although those for NO indicate some structure on the plot, which suggests they the residuals for y_{3t} (NO) are not independent. Comparing them with the credible limits ± 1.96 of the $N(0, 1)$ we observe that there are 6.84% outliers in O_3, 6.55% outliers in NO_2 and 4.56% in NO; this suggests that O_3 and NO_2 slightly underestimate the observation variance, while NO overestimates the respective variance (we would expect to obtain 5% outliers for a perfect fit). This can be further explored by looking

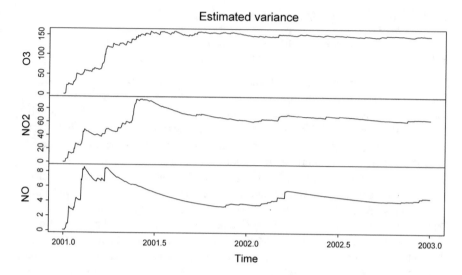

Fig. 5.2 Estimated variances of O_3, NO_2 and NO

at the values of the MSE, MMSE and MAD, which are

$$\mathrm{MSE} = \begin{bmatrix} 243.233 \\ 96.223 \\ 7.819 \end{bmatrix}, \quad \mathrm{MSSE} = \begin{bmatrix} 1.637 \\ 1.895 \\ 1.340 \end{bmatrix}, \quad \mathrm{MAD} = \begin{bmatrix} 11.651 \\ 6.959 \\ 1.574 \end{bmatrix}.$$

Figure 5.2 shows the posterior modes of the observation variances σ_{ii}, where $\Sigma = (\sigma_{ij})_{i,j=1,2,3}$. We observe that after 3 months the variance estimates seem to stabilise to constant values, with O_3 having the largest variation and NO having the smallest variation. The R commands used to extract this plot are given below:

```
# plot of estimated variances
> var1 <- var2 <- var3 <- rep(0,732)
> for(t in 1:732){
+ var1[t] <- fit$ObsVar[[t]][1,1]
+ var2[t] <- fit$ObsVar[[t]][2,2]
+ var3[t] <- fit$ObsVar[[t]][3,3]
+ }
> x <- cbind(var1,var2,var3)
> xts <- ts(x, start=c(2001,0), frequency=365,
+ names=c("O3","NO2","NO"))
> plot.ts(xts, main=expression("Estimated variance"))
```

The cross-correlation of the three time series (O_3, NO_2 and NO) may be explored in the first place by estimating their correlation matrix, which is

$$\begin{bmatrix} 1 & -0.343 & -0.406 \\ -0.343 & 1 & 0.638 \\ -0.406 & 0.638 & 1 \end{bmatrix}$$

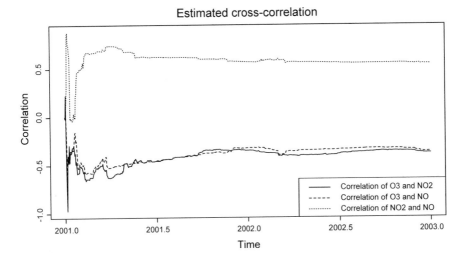

Fig. 5.3 Estimated cross-correlations of O_3, NO_2 and NO

This suggests that O_3 and NO_2 and O_3 and NO are negatively correlated, while NO_2 and NO are positively correlated. Of course such correlation estimates are based on the sample correlations, which fail to take into account the dynamics of the data and in particular the validity of sample statistics is based both on the availability of large amounts of data and on the assumption that this data is stationary. Here, with the availability of 732 observations and with the apparent non-stationarity of the time series data, these assumptions seem to be violated. Thus, one should not rely on sample statistics in order to estimate covariance and correlation matrices in this case. In addition to this the inferred correlations of the MSOP-II algorithm offer the significant advantage of on-line correlation estimates. Figure 5.3 shows the estimated correlations of the observation matrix Σ. We observe that the correlations of NO_2 and NO with O_3 are negative and very close to each other, while the correlation of NO_2 with NO is close to 0.6 mark. These results somehow agree with the sample cross-correlations, but they are more valid as they explicitly make use of the dynamic state space model. The following R code was used to produce Fig. 5.3:

```
> # plot of cross-correlations
> cor1 <- cor2 <- cor3 <- rep(0,732)
> for(t in 1:732){
+ cor1[t] <- fit$ObsVar[[t]][1,2] /
+  sqrt( fit$ObsVar[[t]][1,1] * fit$ObsVar[[t]][2,2] )
+ cor2[t] <- fit$ObsVar[[t]][1,3] /
+   sqrt( fit$ObsVar[[t]][1,1] * fit$ObsVar[[t]][3,3] )
+ cor3[t] <- fit$ObsVar[[t]][2,3] /
+   sqrt( fit$ObsVar[[t]][2,2] * fit$ObsVar[[t]][3,3] )
+ }
```

```
> x <- cbind(cor1,cor2,cor3)
> xts <- ts(x, start=c(2001,0), frequency=365)
> ts.plot(xts[,1],xts[,2],xts[,3],lty=1:3,
+ main=expression("Estimated cross-correlation"),
+ ylab="Correlation")
> smartlegend( x="right", y= "bottom", inset=0,
+ legend =c(expression("Correlation of O3 and NO2"),
+ expression("Correlation of O3 and NO"),
+ expression("Correlation of NO2 and NO")),  lty=c(1,2,3) )
```

5.7   Markov Chain Monte Carlo Inference

5.7.1   Bayesian Inference and the Gibbs Sampler

In this section we introduce the ideas of Markov chain Monte Carlo (MCMC) estimation for a general purpose probability model and Sects. 5.7.2 and 5.7.3 discuss MCMC procedures, specifically tailored to Gaussian state space models.

In most statistical problems we need to estimate probabilities or expectations of some *target* distribution, say $p(x)$, of some random vector X, defined in $\mathbb{R}^k$, for integer $k \geq 1$. Such probabilities and expectations may be expressed as integrals (assuming that X is continuous for simplicity, but replacing Riemann integrals by sums or Lebesgue integrals), e.g. the expectation of X is $E(X) = \int_{\mathbb{R}^k} x p(x) \, dx$ or the probability that X belongs to some set $A \subset \mathbb{R}^k$ is $P(X \in A) = \int_A p(x) \, dx$. The computation of such integrals is usually difficult because typically these will be high dimensional integrals and the distribution $p(\cdot)$ will be complicated. Additional complications in the computation may arise if A is complex. For all those reasons the evaluation of the above integrals will not be available in closed form. However, if we are able to draw a sample from $p(\cdot)$, say $x^{(1)}, x^{(2)}, \ldots, x^{(N)}$, then we can approximate $E(X)$ by $N^{-1} \sum_{i=1}^{N} x^{(i)}$, which in essence is known as Monte Carlo estimation. In words, instead of evaluating the integrals, we approximate them by sample statistics (mean, variance or proportions) of simulated vectors, randomly sampled from the target distribution. From the law of large numbers we know that as $N \to \infty$, $N^{-1} \sum_{i=1}^{N} x^{(i)}$ will converge in probability to the true value of $E(X)$.

Now consider the probability model whereby observations y are generated by a probability distribution $p(y \mid \theta)$, conditional on some parameter vector θ. This distribution is the likelihood function of θ, based on y. Now suppose that we place a prior on θ, $p(\theta)$, and we make use of the Bayes theorem to obtain the posterior distribution of θ, i.e.

$$p(\theta \mid y) \propto p(y \mid \theta)p(\theta) \tag{5.18}$$

(see also the relevant discussion of Sect. 2.4.3). If we can draw a random sample from $p(\theta \mid y)$, then we can use Monte Carlo methods to estimate expectations or probabilities of the posterior distribution of θ. Procedures to simulate a random sample from a given distribution are available, but they are limited to relatively simple forms of distributions. Such forms include most known distributions, some of which are mentioned in Sects. 2.3.2 and 2.3.3, with simulation procedures being based upon simulation of the uniform distribution. For more details on these procedures the reader is referred to Gamerman and Lopes (2006, Chapter 1). For complex posterior distributions, such as those arising from sequential application of the Bayes theorem, the above simulation procedures are not suitable. Moreover, Bayes rule (5.18) suggests that the posterior distribution $p(\theta \mid y)$ is only available up to a proportionality constant $c = 1/p(y)$, which computation involves the integral $p(y) = \int_{\mathbb{R}^k} p(y \mid \theta) p(\theta) \, d\theta$. The computation of this integral is usually difficult or not possible, except in very special cases. Hence drawing a sample from $p(\theta \mid y)$ may not be possible.

The principle idea of Markov chain Monte Carlo (MCMC) estimation is to propose sampling from a *Markov chain* (see definition below), which converges to the posterior $p(\theta \mid y)$; then assuming convergence, the random sample we draw from the chain will also be drawn from $p(\theta \mid y)$. Thus, the task is to construct an *appropriate* Markov chain so that (a) it converges to the posterior distribution $p(\theta \mid y)$ and (b) a random sample can be drawn from the chain. Below we provide a brief discussion of Markov chains, necessary to the development of this book. For a detailed treatment relevant to MCMC methods the reader is referred to Gamerman and Lopes (2006, Chapter 4) and to references therein.

A stochastic process is defined as a collection of random vectors over time; if the random vectors are collected in discrete-time we speak of a sequence of random vectors and this is the situation we will be concerned with in this book; for a detailed treatise of stochastic processes the reader is referred to Doob (1955). For example the time series $(y_t, t \geq 1)$ discussed throughout the book as well as the state process $(\beta_t, t \geq 0)$ are examples of stochastic processes. For the purposes of the Markov chains discussed in this section, each member of the sequence (a random vector) is denoted by $\theta^{(i)}$, $i = 0, 1, \ldots$. Each vector $\theta^{(i)}$ is known as a *state* and the vector space of the values of $\theta^{(i)}$ is known as the *state space* and is denoted by S; the state space should not be confused with the state space model, but below we draw some similarities. This space may be discrete $S = \{1, 2, 3, 4, \ldots\}$ or continuous $S = \mathbb{R}^k$. In what follows we will assume that S is discrete, but similar results apply for continuous state spaces (Gamerman & Lopes, 2006, Chapter 4). A Markov chain is a stochastic process so that, given the present, the future and the past are conditionally independent. This property can be written as

$$P(\theta^{(i+1)} \in A_{i+1} \mid \theta^{(i)} = x, \theta^{(i-1)} \in A_{i-1}, \ldots, \theta^{(0)} \in A_0) =$$

$$P(\theta^{(i+1)} \in A_{i+1} \mid \theta^{(i)} = x),$$

for a particular value x of $\theta^{(i)}$, where A_i is a subset of the state space. This indicates that the event $\{\theta^{(i+1)} \in A_{i+1}\}$ depends only on the state of the Markov chain at time i (observed value $\theta^{(i)} = x$) and not on past values of $\theta^{(i-1)}, \theta^{(i-2)}, \ldots, \theta^{(0)}$. A simple example of a Markov chain is the random walk $\theta^{(i+1)} = \theta^{(i)} + \epsilon_i$, where

ϵ_i is a white noise process. This is a Markov chain as, given $\theta^{(i)}$, $\theta^{(i+1)}$ and $\theta^{(i-s)}$ are conditionally independent, for $0 \le s \le i$. A real-life example of a Markov chain is the price of an asset traded in the stock market: for a given a share price of today, the price of tomorrow will not depend on the price the share had yesterday or the day before. Many more examples of Markov chains are given in Gamerman and Lopes (2006, Chapter 4). The state vectors $\beta_1, \beta_2, \ldots$ of the transition equation (5.1b) define a Markov chain, since given the present vector β_t, the past β_{t-1} is independent of the future β_{t+1}. The same property can be applied to the entire state space model, including the observations in Eq. (5.1a). This provides a justification of the name of the state space model.

Associated with a Markov chain is the transition matrix, which is defined as the matrix whose rs-th element is the probability $p_{rs} = P(\theta^{(i+1)} = s \mid \theta^{(i)} = r)$, which is the probability of the chain moving from state r (at time i) to state s (at time $i+1$). Usually this probability will depend on i, but in this discussion the chain is assumed to be *homogeneous*, in which case p_{rs} does not depend on i, but only on r and s; the transition matrix is denoted by P. One of the main utilities of the transition matrix is that of convergence of the chain. Under certain weak conditions (see e.g. Gamerman and Lopes (2006, Chapter 4)), as $i \to \infty$ the chain reaches its stationary distribution $\pi(\cdot)$ defined as

$$\sum_{r \in S} \pi(r) p_{rs} = \pi(s), \quad \text{for } s \in S.$$

From this equation we can see that $\pi P = \pi$, which suggests that once the chain has marginal distribution π at time n, then π is the marginal distribution of the chain for any time larger than n, hence the convergence. This has important implications for MCMC, because if we know that the chain converges to the posterior distribution $p(\theta \mid y)$, then when the chain has reached its stationary distribution π, then we have $\pi(\theta) \equiv p(\theta \mid y)$. Thus, if we simulate from a Markov chain after it reaches its marginal stationary distribution we have a simulation from the target distribution $p(\theta \mid y)$. The task then reduces to the construction of a suitable Markov chain. We close this introductory section by discussing the most popular class of Markov chains known as Gibbs sampling. This will be used in subsequent Sects. 5.7.2 and 5.7.3 to construct Markov chains for state space models.

The Gibbs sampling MCMC scheme originates from the Gibbs distribution in the context of mechanical statistics, discussed in detail in Geman and Geman (1984). Below we give the Gibbs sampling algorithm; for more details and examples the reader is referred to Gamerman and Lopes (2006, Chapter 5) and Robert (2007) and to references therein. Suppose that $\theta = [\theta_1, \theta_2, \ldots, \theta_k]^\top$ is a k-dimensional parameter vector (typically it will be associated to the multi-dimensional posterior distribution $p(\theta \mid y)$ derived by (5.18)).

Gibbs Sampler

1. Initialise the iteration counter $i = 0$, with
   $\theta^{(0)} = [\theta_1^{(0)}, \theta_2^{(0)}, \ldots, \theta_k^{(0)}]^\top$.

2. Obtain a set of vectors $\theta^{(i)} = [\theta_1^{(i)}, \theta_2^{(i)}, \ldots, \theta_k^{(i)}]^\top$ from $\theta^{(i-1)}$ by sampling one value $\theta_j^{(i)}$ $(j = 1, 2, \ldots, k)$ according to

$$\theta_1^{(i)} \sim p(\theta_1 \mid \theta_2^{(i-1)}, \theta_3^{(i-1)}, \ldots, \theta_{k-1}^{(i-1)}, \theta_k^{(i-1)})$$

$$\theta_2^{(i)} \sim p(\theta_2 \mid \theta_1^{(i)}, \theta_3^{(i-1)}, \ldots, \theta_{k-1}^{(i-1)}, \theta_k^{(i-1)})$$

$$\vdots$$

$$\theta_k^{(i)} \sim p(\theta_k \mid \theta_1^{(i)}, \theta_2^{(i)}, \theta_3^{(i)}, \ldots, \theta_{k-1}^{(i)})$$

3. Set the counter $i = i + 1$ and go to step (2) until convergence.

The algorithm assumes that we can sample from the distributions $p(\theta_j \mid \theta_{-j})$, where θ_{-j} is the $k - 1$-dimensional vector including the elements of θ except θ_j, for $j = 1, 2, \ldots, k$. We note that the simulated vector $\theta^{(i)}$ is a draw from a Markov chain, because by the Gibbs sampler above $\theta^{(i)}$ depends only on $\theta^{(i-1)}$ and not on past iterations $\theta^{(i-2)}, \theta^{(i-3)}, \ldots, \theta^{(0)}$. When convergence is reached we obtain samples from the stationary distribution π of the chain. If the target distribution we want to sample from is the posterior $p(\theta \mid y)$, then we need to condition the distributions in the Gibbs sampler to y, hence at each iteration i we shall sample $\theta_j^{(i)}$ from the conditional distribution $p(\theta_j \mid \theta_1^{(i)}, \ldots, \theta_{j-1}^{(i)}, \theta_{j+1}^{(i-1)}, \ldots, \theta_k^{(i-1)}, y)$.

It is worthwhile to note that we do not need to know the stationary distribution $\pi(\cdot)$ of the chain; instead, convergence is assessed by conducting informal and formal tests, based only on the simulated values $\theta^{(i)}$. It is a usual practice to run the chain for a considerable length of time to train on the data; this length is known as *burn-in* and a typical value is 1000 iterations, but this value will be problem-specific. After the burn-in period it is a usual practice to run the chain for as long as is needed and convergence diagnostics can confirm that the chain has converged. If this is the case we can plot summaries of the simulations, e.g. mode, histograms, or quantiles of $\theta^{(i)}$. Informal convergence diagnostics can be based on generating multiple chains and assess whether simulations from those look similar. Perhaps the simplest way to assess convergence is to judge whether $\theta^{(i)}$ constitute a sample from a stationary distribution (the distribution $\pi(\cdot)$ of the chain). Hence graphical methods of assessing this include the plot of $\theta^{(i)}$ against i, known as the trace plot, and the plot of the autocorrelation function (ACF) against lags $1, 2, 3, \ldots$, known as the correlogram. If the chain has reached convergence, we would expect that the trace plot indicates no structure, i.e. the mean and the variance of $\theta^{(i)}$ not to depend on i; likewise all values of the ACF should lie inside the $\pm 2/\sqrt{N - t_B}$ credible intervals, where N is the total number of iterations and t_B is the burn-in length. The trace plot checks whether blocks or shifted segments of the simulated vectors $\theta^{(i)}$ appear to have the same distribution (at least having the same mean and variance), while the correlogram checks whether the simulated vectors $\theta^{(i)}$, $\theta^{(j)}$ are uncorrelated, for $i \neq j$. More details on informal and formal convergence diagnostics are discussed in Gamerman and Lopes (2006, §5.4).

5.7.2 The Forward Filtering Backward Sampling Scheme

Coming back to the context of multivariate state space models, consider model (5.1a)–(5.1b), where the covariance matrices $\mathbf{\Sigma}$ and $\mathbf{Z}_t = \mathbf{Z}$ (time-invariant) are unknown and subject to estimation. As before the aim is to estimate the state vectors β_t and these covariance matrices, provided observed data $y_{1:n} = \{y_1, \ldots, y_n\}$. As we have seen earlier, this task may be performed by using the EM algorithm, as discussed in Sect. 5.2, but here we are interested in performing Bayesian inference using the Gibbs sampler. In this section we consider that $\mathbf{\Sigma}$ and $\mathbf{Z}$ are known and hence interest is solely focused on the estimation of β_t, given $y_{1:n}$. The general case, which considers estimation of $\mathbf{\Sigma}$ and $\mathbf{Z}$ is discussed in Sect. 5.7.3.

Since $\mathbf{\Sigma}$ and $\mathbf{Z}$ are known the fixed-interval smoothing applies (see p. 212) and so the distribution of β_t, given β_{-t} and $y_{1:n}$ is obtained as multivariate Gaussian. Since we can sample from $\pi(\beta_t \mid \beta_{-t})$, this provides a single step of the Gibbs sampler where it is noted that at time $t = n$ we sample from the posterior $\beta_n \mid y_{1:n}$, which by the Kalman filter (see Theorem 5.1) is again Gaussian. This approach was proposed by Carlin et al. (1992) together with extensions to non-linear and non-Gaussian state space models. Unfortunately, this approach can be very inefficient, because the prior correlation imposed in the system of state vectors $\beta = [\beta_1^\top, \beta_2^\top, \ldots, \beta_n^\top]^\top$ is largely transferred to the posterior state vectors $\beta \mid y_{1:n}$. The aforementioned chain correlation together with the high dimensional state space imposed by the time series length n introduces convergence problems in the Gibbs sampler and slows it down considerably. This issue is explained below and it is motivated by a relevant discussion in Gamerman and Lopes (2006, §5.5.2).

Consider that $\theta = [\theta_1, \theta_2]^\top$ follows a bivariate Gaussian distribution

$$\theta = \begin{bmatrix} \theta_1 \\ \theta_2 \end{bmatrix} \sim N \left\{ \begin{bmatrix} 2 \\ 3 \end{bmatrix}, \begin{bmatrix} 1 & \rho \\ \rho & 1 \end{bmatrix} \right\},$$

where ρ is the correlation of θ_1 and θ_2. It follows that the conditional distributions are

$$\theta_1 \mid \theta_2 \sim N[2 + \rho(\theta_2 - 3), 1 - \rho^2]$$

$$\theta_2 \mid \theta_1 \sim N[3 + \rho(\theta_1 - 2), 1 - \rho^2].$$

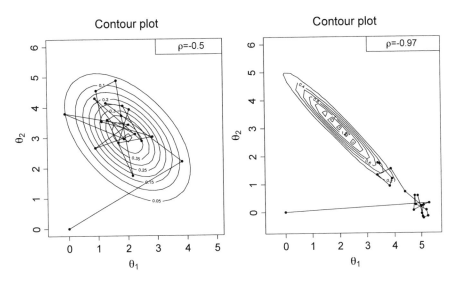

Fig. 5.4 Contour plots of bivariate Gaussian distributions with correlation $\rho = -0.5$ (left panel) and $\rho = -0.97$ (right panel); shown are draws from Gibbs sampler for the first 20 iterations

We initialise the Gibbs sampler at $[0, 0]^\top$, i.e. $\theta^{(0)} = [\theta_1^{(0)} = 0, \theta_2^{(0)} = 0]^\top$ and consequently we successively simulate $\theta_1^{(i)}$ from $\theta_1 \mid \theta_2^{(i-1)}$ and $\theta_2^{(i)}$ from $\theta_2 \mid \theta_1^{(i)}$, for $i = 1, 2, \ldots, 1000$. We repeat this process for two values of the correlation coefficient ρ, first with a moderate correlation $\rho = -0.5$ and second with a correlation very close to -1, $\rho = -0.97$. Figure 5.4 shows contour plots in both cases with the first 20 iterations being plotted. We observe that for $\rho = -0.5$ (left panel of Fig. 5.4) the first iteration is inside the 5% probability ellipsis of the contour, while for $\rho = -0.97$ (right panel of Fig. 5.4) it takes a hefty 15 iterations for $\theta^{(i)}$ to get inside the 5% probability ellipsis. We note that although it has taken about 15 iterations for this chain to reach the 5% probability region, the chain moves much slower towards the mean $[2, 3]^\top$ than in the case of $\rho = -0.5$. This indicates that high absolute correlation between θ_1 and θ_2 introduces significant delays in the convergence.

Returning to the state space model we note that β_t is likely to be highly correlated with β_{t-1}, because in the transition equation (5.1b) the values of $\mathbf{Z}$ are usually small. Indeed, we note that if $\mathbf{Z} = \mathbf{0}$, then β_t is perfectly correlated with β_{t-1}; it is not reasonable to suggest the values of $\mathbf{Z}$ to be large, because this would imply erratic changes in the dynamics of the state vector. Thus, the, relatively, small values of $\mathbf{Z}$ introduce high absolute correlation between β_t and β_{t-1}. It turns out that the conditional distribution $\beta_t \mid \beta_{-t}, y_{1:n}$, from which we need to draw $\beta_t^{(i)}$ in the Gibbs sampling proposal of Carlin et al. (1992) is likely to introduce computational inefficiency and convergence delays.

As a result alternative Gibbs sampling schemes are proposed in the literature; indeed, Carter and Kohn (1994) and Fruhwirth-Schnatter (1994b) independently proposed a *block* application of Gibbs sampling, which is considerably more stable and orders of magnitude faster than the above scheme, as reported in Shephard (1994b). According to this instead of sampling from $\beta_t \mid \beta_{-t}, y_{1:n}$, we can successively sample from just $\beta_t \mid \beta_{t+1}, y_{1:n}$. This is possible because we can write

$$
p(\beta_1, \beta_2, \ldots, \beta_n \mid y_{1:n}) = \prod_{t=1}^{n-1} p(\beta_t \mid \beta_{t+1}, \ldots, \beta_n, y_{1:n}) p(\beta_n \mid y_{1:n})
$$

$$
= \prod_{t=1}^{n-1} p(\beta_t \mid \beta_{t+1}, y_{1:n}) p(\beta_n \mid y_{1:n}) \tag{5.19}
$$

$$
= \prod_{t=1}^{n-1} p(\beta_t \mid \beta_{t+1}, y_{1:t}) p(\beta_n \mid y_{1:n}), \tag{5.20}
$$

where (5.19) is obtained since given β_{t+1}, β_t and $\beta_{t+2}, \ldots, \beta_n$ are conditionally independent (i.e. given the present, the past and the future are conditionally independent). For the same reason given y_t, β_t and $y_{t+1}, \ldots, y_n$ are conditionally independent, hence we have (5.20) (a similar argument was also used in the proof of Theorem 3.4 in Sect. 3.3).

From an application of the Kalman filter, the posterior distribution of β_n at time $t = n$ is $\beta_n \mid y_{1:n} \sim N(\hat{\beta}_{n|n}, \mathbf{P}_{n|n})$, where $\hat{\beta}_{n|n}$ and $\mathbf{P}_{n|n}$ are calculated recursively as in Theorem 5.1. For univariate state space models ($d = 1$) the distribution of $\beta_t \mid \beta_{t+1}, y_{1:t}$ was established in Theorem 3.4; the modification for the multivariate case ($d \geq 1$) are minor. It turns out in the multivariate case the distribution of $\beta \mid \beta_{t+1}, y_{1:t}$ is $\beta_t \mid \beta_{t+1}, y_{1:t} \sim N[\hat{\beta}_{t|t+1}(\beta_{t+1}), \mathbf{P}_{t|t+1}(\beta_{t+1})]$, where

$$
\hat{\beta}_{t|t+1}(\beta_{t+1}) = \hat{\beta}_{t|t} + \mathbf{L}_t(\beta_{t+1} - \hat{\beta}_{t+1|t}), \quad \mathbf{P}_{t|t+1}(\beta_{t+1}) = \mathbf{P}_{t|t} - \mathbf{L}_t \mathbf{P}_{t+1|t} \mathbf{L}_t^\top,
$$

with $\mathbf{L}_t = \mathbf{P}_{t|t} \mathbf{F}_{t+1}^\top \mathbf{P}_{t+1|t}^{-1}$ and $\hat{\beta}_{t|t}$, $\hat{\beta}_{t+1|t}$, $\mathbf{P}_{t|t}$ and $\mathbf{P}_{t+1|t}$ provided by the Kalman filter. Note that here we have made explicit the dependence of $\hat{\beta}_{t|t+1}(\beta_{t+1})$ on β_{t+1}, while $\mathbf{P}_{t|t+1}(\beta_{t+1})$ does not depend on β_{t+1}.

Equation (5.20) suggests that, in order to simulate from $p(\beta_1, \ldots, \beta_n \mid y_{1:n})$, at each iteration i of the Gibbs sampler, we need to draw $\beta_n^{(i)}$ from a $N(\hat{\beta}_{n|n}, \mathbf{P}_{n|n})$ (forward filtering step) and then for $t = n-1, n-2, \ldots, 1$ we need to draw $\beta_t^{(i)}$ from $N[\hat{\beta}_{t|t+1}(\beta_{t+1}^{(i)}), \mathbf{P}_{t|t+1}(\beta_{t+1}^{(i)})]$ (backward sampling step). The algorithm is known as *forward filtering backward sampling* (FFBS). This sampling scheme benefits from sampling β_t at each time t, conditional on the state one time ahead (β_{t+1}) instead of conditioning β_t on the whole β_{-t}. The FFBS algorithm is schematically given below.

Forward Filtering Backward Sampling (Known Covariance Matrices)
In the state space model (5.1a)–(5.1b) with $\boldsymbol{\Sigma}$ and $\mathbf{Z}$ and the prior of β_0 as in
the Kalman filter (Theorem 5.1) the following steps provide Gibbs sampling
of the state vectors:

1. Run the Kalman filter for $t = 1, 2, \ldots, n$ and obtain $\hat{\beta}_{t+1|t}, \hat{\beta}_{t|t}, \mathbf{P}_{t|t-1}$ and
   $\mathbf{P}_{t|t}$.
2. For $i \geq 1$, draw a vector $\beta_n^{(i)}$ from $N(\hat{\beta}_{n|n}, \mathbf{P}_{n|n})$.
3. For each $t = n - 1, n - 2, \ldots, 1$ draw state $\beta_t^{(i)}$ from
   $N[\hat{\beta}_{t|t+1}(\beta_{t+1}^{(i)}), \mathbf{P}_{t|t+1}(\beta_{t+1}^{(i)})]$, where

$$\hat{\beta}_{t|t+1}(\beta_{t+1}) = \hat{\beta}_{t|t} + \mathbf{L}_t(\beta_{t+1} - \hat{\beta}_{t+1|t}),$$

$$\mathbf{P}_{t|t+1}(\beta_{t+1}) = \mathbf{P}_{t|t} - \mathbf{L}_t \mathbf{P}_{t+1|t} \mathbf{L}_t^\top,$$

   and $\mathbf{L}_t = \mathbf{P}_{t|t} \mathbf{F}_{t+1}^\top \mathbf{P}_{t+1|t}^{-1}$.
4. Set the counter $i = i + 1$ and go to step (2) until convergence.

Some comments are in order. Firstly, note that the FFBS algorithm does not
require an initialisation of the state vector $\beta_t^{(0)}$. In other words, the Kalman filter
provides a learned procedure for the initialisation of $\beta_n^{(i)}$. Secondly, the covariance
matrix $\mathbf{P}_{t|t+1}(\beta_{t+1}) = \mathbf{P}_{t|t+1}$ does not depend on β_{t+1} and can be provided by the
Kalman filter in step 1. This can result in significant computational savings, as only
the computation of the mean vector $\beta_{t|t+1}(\beta_{t+1}^{(i)})$ requires to know the simulated
vector $\beta_{t+1}^{(i)}$.

5.7.3 Unknown Variances-Covariances

This section considers the state space model (5.1a)–(5.1b), where now the covari-
ance matrix $\boldsymbol{\Sigma}$ of the innovation vector ϵ_t and the covariance matrix $\mathbf{Z}$ of the
innovation vector ζ_t are unknown and subject to estimation. Initially, the FFBS
algorithm proposed in Carter and Kohn (1994) and Fruhwirth-Schnatter (1994b)
considered univariate time series. Carter and Kohn (1994) placed improper priors on
the observation and transition variances, resulting in proper inverse gamma priors
for these variances; these authors assumed that the transition covariance matrix is
known up to a variance component, which is subject to estimation. On the other
hand, Fruhwirth-Schnatter (1994b) introduced the d-inverse gamma state space
model, whereby the variance of the scalar observation innovation is gamma and
the covariance of the transition innovation vector $\mathbf{Z}$ is diagonal, each element of
its main diagonal independently following a priori an inverse gamma distribution.

The d-inverse gamma state space model is implemented within the dlm package in R, see e.g. Petris et al. (2009, §4.5). Here we describe the more general approach where the priors of both Σ and $\mathbf{Z}$ are inverse Wishart, allowing us to learn for the correlation between elements of ϵ_t and ζ_t, respectively. Gibbs sampling for the d-inverse gamma model is described in Exercise 11 and a summary of the algorithm is given in page 260.

The general Gibbs sampler described in Sect. 5.7.1 can be applied in blocks of parameters, i.e. if we are interested in sampling from the posterior distribution of $[\theta, \psi]^\top$, where θ is a vector of parameters and ψ is a vector or matrix containing hyperparameters, then we need to sample from $\theta \mid \psi, y$ and from $\psi \mid \theta, y$. In order to apply this modification first we set $\theta = [\beta_1^\top, \ldots, \beta_n^\top]^\top$ and $\psi = [\Sigma, \mathbf{Z}]$ and then we note that the FFBS scheme of the previous section provides a sampling scheme from β_t, given the hyperparameters Σ and $\mathbf{Z}$. Therefore from (5.20) we have

$$p(\beta_{1:n} \mid \Sigma, \mathbf{Z}, y_{1:n}) = \prod_{t=1}^{n-1} p(\beta_t \mid \beta_{t+1}, \Sigma, \mathbf{Z}, y_{1:t}) p(\beta_n \mid \Sigma, \mathbf{Z}, y_{1:n}), \qquad (5.21)$$

where we use $\beta_{1:n}$ for the joint state vectors $\beta_{1:n} = \{\beta_1, \beta_2, \ldots, \beta_n\}$.

Moving on to the prior structure of the hyperparameters $\Sigma, \mathbf{Z}$ we assume that a priori they are independent and that each of which has an inverse Wishart prior distribution, i.e.

$$p(\Sigma) = c_2 |\Sigma|^{-(\nu+d+1)/2} \exp\left[-\frac{1}{2}\mathrm{trace}(\mathbf{S}_\Sigma \Sigma^{-1}) \right], \qquad (5.22)$$

$$p(\mathbf{Z}) = c_2 |\mathbf{Z}|^{-(\nu+p+1)/2} \exp\left[-\frac{1}{2}\mathrm{trace}(\mathbf{S}_\mathbf{Z} \mathbf{Z}^{-1}) \right], \qquad (5.23)$$

where $\mathbf{S}_\Sigma$ and $\mathbf{S}_\mathbf{Z}$ are prior scale matrices, ν are the prior degrees of freedom and c_2 is the proportionality constant (the expression of c_2 is given in Sect. 5.5.2 where the inverse Wishart distribution is discussed). These distributions are abbreviated as $\Sigma \sim IW(n, \mathbf{S}_\Sigma)$ and $\mathbf{Z} \sim IW(n, \mathbf{S}_\mathbf{Z})$; for a more detailed discussion on the inverse Wishart distribution see also Sect. 5.5.2. Then the conditional distribution of Σ, given $\mathbf{Z}$, $\beta_{1:n}$ and $y_{1:n}$ is

$$\begin{aligned}
p(\Sigma \mid \mathbf{Z}, \beta_{1:n}, y_{1:n}) &= p(\Sigma, \mathbf{Z}, \beta_{1:n}, y_{1:n}) \\
&= p(y_{1:n} \mid \beta_{1:n}, \Sigma, \mathbf{Z}) p(\beta_{1:n} \mid \Sigma, \mathbf{Z}) p(\Sigma) p(\mathbf{Z}) \\
&= \prod_{t=1}^{n} p(y_t \mid \beta_t, \Sigma) \prod_{t=1}^{n} p(\beta_t \mid \beta_{t-1}, \mathbf{Z}) p(\Sigma) p(\mathbf{Z}) \qquad (5.24) \\
&\propto \prod_{t=1}^{n} |\Sigma|^{-1/2} \exp\left[-\frac{1}{2}\mathrm{trace}\left\{ (y_t - x_t^\top \beta_t)(y_t - x_t^\top \beta_t)^\top \Sigma^{-1} \right\} \right]
\end{aligned}$$

$$\times |\boldsymbol{\Sigma}|^{-(\nu+d+1)/2} \exp\left[-\frac{1}{2}\text{trace}(\mathbf{S}_\Sigma \boldsymbol{\Sigma}^{-1})\right]$$

$$= |\boldsymbol{\Sigma}|^{-(\nu+n+d+1)/2} \exp\left[-\frac{1}{2}\text{trace}\left\{\sum_{t=1}^{n}(y_t - x_t^\top \beta_t)(y_t - x_t^\top \beta_t)^\top + \mathbf{S}_\Sigma\right\}\right],$$

which is proportional to the inverse Wishart distribution

$$\boldsymbol{\Sigma} \mid \mathbf{Z}, \beta_{1:n}, y_{1:n} \sim IW\left[\nu + n, \sum_{t=1}^{n}(y_t - x_t^\top \beta_t)(y_t - x_t^\top \beta_t)^\top + \mathbf{S}_\Sigma\right]. \quad (5.25)$$

We have used the assumption that *a priori* $\boldsymbol{\Sigma}$ and $\mathbf{Z}$ are independent, hence $p(\boldsymbol{\Sigma}, \mathbf{Z}) = p(\boldsymbol{\Sigma})p(\mathbf{Z})$. Also, Eq. (5.24) shows that, given $\beta_{1:n}$ and $y_{1:n}$, the covariances matrices $\boldsymbol{\Sigma}$ and $\mathbf{Z}$ are conditionally independent.

Similarly, we can see that the conditional distribution of $\mathbf{Z}$, given $\beta_{1:n}$, $y_{1:n}$ and $\boldsymbol{\Sigma}$ is

$$p(\mathbf{Z} \mid \boldsymbol{\Sigma}, \beta_{1:n}, y_{1:n}) = p(\mathbf{Z}, \boldsymbol{\Sigma}, \beta_{1:n}, y_{1:n}) \propto \prod_{t=1}^{n} p(\beta_t \mid \beta_{t-1}, \mathbf{Z})p(\mathbf{Z})$$

$$= |\mathbf{Z}|^{-(\nu+n+p+1)/2}$$

$$\times \exp\left[-\frac{1}{2}\text{trace}\left\{\sum_{t=1}^{n}(\beta_t - \mathbf{F}_t\beta_{t-1})(\beta_t - \mathbf{F}_t\beta_{t-1})^\top + \mathbf{S}_Z\right\}\mathbf{Z}^{-1}\right],$$

which is proportional to the inverse Wishart distribution

$$\mathbf{Z} \mid \boldsymbol{\Sigma}, \beta_{1:n}, y_{1:n} \sim IW\left[\nu + n, \sum_{t=1}^{n}(\beta_t - \mathbf{F}_t\beta_{t-1})(\beta_t - \mathbf{F}_t\beta_{t-1})^\top + \mathbf{S}_Z\right].$$

$$(5.26)$$

From the above discussion, the full conditional distributions are given by (5.21), (5.25) and (5.26). Hence the FFBS algorithm for unknown covariance matrices $\boldsymbol{\Sigma}$ and $\mathbf{Z}$ are given below.

Forward Filtering Backward Sampling (Unknown Covariance Matrices)
In the state space model (5.1a)–(5.1b) with the prior of β_0 as in the Kalman filter (Theorem 5.1) and the priors of $\boldsymbol{\Sigma}$ and $\mathbf{Z}$ as in (5.22) and (5.23), the following steps provide Gibbs sampling of the state vectors and the covariance matrices:

(continued)

1. Initialisation: draw $\Sigma^{(0)}$ from (5.22) and $Z^{(0)}$ from (5.23).
2. For iteration $i \geq 1$, set $\Sigma = \Sigma^{(i-1)}$ and $Z = Z^{(i-1)}$.

   a. Run the Kalman filter for $t = 1, 2, \ldots, n$ and obtain $\hat{\beta}_{t+1|t}, \hat{\beta}_{t|t}, P_{t|t-1}$ and $P_{t|t}$.
   b. Draw a vector $\beta_n^{(i)}$ from $N(\hat{\beta}_{n|n}, P_{n|n})$.
   c. For each $t = n - 1, n - 2, \ldots, 1$ draw state $\beta_t^{(i)}$ from $N[\hat{\beta}_{t|t+1}(\beta_{t+1}^{(i)}), P_{t|t+1}(\beta_{t+1}^{(i)})]$, where

   $$\hat{\beta}_{t|t+1}(\beta_{t+1}) = \hat{\beta}_{t|t} + L_t(\beta_{t+1} - \hat{\beta}_{t+1|t}),$$

   $$P_{t|t+1}(\beta_{t+1}) = P_{t|t} - L_t P_{t+1|t} L_t^{\top},$$

   and $L_t = P_{t|t} F_{t+1}^{\top} P_{t+1|t}^{-1}$.
   d. Draw $\Sigma^{(i)}$ from

   $$IW\left[v + n, \sum_{t=1}^{n} (y_t - x_t^{\top} \beta_t^{(i)})(y_t - x_t^{\top} \beta_t^{(i)})^{\top} + S_{\Sigma} \right]$$

   and $Z^{(i)}$ from

   $$IW\left[v + n, \sum_{t=1}^{n} (\beta_t^{(i)} - F_t \beta_{t-1}^{(i)})(\beta_t^{(i)} - F_t \beta_{t-1}^{(i)})^{\top} + S_Z \right]$$

   where $\beta_0^{(i)}$ is drawn from a $N(\hat{\beta}_{0|0}, P_{0|0})$.
3. Set the counter $i = i + 1$ and go to step (2) until convergence.

Example 5.1 (Production Time Series) In this section we consider a bivariate time series consisting of 276 observation vectors $y_t = [y_{1t}, y_{2t}]^{\top}$ of temperatures measured at two components (y_{1t} is the temperature of Component 3 and y_{2t} of Component 4) during the process of a production of a plastic mould. This is part of a larger data set (considering 5 components) that is studied in Pan and Jarrett (2004) and also in Triantafyllopoulos (2006a). Large levels of the temperatures during the production process indicate hazards and may prompt relevant action. The objective of the above studies is to propose a control mechanism that can signal large temperatures using statistical process control methods. Pan and Jarrett (2004) demonstrate that the estimation of the covariance matrices (such as Σ and Z in the context and notation of this book) plays a crucial role in the detection of out of control signals. In this section we use a bivariate local level model for y_t and we

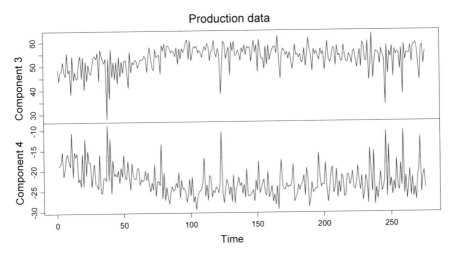

Fig. 5.5 Plot of the production time series $y_t = [y_{1t}, y_{2t}]^\top$

provide smoothed estimates of the level β_t as well as estimates of the observation and transition covariance matrices $\mathbf{\Sigma}$ and $\mathbf{Z}$.

Figure 5.5 plots the data (the top panel shows y_{1t} the temperatures of Component 1 and the lower panel shows the temperatures y_{2t} of Component 2). We observe that each component time series exhibits local variation around the level, which suggests that a local level is a plausible model. Moreover, histograms of each component indicate that the normality assumption of the data is broadly satisfied. Figure 5.5 also indicates that y_{1t} increases steadily for up to about $t = 140$ and then it seems to drop a degree ^{0}C compared to the start of the process, while y_{2t} seems more stable over time. The proposed model is

$$y_t = \beta_t + \epsilon_t \quad \text{and} \quad \beta_t = \beta_{t-1} + \zeta_t,$$

where $\beta_t = [\beta_{1t}, \beta_{2t}]^\top$ is a bivariate state vector and the innovation vectors ϵ_t and ζ_t satisfy the usual conditions, stated in several places in the book, e.g. as in model (5.1a)–(5.1b). Initially we assume that $\beta_0 \sim N([0, 0]^\top, 1000\mathbf{I})$ for a weak prior specification on β_0, while the inverse Wishart priors (5.22) and (5.23) of $\mathbf{\Sigma} = \mathrm{Var}(\epsilon_t)$ and $\mathbf{Z} = \mathrm{Var}(\zeta_t)$ are used, where $v = 10$, $d = p = 2$ and $\mathbf{S}_\Sigma = \mathbf{S}_Z = \mathbf{I}$. These settings suggest weak prior knowledge on $\mathbf{\Sigma}$ and $\mathbf{Z}$, but if prior knowledge is available this may be incorporated easily by specifying $\mathbf{S}_\Sigma$ and $\mathbf{Z}$ as non-diagonal covariance matrices. In our experience, even such a prior information is available, there is little loss, if any, by considering the setting proposed in this section; see also the discussion in Sect. 4.5.

With the above set-up in place we run the forward filtering backward sampling
(FFBS) algorithm described earlier in this section. The following commands are
used to fit the model in R:

```
> # read data
> pro <- read.table("productiondata.txt")
> ypro <- as.matrix(pro[,3:4])
> # fit the local level model
> fit <- bts.ffbs(y=ypro, v=10, m=5000)
```

The Gibbs algorithm runs for 5000 iterations, which was judged enough for convergence and the first 1000 iterations were used for the burn-in and were excluded
from any further computations. This convergence is backed by the trace plots of
the simulated values of $\beta_{1t}^{(i)}$ and $\beta_{2t}^{(i)}$, shown for iterations $i = 1001, \ldots, 5000$
in Fig. 5.6. Indeed, we observe that $\beta_{1t}^{(i)}$ and $\beta_{2t}^{(i)}$ are stationery and they are also
uncorrelated (this can be checked by looking at the correlogram of $\beta_{jt}^{(i)}$, $j = 1, 2$).
This figure is produced using the R code:

```
> # plot of estimated beta's
> par(mfrow=c(2,1))
> ts.plot(ts(fit$beta[276,1,1001:5000],start=1001,
+ frequency=1),main=expression(paste(beta[1])),
+ xlab="Iteration",ylab="")
> ts.plot(ts(fit$beta[276,2,1001:5000],start=1001,
+ frequency=1),main=expression(paste(beta[2])),
+ xlab="Iteration",ylab="")
```

Figure 5.7 shows the histograms of the estimated variances Σ_{11} and Σ_{22}, for $\mathbf{\Sigma} = (\Sigma_{ij})_{i,j=1,2}$ as well as the estimated variances Z_{11} and Z_{22} for $\mathbf{Z} = (Z_{ij})_{i,j=1,2}$.

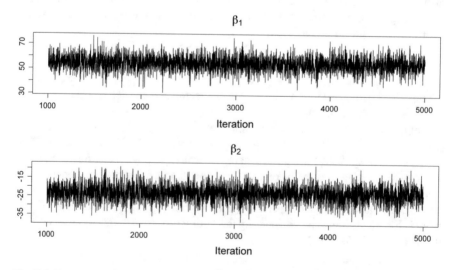

Fig. 5.6 Trace plot of the level $\beta_t = [\beta_{1t}, \beta_{2t}]^\top$ of the time series; shown are the simulated values $\beta_{1t}^{(i)}$ (top panel) and $\beta_{2t}^{(i)}$ (low panel) for iterations $i = 1001, \ldots, 5000$

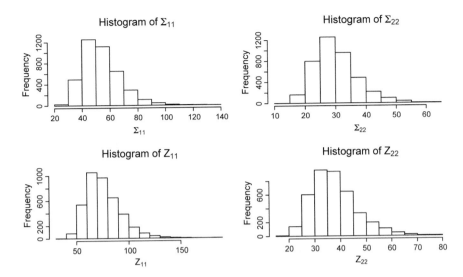

Fig. 5.7 Histograms of the simulated variances Σ_{11}, Σ_{22}, Z_{11} and Z_{22}

The modes of the estimated correlation coefficients in Σ and Z were -0.832 and -0.825, respectively. This indicates significant negative correlation between y_{1t} and y_{2t}. Assuming that Σ and Z are diagonal matrices, as it is the case in the d-inverse gamma state space models of Carter and Kohn (1994) and Fruhwirth-Schnatter (1994b) is not appropriate; here the inverse Wishart priors offer the advantage of estimating the correlation coefficients of ϵ_{1t}, ϵ_{2t} and ζ_{1t}, ζ_{2t}, where $\epsilon_t = [\epsilon_{1t}, \epsilon_{2t}]^\top$ and $\zeta_t = [\zeta_{1t}, \zeta_{2t}]^\top$. The following R code was used for Fig. 5.7:

```
> # plot of simulated variances
> Sigma <- Z <- array(0, dim=c(2,2,5000))
> for (i in 1001:5000){
+ Sigma[,,i] <- fit$Sigma[[i]]
+ Z[,,i] <- fit$Z[[i]]
+ }

> par(mfrow=c(2,2))
> hist(Sigma[1,1,1001:5000],main=expression(paste("Histogram
+ of ", Sigma[11])), xlab=expression(Sigma[11]) )
> hist(Sigma[2,2,1001:5000],main=expression(paste("Histogram
+ of ", Sigma[22])), xlab=expression(Sigma[22]) )
> hist(Z[1,1,1001:5000],main=expression(paste("Histogram
+ of ",Z[11])), xlab=expression(Z[11]) )
> hist(Z[2,2,1001:5000],main=expression(paste("Histogram
+ of ",Z[22])), xlab=expression(Z[22]) )
```

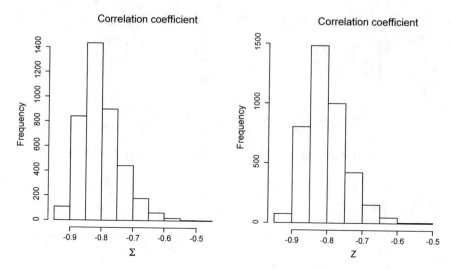

Fig. 5.8 Histograms of the simulated correlations (left panel for **Σ** and right panel for **Z**)

The estimated correlation of ϵ_{1t} and ϵ_{2t} (corresponding to the observation covariance matrix **Σ**) and the correlation of ζ_{1t} and ζ_{2t} (corresponding to the covariance matrix **Z**) is visualised in Fig. 5.8. The R code used here is:

```
> # plot of simulated correlations
> par(mfrow=c(1,2))
> hist( Sigma[1,2,1001:5000] /
+ sqrt( Sigma[1,1,1001:5000] * Sigma[2,2,1001:5000] ),
+ main=expression("Correlation coefficient"),
+ xlab=expression(Sigma) )

> hist( Z[1,2,1001:5000] /
+ sqrt( Z[1,1,1001:5000] * Z[2,2,1001:5000] ),
+ main=expression("Correlation coefficient"), xlab="Z")
```

5.8 Exercises

1. Suppose that you want to forecast a bivariate time series $y_t = [y_{1t}, y_{2t}]^{\top}$ consisting of a linear trend y_{1t} and a local level y_{2t}. Write down two state space models for y_t of the form of model (5.1a)–(5.1b), with design matrices

    a.

    $$\mathbf{x}_t = \begin{bmatrix} 1 & 0 \\ 0 & 1 \end{bmatrix};$$

b.

$$\mathbf{x}_t = \begin{bmatrix} 1 & 0 \\ 0 & 0 \\ 0 & 1 \end{bmatrix}$$

i.e. determine the form of the transition matrices, the observation covariance matrices and the transition covariance matrices.

2. (Aggregate forecasting). Suppose that the d-dimensional time series vector $y_t = [y_{1t}, \ldots, y_{td}]^\top$ is modelled with the multivariate scaled observational precision state space model (5.13a)–(5.13b) of Sect. 5.5.3. Define the aggregate time series of 'totals' $T_t = \sum_{i=1}^{d} y_{it} = 1_d^\top y_t$, where $1_d = [1, \ldots, 1]^\top$ is the d-dimensional vector of units. Show that T_t follows a univariate state space model with observation equation

$$T_t = x_t^\top \gamma_t + \upsilon_t, \quad \upsilon_t \sim N(0, 1_d^\top \Sigma 1_d)$$

and transition equation

$$\gamma_t = \mathbf{F}_t \gamma_{t-1} + \omega_t, \quad \omega_t \sim N(0, 1_d^\top \Sigma 1_d \mathbf{Z}_t),$$

where x_t is the design vector, $\mathbf{F}_t$ is the transition matrix, Σ is the observation covariance matrix and $\mathbf{Z}_t$ is the transition left covariance matrix of the model (5.13a)–(5.13b). Define γ_t, υ_t and ω_t in terms of the state matrix and innovations of the state space model of y_t.

3. A marketing company is running a pricing assessment for 3 of its products A, B and C. The company adopts a multivariate state space model in order to forecast the prices of these three products. Let y_{1t} be the price of product A at time t, y_{2t} the price of product B at time t, y_{3t} the price of product C at time t and write $y_t = [y_{1t}, y_{2t}, y_{3t}]^\top$. The one-step forecast of these products at time $t+1$ is $\hat{y}_{t+1|t} = [3, 2, 5]^\top$ (in £) with associated forecast covariance matrix $\mathbf{Q}_{t+1|t} = \mathrm{diag}(0.5, 2, 1)$. The company asserts that the total price of the three products at time $t+1$ should be $T_{t+1} = 9$. Obtain 99% marginal prediction intervals of y_{1t}, y_{2t}, y_{3t}, given that $T_{t+1} = 9$. Based on these intervals suggest whether the company should revise downwards or upwards the prices of A, B and C as compared to the forecasts 3, 2 and 5 respectively.

4. Consider a $d \times 1$ random vector Y and a $p \times 1$ random vector X having a joint dp-dimensional multivariate normal distribution

$$\begin{bmatrix} X \\ Y \end{bmatrix} \sim N \left\{ \begin{bmatrix} m_X \\ m_Y \end{bmatrix}, \begin{bmatrix} \mathbf{V}_X & \mathbf{C} \\ \mathbf{C}^\top & \mathbf{V}_Y \end{bmatrix} \right\},$$

where m_X is the mean vector of X, m_Y is the mean vector of Y, $\mathbf{V}_X$ is the covariance matrix of X, $\mathbf{C}$ is the covariance of X and Y, and $\mathbf{V}_Y$ is the covariance matrix of Y.

a. Define $\mathbf{K} = \mathbf{CV}_Y^{-1}$ and show that $X - \mathbf{K}Y$ is uncorrelated with Y. Find the distribution of $X - \mathbf{K}Y$.

b. Using (a) show that the conditional distribution of X, given $Y = y$, for a particular value y of Y is

$$X \mid Y = y \sim N[m_X + \mathbf{K}(y - m_Y), \mathbf{V}_X - \mathbf{KV}_Y\mathbf{K}^\top].$$

c. Use this result to prove Theorem 5.1 (Kalman filter).

5. In the context of Exercise 4 suppose that X and Y are only partially specified in terms of their means and variances, without specifying their joint distribution. If we write $X \sim [m_X, \mathbf{V}_X]$ this implies that the mean vector of X is m_X and its covariance matrix is $\mathbf{V}_X$. For the joint partial specification of X and Y we write

$$\begin{bmatrix} X \\ Y \end{bmatrix} \sim \left\{ \begin{bmatrix} m_X \\ m_Y \end{bmatrix}, \begin{bmatrix} \mathbf{V}_X & \mathbf{C} \\ \mathbf{C}^\top & \mathbf{V}_Y \end{bmatrix} \right\},$$

where $m_X, m_Y, \mathbf{V}_X, \mathbf{V}_Y, \mathbf{C}$ are as in Exercise 4.
Two random vectors Z and W are said to be second order independent if

$$E(Z \mid W = w) = E(Z) \quad \text{and} \quad \text{Var}(Z \mid W = w) = \text{Var}(Z),$$

for any value w of W. We write $Z \perp_2 W$ to denote that Z and W are 2nd order independent.

a. Show that if Z and W are mutually independent, then $Z \perp_2 W$. Show that the reverse is not always true.

b. With $\mathbf{K}$ defined as in Exercise 4, show that if $X - \mathbf{K}Y \perp_2 Y$, then the mean vector and the covariance matrix of the conditional distribution of $X \mid Y = y$ are given by

$$X \mid Y = y \sim [m_X + \mathbf{K}(y - m_Y), \mathbf{V}_X - \mathbf{KV}_Y\mathbf{K}^\top].$$

c. Now consider the state space model (5.1a)–(5.1b) where the Gaussian assumption of the innovations ϵ_t and ζ_t is dropped and the model be partially specified via the innovations' mean vectors and covariance matrices, i.e. $\epsilon_t \sim (0, \mathbf{\Sigma})$ and $\zeta_t \sim (0, \mathbf{Z}_t)$. Use (b) to show

$$\beta_t \mid y_{1:t} \sim (\hat{\beta}_{t|t}, \mathbf{P}_{t|t}),$$

where $\hat{\beta}_{t|t}$ and $\mathbf{P}_{t|t}$ are given by updating recurrences exactly the same as those provided by the Kalman filter of Theorem 5.1. Thus extend the Kalman filter in cases that the modeller is reluctant to specify the Gaussian distribution for the innovations.

6. Triantafyllopoulos and Harrison (2008). In the context of Exercise 5 suppose that the covariance matrices $\mathbf{V}_X$ and $\mathbf{V}_Y$ are scaled by an unknown variance σ^2. Hence, X and Y are partially specified by their mean vectors and covariance matrices as

$$\begin{bmatrix} X \\ Y \end{bmatrix} \sim \left\{ \begin{bmatrix} m_X \\ m_Y \end{bmatrix}, \sigma^2 \begin{bmatrix} \mathbf{V}_X & \mathbf{C} \\ \mathbf{C}^\top & \mathbf{V}_Y \end{bmatrix} \right\},$$

where $m_X, m_Y, \mathbf{V}_X, \mathbf{V}_Y, \mathbf{C}$ are as in Exercise 5.

a. Now with $\perp_2$ denoting second order independence as defined in Exercise 5 above, assuming $X - \mathbf{K}Y \perp_2 Y \mid \sigma^2$, show

$$X \mid \sigma^2, Y = y \sim [\mu_X + \mathbf{K}(y - \mu_Y), \sigma^2(\mathbf{V}_X - \mathbf{K}\mathbf{V}_Y\mathbf{K}^\top)],$$

where $\mathbf{K} = \mathbf{C}\mathbf{V}_Y^{-1}$.

b. Let T be a non-linear function of Y, often taken as

$$T = (Y - \mu_Y)^\top \mathbf{V}_Y^{-1}(Y - \mu_Y).$$

Define κ to be α times the variance of $T \mid \sigma^2$, for some $\alpha > 0$. Assume that $\sigma^2 - LT \perp_2 Y, \kappa$, with

$$T \mid \sigma^2, \kappa \sim (\sigma^2, \kappa/\alpha) \quad \text{and} \quad \text{Cov}(T, \sigma^2 \mid \kappa) = \text{Var}(\sigma^2 \mid \kappa),$$

where $L = \alpha/(\eta + \alpha)$ and $\sigma^2 \mid \kappa \sim (\hat\sigma^2, \kappa/\eta$, which is η/α times as precise as the conditional distribution of T, for some known $\hat\sigma^2, \alpha$ and η.

Show that

$$\begin{bmatrix} \sigma^2 \\ T \end{bmatrix} \sim \left\{ \begin{bmatrix} \hat\sigma^2 \\ \hat\sigma^2 \end{bmatrix}, \frac{\kappa}{\eta} \begin{bmatrix} 1 & 1 \\ 1 & (\eta + \alpha)/\alpha \end{bmatrix} \right\}.$$

Using $\sigma^2 - LT \perp_2 Y \mid \kappa$, with L as above, show

$$\sigma^2 \mid \kappa, T = \tau \sim \left(\frac{\eta\hat\sigma^2 + \alpha\tau}{\eta + \alpha}, \frac{\kappa}{\eta + \alpha} \right).$$

Using the tower law of expectations show that

$$X \mid Y = y \sim \left[\mu_X + \mathbf{K}(y - \mu_Y), \frac{\eta\hat\sigma^2 + \alpha\tau}{\eta + \alpha}(\mathbf{V}_X - \mathbf{K}\mathbf{V}_Y\mathbf{K}^\top) \right],$$

where $\tau = (y - \mu_Y)\mathbf{V}_Y^{-1}(y - \mu_Y)$.

c. Now suppose that the joint distribution of X and Y is Gaussian, so that

$$\begin{bmatrix} X \\ Y \end{bmatrix} \sim N \left\{ \begin{bmatrix} m_X \\ m_Y \end{bmatrix}, \sigma^2 \begin{bmatrix} \mathbf{V}_X & \mathbf{C} \\ \mathbf{C}^\top & \mathbf{V}_Y \end{bmatrix} \right\},$$

and assume that vs/σ^2 follows a chi-square distribution with v degrees of freedom, i.e. $vs/\sigma^2 \sim \chi^2_v$, for some $v, s > 0$.
Writing $T = (Y - \mu_y)^\top \mathbf{V}_Y^{-1} (Y - \mu_Y)$, show

$$X \mid Y = y \sim t \left[v + p, \mu_x + \mathbf{K}(y - \mu_Y), \frac{vs + \tau}{v + p} \left(\mathbf{V}_X - \mathbf{K} \mathbf{V}_Y \mathbf{K}^\top \right) \right],$$

$$\frac{vs + \tau}{\sigma^2} \mid Y = y \sim \chi^2_{v+p}, \quad \tau = (y - \mu_Y)^\top \mathbf{V}_Y^{-1} (y - \mu_Y),$$

for some y observed vector of Y and $t(\cdot)$ denotes a Student t distribution.
d. Compare the posterior mean and variance of σ^2 of approaches (b) and (c) above. In particular find a value for α in (b) so that $E(X \mid Y = y)$ and $\text{Var}(X \mid Y = y)$ are the same under (b) and (c).
Adopting this value of α show that

$$E(\sigma^2 \mid Y = y, \text{approach (c) }) - E(\sigma^2 \mid Y = y, \text{approach (b) })$$

$$= \frac{(p - 1)vs}{(v - 2)(v + p - 2)},$$

hence establish that the posterior means of σ^2 under (b) and (c) are different. Explore further the impact of the differences of the estimation of σ^2 by considering the posterior variances of σ^2 under (b) and (c).
e. In the context of state space models, set $X = \beta_t$ (state vector at t) and $Y = y_t$ (observation vector at t) and obtain a version of the results in (b) and (c) for state space models. Write down explicitly your state space model under (b) and under (c) and comment on their differences.

7. New York air quality (source Chambers et al. (1983)). Measurements of wind speed and temperature are observed in daily frequency over the period 1 May to 30 September 1973 in New York. The data is available in R by executing the following commands

```
> library(MASS)
> data(airquality)
> y <- airquality[,3:4]
```

It is suggested that air quality reflects the levels of these two variables. Use a state space model to jointly model the effects and provide prediction of the wind speed and the temperature. Your analysis should include estimating the correlation of the two variables as well as assessing the goodness of fit by considering appropriate error analysis.

8. Longley's macroeconomic data set (source Longley (1967)). The data set comprise of annual measurements on six economic variables: gross national product (GNP) deflator, GNP, number of unemployed, number of people in the armed forces, 'noninstitutionalized' population ≥ 14 years of age and number of employed, from 1947 to 1962. The data is available in R by executing the following commands

```
> library(MASS)
> data(longley)
> y <- longley
```

Discuss the data commenting on any interesting features they provide. Consider the four variables: variables GNP, number of unemployed, number of people in the armed forces and number of employed and suggest a state space model to describe these four variables. Fit the model in R and estimate the observation covariance matrix.

9. Petroleum rock samples (source BP Research, image analysis by Ronit Katz). 48 rock measurements on 3 variables from a petroleum reservoir are made. The variables are: (1) area of pores space, in pixels out of 256 by 256, (2) perimeter in pixels and (3) perimeter per sqrt(area). The data are sampled in 12 main time-occasions and 4 cross-sections. The time units are not known, but are believed to be equally spaced and of high frequency. The data i is available in R by executing the following commands

```
> library(MASS)
> data(rock)
> y <- rock[,1:3]
```

Fit a trivariate state space model to the data. Estimate the 3×3 observation and transition covariance matrices using a suitable Gibbs sampling procedure.

10. Prove the EM algorithm for multivariate state space models, which is summarised on page 215.

11. d-inverse gamma state space model. In the state space model (5.1a)–(5.1b) suppose that the observation covariance matrix Σ and the time-invariant transition covariance matrix $Z_t = Z$ are diagonal, written as $\Sigma = \text{diag}(\sigma_{11}, \sigma_{22}, \ldots, \sigma_{dd})$ and $Z = \text{diag}(z_{11}, z_{22}, \ldots, z_{pp})$, for some variances σ_{ii} and z_{jj} and $i = 1, 2, \ldots, d$; $j = 1, 2, \ldots, p$.
We assume that σ_{ii} is independent of σ_{kk}, for $i \neq k$, z_{jj} is independent of z_{ll}, for $j \neq l$ and σ_{ii} is independent of z_{jj}, for any i, j. The prior of β_0 is as in the Kalman filter (Theorem 5.1) and inverse gamma priors are assumed for σ_{ii} and z_{jj}, i.e.

$$\sigma_{ii} \sim IG(v_i, S_{\Sigma,ii}) \quad \text{and} \quad z_{jj} \sim IG(w_j, S_{Z,jj}), \qquad (5.27)$$

for some parameters v_i, $S_{\Sigma,ii}$, w_i, $S_{Z,jj}$.

In the context of forward filtering backward sampling scheme, show that the conditional posterior distributions of σ_{ii} and z_{jj} are

$$\sigma_{ii} \mid \mathbf{Z}, \beta_{1:n}, y_{1:n} \sim IG\left[\frac{n}{2} + v_i, \frac{1}{2}\sum_{t=1}^{n}(y_{it} - x_{it}^\top \beta_t)^2 + S_{\Sigma,ii}\right],$$

$$z_{jj} \mid \mathbf{\Sigma}, \beta_{1:n}, y_{1:n} \sim IG\left[\frac{n}{2} + w_j, \frac{1}{2}\sum_{t=1}^{n}(\beta_{jt} - F_{jt}^\top \beta_{t-1})^2 + S_{Z,jj}\right],$$

where $y_t = [y_{1t}, \ldots, y_{dt}]^\top$, $\mathbf{x}_t^\top = [x_{1t}, \ldots, x_{pt}]$, $\beta_t = [\beta_{1t}, \ldots, \beta_{pt}]^\top$ and $\mathbf{F}_t^\top = [F_{1t}, \ldots, F_{pt}]$.

Hence obtain the forward filtering backward sampling algorithm for the d-inverse gamma state space model, given below:

Forward Filtering Backward Sampling (d-Inverse Gamma Model)
In the state space model (5.1a)–(5.1b) with the prior of β_0 as in the Kalman filter (Theorem 5.1) and the priors of σ_{ii} and z_{jj} as in (5.27) the following steps provide Gibbs sampling of the state vectors and the variances:

a. Initialisation: draw $\sigma_{ii}^{(0)}$ and $z_{jj}^{(0)}$ from (5.27) and obtain $\mathbf{\Sigma}^{(0)}$ and $\mathbf{Z}^{(0)}$.
b. For iteration $k \geq 1$, set $\mathbf{\Sigma} = \mathbf{\Sigma}^{(k-1)}$ and $\mathbf{Z} = \mathbf{Z}^{(k-1)}$.

  i. Run the Kalman filter for $t = 1, 2, \ldots, n$ and obtain $\hat{\beta}_{t+1|t}$, $\hat{\beta}_{t|t}$, $\mathbf{P}_{t|t-1}$ and $\mathbf{P}_{t|t}$.
  ii. Draw a vector $\beta_n^{(i)}$ from $N(\hat{\beta}_{n|n}, \mathbf{P}_{n|n})$.
  iii. For each $t = n - 1, n - 2, \ldots, 1$ draw state $\beta_t^{(i)}$ from $N[\hat{\beta}_{t|t+1}(\beta_{t+1}^{(i)}), \mathbf{P}_{t|t+1}(\beta_{t+1}^{(i)})]$, where

$$\hat{\beta}_{t|t+1}(\beta_{t+1}) = \hat{\beta}_{t|t} + \mathbf{L}_t(\beta_{t+1} - \hat{\beta}_{t+1|t}),$$

$$\mathbf{P}_{t|t+1}(\beta_{t+1}) = \mathbf{P}_{t|t} - \mathbf{L}_t \mathbf{P}_{t+1|t} \mathbf{L}_t^\top,$$

  and $\mathbf{L}_t = \mathbf{P}_{t|t} \mathbf{F}_{t+1}^\top \mathbf{P}_{t+1|t}^{-1}$.
  iv. For $i = 1, 2, \ldots, d$, draw $\sigma_{ii}^{(k)}$ from

$$IG\left[\frac{n}{2} + v_i, \frac{1}{2}\sum_{t=1}^{n}(y_{it} - x_{it}^\top \beta_t^{(k)})^2 + S_{\Sigma,ii}\right]$$

(continued)

and for $j = 1, 2, \ldots, p$, draw $z_{jj}^{(k)}$ from

$$IG\left[\frac{n}{2} + w_j, \frac{1}{2}\sum_{t=1}^{n}(\beta_{jt}^{(k)} - F_{jt}^{\top}\beta_{t-1}^{(k)})^2 + S_{Z,jj}\right]$$

where $\beta_0^{(k)}$ is drawn from a $N(\hat{\beta}_{0|0}, \mathbf{P}_{0|0})$. Set

$$\mathbf{\Sigma}^{(k)} = \text{diag}\left[\sigma_{11}^{(k)}, \sigma_{22}^{(k)}, \ldots, \sigma_{dd}^{(k)}\right],$$

$$\mathbf{Z}^{(k)} = \text{diag}\left[z_{11}^{(k)}, z_{22}^{(k)}, \ldots, z_{pp}^{(k)}\right].$$

c. Set the counter $k = k + 1$ and go to step (b) until convergence.

Chapter 6
Non-Linear and Non-Gaussian State Space Models

This chapter discusses Bayesian inference for non-linear and non-Gaussian state space models. The chapter considers a general model formulation, which includes as special case the linear and Gaussian state space models, discussed in Chaps. 3–5, and allows for generalisation aimed at wide application. Dynamic generalised linear models and conditionally Gaussian models are discussed in some detail. The power local level model is presented for historical reasons as a relatively limited modelling setup which allows conjugate Bayesian inference. Moving away from its limitations we discuss approximate inference and the extended Kalman filter. Simulation-based estimation is considered in detail focusing upon sequential Monte Carlo methods and Markov chain Monte Carlo methods.

Section 6.1 introduces the general models and provides their motivation starting from linear Gaussian state space models. Dynamic generalised linear models are described in detail in Sect. 6.2. Special cases include models for counts, categorical data and continuous proportions. Other non-linear and non-Gaussian models, such as stochastic volatility and bearings-only tracking models described in Sect. 6.3. The power local level models are discussed in Sect. 6.5; this is a class of models allowing for conjugate inference, albeit is a very restricted class of models. Special interesting cases include the Poisson-gamma and the exponential-gamma models. The extended Kalman filter is described for approximate inference in Sect. 6.6. Essentially a Taylor series expansion is used and the Kalman filter is used to provide approximation of the posterior distribution of the states. Sequential Monte Carlo methods are discussed in detail in Sect. 6.7. In detail there is consideration of the choice of the importance function (Sect. 6.7.4) and of the problem of the estimation of static-parameters (Sect. 6.7.8). Various examples are used to illustrate the algorithms including bearings-only tracking data, simulations from a conditionally Gaussian model and forecasting of counts of schoolchildren suffering from asthma. Section 6.8 discusses MCMC inference for dynamic generalised linear models. The chapter concludes by discussing dynamic survival models in Sect. 6.9.

6.1   General Model Formulation

So far in this book we have studied state space models, which are Gaussian (the distributions of the observations given the states and the distribution of the states given past states are Gaussian) and the observations are linear functions of the states. This is formally described by the two conditional distributions

$$p(y_t \mid \beta_t) \equiv N(\mathbf{x}_t^\top \beta_t, \boldsymbol{\Sigma}) \quad \text{and} \quad p(\beta_t \mid \beta_{t-1}) \equiv N(\mathbf{F}_t \beta_{t-1}, \mathbf{Z}_t), \tag{6.1}$$

which are immediately obtained from the observation and transition equations (5.1a)–(5.1b) with the innovation distributions (5.2), where the components $\mathbf{x}_t$, $\mathbf{F}_t$, $\boldsymbol{\Sigma}$ and $\mathbf{Z}_t$ are described in Sect. 5.1. The model is complete when we specify the prior distribution of β_0, which is again Gaussian, i.e. $\beta_0 \sim N(\hat{\beta}_{0|0}, \mathbf{P}_{0|0})$, for some known prior mean vector and prior covariance matrix $\hat{\beta}_{0|0}$ and $\mathbf{P}_{0|0}$).

However, as we have already seen in Chap. 1 many data exhibit a non-linear relationship between the observations and the states and are not supported by the linear structure implicit in the Gaussian distributions of (6.1). In the bearings only ship tracking example of Sect. 1.3.2 the distribution of the observations given the states is non-linear, see Eq. (1.12); in the stochastic volatility example discussed in Sect. 1.3.3 again observation and state are linked via a non-liner relationship.

These models can be described by the so-called *conditionally Gaussian* state space models, that is given the states, the distribution of the observations is Gaussian and given past states the distribution of the present state is Gaussian, but now the linearity is dropped. This can be expressed as

$$y_t = f(\beta_t) + \epsilon_t \quad \text{and} \quad \beta_t = g(\beta_{t-1}) + \zeta_t, \tag{6.2}$$

where $f(\cdot)$ is generally a non-linear function on β_t and $g(\cdot)$ is a non-linear function on β_{t-1} and the innovations ϵ_t and ζ_t follow the usual assumptions (each of ϵ_t and ζ_t is a white noise and ϵ_t is independent of ζ_t). It is also assumed that both ϵ_t and ζ_t are independent of β_0, which follows some Gaussian distribution. Clearly, if we set $f(\beta_t) = \mathbf{x}_t^\top \beta_t$ and $g(\beta_{t-1}) = \mathbf{F}_t \beta_{t-1}$, then we obtain the usual linear Gaussian state space model (6.1).

The state space model (6.2) is known as conditionally Gaussian, because given β_t the distribution of y_t is Gaussian and also given β_{t-1}, the distribution of β_t is Gaussian, i.e.

$$p(y_t \mid \beta_t) \equiv N[f(\beta_t), \boldsymbol{\Sigma}] \quad \text{and} \quad p(\beta_t \mid \beta_{t-1}) \equiv N[g(\beta_{t-1}), \mathbf{Z}_t],$$

where $\boldsymbol{\Sigma}$ and $\mathbf{Z}_t$ are the covariance matrices of ϵ_t and ζ_t, respectively.

The class of non-linear state space models described above assumes that the distributions of $y_t \mid \beta_t$ and $\beta_t \mid \beta_{t-1}$ are Gaussian. However, there are situations where these assumptions are violated, in particular the distribution of y_t, given β_t (sometimes referred to as the *response distribution*) may not be Gaussian.

For example y_t may be discrete-valued, describing counts or observations of a categorical variable, or continuous-valued, describing proportions or positive processes. In all these examples the support of y_t is not the real-hyperplane $\mathbb{R}^d$, and as a result the d-dimensional Gaussian distribution for $y_t \mid \beta_t$ is inappropriate. Motivating examples of such time series data may be medical counts of patients that visit their family doctor observed over time, proportions of world car manufacturing volume over time and so forth. Such models may exhibit a linear or non-linear relationship between y_t and β_t. As a result we can extend the model definition of (6.2) to account for non-Gaussian distributions. The general non-linear and non-Gaussian state space model is defined by specifying the distributions

$$p(y_t \mid \beta_t), \quad p(\beta_t \mid \beta_{t-1}) \quad \text{and} \quad p(\beta_0). \tag{6.3}$$

Thus, this general state space model specifies the distribution of $y_t \mid \beta_t$ (response distribution), the distribution of $\beta_t \mid \beta_{t-1}$ (transition distribution) and the distribution of β_0 (initial or prior distribution). In addition to the distributional specification (6.3) we will assume that given the present β_t, the past β_{t-i} and the future β_{t+j} are conditionally independent (i.e. that β_t carries all information relevant to the present); this is known as the Markov property of the state space model. This assumption together with (6.3) defines a very wide class of models. Sections 6.2 and 6.3 aim to give some particular examples of possible general state space models, but there might be other models not discussed here.

6.2 Dynamic Generalised Linear Models

6.2.1 Model Definition

The generalised linear model (GLM), proposed originally by Nelder and Wedderburn (1972) as a generalisation to linear models, including regression and analysis of variance, in order to deal with non-Gaussian observations. The framework of the GLM revolutionised statistical inference and application as evidenced in the monographs (McCullagh & Nelder, 1989; Fahrmeir & Tutz, 2001). The GLM framework basically considers that observations are generated by a response distribution belonging to the natural exponential family of distributions. Then the *mean response* (the mean of the observations given some parameters) is transformed to the linear predictor by the so-called *link function*. This achieves to transform the mean of the observations (which distribution is non-Gaussian and can be discrete or continuous distribution) into the linear predictor, which takes values in $\mathbb{R}^d$ and can be assumed to follow a multivariate Gaussian distribution. Hence inference can follow parallel estimation steps to the standard linear models, for which the link function is the identity function and the observations follow a Gaussian distribution.

The dynamic generalised linear model (DGLM) is an extension of the GLM by assuming that the parameters of the exponential family as well as the regression coefficients that take part in the linear predictor are time-varying. In particular, the DGLM considers that the response distribution is a member of the exponential family with time-varying parameters and the mean response is mapped via the link function to the linear predictor, which in turn follows a state space model (see below for a technical description). This representation was first considered in West et al. (1985) and later on adopted in Fahrmeir (1992) and is described below.

Let $\{y_t\}$ be a d-dimensional time series, generated by the multivariate exponential family of distributions

$$p(y_t \mid \gamma_t) = \exp \left\{ \frac{1}{a_t} \left[\gamma_t^\top z(y_t) - b(\gamma_t) \right] + c(y_t) \right\}, \quad \text{(observation model)}$$

(6.4)

where γ_t is the natural d-dimensional parameter vector, a_t is a possible time-varying parameter not depending on y_t or γ_t, $b(\cdot)$ is a twice differentiable function, $c(\cdot)$ is a function that depends on y_t, but not on γ_t. The function $z(\cdot)$, usually taken as the identity function $z(y_t) = y_t$, is a deterministic function of y_t, necessary to consider to express several distributions in the form (6.4) (see the examples below). Distribution (6.4) may depend on some other hyperparameters θ; this will become clear in a case by case situation as the examples that follow illustrate.

Given γ_t, we can express the mean response vector and the covariance matrix of $z(y_t)$ as derivatives of the function $b(\cdot)$. For the mean, first observe that $\int_{\mathbb{R}^d} p(y_t \mid \gamma_t) \, dy_t = 1$ and take partial derivatives in both sides:

$$0 = \int_{\mathbb{R}^d} \frac{\partial}{\partial \gamma_t} \exp \left\{ \frac{1}{a_t} \left[\gamma_t^\top z(y_t) - b(\gamma_t) \right] + c(y_t) \right\} \, dy_t$$

$$= \frac{1}{a_t} \int_{\mathbb{R}^d} \left[z(y_t) - \frac{\partial b(\gamma_t)}{\partial \gamma_t} \right] \exp \left\{ \frac{1}{a_t} \left[\gamma_t^\top z(y_t) - b(\gamma_t) \right] + c(y_t) \right\},$$

which implies $E[z(y_t) \mid \gamma_t] = \partial b(\gamma_t)/\partial \gamma_t$.

For the covariance matrix of $z(y_t)$ we start from $\int_{\mathbb{R}^d} p(y_t \mid \gamma_t) \, dy_t = 1$ and we obtain

$$0 = \frac{\partial^2}{\partial \gamma_t \partial \gamma_t^\top} \int_{\mathbb{R}^d} \exp \left\{ \frac{1}{a_t} \left[\gamma_t^\top z(y_t) - b(\gamma_t) \right] + c(y_t) \right\} \, dy_t$$

$$= \frac{\partial}{\partial \gamma_t^\top} \left\{ \frac{1}{a_t} \int_{\mathbb{R}^d} \exp \left[\frac{1}{a_t} \left(\gamma_t^\top z(y_t) - b(\gamma_t) \right) + c(y_t) \right] \, dy_t - \frac{1}{a_t} \frac{\partial b(\gamma_t)}{\partial \gamma_t} \right\}$$

$$= \frac{1}{a_t} \int_{\mathbb{R}^d} \left\{ z(y_t)z(y_t)^\top - z(y_t)\frac{\partial b(\gamma_t)}{\partial \gamma_t} \right\} \exp\left\{ \frac{1}{a_t}\left[\gamma_t^\top z(y_t) - b(\gamma_t) \right] + c(y_t) \right\} dy_t$$

$$-\frac{1}{a_t}\frac{\partial^2 b(\gamma_t)}{\partial \gamma_t \partial \gamma_t^\top}$$

$$= \frac{1}{a_t}\mathrm{E}[z(y_t)z(y_t)^\top] - \frac{1}{a_t}\frac{\partial b(\gamma_t)}{\partial \gamma_t}\mathrm{E}[z(y_t)]^\top - \frac{1}{a_t}\frac{\partial^2 b(\gamma_t)}{\partial \gamma_t \partial \gamma_t^\top},$$

which implies

$$\mathrm{E}[z(y_t)z(y_t)^\top] = \left[\frac{\partial b(\gamma_t)}{\partial \gamma_t} \right]^2 + \frac{\partial^2 b(\gamma_t)}{\partial \gamma_t \partial \gamma_t^\top}.$$

Hence, given γ_t, the covariance matrix of $z(y_t)$ is

$$\mathrm{Var}[z(y_t) \mid \gamma_t] = \mathrm{E}[z(y_t)z(y_t)^\top \mid \gamma_t] - \mathrm{E}[(z(y_t) \mid \gamma_t]\mathrm{E}[z(y_t) \mid \gamma_t]^\top = \frac{\partial^2 b(\gamma_t)}{\partial \gamma_t \partial \gamma_t^\top}.$$

Returning to the definition of the DGLM, the mean response vector $\mu_t = \mathrm{E}(y_t \mid \gamma_t)$ is mapped into the linear predictor η_t via the link function

$$g(\mu_t) = \eta_t \tag{6.5}$$

and the linear predictor η_t is assumed to follow a state space model

$$\eta_t = \mathbf{x}_t^\top \beta_t \quad \text{and} \quad \beta_t = \mathbf{F}_t\beta_{t-1} + \zeta_t, \quad \zeta_t \sim N(0, \mathbf{Z}_t), \tag{6.6}$$

where $\mathbf{x}_t$ is a $p \times d$ design matrix, β_t is a p-dimensional state vector, $\mathbf{F}_t$ is a $p \times p$ transition matrix and ζ_t is a white noise vector with covariance matrix $\mathbf{Z}_t$. In words the mean response vector μ_t which domain is a restricted subset of $\mathbb{R}^d$ is mapped into the unrestricted real field $\mathbb{R}^d$ via the link function $g(\cdot)$.

Furthermore, it is assumed that ζ_t is independent of the initial or prior state β_0, which follows a Gaussian distribution:

$$\beta_0 \sim N(\hat{\beta}_{0|0}, \mathbf{P}_{0|0}), \tag{6.7}$$

for some prior mean vector $\hat{\beta}_{0|0}$ and covariance matrix $\mathbf{P}_{0|0}$. The link function $g(\cdot)$ is assumed to be a bijection, so that $g^{-1}(\cdot)$ the inverse of $g(\cdot)$ exists and (6.5) implies $\mu_t = g^{-1}(\eta_t)$. To summarise the DGLM is defined by Eqs. (6.4)–(6.7).

Note that conditional on the state β_t, the mean response $\mu_t = g^{-1}(\mathbf{x}_t^\top \beta_t)$ is precisely known and so we can write the response distribution as $p(y_t \mid \beta_t)$, because the natural parameter vector γ_t is given as a deterministic function of β_t. Hence, we may express (6.4) as $p(y_t \mid \beta_t)$ and the state distribution as $p(\beta_t \mid \beta_{t-1}) \equiv N(\mathbf{F}_t\beta_{t-1}, \mathbf{Z}_t)$, which together with the prior (6.7) are in the form of the general state space model (6.3). Below we provide example of specific DGLMs, which are suitable of modelling frequently met non-Gaussian data.

6.2.2   Count Time Series

Count time series data are met in many fields: in medicine they may be count of medical contacts of patients recorded over time, in economics they may be counts of passengers carried in an airline over time and so forth. The two most basic distributions describing count data are the Poisson and the negative binomial distributions, for a description of which the reader is referred to Sect. 2.3.2. A more detailed treatment of count time series is discussed in Kedem and Fokianos (2002).

Suppose that $\{y_t\}$ is generated by a Poisson distribution with rate $\lambda_t > 0$, i.e.

$$p(y_t \mid \lambda_t) = \exp(-\lambda_t)\frac{\lambda_t^{y_t}}{y_t!}, \quad y_t = 0, 1, 2, \dots . \tag{6.8}$$

We can write this distribution as

$$p(y_t \mid \lambda_t) = \exp(y_t \gamma_t - \lambda_t - \log y_t!),$$

which is in the form of (6.4), with $a = 1$, $z(y_t) = y_t$, $\gamma_t = \log \lambda_t$, $b(\gamma_t) = \lambda_t = \exp(\gamma_t)$ and $c(y_t) = -\log y_t!$.

The canonical link $g(\mu_t) = \gamma_t = \log \lambda_t$ together with the transition equation provide

$$\eta_t = \log \lambda_t = x_t^\top \beta_t \tag{6.9}$$

$$\beta_t = F_t \beta_{t-1} + \zeta_t, \quad \zeta_t \sim N(0, Z_t), \tag{6.10}$$

where β_t is a p-dimensional state vector, F_t is a $p \times p$ transition matrix, ζ_t is a p-dimensional innovation vector following a Gaussian distribution with zero mean vector and covariance matrix Z_t. As usual a Gaussian distribution is assumed for the initial state β_0. Thus, the model consists of Eqs. (6.8)–(6.10) together with the prior distribution of β_0.

The Poisson model above imposes the assumption that, given λ_t, the mean and the variance of y_t are equal, i.e. $E(y_t \mid \lambda_t) = \text{Var}(y_t \mid \lambda_t) = \lambda_t$. In many situations this is not the case, in particular interest is placed when the variance of the data is larger than the mean. In such a situation the data is known to be over-dispersed and can be better modelled with the negative binomial distribution as detailed below; the negative binomial distribution is discussed in Sect. 2.3.2.

Suppose that that $\{y_t\}$ is generated by a negative binomial distribution

$$p(y_t \mid \pi_t) = \binom{y_t + \lambda_t - 1}{y_t}(1 - \pi_t)^{\lambda_t}\pi_t^{y_t}, \quad y_t = 0, 1, 2, \ldots, \tag{6.11}$$

where $\lambda_t > 0$ the number of failures before the y_tth success is assumed known and π_t is the probability of success at time t; see also the description of the negative binomial distribution in Sect. 2.3.2.

The above model can be written as

$$p(y_t \mid \pi_t) = \exp\left[y_t\gamma_t + \lambda_t \log(1 - \pi_t) + \log\binom{y_t + \lambda_t - 1}{y_t}\right],$$

where $z(y_t) = y_t$, the natural parameter is $\gamma_t = \log \pi_t$ and $b(\gamma_t) = -\lambda_t \log(1 - \pi_t) = -\lambda_t \log(1 - e^{\gamma_t})$.

The mean response is

$$\mu_t = \frac{\partial b(\gamma_t)}{\partial \gamma_t} = \frac{\lambda_t e^{\gamma_t}}{1 - e^{\gamma_t}} = \frac{\lambda_t \pi_t}{1 - \pi_t},$$

confirming the mean of the negative binomial distribution, given in Sect. 2.3.2.

Thus, the logarithmic link can be applied to map $\mu_t > 0$ to the real line. This results in the link and transition equations

$$\eta_t = \log\left(\frac{\lambda_t \pi_t}{1 - \pi_t}\right) = x_t^\top \beta_t \tag{6.12}$$

$$\beta_t = F_t \beta_{t-1} + \zeta_t, \quad \zeta_t \sim N(0, Z_t), \tag{6.13}$$

with x_t, β_t, F_t, ζ_t and Z_t as defined for the Poisson model above.

The negative binomial distribution is obtained as a mixture of a Poisson distribution when the rate λ follows a gamma distribution. For example, dropping the time index t, for convenience, the probability mass function of the negative binomial distribution is obtained from the gamma mixture

$$p_{NB}(y) = \int_0^\infty p_{Pois}(y; \lambda) p_{Ga}(\lambda; \alpha, \beta)\, d\lambda$$

$$= \int_0^\infty \exp(-\lambda)\frac{\lambda^y}{y!}\frac{\beta^\alpha}{\Gamma(\alpha)}\lambda^{\alpha-1}\exp(-\beta\lambda)\, d\lambda$$

$$= \frac{\Gamma(y+\alpha)}{y!\Gamma(\alpha)}\frac{\beta^\alpha}{(\beta+1)^{y+\alpha}} = \binom{y+\alpha-1}{y}(1-\pi)^\alpha \pi^y,$$

where $p_{Pois}(y; \lambda)$ denotes the probability mass function of the Poisson distribution with rate λ, $p_{Ga}(\lambda; \alpha, \beta)$ denotes the density function of the gamma distribution with parameters α, $\beta = (1 - \pi)\pi^{-1}$ and $\Gamma(\cdot)$ denotes the gamma function.

Hence one can regard the negative binomial model as a Poisson model when $\lambda \sim \text{Ga}[\alpha, (1-\pi)\pi^{-1}]$. It also follows that for large α the Poisson distribution is obtained by the negative binomial, as in this case the mean and the variance of the gamma distribution are approximately the same. For more details on the many Poisson mixture models the reader is referred to Karlis and Xekalaki (2005) and to references therein.

6.2.3 Categorical Time Series

Suppose that the series $\{y_t\}$ follows a binomial distribution (see also Sect. 2.3.2) with probability mass function

$$p(y_t|\pi_t) = \binom{\lambda_t}{y_t} \pi_t^{y_t} (1-\pi_t)^{\lambda_t - y_t}, \quad y_t = 0, 1, \ldots, \lambda_t, \tag{6.14}$$

where π_t is a random variable which denotes the probability of success at time t and λ_t is a known positive integer at time t. Then

$$p(y_t|\pi_t) = \exp\left[y_t \log\left(\frac{\pi_t}{1-\pi_t}\right) + \lambda_t \log(1-\pi_t) + \log\binom{\lambda_t}{y_t} \right]$$

$$= \exp\left\{ \lambda_t \left[\frac{y_t}{\lambda_t} \gamma_t - \log(1 + e^{\gamma_t}) \right] + \log\binom{\lambda_t}{y_t} \right\}.$$

This implies that $z(\cdot)$ is the proportion $z(y_t) = y_t/\lambda_t$ and that the natural parameter $\gamma_t = \log\{\pi_t/(1-\pi_t)\}$. In this case $b(\gamma_t) = \log(1 + e^{\gamma_t})$, confirming that $\pi_t = \text{E}[z(y_t)|\pi_t] = b'(\gamma_t) = e^{\gamma_t}/(1+e^{\gamma_t}) = \lambda_t \pi_t$. The link function is $g(\mu_t) = \gamma_t$ and so the canonical link and transition equations are

$$\eta_t = \log\left(\frac{\pi_t}{1-\pi_t}\right) = x_t^\top \beta_t, \tag{6.15}$$

$$\beta_t = \mathbf{F}_t \beta_{t-1} + \zeta_t, \quad \zeta_t \sim N(0, \mathbf{Z}_t). \tag{6.16}$$

The above binomial model can be generalised in order to accommodate for multicategorical data. Considering $k \geq 2$ categories, let y_{it} denote the count or total measurement of a quality characteristic observed in category $i = 1, 2, \ldots, k$ at time t. Denote with π_{it} the *cell probability* that the random variable y_{it} is equal to the observed count. Fix $\lambda_t = y_{1t} + \cdots + y_{kt}$ to be total count and $\pi_{1t} + \cdots + \pi_{kt} = 1$, for some known positive integer λ_t. The joint p.m.f. of $y_t = [y_{1t}, \ldots, y_{k-1,t}]^\top$ is

$$p(y_t) = \frac{\lambda_t!}{\prod_{i=1}^k y_{it}!} \prod_{i=1}^k \pi_{it}^{y_{it}}, \tag{6.17}$$

which defines the *multinomial distribution* and is discussed in some detail in Chapter 3 of Fahrmeir and Tutz (2001); see also the relevant discussion in Sect. 2.3.2.

The above p.m.f. may be written as

$$
p(y_t) = \exp\left[\sum_{i=1}^{k-1} y_{it} \log \pi_{it} + \left(\lambda_t - \sum_{i=1}^{k-1} y_{it}\right) \log\left(1 - \sum_{i=1}^{k-1} \pi_{it}\right)\right] \frac{\lambda_t!}{\prod_{i=1}^{k} y_{it}!}
$$

$$
= \exp\left\{\lambda_t \left[\sum_{i=1}^{k-1} y_{it} \log \frac{\pi_{it}}{1 - \sum_{i=1}^{k-1} \pi_{it}} + \log\left(1 - \sum_{i=1}^{k-1} \pi_{it}\right)\right]\right\} \frac{\lambda_t!}{\prod_{i=1}^{k} y_{it}!},
$$

which is in the form of (6.4), with $a_t = \lambda_t^{-1}$, $z(y_t) = \lambda_t^{-1} y_t$, $\gamma_t = [\gamma_{1t}, \ldots, \gamma_{k-1,t}]^\top$, $\gamma_{it} = \log\left[\pi_{it}\left(1 - \sum_{i=1}^{k-1} \pi_{it}\right)^{-1}\right]$ and $b(\gamma_t) = \log\left(1 - \sum_{i=1}^{k-1} e^{\gamma_{it}}\right)$.

The $k - 1$ dimensional linear predictor vector η_t maps the mean vector $E(y_t)$ via the canonical link and transition equations

$$
\eta_t = \begin{bmatrix} \log \frac{\pi_{1t}}{1-\sum_{i=1}^{k-1} \pi_{it}} \\ \log \frac{\pi_{2t}}{1-\sum_{i=1}^{k-1} \pi_{it}} \\ \vdots \\ \log \frac{\pi_{k-1,t}}{1-\sum_{i=1}^{k-1} \pi_{it}} \end{bmatrix} = \mathbf{x}_t^\top \beta_t, \tag{6.18}
$$

$$
\beta_t = \mathbf{F}_t \beta_{t-1} + \zeta_t, \quad \zeta_t \sim N(0, \mathbf{Z}_t). \tag{6.19}
$$

Here $\mathbf{x}_t$ is a $p \times (k - 1)$ design matrix, β_t is the p-dimensional state vector, $\mathbf{F}_t$ is the $p \times p$ transition matrix and $\mathbf{Z}_t$ is the $p \times p$ transition covariance matrix. We note that for two categories $k = 2$ the multinomial model (6.17)–(6.19) reduces to the binomial model (6.14)–(6.16). Hence the multinomial model is a direct generalisation of the binomial model and is suitable for modelling multi-categorical data.

6.2.4 Continuous Proportions

In Sect. 6.2.3 we discuss models for categorical data, which may be used to make inference for proportions, e.g. by considering the proportion y_t/λ_t in the binomial model (6.14). Sometimes this is referred to as *discrete proportion*, because it is based on discrete observations y_t and the totals λ_t. In this set-up the total λ_t is assumed known and the focus of the inference is placed on the count y_t. However, we may wish to model directly the proportion, for example if such a proportion is observed rather than the counts. In other situations we may wish to make inference regarding some probability, not necessarily being a proportion, or any other measurable quantity which takes values in the interval [0, 1]. For convenience such measurements are known as *continuous proportions*, although as noted above, strictly speaking, they may not be restricted to proportions.

Consider that y_t denotes observations taking values in the interval [0, 1], for any time t. A natural distribution to describe such observations is the beta distribution, hence the observation model is

$$p(y_t \mid \alpha_{1t}, \alpha_{1t}) = \frac{\Gamma(\alpha_{1t} + \alpha_{2t})}{\Gamma(\alpha_{1t})\Gamma(\alpha_{2t})} y_t^{\alpha_{1t}-1} (1 - y_t)^{\alpha_{2t}-1}, \tag{6.20}$$

where $\alpha_{1t}, \alpha_{2t} > 0$ are time-varying parameters and $\Gamma(\cdot)$ is the gamma function; the beta distribution is discussed in Sect. 2.3.3.

Without loss in clarity, we omit the conditioning on α_{it} in the density (6.20). This beta distribution can be written as

$$p(y_t) = \exp\left[\alpha_{1t} \log y_t - \log \frac{\Gamma(\alpha_{1t} + \alpha_{2t})}{\Gamma(\alpha_{1t})\Gamma(\alpha_{2t})} + (\alpha_{2t} - 1)\log(1 - y_t) - \log y_t\right],$$

which is in the form of (6.4), with $a_t = 1$, $z(y_t) = \log y_t$, $\gamma_t = \alpha_{1t}$, $b(\gamma_t) = -\log[\Gamma(\gamma_t)^{-1}\Gamma(\alpha_{2t})^{-1}\Gamma(\gamma_t + \alpha_{2t})]$, $c(y_t) = (\alpha_{2t} - 1)\log(1 - y_t) - \log y_t$.

Given α_{1t}, α_{2t} the mean of y_t is $\mu_t = E(y_t \mid \alpha_{1t}, \alpha_{2t}) = \alpha_{1t}(\alpha_{1t} + \alpha_{2t})^{-1}$ and so the logarithmic link can be used, leading to the link and transition equations

$$\eta_t = \log\left(\frac{\alpha_{1t}}{\alpha_{1t} + \alpha_{2t}}\right) = x_t^\top \beta_t \tag{6.21}$$

$$\beta_t = F_t \beta_{t-1} + \zeta_t, \quad \zeta_t \sim N(0, Z_t), \tag{6.22}$$

with the usual definitions of β_t, ζ_t, F_t and Z_t.

The state space model (6.20)–(6.22) can be generalised to account for several correlated proportions. Considering $k \geq 2$ categories, let $y_t = [y_{1t}, \dots, y_{k-1,t}]^\top$ be a vector of $k - 1$ continuous proportions and $\alpha_t = [\alpha_{1t}, \dots, \alpha_{kt}]^\top$ be a vector of k positive parameters. The joint effects of $y_{1t}, \dots, y_{k-1,t}$ can be described by the *Dirichlet distribution*, which is a generalisation of the beta distribution and its density is

$$p(y_t \mid \alpha_t) = \frac{1}{D(\alpha_t)} \prod_{i=1}^{k-1} y_{it}^{\alpha_{it}-1} \left(1 - \sum_{i=1}^{k-1} y_{it}\right)^{\alpha_{kt}-1}, \tag{6.23}$$

where $D(\alpha_t)$ denotes the Dirichlet function, which can be expressed as functions of gamma functions

$$D(\alpha_t) = \frac{\prod_{i=1}^{k} \Gamma(\alpha_{it})}{\Gamma\left(\sum_{i=1}^{k} \alpha_{it}\right)}.$$

This distribution can be expressed as

$$p(y_t \mid \alpha_t) = \exp\left[\sum_{i=1}^{k-1} \alpha_{it} \log y_{it} - \log D(\alpha_t) + (\alpha_{kt} - 1) \log\left(1 - \sum_{i=1}^{k-1} y_{it}\right) \right.$$
$$\left. - \sum_{i=1}^{k-1} \log y_{it} \right],$$

which is in the form of (6.4), with $a_t = 1$, $z(y_t) = [\log y_{1t}, \ldots, \log y_{k-1,t}]^\top$, $\gamma_t = [\alpha_{1t}, \ldots, \alpha_{k-1,t}]^\top$, $b(\gamma_t) = \log D(\alpha_t)$ and $c(y_t) = (\alpha_{kt} - 1) \log\left(1 - \sum_{i=1}^{k-1} y_{it}\right) - \sum_{i=1}^{k-1} \log y_{it}$.

It can be shown that, conditional on α_t, the mean of y_t is $\mu_t = [\mu_{1t}, \ldots, \mu_{k-1,t}]^\top$, with $\mu_{it} = \alpha_{it}(\alpha_{1t} + \cdots + \alpha_{kt})^{-1}$. As a result the logarithmic link can be used to map the mean μ_{it} to the real line, leading to the link and transition equations

$$\eta_t = \begin{bmatrix} \log \frac{\alpha_{1t}}{\sum_{i=1}^{k} \alpha_{it}} \\ \log \frac{\alpha_{2t}}{\sum_{i=1}^{k} \alpha_{it}} \\ \vdots \\ \log \frac{\alpha_{k-1,t}}{\sum_{i=1}^{k} \alpha_{it}} \end{bmatrix} = \mathbf{x}_t^\top \beta_t \qquad (6.24)$$

$$\beta_t = \mathbf{F}_t \beta_{t-1} + \zeta_t, \quad \zeta_t \sim N(0, \mathbf{Z}_t). \qquad (6.25)$$

We note that for $k = 2$ categories the above model reduces to the beta model (6.20)–(6.22).

6.2.5   Decomposition of Dynamic Generalised Linear Models

Decomposition of linear and Gaussian state space models is discussed in Sect. 4.2.2. Based on the assumption of observability, the mean response μ_t of the model (3.10a)–(3.10b) is decomposed into a sum of component state space models. This approach can be extended to the class of dynamic generalised model (6.4)–(6.6) (see Sect. 6.2.1). We will assume that the model is observable, or that by extending the definition of linear state space models (see Sect. 3.5.1) that the rank of the $p \times p$ observability matrix

$$
O = \begin{bmatrix} x^\top \\ x^\top \mathbf{F} \\ \vdots \\ x^\top \mathbf{F}^{p-1} \end{bmatrix}
$$

is p. With this assumption in place we have that the linear predictor η_t is decomposed into a sum of ℓ component state space models

$$
\eta_t = \chi_t^{(1)} + \chi_t^{(2)} + \cdots + \chi_t^{(\ell)}, \tag{6.26}
$$

$$
\chi_t^{(j)} = e_1^\top \gamma_{jt}, \tag{6.27}
$$

$$
\gamma_{jt} = \mathbf{C}(\Phi_j)\gamma_{j,t-1} + \xi_{jt}, \quad j = 1, \ldots, \ell, \tag{6.28}
$$

with model components γ_{jt}, e_1, $\mathbf{C}(\Phi_j)$ and ξ_{jt} defined as in Sect. 4.2.2. The decomposition of the Gaussian state space of Sect. 4.2.2 is obtained as a special case of the above decomposition, since in that case $\eta_t = \mu_t$ (the link function $g(\mu_t)$ is the identity function).

The decomposition (6.26)–(6.28) implies that

$$
E(y_t \mid \gamma_t) = \mu_t = g^{-1}\left(\chi_t^{(1)} + \chi_t^{(2)} + \cdots + \chi_t^{(\ell)}\right).
$$

For example if the logarithmic link $g(\mu_t) = \log \mu_t$ is used (Poisson, gamma, exponential, inverse gamma), then

$$
E(y_t \mid \gamma_t) = \prod_{j=1}^{\ell} \exp\left(\chi_t^{(j)}\right)
$$

and the decomposition can be thought of as multiplicative.

6.3 Other Non-Gaussian and Non-linear Models

In Sects. 6.2.2, 6.2.3 and 6.2.4 dynamic generalised linear models are described for specific types of non-Gaussian time series data, i.e. counts, categorical observations and proportions. The list is easily extended to other types of data, in which observation model can be described by any member of the exponential family (6.4). Some further examples are provided in the exercises in the end of the chapter. However, data may not be described by a distribution that belongs to the exponential family. We have already met two such examples in Chap. 1. In particular, the state space model (1.15)–(1.16) of Sect. 1.3.3 is a conditional Gaussian state space model, since, conditional on the states β_t, the return series y_t follows a Gaussian distribution

$$y_t \mid \beta_t \sim N[0, \exp([1, 0]\beta_t)], \tag{6.29}$$

where the bivariate state vector is $\beta_t = [h_t, 1]^\top$ and h_t is the logarithm of the volatility; see Sect. 1.3.3 for details. Conditionally on the past states β_{t-1}, the distribution of β_t is Gaussian: $\beta_t \mid \beta_{t-1} \sim N(\mathbf{F}\beta_{t-1}, \mathbf{Z})$, for some transition matrix $\mathbf{F}$ and covariance matrix $\mathbf{Z}$. In Sect. 1.3.3 the motivation for this model is discussed in some detail.

Another conditionally Gaussian state space model is described in Sect. 1.3.2. In particular, given states β_t, the observations z_t (angular process) follow a Gaussian distribution

$$z_t \mid \beta_t \sim N\left[\arctan\left(\frac{[0, 1, 0, 0]\beta_t}{[1, 0, 0, 0]\beta_t}\right), \sigma^2\right], \tag{6.30}$$

for some variance σ^2. Given past states β_{t-1}, β_t follows a Gaussian distribution: $\beta_t \mid \beta_{t-1} \sim N(\mathbf{F}\beta_{t-1}, \mathbf{Z})$, for some transition matrix $\mathbf{F}$, which is provided in (1.11) and some covariance matrix $\mathbf{Z}$.

Finally we give an example of a non-Gaussian and non-linear time series model, which is not conditionally Gaussian and does not belong to the class of dynamic generalised linear models (DGLM). Consider the above stochastic volatility model, but replace the Gaussian distribution in Eq. (6.29) by a Student t distribution, so that

$$y_t \mid \beta_t \sim t[\nu, 0, \exp([1, 0]\beta_t)], \tag{6.31}$$

where $\nu > 0$ denotes the degrees of freedom, and the evolution of β_t follows the usual linear and Gaussian law, described above, i.e. $\beta_t = \mathbf{F}\beta_{t-1} + \zeta_t$. This state space model is not conditionally Gaussian, since given β_t, the distribution of y_t is a Student t distribution, and it is not a DGLM, since the Student t distribution is not a member of the exponential family of distributions. Nevertheless, this is a more plausible stochastic volatility model than the one described earlier and in Sect. 1.3.3, because it enables y_t to have heavy tails (depending on the value of ν), which is a desirable property, since financial returns typically exhibit heavier tails than the Gaussian distribution.

In the following sections we describe inference for non-linear and non-Gaussian state space models; the list is not exhaustive and it covers exact inference for a restricted class of models (Sect. 6.5), approximate inference for a wide class of models (Sect. 6.6) and simulation-based inference (Sect. 6.7). We start by describing first the general formulation for estimation and forecasting.

6.4 Inference for the General State Space Model

Consider the general non-linear and non-Gaussian state space model (6.3) and let $y_{1:t} = (y_1, \ldots, y_t)$ be a collection of observations. Suppose that at time $t - 1$ the posterior distribution $p(\beta_{t-1} \mid y_{1:t-1})$ of β_{t-1} is known. Then the prior distribution of β_t is

$$p(\beta_t \mid y_{1:t-1}) = \int_A p(\beta_t \mid \beta_{t-1}) p(\beta_{t-1} \mid y_{1:t-1}) \, d\beta_{t-1}, \qquad (6.32)$$

where A, the domain of β_t, is a subset of $\mathbb{R}^p$. The distribution of $p(\beta_t \mid \beta_{t-1})$ is the transition model in (6.3) and the distribution $p(\beta_{t-1} \mid y_{1:t-1})$ is assumed to be known. Hence the integral may be computed.

Given information $y_{1:t-1}$, the one-step forecast distribution of y_t is provided by the integral

$$p(y_t \mid y_{1:t-1}) = \int_B p(y_t \mid \beta_t) p(\beta_t \mid y_{1:t-1}) \, d\beta_{t-1}, \qquad (6.33)$$

where B is the domain of y_t. The distribution $p(y_t \mid \beta_t)$ is provided by the observation model in (6.3) and the distribution $p(\beta_t \mid y_{1:t-1})$ is provided by (6.32). Hence $p(y_t \mid y_{1:t-1})$ may be computed by the above integral.

When y_t is observed, the collection of data $y_{1:t-1}$ is updated to include y_t, i.e. $y_{1:t} = (y_1, \ldots, y_t)$ and by an application of the Bayes theorem

$$p(\beta_t \mid y_{1:t}) = \frac{p(y_t \mid \beta_t) p(\beta_t \mid y_{1:t-1})}{p(y_t \mid y_{1:t-1})}, \qquad (6.34)$$

the posterior distribution of β_t can be obtained. Starting at time $t = 0$ with the assumed prior $p(\beta_0)$, this process is repeated sequentially to obtain the posterior distributions $p(\beta_t \mid y_{1:t})$, for any $t = 1, 2, \ldots$. For example in the linear and Gaussian state space model (3.10a)–(3.10b) integrals (6.32) and (6.33) are obtained in closed form and together with (6.34) provide the Kalman filter recursions of Theorem 3.2.

However, moving away from the linear and Gaussian state space model, integrals (6.32) and (6.33) are not available in closed form and this prevents the calculation of the posterior distribution of β_t in (6.34). There might be only special and limited models for which the above calculations will be obtained in closed form. For the general and often most interesting cases we shall resort to approximations or to simulation-based inference.

6.5 Power Local Level Models

6.5.1 Motivation and Main Model Structure

In this section we briefly describe the class of the so-called *power local level models*, which were first proposed by Smith (1979) and Harvey and Fernandes (1989) and primarily initiated research efforts on non-Gaussian state space modelling. These models are further developed in Smith (1981) and Smith and Miller (1986), but their applicability is somewhat limited to state transitions that resemble a random walk process. These models are similar in nature with the work of Gamerman et al. (2013) who propose a new class of non-Gaussian state space models constructed in such a way to allow exact computation of the marginal likelihood. This approach deploys the gamma-beta conjugacy, used also in Smith and Miller (1986) and in Shephard (1994a), but extends it in order to accommodate a wide class of response distributions.

The idea of Bayesian inference with a prior-posterior facility for non-Gaussian models is developed in the late 1970s with the path-breaking work of Diaconis and Ylvisaker (1979). These authors adopt the so-called *conjugate prior* for independent observations generated by a response distribution in the exponential family.

Considering the general state space model (6.3) Smith (1979) motivates transition laws based on the random walk of the Gaussian linear state space model, defined by

$$y_t = \beta_t + \epsilon_t \quad \text{and} \quad \beta_t = \beta_{t-1} + \zeta_t,$$

where $\epsilon_t \sim N(0, \sigma^2)$ and $\zeta_t \sim N(0, Z_t)$, for some variances σ^2 and Z_t and remaining assumptions as in the state space model (3.10a)–(3.10b). For this model, adopting the discounting approach to specify Z_t as in Sect. 4.3, the prior distribution of β_{t+1} in the Kalman filter (see Theorem 3.2) can be written as

$$p(\beta_t = \beta \mid y_{1:t-1}) = \frac{1}{\sqrt{2\pi P_{t|t-1}}} \exp\left\{ -\frac{(\beta - \hat{\beta}_{t|t-1})^2}{2P_{t|t-1}} \right\}$$

$$= c_t \left[\frac{1}{\sqrt{2\pi P_{t-1|t-1}}} \exp\left\{ -\frac{(\beta - \hat{\beta}_{t-1|t-1})^2}{2P_{t-1|t-1}} \right\} \right]^{\delta}$$

$$= c_t \, p(\beta_{t-1} = \beta \mid y_{1:t-1})^{\delta}, \tag{6.35}$$

where $P_{t|t-1} = P_{t-1|t-1}/\delta$ is used, for a discount factor $0 < \delta \le 1$, and the constant $c_t = \delta^{1/2}(2\pi P_{t-1|t-1})^{-(1-\delta)/2}$ does not depend on β. Equation (6.35) indicates that, given information $y_{1:t-1}$, the prior distribution of β_t is flatter than the posterior distribution of β_{t-1}, which reflects on the increased uncertainty, i.e. $\text{Var}(\beta_t \mid y_{1:t-1}) = P_{t-1|t-1} + Z_t = P_{t-1|t-1}/\delta \ge P_{t-1|t-1} = \text{Var}(\beta_t - 1 \mid y_{1:t-1})$.

Moving on to the non-Gaussian model (6.3), Smith (1979) defines the power local level models as a dynamic model whose transition density $p(\beta_t \mid \beta_{t-1})$ satisfies

$$p(\beta_t = \beta \mid y_{1:t-1}) = c_t p(\beta_{t-1} = \beta \mid y_{1:t-1})^\delta, \qquad (6.36)$$

for some deterministic (constant) c_t and a discount factor $0 < \delta \le 1$. The law (6.36) suggests that, conditional on information $y_{1:t-1}$, the prior of β_t is flatter than the posterior of β_{t-1}. It turns out that for a wide class of observation models $p(y_t \mid \beta_t)$ of (6.3), one does not need to know precisely the value of the constant c_t, as this is implied by the structure of (6.36). Such a modelling framework is discussed here by considering that $p(y_t \mid \beta_t)$ is a member of the exponential family (6.4) and is detailed as follows.

First note that through the link function $g(\cdot)$ of Eq. (6.5) the prior and posterior distributions of β_t yield prior and posterior distributions of the natural parameter γ_t. Hence we work with γ_t, as it is more convenient to form the prior distribution. Suppose that at time $t-1$, the posterior distribution of γ_{t-1} is

$$p(\gamma_{t-1} = \gamma \mid y_{1:t-1}) = \kappa(r_{t-1}, s_{t-1}) \exp[r_{t-1}^\top \gamma - s_{t-1} b(\gamma)], \qquad (6.37)$$

for some parameter vector r_{t-1} and scalar s_{t-1}, where $b(\gamma)$ is provided in the observation model equation (6.4) and $\kappa(r_{t-1}, s_{t-1})$ is the proportionality constant, i.e.

$$\kappa(r_{t-1}, s_{t-1}) = \left[\int_A \exp\{ r_{t-1}^\top \gamma - s_{t-1} b(\gamma) \} \, d\gamma \right]^{-1}$$

and A being the domain of γ.

At time t, the prior of γ_t is obtained by adopting prior (6.36) with the posterior (6.37) as

$$p(\gamma_t = \gamma \mid y_{1:t-1}) = c_t \kappa(r_{t-1}, s_{t-1})^\delta \exp[\delta r_{t-1}^\top \gamma - \delta s_{t-1} b(\gamma)].$$

This prior distribution is known as *conjugate prior* having the property that prior $p(\gamma_t = \gamma \mid y_{1:t-1})$ and posterior $p(\gamma_{t-1} = \gamma \mid y_{1:t-1})$ belong to the same family of distributions; we shall write $\gamma_t \mid y_{1:t-1} \sim CP(\delta r_{t-1}, \delta s_{t-1})$ to denote this distribution. Conjugate prior distributions for the exponential family were introduced in Diaconis and Ylvisaker (1979) and are discussed in Robert (2007).

At time t as the observation y_t becomes available, the posterior distribution of γ_t is updated by an application of the Bayes theorem

$$p(\gamma_t \mid y_{1:t}) \propto p(y_{t+1} \mid \gamma_t) p(\gamma_t \mid y_{1:t-1})$$

$$\propto \exp\left\{ \left(\frac{z(y_t)}{a_t} + \delta r_{t-1} \right)^\top \gamma_t - \left(\frac{1}{a_t} + \delta s_{t-1} \right) b(\gamma_t) \right\}$$

$$= \exp[r_t^\top \gamma_t - s_t b(\gamma_t)].$$

This establishes that $r_t = a_t^{-1} z(y_t) + \delta r_{t-1}$, $s_t = a_t^{-1} + \delta s_{t-1}$, so that $\gamma_t \mid y_{1:t} \sim CP(r_t, s_t)$. Assuming some prior values for r_0, s_0, the recursions of r_t and s_t above, together with the prior (6.36) and the posterior (6.37) provide a sequential algorithm over time $t = 1, 2, \ldots$. In the next section two specific models are used to illustrate the application of this algorithm.

We close this section by discussing a key result of the power local level model.

Theorem 6.1 *Let the time series $\{y_t\}$ follow the power local level model governed by (6.36). Let the prior and the posterior distributions of the state vector β_t be differentiable and unimodal. Then, given information $y_{1:t}$, the mode of β_{t+1} is the same as the mode of β_t.*

Proof Let $\hat{\beta}_t$ be the mode of β_t, given $y_{1:t}$. From Eq. (6.36) we obtain

$$\frac{\partial \log p(\beta_{t+1} = \beta)}{\partial \beta} = \delta \frac{\partial \log p(\beta_t = \beta)}{\partial \beta}$$

and since $\hat{\beta}_t$ is the mode of β_t we get

$$\left. \frac{\partial \log p(\beta_{t+1} = \beta)}{\partial \beta} \right|_{\beta = \hat{\beta}_t} = 0,$$

hence $\hat{\beta}_t$ is also the mode of β_{t+1}.   □

Some comments are in order. Theorem 6.1 provides a key property of the power local level model. Basically is suggests that for a wide class of differentiable and unimodal prior and posterior distributions, the mode is invariant going from time t to $t+1$ with the same information. This proposes a local level evolution of the states β_t, which resembles that of the random walk in the Gaussian local level model, see Sect. 3.1.3. The name *local level* originates by this local level evolution of the states, while the word *power* reflects the power law in (6.36). In the Gaussian local level model the mode invariance is the same as a mean invariance, written as $E(\beta_{t+1} \mid y_{1:t}) = E(\beta_t \mid y_{1:t})$ and the power law (6.36) in this case is given by (6.35). Below we discuss a power local level for count data and we verify Theorem 6.1 for that model.

6.5.2   Poisson-Gamma and Exponential-Gamma Models

Consider the Poisson model (6.8) of Sect. 6.2.2. Suppose that at time t the posterior of λ_t is a gamma distribution $\lambda_{t-1} \mid y_{1:t-1} \sim G(\alpha_{t-1}, \beta_{t-1})$, for some $\alpha_{t-1}, \beta_{t-1} > 0$; here β_{t-1} is just the second parameter (known as *rate*) of the gamma distribution and should not be confused with the states of the state space model. Then, applying the power law (6.36), we get that the prior of λ_t is

$$p(\lambda_t = \lambda \mid y_{1:t-1}) = c_t \, p(\lambda_{t-1} = \lambda \mid y_{1:t-1})^\delta$$

$$= c_t \left[\frac{\beta_{t-1}^{\alpha_{t-1}}}{\Gamma(\alpha_{t-1})} \lambda^{\alpha_{t-1}-1} \exp(-\beta_{t-1}\lambda) \right]^\delta$$

$$\propto \lambda^{\delta\alpha_{t-1}+1-\delta-1} \exp(-\delta\beta_{t-1}\lambda),$$

so that the prior λ_t is the gamma distribution $\lambda_t \mid y_{1:t-1} \sim G(\delta\alpha_{t-1}+1-\delta, \delta\beta_{t-1})$. Upon observing y_t, the posterior distribution of λ_t is revised using Bayes theorem as

$$p(\lambda_t \mid y_{1:t}) \propto p(y_t \mid \lambda_t) p(\lambda_t \mid y_{1:t-1})$$

$$\propto \lambda_t^{\delta\alpha_{t-1}+1-\delta+y_t-1} \exp\{(\delta\beta_{t-1}+1)\lambda_t\},$$

so that $\lambda_t \mid y_{1:t} \sim G(\alpha_t, \beta_t)$, with

$$\alpha_t = \delta\alpha_{t-1} + 1 - \delta + y_t \quad \text{and} \quad \beta_t = \delta\beta_{t-1} + 1.$$

Before we proceed with forecasting, we verify Theorem 6.1 for this model. By noting that the mode of a random variable following the gamma distribution $G(\alpha, \beta)$ is $(\alpha - 1)/\beta$, we have

$$\text{mode}(\lambda_{t+1} \mid y_{1:t}) = \frac{\delta(\alpha_t - 1)}{\delta\beta_t} = \frac{\alpha_t - 1}{\beta_t} = \text{mode}(\lambda_t \mid y_{1:t}),$$

which establishes the mode invariance as we are moving from time t to $t + 1$ with the same information $y_{1:t}$. Note that the conditions of Theorem 6.1 are met as the density of the gamma distribution is differentiable and unimodal.

With the posterior distribution of λ_t in place, the one-step forecast distribution of y_{t+1} is

$$p(y_{t+1} \mid y_{1:t}) = \int_0^\infty p(y_{t+1} \mid \lambda_{t+1}) p(\lambda_{t+1} \mid y_{1:t}) \, d\lambda_{t+1}$$

$$= \frac{(\delta\beta_t)^{\delta\alpha_t+1-\delta}}{\Gamma(\delta\alpha_t + 1 - \delta) y_{t+1}!} \int_0^\infty \lambda^{\delta\alpha_t+1-\delta-1} \exp[-(\delta\beta_t + 1)\lambda_{t+1}] \, d\lambda_{t+1}$$

$$= \frac{(\delta\beta_t)^{\delta\alpha_t+1-\delta} \Gamma(\delta\alpha_t + 1 - \delta + y_{t+1})}{\Gamma(\delta\alpha_t + 1 - \delta) y_{t+1}! (\delta\beta_t + 1)^{\delta\alpha_t+1-\delta+y_{t+1}}}$$

$$= \binom{n + y - 1}{y} \left(\frac{a}{1+a} \right)^n \left(\frac{1}{1+a} \right)^y,$$

with $n = \delta\alpha_t + 1 - \delta$, $y = y_{t+1}$ and $a = \delta\beta_t$. Here we have made use of the gamma integral of Eq. (4.60) and

$$\frac{\Gamma(n + y)}{y!\,\Gamma(n)} = \binom{n + y - 1}{y},$$

see also Exercises 7 and 8. Hence the one-step forecast distribution of y_{t+1} is a negative binomial distribution $y_{t+1} \mid y_{1:t} \sim \text{NegBinom}(n, a)$.

So far we have adopted the power law (6.36), which implies a local-level type evolution of λ_t, but an explicit transition of λ_t is not derived. Such evolution of λ_t is supported (see Exercise 5) by the multiplicative evolution

$$\lambda_t = \frac{\lambda_{t-1}\xi_t}{\delta}, \tag{6.38}$$

where $\lambda_{t-1} \mid y_{1:t-1} \sim G(\alpha_{t-1}, \beta_{t-1})$, ξ_t is a random variable, which is independent of λ_{t-1} and follows the beta distribution $\xi_t \sim \text{Beta}[\delta\alpha_{t-1}, (1 - \delta)\alpha_{t-1}]$. This evolution was first proposed in Smith and Miller (1986) and further deployed in Shephard (1994a) for a non-Gaussian state space model with exponential or gamma responses; a similar model is described below.

Suppose that non-negative observations are collected over time. Such data may be records of athletes, such as those discussed in Smith and Miller (1986), which are non-negative and take values in $[0, +\infty)$. We can also motivate data of this sort from many fields, such as environmetrics, where observations may represent pollutant or temperature readings, or economics, where observations may represent volatility (all this data take non-negative values). Returning to the records, a plausible model to describe the non-negative time series y_t is the exponential model, so that given a state λ_t, the response y_t is assumed to follow an exponential distribution with rate λ_t, or

$$p(y_t \mid \lambda_t) = \lambda_t \exp(-\lambda_t y_t),$$

for some $\lambda_t > 0$. Assuming that at time $t - 1$ the posterior distribution of λ_{t-1} is a gamma distribution, $\lambda_{t-1} \mid y_{1:t-1} \sim G(\alpha_{t-1}, \beta_{t-1})$, for some known parameters $\alpha_{t-1}, \beta_{t-1} > 0$, we can apply the power law as above and obtain exactly the same expressions of the prior and posterior distribution of λ_t in the Poisson model above, i.e. the posterior distribution of λ_t is $\lambda_t \mid y_{1:t} \sim G(\alpha_t, \beta_t)$, with $\alpha_t = \delta\alpha_{t-1} + y_t$ and $\beta_t = \delta\beta_{t-1} + 1$. This is because the same gamma distribution for λ_t is applied in both models. Consequently, the transition law (6.38) is established, as is discussed in Smith and Miller (1986); see also Exercise 5. With the posterior of λ_t in place, we can derive the one-step ahead forecast distribution of y_{t+1} as

$$p(y_{t+1} \mid y_{1:t}) = \int_0^\infty p(y_{t+1} \mid \lambda_{t+1}) p(\lambda_{t+1} \mid y_{1:t}) \, d\lambda_{t+1}$$

$$= \frac{(\delta \beta_t + y_{t+1})^{\delta \alpha_t + 2 - \delta}}{\Gamma(\delta \alpha_t + 2 - \delta)},$$

the full derivation of which is left to the reader as an exercise.

6.6 Approximate Inference

6.6.1 Motivation and Methodology

In this section we discuss approximate inference for a conditional Gaussian state space model (6.2). The non-linear functions $f(\cdot)$ and $g(\cdot)$ of model (6.2) are approximated by a first order Taylor expansion, and hence approximating the non-linear model (6.2) as a linear and Gaussian one.

Consider the state space model (6.2) where the observation covariance matrix $\mathrm{Var}(\epsilon_t) = \Sigma$ and the transition covariance matrix $\mathrm{Var}(\zeta_t) = Z_t$ are assumed known. Suppose that at time $t-1$ the posterior mean vector of β_{t-1} is approximated as $\hat{\beta}_{t-1|t-1} \approx \mathrm{E}(\beta_{t-1} \mid y_{1:t-1})$ and it is available. The function $g(\beta_{t-1})$ can be approximated using a Taylor expansion around the vector $\hat{\beta}_{t-1|t-1}$ as

$$g(\beta_{t-1}) \approx g(\hat{\beta}_{t-1|t-1}) + \left. \frac{\partial g(\beta_{t-1})}{\partial \beta_{t-1}} \right|_{\beta_{t-1} = \hat{\beta}_{t-1|t-1}} (\beta_{t-1} - \hat{\beta}_{t-1|t-1})$$

after ignoring second and higher order terms. Thus, the transition equation of β_t in (6.2) is approximated as

$$\beta_t \approx \left. \frac{\partial g(\beta_{t-1})}{\partial \beta_{t-1}} \right|_{\beta_{t-1} = \hat{\beta}_{t-1|t-1}} \beta_{t-1} + v_t + \zeta_t, \tag{6.39}$$

where $v_t = g(\hat{\beta}_{t-1|t-1}) - [\partial g(\beta_{t-1})/\partial \beta_{t-1}]_{\beta_{t-1} = \hat{\beta}_{t-1|t-1}} \hat{\beta}_{t-1|t-1}$. Now we can see that the prior mean vector of β_t is approximated as $\hat{\beta}_{t|t-1} = \mathrm{E}(\beta_t \mid y_{1:t-1}) \approx [\partial g(\beta_{t-1})/\partial \beta_{t-1}]_{\beta_{t-1} = \hat{\beta}_{t-1|t-1}} \hat{\beta}_{t-1|t-1} + v_t = g(\hat{\beta}_{t-1|t-1})$.

Provided the availability of $\hat{\beta}_{t|t-1}$ as above, we use a second Taylor expansion around $\hat{\beta}_{t|t-1}$ to approximate $f(\beta_t)$ in the observation equation as

$$f(\beta_t) \approx f(\hat{\beta}_{t|t-1}) + \left. \frac{\partial f(\beta_t)}{\partial \beta_t} \right|_{\beta_t = \hat{\beta}_{t|t-1}}^{\mathsf{T}} (\beta_t - \hat{\beta}_{t|t-1}),$$

after ignoring second and higher order terms as before. Thus, the observation equation of model (6.2) is approximated as

$$y_t = \frac{\partial f(\beta_t)}{\partial \beta_t}\bigg|_{\beta_t = \hat{\beta}_{t|t-1}}^{\top} \beta_t + \mu_t + \epsilon_t, \tag{6.40}$$

where $\mu_t = f(\hat{\beta}_{t|t-1}) - [\partial f(\beta_t)/\partial \beta_t]_{\beta_t = \hat{\beta}_{t|t-1}}^{\top} \hat{\beta}_{t|t-1}$.

The linear state space model (6.39)–(6.40) is an approximation of the non-linear model (6.2). Indeed we can write

$$y_t \approx \mathbf{x}_t^{\top} \beta_t + \varepsilon_t \quad \text{and} \quad \beta_t \approx \mathbf{F}_t \beta_{t-1} + \eta_t, \tag{6.41}$$

where the design matrix $\mathbf{x}_t$ and the transition matrix $\mathbf{F}_t$ are equal to

$$\mathbf{x}_t = \frac{\partial f(\beta_t)}{\partial \beta_t}\bigg|_{\beta_t = \hat{\beta}_{t|t-1}} \quad \text{and} \quad \mathbf{F}_t = \frac{\partial g(\beta_{t-1})}{\partial \beta_{t-1}}\bigg|_{\beta_{t-1} = \hat{\beta}_{t-1|t-1}},$$

and the innovations ε_t and η_t have mean vectors μ_t and ν_t and covariance matrices Σ and $\mathbf{Z}_t$, respectively, i.e. $\varepsilon_t \sim N(\mu_t, \Sigma)$ and $\eta_t \sim N(\nu_t, \mathbf{Z}_t)$. The state space model (6.41) is very similar to the multivariate models of Chap. 5 (see model (5.1a)–(5.1b)), the only difference being the non-zero mean vectors of ε_t and η_t. The Kalman filter used to estimate the state vectors β_t in Chap. 5 can be updated to accommodate for a non-zero mean of the innovations ε_t and η_t. It is relatively easy to verify that the Kalman filter recursions of Theorem 5.1 are still applied if we modify $\hat{\beta}_{t|t-1}$ and $\hat{y}_{t|t-1}$ as

$$\hat{\beta}_{t|t-1} = \mathbf{F}_t \hat{\beta}_{t-1|t-1} + \nu_t \quad \text{and} \quad \hat{y}_{t|t-1} = \mathbf{x}_t^{\top} \hat{\beta}_{t|t-1} + \mu_t,$$

where μ_t and ν_t are the mean vectors of ε_t and η_t defined above. Thus, with the linearised state space model (6.41) we can obtain the posterior distribution of β_t given $y_{1:t}$ as well as we can routinely forecast future values of y_{t+h}, for some $h \geq 1$.

Some comments are in order. We observe that in the linear case $f(\beta_t) = \mathbf{x}_t^{\top} \beta_t$ and $g(\beta_{t-1}) = \mathbf{F}_t \beta_{t-1}$. In this case the derivatives of $f(\beta_t)$ and $g(\beta_{t-1})$ yield just $\mathbf{x}_t$ and $\mathbf{F}_t$, respectively, and hence $\mu_t = 0$, $\nu_t = 0$. This implies that the approximations are exact as expected.

When at least one of $f(\beta_t)$ and $g(\beta_{t-1})$ are non-linear the above approximation comes into play. In most practical situations $g(\beta_{t-1})$ will still be linear, as it describes a viable Markov evolution which may well be linear. As a result we may have $f(\beta_t)$ to be non-linear on β_t, but $g(\beta_{t-1}) = \mathbf{F}_t \beta_{t-1}$ being linear on β_{t-1}. The empirical motivation here could be that it is hard to anticipate or specify a priori a non-linear relationship of the unobserved state evolution $\{\beta_t\}$, while we shall have more information on the observation linking the data $\{y_t\}$ with the states $\{\beta_t\}$.

6.6.2   Tracking a Ship

In this section we consider the bearings-only tracking problem, described in some detail in Sect. 1.3.2. In Sect. 6.3, it is shown that z_t, the time series defined as $z_t = \arctan(x_t/y_t)$, where x_t is the position of the ship in the x-axis at time t and y_t its position in the y-axis at time t, follows a conditionally Gaussian state space model (6.2) with

$$f(\beta_t) = \arctan\left(\frac{[0, 1, 0, 0]\beta_t}{[1, 0, 0, 0]\beta_t}\right)$$

and the four-dimensional state vector β_t, which is defined in Sect. 6.3, follows a random walk so that $\mathbf{F}_t = \mathbf{I}$.

In order to use the approximate inference of Sect. 6.6.1 above, we need the partial derivative of $f(\beta_t)$ with respect to β_t. By using the chain rule of differentiation we obtain

$$\frac{\partial f(\beta_t)}{\partial \beta_t} = \left[1 + \left(\frac{[0, 1, 0, 0]\beta_t}{[1, 0, 0, 0]\beta_t}\right)^2\right]^{-1} ([1, 0, 0, 0]\beta_t)^{-1}$$

$$\times \left(\frac{\partial [0, 1, 0, 0]\beta_t}{\partial \beta_t} - \frac{\partial [1, 0, 0, 0]\beta_t}{\partial \beta_t}[0, 1, 0, 0]\beta_t\right)$$

$$= \left\{([1, 0, 0, 0]\beta_t)^2 + ([0, 1, 0, 0]\beta_t)^2\right\}^{-1} \begin{bmatrix} 0 & -1 & 0 & 0 \\ 1 & 0 & 0 & 0 \\ 0 & 0 & 0 & 0 \\ 0 & 0 & 0 & 0 \end{bmatrix} \beta_t$$

$$= (2\beta_t^\top \beta_t)^{-1} \begin{bmatrix} 0 & -1 & 0 & 0 \\ 1 & 0 & 0 & 0 \\ 0 & 0 & 0 & 0 \\ 0 & 0 & 0 & 0 \end{bmatrix} \beta_t$$

Thus, using the first order Taylor expansion of $f(\beta_t)$ around $\hat{\beta}_{t|t-1}$, the observation equation is approximated as

$$z_t \approx \frac{\hat{\beta}_{t|t-1}^\top}{2\hat{\beta}_{t|t-1}^\top \hat{\beta}_{t|t-1}} \begin{bmatrix} 0 & 1 & 0 & 0 \\ -1 & 0 & 0 & 0 \\ 0 & 0 & 0 & 0 \\ 0 & 0 & 0 & 0 \end{bmatrix} \beta_t + \varepsilon_t, \tag{6.42}$$

where $\varepsilon_t \sim N(\mu_t, \sigma^2)$ and

$$\mu_t = \arctan\left(\frac{[0, 1, 0, 0]\hat{\beta}_{t|t-1}}{[1, 0, 0, 0]\hat{\beta}_{t|t-1}}\right) - (2\hat{\beta}_{t|t-1}^\top \hat{\beta}_{t|t-1})^{-1}\hat{\beta}_{t|t-1}^\top \begin{bmatrix} 0 & 1 & 0 & 0 \\ -1 & 0 & 0 & 0 \\ 0 & 0 & 0 & 0 \\ 0 & 0 & 0 & 0 \end{bmatrix} \hat{\beta}_{t|t-1}$$

The random walk transition is $\beta_t = \beta_{t-1} + \zeta_t$, where $\zeta_t \sim N(0, \mathbf{Z})$, for some covariance matrix $\mathbf{Z}$. This transition together with the observation equation (6.42) provide the linearised state space model of the non-linear model (1.10)–(1.11), where we have used $\mathbf{F} = \mathbf{I}$. In this model σ^2 and $\mathbf{Z}$ are assumed known and an initial state $\beta_0 \sim N(\hat{\beta}_{0|0}, \mathbf{P}_{0|0})$ is considered. With this setting in place the Kalman filter applies providing an approximate posterior mean vector $\hat{\beta}_{t|t}$ and covariance matrix $\mathbf{P}_{t|t}$. This approximate estimation algorithm benefits from the rich availability of estimation in linear state space models, e.g. σ^2 and $\mathbf{Z}$ can be estimated using the forward filtering backward sampling MCMC scheme, described in Sect. 5.7.

6.6.3 The Extended Kalman Filter

Historical note. The extended Kalman filter (EKF) was first developed in the significant work of Stanley F. Schmidt at the Ames Research Centre at NASA. In 1960 Kalman visited Schmidt to present his approach of sequential filtering (Kalman, 1960). Schmidt and his colleagues were interested in Kalman's work because the new methodology was able to solve sequentially the filtering problem, without relying on stationarity. However, the assumption of normality seemed to be an obstacle for the target application, part of the Apollo project. Schmidt and his colleagues modified the Kalman filter to what is now known as the extended Kalman filter. McGee and Schmidt (1985) and Schmidt (1981) discuss the story around the discovery of the EKF and how it was applied for the Apollo project. From dynamic systems point of view the EKF is discussed in Sect. 8.4.4; see also Grewal and Andrews (2010).

Conditionally Gaussian state space models (6.2), as described above, is a large class of non-linear state models. However, this class of models is limited to Gaussian innovations forcing the response y_t to be conditionally Gaussian. Such an assumption restricts the applicability of these models, as the distribution of many time series exhibit departure from normality. For example, in Sect. 6.2 we describe dynamic generalised linear models (DGLM), with response belonging to the exponential family of distributions. It is therefore necessary to consider non-Gaussian time series analysis; the power local level models of Sect. 6.5 is a first attempt towards this direction. However, the power local level is restricted to transitions of the states around a local level and do not permit more general transitions, such as those considered in the previous chapters for the Gaussian models or those for the DGLM. In Sect. 6.7 we discuss modern Monte Carlo inference for such models; in this section we provide a simple solution to the non-Gaussian and non-linear state space model estimation, which is known as *extended Kalman filtering*. EKF is described and further applied in Fahrmeir (1992) and in Fruhwirth-Schnatter (1994a) among other studies; a book length discussion can be found in Fahrmeir and Tutz (2001).

Consider the general state space model (6.3), where for simplicity we shall assume that the transition equation is linear and Gaussian, i.e.

$$\beta_t = \mathbf{F}_t\beta_{t-1} + \zeta_t, \quad \zeta_t \sim N(0, \mathbf{Z}_t), \tag{6.43}$$

exactly as in the Gaussian case discussed in detail in the previous chapters.

We shall be interested in observations generated by a non-linear and non-Gaussian model and in particular we consider that conditionally on a state β_t, y_t follows some distribution with mean vector $f(\beta_t)$ and some covariance matrix $\mathbf{V}_t$, which may depend on β_t. Extended Kalman filtering considers a rough approximation of that distribution by a Gaussian distribution with mean vector $f(\beta_t)$ and covariance matrix $\mathbf{V}_t$. Hence, the response distribution approximation states that

$$y_t \approx f(\beta_t) + \epsilon_t, \quad \epsilon_t \sim N(0, \mathbf{V}_t), \tag{6.44}$$

so that

$$E(y_t \mid \beta_t) = f(\beta_t) \quad \text{and} \quad \text{Var}(y_t \mid \beta_t) = \mathbf{V_t}.$$

The working model is the conditional Gaussian state space model (6.43)–(6.44), for which the theory of Sect. 6.6.1 may be applied. As a result the extended Kalman filter (EKF) proposes two approximations: in the first the non-Gaussian response distribution is approximated by a Gaussian distribution matching the mean vector and the covariance matrix of the non-Gaussian one; in the second approximation the non-linear function $f(\beta_t)$ is approximated by a first order Taylor series in order to linearise the model and hence the Kalman filter may then be applied. Care must be applied as non-linearities in $f(\beta_t)$ and poor approximation of the distribution of $y_t\beta_t$ by its first two moments might result in significant errors and poor performance of the Kalman filter. Below we give a basic example illustrating the applicability of EKF.

For example consider the Poisson time series (6.8) described in Sect. 6.2.2. From the Poisson distribution (see also Sect. 2.3.2) we have

$$E(y_t \mid \beta_t) = f(\beta_t) = \exp(x_t^\top \beta_t) \quad \text{and} \quad V_t = \exp(x_t^\top \beta_t)$$

where x_t is the design vector and β_t is the state vector following the transition equation $\beta_t = \mathbf{F}\beta_{t-1} + \zeta_t$ and the usual assumptions and component definitions described in (6.8). Here, as y_t is scalar, x_t is a column vector and V_t, which depends implicitly on β_t, is a variance. Thus, the Poisson model (6.8) is approximated by the conditionally Gaussian state space model

$$y_t \approx f(\beta_t) + \epsilon_t = \exp(x_t^\top \beta_t) + \epsilon_t, \quad \epsilon_t \sim N(0, V_t),$$

and $\beta_t = \mathbf{F}\beta_{t-1} + \zeta_t$. Proceeding to the approximation of $f(\beta_t)$ as described in Sect. 6.6.1 we have

$$\frac{\partial f(\beta_t)}{\partial \beta_t} = \exp(x_t^\top \beta_t)x_t$$

Following the theory of conditionally Gaussian in Sect. 6.6.1 suppose that at time $t-1$ the posterior distribution of β_{t-1} is approximated by a $N(\hat{\beta}_{t-1|t-1}, \mathbf{P}_{t|t-1})$, for some mean vector $\hat{\beta}_{t-1|t-1}$ and covariance matrix $\mathbf{P}_{t|t-1}$. From the transition equation of β_t we obtain the approximate prior mean vector $\hat{\beta}_{t|t-1}$ and covariance matrix $\mathbf{P}_{t|t-1}$ of β_t, from the Kalman filter recursions (see Theorem 5.1 of Chap. 5).

With these moments in place, y_t is approximated by the Gaussian linear state space model

$$y_t \approx \exp(x_t^\top \hat{\beta}_{t|t-1})x_t^\top \beta_t + \varepsilon_t, \tag{6.45}$$

where $\varepsilon_t \sim N(\mu_t, \hat{V}_t)$, with

$$\mu_t = \exp(x_t^\top \hat{\beta}_{t|t-1})(1 - x_t^\top \hat{\beta}_{t|t-1}) \quad \text{and} \quad \hat{V}_t = \exp(x_t^\top \hat{\beta}_{t|t-1})$$

and β_t follows the linear transition $\beta_t = \mathbf{F}\beta_{t-1} + \zeta_t$. Model (6.45) together with the transition of β_t provide a linear and Gaussian state space model, an approximation of the non-linear and non-Gaussian state space model (6.8).

6.6.4 The Unscented Kalman Filter

The extended Kalman filter described above is able to deal with non-linear filtering, but it is reported to have large cumulative state and forecast errors, which may lead to poor performance overall (Wan & Van Der Merwe, 2000). This is particularly prevalent when complex non-linearities in the function $f(\cdot)$ are observed, in a way that first-order Taylor series approximation is a poor approximation of the system.

Efforts to deal with the non-linear modelling, but go beyond the extended Kalman filter, usually involve non-linear filter heuristics. These methods usually deploy either the standard Kalman filter or the extended Kalman filter, coupled with heuristic features in order to deal with system non-linearities, see e.g. Saab (2004). One of the standard approaches is the introduction of so-called *sigma points* and associated *weights*, which are chosen in such a way so that to concentrate around the high-probability regions of the posterior distribution of the states. The choice of these points and weights is very much the topic of current research and computational implementation efforts, see e.g. Saab (2004), Ponomareva and Date (2013), Radhakrishnan et al. (2018), Pakrashi and Namee (2019) and references therein. Since its discovery the unscented transformation and unscented Kalman filter have been extended and enriched by improving its performance (usually by making a clever determination of the sigma points), see e.g. Julier (2002), Julier and Uhlmann (2004) and references mentioned above. The UKF is used in sequential Monte Carlo (see Sect. 6.7 below) combined with Markov chain Monte Carlo steps in order to choose more accurately the importance function (Van Der Merwe et al., 2001).

The *unscented Kalman filter* (UKF) aims to improve on the second moment approximation (mean and variance) of the EKF described above. The UKF, introduced by Julier and Uhlmann (1997), uses the so-called *unscented transformation* in order to calculate approximations of the mean vector and covariance matrix of the posterior distribution of the states. While, the EKF uses a first order Taylor expansion to approximate the non-linear state space by a linear one, the UKF introduces sigma points and weights from a high probability region of the posterior distribution of the states in close proximity of the posterior mean of the states.

In its original version Julier and Uhlmann (1997) consider first the unscented transformation, which is basically approximating the mean and variance of a random variable, which undergoes a non-linear transformation. the UT is achieved in such a way that the sample mean and sample variance of the transformed random variable match the true mean and variance. Suppose that the p-dimensional column random vector x has sample mean vector $\bar{x}$ and sample covariance matrix $\mathbf{V}_x$. We wish to approximate the mean vector and covariance matrix of the random vector $y = f(x)$, where $f(\cdot)$ is a non-linear function. The random vector x may be approximated by a cloud of $2n + 1$ points

$$x^{(0)} = \bar{x}, \quad x^{(i)} = \bar{x} + \sqrt{n + \kappa}\mathbf{V}_{x,i}^{1/2}, \quad x^{(i+n)} = \bar{x} - \sqrt{n + \kappa}\mathbf{V}_{x,i}^{1/2},$$

with associated weights

$$w_0 = \frac{\kappa}{n + \kappa}, \quad w_i = w_{i+n} = \frac{1}{2(n + \kappa)},$$

where $\mathbf{V}_{x,i}^{1/2}$ denotes the i-th column of the symmetric square root matrix of $\mathbf{V}_x$ ($i = 1, \ldots, n$), for some $\kappa \in \mathbb{R}$.

Then for the random vector y, cloud points, mean and variance approximations are computed as

$$y^{(j)} = f[(x^{(j)}], \quad j = 1, \ldots, 2n,$$

$$\bar{y} = \sum_{j=0}^{2n} w_j y^{(j)}, \tag{6.46}$$

$$\mathbf{V}_y = \sum_{j=0}^{2n} w_j [y^{(j)} - \bar{y}][y^{(j)} - \bar{y}]^\top.$$

Some comments are in order.

1. Assuming that the cloud values $x^{(j)}$ accurately describe the distribution of x and the sample mean $\bar{x}$ is close to $E(x)$, then approximations of the mean vector and covariance matrix of y are accurate up to some degree. The algorithm is designed so that $\bar{y}$ matches the sample mean of y (evaluated at the sigma points) and hence in comparison to EKF the mean estimation is more accurate.

2. The above algorithm avoids approximating $f(\cdot)$ as the EKF does. Instead it evaluates the sigma points from the distribution of x and hence it avoids making approximation errors of $f(\cdot)$ such as those in EKF.

3. In the description above we have used the symmetric square root for $\mathbf{V}_x^{1/2}$. Any other suitable matrix square root may be used, such as based on the Choleski decomposition.

4. If $f(\cdot)$ is a linear function, say $y = \mathbf{A}x$, for some matrix A, then both the mean vector and the covariance matrix of y are exact, or $\bar{y} = A\bar{x}$ and $\mathbf{V}_y = A\mathbf{V}_x A^\top$.

5. The algorithm requires fine-tuning of the parameter κ. If $\kappa < 0$, then the algorithm can return a negative definite matrix $\mathbf{V}_y$. In this case modifications of the above algorithm is required to ensure that $\mathbf{V}_y$ is non-negative definite; for more details the reader is referred to Julier and Uhlmann (1997, 2004).

The UKF considers the above unscented transformation applied at the states at each point of time. We shall consider the conditionally Gaussian state space model (6.2), although it is possible to describe the algorithm for the more general non-linear model (6.3) of Sect. 6.1.

Suppose that at time $t-1$ the posterior mean vector $\hat{\beta}_{t-1|t-1}$ and the posterior covariance matrix $\mathbf{P}_{t-1|t-1}$ are available. We apply the above UT (6.46), with $x = \beta_{t-1}, \bar{x} = \hat{\beta}_{t-1|t-1}, \mathbf{V}_x = \mathbf{P}_{t-1|t-1}$. Hence we generate $2n$ points $\beta_{t-1}^{(j)}$.

In the prediction step we approximate the mean vector and the covariance matrix of β_t (suing the non-linear function $g(\beta_{t-1})$ of the state equation) and of y_t (using the non-linear function $f(\cdot)$ of the observation equation). So for β_t we have

$$\beta_{t|t-1}^{(j)} = g[\beta_{t-1}^{(j)}], \quad \hat{\beta}_{t|t-1} = \sum_{j=0}^{2n} w_i \beta_{t|t-1}^{(j)},$$

$$\mathbf{P}_{t|t-1} = \sum_{j=0}^{2n} w_i \left[\beta_{t|t-1}^{(j)} - \hat{\beta}_{t|t-1}\right]\left[\beta_{t|t-1}^{(j)} - \hat{\beta}_{t|t-1}\right]^\top$$

and for y_t we have

$$y_{t|t-1}^{(j)} = f[\beta_{t|t-1}^{(j)}], \quad \hat{y}_{t|t-1} = \sum_{j=0}^{2n} w_i y_{t|t-1}^{(j)},$$

$$\mathbf{Q}_{t|t-1} = \sum_{j=0}^{2n} w_i \left[y_{t|t-1}^{(j)} - \hat{y}_{t|t-1} \right] \left[y_{t|t-1}^{(j)} - \hat{y}_{t|t-1} \right]^{\mathsf{T}}$$

The covariance between β_t and y_t is approximated as

$$\mathbf{C}_t = \mathrm{Cov}(\beta_t, y_t) = \sum_{j=0}^{2n} w_i \left[\beta_{t|t-1}^{(j)} - \hat{\beta}_{t|t-1} \right] \left[y_{t|t-1}^{(j)} - \hat{y}_{t|t-1} \right]^{\mathsf{T}}.$$

When observation y_t becomes available, the algorithm updates the mean vector $\hat{\beta}_{t|t}$ and the covariance matrix $\mathbf{P}_{t|t}$ as

$$\hat{\beta}_t = \hat{\beta}_{t|t-1} + \mathbf{K}_t(y_t - \hat{y}_{t|t-1})$$

and $\mathbf{P}_{t|t} = \mathbf{P}_{t|t-1} - \mathbf{K}_t \mathbf{Q}_{t|t-1} \mathbf{K}_t^{\mathsf{T}}$, where the UKF gain $\mathbf{K}_t = \mathbf{C}_t \mathbf{Q}_{t|t-1}^{-1}$. Note that recursions of $\hat{\beta}_{t|t}$ and $\mathbf{P}_{t|t}$ are very similar to the standard Kalman filter, except that here $\hat{\beta}_{t|t-1}$, $\hat{y}_{t|t-1}$, $\mathbf{P}_{t|t-1}$ and $\mathbf{K}_t$ are computed using the sigma points. It should be noted that the above recursions starting from $\hat{\beta}_{t-1|t-1}$, complete an iteration of the UKF. This suggests a sequential algorithm starting from the prior β_0, with given mean vector $\hat{\beta}_{0|0}$ and covariance matrix $P_{0|0}$.

One of the advantages of the UKF is that the sigma points generated at each point of time are deterministic and easy to set for high dimensions, hence EKF has been proposed for high dimensional studies of non-linear systems. Monte Carlo methods and sequential Monte Carlo (see Sect. 6.7 below) at each point of time simulate a random sample and these methods are known to diverge for high dimensional data, for a discussion see Petris et al. (2009).

6.7 Sequential Monte Carlo Inference

6.7.1 Monte Carlo Integration

Monte Carlo is a popular yet simple procedure for the approximation of an integral of a given continuous function. Suppose we wish to compute the integral

$$I = \int_A f(x) p(x) \, dx,$$

where $f(\cdot)$ is a continuous function and $p(\cdot)$ is a density function defined on some domain A. We can regard the above integral as the expectation $E[f(X)]$, where X is the random vector having density function $p(x)$. If we are able to generate a sample from $p(\cdot)$, say $x^{(i)}$, for $i = 1, \ldots, N$, then we can approximate $I = E[f(X)]$ as

$$\hat{I} = \frac{1}{N} \sum_{i=1}^{N} f(x^{(i)}).$$

From the central limit theorem, we know that $\hat{I}$ converges almost surely to the true expectation $I = E[f(X)]$. In many applications, it will be difficult to simulate from $p(x)$, as $p(x)$ may be too complex. Another issue arises in Bayesian inference whereby $p(x)$ is a posterior distribution of X, given some data y and typically is available only up to a proportionality constant c. Indeed writing $p(x \mid y)$ for this posterior density, by applying Bayes theorem we have

$$p(x \mid y) = cp(y \mid x)\pi(x),$$

where $p(y \mid x)$ is the likelihood of X with data y, $\pi(x)$ is the prior distribution of X and

$$c = \left[\int_A p(y \mid x)\pi(x) \right]^{-1}.$$

For most applications c will not be available in closed form. As a result, $p(x \mid y)$ is only available up to a proportionality constant and obtaining a sample from this posterior is even harder. This poses additional obstacles for simulating directly from $p(\cdot)$.

6.7.2   *Importance Sampling*

In order to overcome the above difficulties a procedure known as *importance sampling* is deployed. The basic idea is that instead of simulating from $p(x)$ or $p(x \mid y)$, which might be difficult or even impossible as described above, we simulate from a convenient distribution $g(x)$, known as *importance density* or *importance function* and then we calculate some weights to make the necessary adjustment. To detail the computations we can write I as

$$I = \int_A \left[f(x) \frac{p(x)}{g(x)} \right] g(x)\, dx = E[w(x)f(x)],$$

where the weight function $w(x) = p(x)/g(x)$ and it is assumed that $g(\cdot)$ has the same domain as $p(\cdot)$ and that $g(x) \neq 0$, for $x \in A$. Following the ideas of Monte

Carlo approximation in Sect. 6.7.1 we can approximate I by

$$\hat{I} = \frac{1}{N} \sum_{i=1}^{N} w^{(i)} f(x^{(i)}), \qquad (6.47)$$

where $w^{(i)} = p(x^{(i)})/g(x^{(i)})$, for a sample $x^{(1)}, \ldots, x^{(N)}$ from $g(x)$.

The above approximation $\hat{I}$ relies on the availability of $p(\cdot)$ or $p(x \mid y)$, as $p(\cdot)$ will usually represent a posterior distribution. As it is common in Bayesian inference $p(\cdot)$ can be readily known only up to a proportionality constant (see also the discussion in Sect. 6.7.1), the weights $w^{(i)}$ may not be available. However, it turns out that the Monte Carlo approximation $\hat{I}$ can accommodate this, hence not requiring to compute the proportionality constant.

Suppose that $p(x) = cq(x)$, where c is the proportionality constant and $q(x)$ is a known function. Redefine the weights as $w(x) = q(x)/g(x)$, so that the weights of (6.47) are now equal to $p(x)/g(x) = cq(x)/g(x) = cw(x)$. Noting this and setting $f(x) = 1$ in (6.47) we obtain

$$1 = \frac{1}{N} \sum_{i=1}^{N} cw^{(i)} \quad \text{or} \quad c \sum_{i=1}^{N} w^{(i)} = N.$$

Now from (6.47) for any $f(x)$ we have

$$\hat{I} = \frac{\sum_{i=1}^{N} cw^{(i)} f(x^{(i)})}{\sum_{i=1}^{N} cw^{(i)}} = \sum_{i=1}^{N} \tilde{w}^{(i)} f(x^{(i)}), \qquad (6.48)$$

where $\tilde{w}^{(i)}$ are known as the standardised weights and are defined as

$$\tilde{w}^{(i)} = \frac{w^{(i)}}{\sum_{i=1}^{N} w^{(i)}}, \quad i = 1, \ldots, N, \qquad (6.49)$$

having the property $\sum_{i=1}^{N} \tilde{w}^{(i)} = 1$. The above proposes an algorithm combining importance sampling and Monte Carlo. In brief, for the approximation of I,

- Simulate N values $x^{(1)}, \ldots, x^{(N)}$ from the *importance* density $g(x)$
- Compute the (non-standardised) weights $w^{(i)} = q(x^{(i)})/g(x^{(i)})$
- Get standardised weights $\tilde{w}^{(i)}$ using (6.49)
- Approximate I using (6.48)

By picking appropriate functions $f(\cdot)$ we can approximate mean, variance and other statistics of X. For example for $f(x) = x$ the algorithm approximates $\hat{x}$ the mean of X, while for $f(x) = (x - \hat{x})^2$, the algorithm approximates the variance of X. The distribution of X can be approximated by the *Dirac delta* function, some details of which are given below.

The *Dirac delta* (or δ) is function defined on the real line which is zero everywhere except at point zero and has integral over the real line equal to one. We can think of $\delta(\cdot)$ as a density function with an infinitely high spike around zero and area below this spike equal to one, while everywhere else it is zero. This is the reason why δ is usually referred to as a *point mass* distribution. Formally $\delta(x)$ is defined as

$$\delta(x) = \begin{cases} +\infty, & x = 0 \\ 0, & x \neq 0 \end{cases}$$

and can be regarded as the limit of a sequence of $N(0, \sigma^2)$ densities when $\sigma \to 0$, i.e.

$$\delta(x) = \lim_{\sigma \to 0} \frac{1}{\sqrt{2\pi}\sigma} \exp\left(-\frac{x^2}{2\sigma^2}\right).$$

Suppose we simulate N independent particles $x^{(1)}, x^{(2)}, \ldots, x^{(N)}$ from some distribution $p(x)$. Then an empirical distribution of x is given by

$$\hat{p}(x) = \frac{1}{N} \sum_{i=1}^{N} \delta(x - x^{(i)}).$$

This basically suggest that $p(x)$ is approximated by the sample mean of the Dirac point mass at each particle; for more details see Doucet et al. (2001, Chapter 1).

6.7.3 Sequential Importance Sampling

Consider the general non-linear and non-Gaussian model formulation (6.3). Our aim is to apply importance sampling sequentially over time, in order to approximate the posterior distribution $p(\beta_t \mid y_{1:t})$. In parallel with the definition of $y_{1:t} = \{y_1, \ldots, y_t\}$, it is convenient to define $\beta_{1:t} = \{\beta_1, \ldots, \beta_t\}$ in order to include the history of all state vectors up to and including time t. Denote by $p(\beta_{1:t} \mid y_{1:t})$ the posterior density function of $\beta_{1:t}$ given $y_{1:t}$ and note the required $p(\beta_t \mid y_{1:t})$ is just the marginal distribution of β_t and can be extracted from $p(\beta_{1:t} \mid y_{1:t})$. By using Bayes theorem we have

$$\begin{aligned} p(\beta_{1:t} \mid y_{1:t}) &= p(\beta_t \mid \beta_{1:t-1}, y_{1:t}) p(\beta_{1:t-1} \mid y_{1:t}) \\ &\propto p(y_t \mid \beta_t, \beta_{1:t-1}, y_{1:t-1}) p(\beta_t \mid \beta_{1:t-1}, y_{1:t-1}) \\ &\quad \times p(y_t \mid \beta_{1:t-1}, y_{1:t-1}) p(\beta_{1:t-1} \mid y_{1:t-1}) \\ &\propto p(y_t \mid \beta_t) p(\beta_t \mid \beta_{t-1}) p(\beta_{1:t-1} \mid y_{1:t-1}), \end{aligned} \qquad (6.50)$$

The last line (6.50) is obtained, because given β_t, past states and observations $\beta_{1:t-1}$ and $y_{1:t-1}$ are conditionally independent of the present y_t, hence $p(y_t \mid \beta_t, \beta_{1:t-1}, y_{1:t-1}) = p(y_t \mid \beta_t)$. Likewise given β_{t-1}, the present state β_t is conditionally independent of the past history $\beta_{1:t-2}$ and $y_{1:t-1}$ and so we have $p(\beta_t \mid \beta_{1:t-1}, y_{1:t-1}) = p(\beta_t \mid \beta_{t-1})$. This follows from the basic Markovian property of the state space model (6.3), i.e. that given the present, the past and the future are conditionally independent.

In general we will not be able to sample from the posterior $p(\beta_{1:t} \mid y_{1:t})$ for the reasons outlined in Sect. 6.7.1. Hence, we can apply importance sampling as described in Sect. 6.7.2, appropriately modified to cater for sequential application. Following the ideas of importance sampling, we shall sample the states from a importance function or density $g(\cdot)$, defined on the same domain as the required posterior $p(\beta_{1:t} \mid y_{1:t})$. This importance function may not necessarily be a probability density function, but in most practical cases and for the purposes of this book we shall assume that $g(\cdot)$ is a density function, and hence it satisfies

$$
\begin{aligned}
g(\beta_{1:t} \mid y_{1:t}) &= g(\beta_t \mid \beta_{1:t-1}, y_{1:t}) g(\beta_{1:t-1} \mid y_{1:t}) \\
&\propto g(\beta_t \mid \beta_{t-1}, y_t) g(\beta_{1:t-1} \mid y_{1:t-1}),
\end{aligned} \tag{6.51}
$$

where we have assumed that $g(\beta_t \mid \beta_{1:t-1}, y_{1:t}) = g(\beta_t \mid \beta_{t-1}, y_t)$. We remark that this equality does not follow from the Markovian property of the state space model, because now the importance function $g(\cdot)$ is a density outside the definition of the state space model and hence is not required to satisfy that equality.

The next step is to define the weights and to recover a sequential calculation from $t-1$ to t, for each time $t = 1, 2, , \ldots$. Recall from Sect. 6.7.2 that the *importance weights* defined as the ratio of the posterior density over the importance function, or

$$
w_t = \frac{p(\beta_{1:t} \mid y_{1:t})}{g(\beta_{1:t} \mid y_{1:t})},
$$

which by using Eqs. (6.50) and (6.51) results in

$$
\begin{aligned}
w_t &\propto \frac{p(y_t \mid \beta_t) p(\beta_t \mid \beta_{t-1})}{g(\beta_t \mid \beta_{t-1}, y_t)} \frac{p(\beta_{1:t-1} \mid y_{1:t-1})}{g(\beta_{1:t-1} \mid y_{1:t-1})} \\
&= \frac{p(y_t \mid \beta_t) p(\beta_t \mid \beta_{t-1})}{g(\beta_t \mid \beta_{t-1}, y_t)} w_{t-1}.
\end{aligned}
$$

This formula suggests calculating the sampled importance weights $w_t^{(i)}$ at time t as

$$
w_t^{(i)} = \frac{p(y_t \mid \beta_t^{(i)}) p(\beta_t^{(i)} \mid \beta_{t-1}^{(i)})}{g(\beta_t^{(i)} \mid \beta_{t-1}^{(i)}, y_t)} w_{t-1}^{(i)}, \quad i = 1, \ldots, N, \tag{6.52}
$$

provided that we have sampled $\beta_t^{(i)}$ from $g(\beta_t \mid \beta_{t-1}^{(i)}, y_t)$ and that $\beta_{t-1}^{(i)}$ and $w_{t-1}^{(i)}$ are sampled at time $t - 1$. The density $p(y_t \mid \beta_t)$ is the likelihood of β_t using the single observation y_t and $p(\beta_t \mid \beta_{t-1})$ is the prior of β_t (prior distribution of β_t, given β_{t-1}), both of which are available from the state space model definition (6.3). Hence, with the availability of the sampled states $\beta_t^{(i)}, \beta_{t-1}^{(i)}$, the past sample weights $w_{t-1}^{(i)}$ and the importance function, the sampled weights $w_t^{(i)}$ may be calculated using (6.52).

Once $w_t^{(i)}$ are available the standardised weights may be computed by

$$\tilde{w}_t^{(i)} = \frac{w_t^{(i)}}{\sum_{i=1}^N w_t^{(i)}}, \quad i = 1, \dots, N.$$

For each time t, once the standardised weights are obtained we approximate the posterior density $p(\beta_{1:t} \mid y_{1:t})$ by a weighted sum of Dirac functions (see Sect. 6.7.2 for its definition)

$$\hat{p}(\beta_{1:t} \mid y_{1:t}) = \sum_{i=1}^N \tilde{w}_t^{(i)} \delta(\beta_{1:t} - \hat{\beta}_{1:t}). \tag{6.53}$$

In a sequential application at each time t interest is focused on the state β_t, rather than the entire past of states $\beta_{1:t}$. Hence the posterior (marginal) distribution of β_t given $y_{1:t}$ is approximated as

$$\hat{p}(\beta_t \mid y_{1:t}) = \sum_{i=1}^N \tilde{w}_t^{(i)} \delta(\beta_t - \hat{\beta}_t),$$

where

$$\hat{\beta}_t = \sum_{i=1}^N \tilde{w}_t^{(i)} \beta_t^{(i)}.$$

It follows that from the sample of the states $\beta_t^{(1)}, \dots, \beta_t^{(N)}$ we can obtain any statistics we wish, e.g. its mode, median, quantiles or the empirical distribution.

The above discussion suggests a sequential algorithm: at each point of time t, a sample $\beta_t^{(1)}, \dots, \beta_t^{(N)}$ is drawn from the importance density $g(\beta_t \mid \beta_{t-1}^{(i)}, y_t)$, the weights are computed by (6.52) and then are normalised and finally the posterior distribution of β_t is approximated according to (6.53). However, in application it is usually observed that only a small number of particles have positive weights, which makes for poor Monte Carlo estimation as it is based on a few particles only. To alleviate for this issue researches have proposed a resampling step. The effective sample size, defined as

$$N_{\text{eff}} = \frac{1}{\sum_{i=1}^{N} (\tilde{w}_t^{(i)})^2},$$

is used to decide whether resampling is needed. In one extreme if all particles have equal weight $1/N$, then $N_{\text{eff}} = N$ (in this case no resampling is needed as all weights are positive); in the other extreme if only one particle has weight 1 and the rest are equal to 0, then $N_{\text{eff}} = 1$ (in this case resampling is needed). Hence $1 \leq N_{\text{eff}} \leq N$ and the closer N_{eff} is to N the more particles have non-zero weights and participate in the Monte Carlo estimation of the states. As a result a threshold N_0 may be picked, so that after the calculation of N_{eff} resampling is applied if $N_{\text{eff}} < N_0$; typical values for N_0 include $N_0 = N/2$ or $N_0 = N/3$.

The most common resampling strategy is known as *multinomial resampling* and is briefly described below. Assume that at time t we have sampled the states $\beta_t^{(1)}, \ldots, \beta_t^{(N)}$ and have computed the standardised weights $\tilde{w}_t^{(i)}$. Suppose we have decided to move to a resampling step (assuming $N_{\text{eff}} < N$ as discussed above). First we draw a sample $i_1, i_2, \ldots, i_N$ of size N from the discrete distribution $P(\beta_t = \beta_t^{(i)}) = \tilde{w}_t^{(i)}$ and then we relabel the sample $\beta_t^{(i)} = \beta_t^{(i_j)}$, for $i = 1, 2, \ldots, N$. Finally, the weights are updated to equal weights by $\tilde{w}_t^{(i)} = N^{-1}$ and the algorithm continues to the next time point $t + 1$. There are various other schemes available for resampling, the most popular being residual, stratified and systematic resampling; for more information the reader is referred to the review of Douc et al. (2005) and Chopin and Papaspiliopoulos (2020) and to references therein. The above mentioned sequential algorithm, combining sequential importance sampling with resampling, is known as *sequential Monte Carlo* or *particle filtering*. For the former, Monte Carlo is discussed earlier in Sect. 6.7.1 and its relationship with sequential importance sampling (SIS) discussed in Sect. 6.7.3 is apparent. The latter originates by the signal processing literature where at a given time t a particle is generated $\beta^{(i)}$; each simulated state is seen as a particle and the SIS algorithm in Sect. 6.7.3 proposes the framework of updating the particles over time. Below a summary of the basic particle filter algorithm is given.

Particle Filter Algorithm I (PF-I)

In the state space model (6.3) for each $t = 1, 2, \ldots, n$ the following apply:

1. Simulate N particles $\beta_0^{(1)}, \ldots, \beta_0^{(N)}$ from the prior $p(\beta_0)$.
2. a. For any $t = 1, 2, \ldots, n$ simulate $\beta_t^{(1)}, \ldots, \beta_t^{(N)}$ from the importance function $g(\beta_t \mid \beta_{t-1}^{(i)}, y_t)$.

(continued)

b. Calculate the weights:

$$w_t^{(i)} = \frac{p(y_t \mid \beta_t^{(i)})p(\beta_t^{(i)} \mid \beta_{t-1}^{(i)})}{g(\beta_t^{(i)} \mid \beta_{t-1}^{(i)}, y_t)} w_{t-1}^{(i)}, \quad i = 1, \ldots, N.$$

c. Standardise the weights:

$$\tilde{w}_t^{(i)} = \frac{w_t^{(i)}}{\sum_{i=1}^{N} w_t^{(i)}}, \quad i = 1, \ldots, N.$$

d. Resampling step. Calculate the effective sample size

$$N_{\text{eff}} = \frac{1}{\sum_{i=1}^{N}(\tilde{w}_t^{(i)})^2}.$$

Set threshold $N_0 = N/2$ or $N_0 = N/3$. If $N_{\text{eff}} < N_0$, then resample.
Multinomial resampling: Draw a sample $i_1, i_2, \ldots, i_N$ of size N
from the discrete distribution $P(\beta_t = \beta_t^{(i)}) = \tilde{w}_t^{(i)}$ and then relabel
the sample $\beta_t^{(i)} = \beta_t^{(i_j)}$, for $i = 1, 2, \ldots, N$. Finally, the weights are
updated to equal weights by $\tilde{w}_t^{(i)} = N^{-1}$.

3. Approximate the posterior $p(\beta_t \mid y_{1:t})$ by

$$\hat{p}(\beta_t \mid y_{1:t}) = \sum_{i=1}^{N} \tilde{w}_t^{(i)} \delta(\beta_t - \hat{\beta}_t),$$

where

$$\hat{\beta}_t = \sum_{i=1}^{N} \tilde{w}_t^{(i)} \beta_t^{(i)}.$$

6.7.4   Choice of the Importance Function

The particle filter algorithm of the previous section depends on particles been
generated from the importance density $g(\beta_t \mid \beta_{t-1}^{(i)}, y_t)$. Hence, before the algorithm
may be applied, a choice of this density needs to be made. The main requirement for
the density $g(\cdot)$ is to have the same support as the posterior distribution $p(\beta_t \mid y_{1:t})$.
Since $\beta_t^{(i)}$ is simulated from $g(\cdot)$ it is natural to think that the domains of $g(\cdot)$ and

$p(\cdot)$ must match. Two choices for $g(\cdot)$ stand out: the *suboptimal importance density* and the *optimal importance density* and are described below.

Suboptimal Importance Density (Bootstrap Filter)   This is perhaps the simplest choice for the importance function, according to which $g(\beta_t \mid \beta_{t-1}, y_t) = p(\beta_t \mid \beta_{t-1})$, so that the importance function is just the prior of β_t. With this choice of the importance density, the weights of (6.52) are updated as $w_t^{(i)} = p(y_t \mid \beta_t^{(i)}) w_{t-1}^{(i)}$; the resulting particle filter is known as *bootstrap filter* and is discussed in Gordon et al. (1993) and in Doucet et al. (2001). The advantage of using the bootstrap filter is simplicity, as it is usually easy to simulate from the prior $p(\beta_t \mid \beta_{t-1})$. The disadvantage is that $g(\cdot)$ does not depend on the observation y_t, but only on the states; hence the name *suboptimal importance density*. This choice will be suitable for many non-linear state space models, but it may not be appropriate for highly non-linear systems.

Optimal Importance Density   Out of all possible importance functions that take into account states and observations, there is one which stands out and this is the probability density of β_t, given β_{t-1} and y_t, so that $g(\beta_t \mid \beta_{t-1}, y_t) = p(\beta_t \mid \beta_{t-1}, y_t)$. This importance function, introduced in Zaritskii et al. (1975), is optimal in a sense of minimising the variance of the importance weights over the set of all importance functions. This optimality due to Doucet et al. (2000) is established in the following theorem.

Theorem 6.2 *Conditionally upon $\beta_{t-1}^{(i)}$ and $y_{1:t}$, the importance function $p(\beta_t \mid \beta_{t-1}^{(i)}, y_t)$ minimises the variance of the importance weights $w_t^{(i)}$, in the set of all importance functions $g(\beta_t \mid \beta_{t-1}^{(i)}, y_t)$.*

Proof   We will show that for the importance function $p(\beta_t \mid \beta_{t-1}^{(i)}, y_t)$ the importance weights have zero variance.

Using (6.52) the variance of $w_t^{(i)}$ over all importance functions $g(\beta_t \mid \beta_{t-1}^{(i)}, y_t)$ is

$$
\begin{aligned}
\mathrm{Var}(w_t^{(i)}) &= \mathrm{E}\left(w_t^{(i)2}\right) - \mathrm{E}\left(w_t^{(i)}\right)^2 \\[2mm]
&= \int_A \left(w_t^{(i)}\right)^2 g(\beta_t \mid \beta_{t-1}^{(i)}, y_t)\, d\beta_t - \left[\int_A w_t^{(i)} g(\beta_t \mid \beta_{t-1}^{(i)}, y_t)\, d\beta_t\right]^2 \\[2mm]
&= \left(w_{t-1}^{(i)}\right)^2 \left[\int_A \frac{p(y_t \mid \beta_t)^2 p(\beta_t \mid \beta_{t-1}^{(i)})^2}{g(\beta_t \mid \beta_{t-1}^{(i)}, y_t)}\, d\beta_t - p(y_t \mid \beta_{t-1}^{(i)})^2\right],
\end{aligned}
$$

$$(6.54)$$

where A is the domain of β_t.

If we now choose $g(\beta_t \mid \beta_{t-1}, y_t) = p(\beta_t \mid \beta_{t-1}, y_t)$, then we have

$$\int_A \frac{p(y_t \mid \beta_t)^2 p(\beta_t \mid \beta_{t-1}^{(i)})^2}{p(\beta_t \mid \beta_{t-1}^{(i)}, y_t)} \, d\beta_t = \int_A p(y_t \mid \beta_t) p(\beta_t \mid \beta_{t-1}^{(i)}) p(y_t \mid \beta_{t-1}^{(i)}) \, d\beta_t$$

$$= p(y_t \mid \beta_{t-1}^{(i)})^2$$

and so from (6.54) we have $\mathrm{Var}(w_t^{(i)}) = 0$, for the importance function $p(\beta_t \mid \beta_{t-1}, y_t)$. $\qquad\square$

The optimal importance function discussed above is deployed in Chen and Liu (1996), Doucet et al. (2001), Harvey et al. (2004), Chopin and Papaspiliopoulos (2020) and in references therein. For the optimal importance function to work one needs to be able to draw a sample from it. In many state space models, this is not possible because if we are able to sample from $p(\beta_t \mid \beta_{t-1}, y_t)$ usually we should be able to sample from the posterior $p(\beta_t \mid y_{1:t})$. Between the suboptimal importance function (the prior distribution of β_t) and the optimal importance function (the conditional distribution of β_t, given β_{t-1} and y_t), there are other importance functions that can be used. These functions will be easier to sample than the optimal importance function, and will improve on the suboptimal importance function by incorporating observations y_t in the importance density. Below we briefly describe a popular choice.

Consider the conditionally Gaussian state space model

$$y_t = f(\beta_t) + \epsilon_t \quad \text{and} \quad \beta_t = \mathbf{F}\beta_{t-1} + \zeta_t, \tag{6.55}$$

where $\epsilon_t \sim N(0, \sigma^2)$ and $\zeta_t \sim N(0, \mathbf{Z})$, for some known observation variance σ^2 and some transition covariance matrix $\mathbf{Z}$ and the function $f(\cdot)$ is known. Our objective is to obtain an approximation of the optimal importance function $p(\beta_t \mid \beta_{t-1}, y_t)$.

Following a similar approach as that of Sect. 6.6.1, we use a first order Taylor approximation of $f(\beta_t)$ around the state vector β_{t-1} in order to linearise the state space model (6.55) as

$$y_t \approx x_t^\top \beta_t + \mu_t + \epsilon_t \quad \text{and} \quad \beta_t = \mathbf{F}\beta_{t-1} + \zeta_t,$$

where $\mu_t = f(\mathbf{F}\beta_{t-1}) - [\partial f(\beta_t)/\partial \beta_t]_{\beta_t=\mathbf{F}\beta_{t-1}}^\top \mathbf{F}\beta_{t-1}$ and

$$x_t = \left. \frac{\partial f(\beta_t)}{\partial \beta_t} \right|_{\beta_t=\mathbf{F}\beta_{t-1}}.$$

Note that conditional on β_{t-1}, μ_t is a known function.

Write down the approximate conditional distribution of β_t and y_t, given β_{t-1}, as

$$\begin{bmatrix} \beta_t \\ y_t \end{bmatrix} \Big| \beta_{t-1} \sim N \left\{ \begin{bmatrix} \mathbf{F}\beta_{t-1} \\ x_t^\top \mathbf{F}\beta_{t-1} + \mu_t \end{bmatrix}, \begin{bmatrix} \mathbf{Z} & \mathbf{Z}x_t \\ x_t^\top \mathbf{Z} & x_t^\top \mathbf{Z}x_t + \sigma^2 \end{bmatrix} \right\}.$$

Hence the approximate posterior distribution of β_t, given β_{t-1} and y_t is

$$\beta_t \mid \beta_{t-1}, y_t \sim N \left[\mathbf{F}\beta_{t-1} + \frac{\mathbf{Z}x_t(y_t - x_t^\top \mathbf{F}\beta_{t-1} - \mu_t)}{x_t^\top \mathbf{Z}x_t + \sigma^2}, \mathbf{Z} - \frac{\mathbf{Z}x_t x_t^\top \mathbf{Z}}{\sigma^2} \right].$$

$$(6.56)$$

Considering the conditionally Gaussian model (6.55) we can use the particle filter algorithm PF-I (see p. 296) where the importance function $g(\beta_t \mid \beta_{t-1}, y_t)$ is the Gaussian density (6.56). This importance function is expected to a better choice than the prior $p(\beta_t \mid \beta_{t-1})$ and may well approximate the optimal importance function $p(\beta_t \mid \beta_{t-1}, y_t)$. With the breadth of models incorporated within (6.55), this choice of the importance function is popular, in particular provided that it is easy to simulate the particles from a multivariate normal distribution.

6.7.5   Example 1: Multinomial Time Series

We consider the multinomial dynamic model (6.17)–(6.19) of Sect. 6.2.3. We simulate 100 bivariate states $\beta_t = [\beta_{1t}, \beta_{2t}]^\top$ from a random walk $\beta_t = \beta_{t-1} + \zeta_t$, with $\zeta_t \sim N(0, \mathbf{I})$, which result in 100 simulated probabilities

$$\log \frac{\pi_{1t}}{1 - \pi_{1t} - \pi_{2t}} = \beta_{1t} \quad \text{and} \quad \log \frac{\pi_{2t}}{1 - \pi_{1t} - \pi_{2t}} = \beta_{2t}$$

and $\pi_{3t} = 1 - \pi_{1t} - \pi_{2t}$. These probabilities are used to simulate 100 observation vectors $y_1, \ldots, y_{100}$ from the multinomial distribution (6.17), with size $\lambda_t = 10$. These observations are used in order to estimate the probabilities π_{it} and hence illustrate the performance of the bootstrap filter, for this model when we know the true model.

```
# simulate 100 observations from the model
# Generate random walk
> y <- sim.multinom(100, d=3, size=10, sd=1)
```

We have used the bootstrap filter with 1000 particles and multinomial resampling. Figure 6.1 shows the probabilities π_{it} together with posterior modes of the approximation $\hat{p}(\pi_{it})$, $i = 1, 2, 3$ and $t = 1, \ldots, 100$. We remark that the posterior modes are very close to the original simulated probabilities. This is remarkable, considering the simplicity and speed of the bootstrap filter.

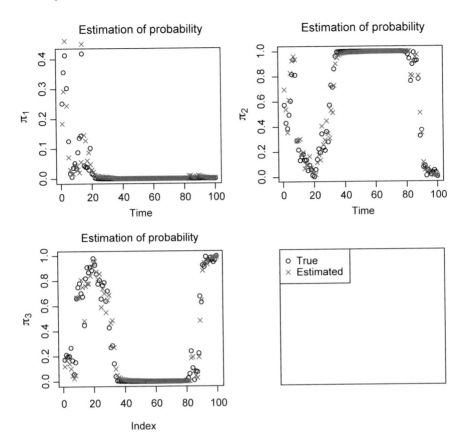

Fig. 6.1 Posterior modes (crosses) of $\hat{p}(\pi_{it})$ in the multinomial dynamic model (6.17)–(6.19), together with the original simulated probabilities (circles)

```
# run the bootstrap filter
> fit <- bts.multinom(y$obs, size=10, N=1000, sd=1)
# extract estimated probabilities
> estimated.probs <- fit$prob
```

The following R code is used to make the plot of Fig. 6.1.

```
# Plot of estimated probabilities against true probabilities
> par(mfrow=c(2,2))
> plot(probabilities[,1],xlab="Time",ylab=expression(pi[1]),
+ main=expression("Estimation of probability"))
> points(estimated.probs[,1],col="red",pch=4) # bootstrap estimates
> plot(probabilities[,2],xlab="Time",ylab=expression(pi[2]),
+ main=expression("Estimation of probability"))
> points(estimated.probs[,2],col="red",pch=4) # bootstrap estimates
> plot(probabilities[,3],ylab=expression(pi[3]),
+ main=expression("Estimation of probability"))
> points(estimated.probs[,3],xlab="Time",col="red",pch=4)
```

```
> plot(1,1,type="n",xaxt="n",yaxt="n",ylab="",xlab="")
> legend("topleft",c("True", "Estimated"),pch=c(1,4),
+ col=c("black","red"))
```

6.7.6 Example 2: Bearings-Only Tracking Revisited

In this section we revisit the bearings-only tracking example discussed in
Sects. 1.3.2 and 6.3. We aim to track a moving target (a ship) in the $x - y$
plane. Observations $z_t = \arctan(y_t/x_t) + \epsilon_t$ are generated by Eq. (1.10), while
the states $\beta_t = (x_t, y_t, \dot{x}_t, \dot{y}_t)^\top$ follow the Markov process (1.11). For full details
and the motivation of this model the reader is referred to Sect. 1.3.2 . We remark
that the bearings-only tracking problem discussed above has notable similarities
to object-tracking, which within the signal processing community has received
considerable attention. This includes single or multiple tracking, spatial tracking and
is closely related to GPS tracking and video processing, with applications to video
surveillance, sport events, forensic science drone and air traffic control, see e.g.
Gordon et al. (1993), Mihaylova et al. (2014) and references therein. For this kind
of problems, Bayesian inference based on simulation has been proven successful
and popular, as is evidenced in Hue et al. (2002), Angelova and Mihaylova (2008),
Andrieu et al. (2010) and in the many references of the overview of Punchihewa
et al. (2018). In the sequel we simulate a simple data set on our bearings-only
tracking problem in order to illustrate the Bootstrap filter discussed earlier.

We simulate two trajectories of the ship from the model of Sect. 1.3.2; we
generate 50 (x_t, y_t) vectors from model (1.10)–(1.11), with

$$\text{Var}(\epsilon_t) = \sigma_\epsilon^2 \quad \text{and} \quad \mathbf{Z} = \text{Var}(\zeta_t) = \begin{bmatrix} 0 & 0 & 0 & 0 \\ 0 & 0 & 0 & 0 \\ 0 & 0 & \sigma_\zeta^2 & 0 \\ 0 & 0 & 0 & \sigma_\zeta^2 \end{bmatrix}.$$

The first trajectory is simulated with $\sigma_\epsilon^2 = 2$ and $\sigma_\zeta^2 = 0.0000001$ and the second
trajectory is simulated with $\sigma_\epsilon^2 = 0.1$ and $\sigma_\zeta^2 = 0.00001$. These settings are similar
to those used in Gilks and Berzuini (2001) and Fearnhead (2002). The R code for
trajectory 1 is

```
# simulate trajectory 1
> z1 <- sim.tracking(50, eta=2, tao=0.0000001)
# fit the model
> fit1 <- bts.tracking(z$z, N=100, eta=2, tao=0.0000001)
# compute the modes
> beta1 <- beta2 <- rep(0, 50)
> for(t in 1:50){ beta1[t] <- Mode(fit1$beta[,1,t])}
> for(t in 1:50){ beta2[t] <- Mode(fit1$beta[,2,t])}
```

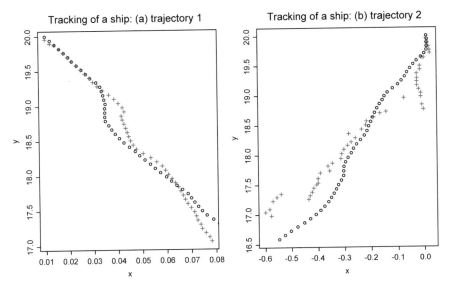

Fig. 6.2 Tracking of a ship. Shown are true positions x_t, y_t (circles) of the ship together with their respective posterior modes (crosses), for each trajectory

The R code for trajectory 2 is

```
# simulate trajectory 2
> z2 <- sim.tracking(50, eta=0.1, tao=0.00001)
> fit2 <- bts.tracking(z4$z, N=100, eta=0.1, tao=0.00001)
# compute the modes
beta3 <- beta4 <- rep(0, 50)
for(t in 1:50){ beta3[t] <- Mode(fit2$beta[,1,t]) }
for(t in 1:50){ beta4[t] <- Mode(fit2$beta[,2,t]) }
```

The bootstrap filter is applied, so that at each time t we simulate 100 particles from the prior $p(\beta_t \mid \beta_{t-1}) \equiv N(\mathbf{F}\beta_{t-1}, \mathbf{Z})$, where $\mathbf{F}$ is the transition matrix of Eq. (1.11). Multinomial resampling is used when the effective sample size is $N_{\text{eff}} < 50/2 = 25$; see Sect. 6.7.4 for a discussion of the bootstrap filter. Figure 6.2 shows the posterior mode of $(x_t, y_t)^\top$ plotted against the true simulated values of $(x_t, y_t)^\top$, for each trajectory. The true values are depicted by a circle, while the estimated values are depicted by a cross. We observe that the estimated values are quite close to the true simulated values, indicating the tracking performance of the particle filter. It is possible to plot empirical densities and credible intervals on the x_t, y_t positions. The R code for the plot of Fig. 6.2 is given below.

```
# plot of x-y trajectories with estimates
> par(mfrow=c(1,2))
>
> plot(z1$x, z1$y, xlab="x", ylab="y",
+ main=expression("Tracking of a ship: (a) trajectory 1"),
+ xlim=c(min(beta1),max(beta1)), ylim=c(min(beta2),max(beta2)) )
```

```
> points(beta1, beta2, col=2, pch=3)
>
> plot(z2$x, z2$y, xlab="x", ylab="y",
+ main=expression("Tracking of a ship: (b) trajectory 2"),
+ xlim=c(min(beta3,z2$x),max(beta3,z2$x)),
+ ylim=c(min(beta4,z2$y),max(beta4,z2$y)) )
> points(beta3, beta4, col=2, pch=3)
```

6.7.7 Example 3: Non-Linear Time Series

In this section we consider filtering from a non-Gaussian and non-linear time series model. This model, which was introduced in Kitagawa (1987), was subsequently used extensively in order to illustrate filtering algorithms for non-Gaussian time series; among other references the reader can find discussions of this model in Carlin et al. (1992), Kitagawa (1998), Godsill et al. (2004) and Andrieu et al. (2010).

The state space model is generated by

$$y_t = \frac{\beta_t^2}{20} + \epsilon_t \quad \text{and} \quad \beta_t = \alpha\beta_{t-1} + \frac{\gamma\beta_{t-1}}{1 + \beta_{t-1}^2} + \delta\cos[1.2(t-1)] + \zeta_t,$$

where ϵ_t and ζ_t independently follow Gaussian distributions, i.e. $\epsilon_t \sim N(0, \sigma^2)$ and $\zeta_t \sim N(0, Z)$; it is also assumed that ϵ_t, ζ_t are independent of the initial state β_0, which is assumed to follow a Gaussian distribution too, i.e. $\beta_0 \sim N(0, 10)$.

This model is conditionally Gaussian as we can see

$$y_t \mid \beta_t \sim N\left(\frac{\beta_t^2}{20}, \sigma^2\right),$$

$$\beta_t \mid \beta_{t-1} \sim N\left[\alpha\beta_{t-1} + \frac{\gamma\beta_{t-1}}{1 + \beta_{t-1}^2} + \delta\cos(1.2t - 1.2), Z\right].$$

Following Godsill et al. (2004) we simulate 50 states $\beta_1, \ldots, \beta_{50}$ and observations $y_1, \ldots, y_{50}$ from this model with $\alpha = 0.5$, $\gamma = 25$, $\delta = 15$, $\sigma^2 = 8$ and $Z = 10$. The bootstrap filter is applied, so that, at each time $t \geq 1$, states $\beta_t^{(i)}$ are simulated from the prior $\beta_t \mid \beta_{t-1}^{(i)} \sim N[0.5\beta_{t-1}^{(i)} + 25\beta_{t-1}^{(i)}(1 + \beta_{t-1}^{(i)2})^{-1} + 15\cos(1.2t - 1.2), 10]$, with $i = 1, \ldots, N$ (here we have used $N = 1000$ particles and resampling is applied if the effective sample size is smaller than 500 particles). Figure 6.3 shows the true simulated states β_t (solid line and solid dots) together with posterior modes (dashed line and crosses), together with 95% posterior credible intervals. We observe that 94% of the simulated states fall within the credible intervals; there are only three points lying outside them: these are states $\beta_5 = -24.823$, $\beta_{45} = -22.953$ and $\beta_{50} = -9.310$. The posterior modes are generally close to the simulated states, with the exception of some points of time towards the

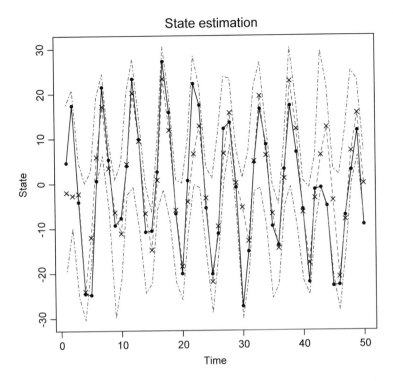

Fig. 6.3 Non-linear state estimation. Shown are the simulated states together with posterior mode and 95% credible intervals

end of the series (times $t = 43, 44, 45$). This fit provides better estimation than the extended Kalman filter of Sect. 6.6.3, and can be further improved if instead of the bootstrap filter, the particles are generated from an approximation of the optimal importance function. Exercise 13 discusses one option of such an approximation.

```
# simulate the data
> obs <- sim.nonlinear(50, alpha=0.5,delta=15, sigma=8, Z=10)
# fit the model
> fit <- bts.nlm.filter(obs$obs,alpha=0.5,delta=15,N=1000,
+ sigma=8, Z=10)
# prepare for plotting
> model <- UQ <- LQ <- rep(0,50)
> for(t in 1:50){
>    model[t] <- Mode(fit1$state[t,])
>.   UQ[t] <- quantile(fit1$state[t,], probs=0.975)
>.   LQ[t] <- quantile(fit1$state[t,], probs=0.025)
> }
# plot the data and the estimates
> ts.plot(ts(obs1$state), ts(model), ts(UQ), ts(LQ),
+ lty=c(1,2,4,4), col=c(1,2,4,4),
+ main=expression("State estimation"), ylab="State")
> points(obs1$state, pch=20)
> points(model, pch=4)
```

6.7.8   Static Parameter Estimation

6.7.8.1   Introduction and Initial Studies

The basic particle filter algorithm (PF-I algorithm on p. 296) assumes that all unknown parameters subject to estimation are placed into the state vector β_t, which is time-dependent. Hence the algorithm cannot treat static parameters, such as variances or other time-invariant hyperparameters. Let θ denote the vector of such static hyperparameters. In order to allow inference for θ, one possibility is to introduce a "static" evolution $\theta_t = \theta_{t-1}$, with $\theta_1 = \theta$ and to incorporate θ_t into the state vector β_t. This has the flaw that sampling from θ_t is essentially the same as sampling from θ_1, hence we do not learn as time increases. For example observe that sampling $\theta_t^{(i)}$ from the importance function $p(\theta_t \mid \theta_{t-1}^{(i)}, y_t)$ implies that the sampled value of $\theta_t^{(i)}$ is necessarily equal to $\theta_{t-1}^{(i)}$, for all t (as $\theta_t = \theta_{t-1}$), hence $\theta_t^{(i)} = \theta_1^{(i)}$ (essentially we sample θ from the prior).

In order to overcome this problem Gordon et al. (1993) propose that θ_t follows an artificial evolution, such that

$$\theta_t = \theta_{t-1} + \eta_t, \qquad\qquad (6.57)$$

where $\eta_t \sim N(0, \mathbf{W}_t)$, for some covariance matrix $\mathbf{W}_t$. As before, θ_t is incorporated in β_t. In this setting at time t we are sampling from θ_t and hence the above problem is overcome. However, we have now introduced an artificial evolution of the static parameters θ. In Gordon et al. (1993) it is proposed that a small covariance matrix of η_t will result in a slow evolution for θ_t, i.e. $\theta_t \approx \theta_{t-1}$. Hence, we may set $\mathbf{W}_t = c\mathbf{I}$, where the constant c should be close to 0 and can be set using discount factors (see e.g. the discussion in Sect. 4.3.2). Nevertheless, this approach has the disadvantage that we are choosing to model static parameters as slowly varying time-varying parameters.

Storvik (2002), considering this problem, proposes that θ is sampled by a conjugate Bayesian analysis and sufficient statistics. This algorithm is somewhat limited in the sense that it requires the existence of sufficient statistics. Gilks and Berzuini (2001) and Fearnhead (2002) discuss MCMC-based particle filter algorithms with leading application the bearings-only tracking problem. Here, we describe the Liu and West filter (Liu & West, 2001), which is simpler and faster than the above MCMC algorithms and is a general-purpose algorithm dealing with the static-parameter problem discussed above.

6.7.8.2   Liu and West Particle Filter

Consider first the case of known static parameters θ. The algorithm makes use of the so-called *auxiliary particle filter*, proposed by Pitt and Shephard (1999). Next we describe the auxiliary particle filter. In the general model (6.3) suppose that at time $t - 1$ the posterior $p(\beta_{t-1} \mid y_{1:t-1})$ is approximated by $\hat{p}(\beta_{t-1} \mid y_{1:t-1}) = \sum_{i=1}^{N} \tilde{w}_{t-1}^{(i)} \delta(\beta_{t-1} - \hat{\beta})$, where $\hat{\beta}_{t-1} = N^{-1} \sum_{i=1}^{n} \tilde{w}_{t-1}^{(i)} \beta_{t-1}^{(i)}$ is the Monte Carlo mean. With the availability of information y_t, the posterior of β_t is approximated as

$$p(\beta_t \mid y_{1:t}) \propto p(y_t \mid \beta_t) p(\beta_t \mid y_{1:t-1})$$

$$= p(y_t \mid \beta_t) \int_A p(\beta_t \mid \beta_{t-1}) p(\beta_{t-1} \mid y_{1:t-1}) \, d\beta_{t-1}$$

$$\approx \sum_{i=1}^{N} \tilde{w}_{t-1}^{(i)} p(\beta_t \mid \beta_{t-1}^{(i)}) p(y_t \mid \beta_t),$$

where A is the domain of β_t.

For each $i = -1, 2, \ldots, N$ we select $\mu_t^{(i)}$ a prior estimate of β_t; this can be the prior mode or prior mean of β_t, given $\beta_{t-1}^{(i)}$. If $\mu_t^{(i)}$ is a good estimate of β_t, then the weight

$$g_t^{(i)} \propto w_{t-1}^{(i)} p(y_t \mid \mu_t^{(i)})$$

should be large, suggesting that $\beta = \mu_t^{(i)}$ is consistent with the datum y_t. The algorithm proceeds by first sampling an indicator variable j with probability proportional to $g_t^{(i)}$ and then sampling a state $\beta_t^{(j)}$ from the prior $p(\beta_t \mid \beta_{t-1}^{(j)})$. Based on these *auxiliary* variables a new weight is computed as

$$w_t^{(j)} = \frac{p(y_t \mid \beta_t^{(j)})}{p(y_t \mid \mu_t^{(j)})}$$

and the algorithm proceeds to the next j. This creates a set of simulated states $\beta_t^{(j_1)}, \ldots, \beta_t^{(j_N)}$. More details of the algorithm are to be found in Pitt and Shephard (1999) and in the many articles which cite it.

Considering now the static parameters θ, let $p(\theta)$ denote the prior distribution of θ. In what follows we shall assume that θ is scalar, but the extension of θ being a vector is trivial. We shall be interested in the approximation of the joint posterior distribution of $(\beta_t, \theta)^\top$. The prior distribution of the states and the likelihood are conditional on θ. The joint posterior of β_t and θ at time t is

$$p(\beta_t, \theta \mid y_{1:t}) \propto p(y_t \mid \beta_t, \theta) p(\beta_t, \theta \mid y_{1:t-1})$$

$$= p(y_t \mid \beta_t, \theta) p(\beta_t \mid \theta, y_{1:t-1}) p(\theta \mid y_{1:t-1}).$$

Hence we have to deal with the posterior of θ at time $t - 1$.

Liu and West (2001) consider first Kernel density estimation for the posterior of θ given $y_{1:t-1}$. At time $t - 1$ and with information $y_{1:t-1}$, suppose we have obtained Monte Carlo samples $\theta_t^{(t-1,j)}$ with weights $w_{t-1}^{(j)}$ which approximate the density

$p(\theta \mid y_{1:t-1}$' the superscript $t-1$ is included to make explicit the dependence of $\theta^{(j)}$ to $t-1$. Write $\bar{\theta}$ and $\mathbf{V}_{t-1}$ the Monte Carlo vector mean and covariance matrix of θ given $y_{1:t-1}$. The smooth Kernel density we consider here is a mixture of normal distributions with mixing weights $w_{t-1}^{(j)}$ so that

$$p(\theta \mid y_{1:t-1}) \approx \sum_{j=1}^{N} w_{t-1}^{(j)} f_{\theta}(m_{t-1}^{(j)}, h^2 \mathbf{V}_{t-1}), \tag{6.58}$$

where $f_{\theta}(m, \mathbf{V})$ denotes the density of a multivariate normal distribution with mean vector m and covariance matrix $\mathbf{V}$.

In order to specify m_{t-1} one idea is to set $m_{t-1}^{(j)} = \theta^{(t-1,j)}$, i.e. to centre the location of f_{θ} around the simulated value $\theta^{(t-1,j)}$. However, this has the disadvantage of an over-dispersed density, with $\mathrm{Var}(\theta \mid y_{1:t-1}) = (h^2 + 1)\mathbf{V}_{t-1}$. Indeed

$$\mathrm{E}(\theta \mid y_{1:t-1}) = \sum_{j=1}^{N} w_{t-1}^{(j)} m_{t-1}^{(j)} = \sum_{j=1}^{N} w_{t-1}^{(j)} \theta^{(t-1,j)} = m_{t-1}$$

and

$$\mathrm{Var}(\theta \mid y_{1:t-1}) = \sum_{j=1}^{N} w_{t-1}^{(j)} \left[\mathrm{Var}(\theta \mid m_{t-1}^{(j)}, h^2 \mathbf{V}_{t-1}) + \theta^{(t-1,j)}(\theta^{(t-1,j)})^{\mathsf{T}} \right.$$

$$\left. - \mathrm{E}(\theta \mid y_{1:t-1})\mathrm{E}(\theta \mid y_{1:t-1})^{\mathsf{T}} \right]$$

$$= \sum_{j=1}^{N} w_{t-1}^{(j)} h^2 \mathbf{V}_{t-1} + \sum_{j=1}^{N} w_{t-1}^{(j)} (\theta^{(t-1,j)}(\theta^{(t-1,j)})^{\mathsf{T}} - m_{t-1} m_{t-1}^{\mathsf{T}})$$

$$= (h^2 + 1)\mathbf{V}_{t-1}.$$

In order to resolve this problem Liu and West propose replacing $m_{t-1}^{(j)} = \theta^{(t-1,j)}$ by

$$m_{t-1}^{(j)} = \alpha \theta^{(t-1,j)} + (1-\alpha)\bar{\theta}, \tag{6.59}$$

where $\alpha = \sqrt{1-h^2}$.

To see this note that

$$\mathrm{E}(\theta \mid y_{1:t-1}) = \sum_{j=1}^{N} w_{t-1}^{(j)} m_{t-1}^{(j)} = \alpha \sum_{j=1}^{N} w_{t-1}^{(j)} \theta^{(t-1,j)} + (1-\alpha) \sum_{j=1}^{N} w_{t-1}^{(j)} \bar{\theta} = \bar{\theta}$$

and

$$\text{Var}(\theta \mid y_{t-1}) = \sum_{j=1}^{N} w_{t-1}^{(j)} h^2 \mathbf{V}_{t-1} + \sum_{j=1}^{N} w_{t-1}^{(j)} [m_{t-1}^{(j)} (m_{t-1}^{(j)})^{\top} - \bar{\theta}\bar{\theta}^{\top}]$$

$$= h^2 \mathbf{V}_{t-1} + \sum_{j=1}^{N} w_{t-1}^{(j)} [\alpha \theta^{(t-1,j)} + (1-\alpha)\bar{\theta}][\alpha \theta^{(t-1,j)} + (1-\alpha)\bar{\theta}]^{\top}$$

$$-\bar{\theta}\bar{\theta}^{\top}$$

$$= (\alpha^2 + h^2) \mathbf{V}_{t-1} = \mathbf{V}_{t-1},$$

using $\alpha^2 + h^2 = 1$.

Hence, with $m_{t-1}^{(j)}$ as in (6.59) the Monte Carlo mean vector $\bar{\theta}$ and covariance matrix $\mathbf{V}_{t-1}$ are preserved in the mixture density. The kernel location $m_{t-1}^{(j)}$ is a exponentially weighted average of the simulated $\theta^{(t-1,j)}$ and the Monte Carlo mean $\bar{\theta}$; the smoothing parameter h controls how close $m_{t-1}^{(j)}$ is to $\bar{\theta}$ and can be chosen using discount factors.

In order to proceed to the posterior $p(\theta \mid y_{1:t})$ Liu and West adopt the artificial evolution (6.57), but modify it appropriately in order to deal with the loss of information incurred by the covariance of η_t. First notice that from the evolution (6.57) the Monte Carlo approximation of the density $p(\theta_t \mid y_{1:t-1})$ is a kernel density

$$p(\theta_t \mid y_{1:t-1}) \approx \sum_{j=1}^{N} w_{t-1}^{(j)} f_{\theta_t}(\theta_{t-1}^{(j)}, \mathbf{W}_t)$$

where $f_{\theta_t}(m, \mathbf{V})$ denotes a Gaussian density with mean vector m and covariance matrix $\mathbf{V}$ and $\theta_{t-1}^{(j)}$ is the sample obtained at time $t-1$ from the density $p(\theta_{t-1} \mid y_{1:t-1})$.

The above-mentioned loss of information is the result of (6.57) and with the usual assumption that θ_{t-1} and η_t are independent, $\text{Var}(\theta_t \mid y_{1:t-1}) = \text{Var}(\theta_{t-1} \mid y_{t-1}) + \mathbf{W}_t$, hence there is a loss of information coming from $ti1$ to t with information $y_{1:t-1}$. This loss of information is depicted by the increased variance $\text{Var}(\theta_t \mid y_{1:t-1}) \geq \text{Var}(\theta_{t-1} \mid y_{t-1})$ and is quantified by the covariance matrix $\mathbf{W}_t$. Hence $\mathbf{W}_t = \mathbf{0}$ corresponds to a static $\theta_t = \theta_{t-1} = \theta$, as $\eta_t = 0$ in (6.57), with probability 1. However, as we wish to keep the evolution (6.57) so that we can update the particle filter from one time to another, we can modify the assumption of independence between θ_{t-1} and η_t in order to cater for the time-invariance of θ. Let us assume that θ_{t-1} and η_t have a non-zero covariance $\text{Cov}(\theta_{t-1}, \eta_t)$; we can set this covariance so that $\text{Var}(\theta_t \mid y_{1:t-1}) = \text{Var}(\theta_{t-1} \mid y_{1:t-1})$ in order to accommodate for the loss of information described above. Indeed

$$\text{Var}(\theta_t \mid y_{1:t-1}) = \text{Var}(\theta_{t-1} \mid y_{1:t-1}) + \mathbf{W}_t + 2\text{Cov}(\theta_{t-1}, \eta_t)$$

Hence, the specification $\text{Cov}(\theta_{t-1}, \eta_t) = -2^{-1}\mathbf{W}_t$ ensures the property $\text{Var}(\theta_t \mid y_{1:t-1}) = \text{Var}(\theta_{t-1} \mid y_{1:t-1}) = \mathbf{V}_{t-1}$ required for the time-invariance of the parameter vector θ. With this in place we can write down the join distribution of θ_t, given θ_{t-1} and $y_{1:t-1}$ as

$$\begin{bmatrix} \theta_t \\ \theta_{t-1} \end{bmatrix} \mid y_{1:t-1} \sim N \left\{ \begin{bmatrix} \bar{\theta} \\ \bar{\theta} \end{bmatrix}, \begin{bmatrix} \mathbf{V}_{t-1} & \mathbf{A}_t \mathbf{V}_{t-1} \\ \mathbf{V}_{t-1} \mathbf{A}_t^\top & \mathbf{V}_{t-1} \end{bmatrix} \right\},$$

where $\bar{\theta}$ and $\mathbf{V}_{t-1}$ are the Monte Carlo mean vector and covariance matrix of θ_{t-1} and from the covariance $\mathbf{A}_t \mathbf{V}_{t-1} = \text{Cov}(\theta_t, \theta_{t-1} \mid y_{1:t-1}) = \text{Cov}(\theta_{t-1} + \eta_t, \theta_{t-1} \mid y_{1:t-1}) = \mathbf{V}_{t-1} - \frac{1}{2}\mathbf{W}_t$, the matrix $\mathbf{A}_t$ is determined as $\mathbf{A}_t = \mathbf{I} - \frac{1}{2}\mathbf{W}_t \mathbf{V}_{t-1}^{-1}$. From the above joint distribution of θ_t and θ_{t-1}, we deduce that the condition distribution of θ_t, given θ_{t-1} is

$$\theta_t \mid \theta_{t-1} \sim N(\bar{\theta} + \mathbf{A}_t(\theta_t - \bar{\theta}), \mathbf{V}_{t-1} - \mathbf{A}_t \mathbf{V}_{t-1} \mathbf{A}_t^\top)$$

Consequently, we can choose a discount or forgetting factor δ in order to specify $\mathbf{W}_t$, which is then used to propose a specification for $\mathbf{A}_t$. With the discounting approach discussed in detail in Sect. 4.3.2, we can set $\mathbf{W}_t = (1-\delta)\delta^{-1}\mathbf{V}_{t-1}$, which results in

$$\mathbf{A}_t = \mathbf{I} - \delta^{-1}(1-\delta)\mathbf{V}_{t-1}\mathbf{V}_{t-1}^{-1}/2 = \frac{3\delta - 1}{2\delta}\mathbf{I} = \alpha \mathbf{I},$$

Hence the shrinkage parameter h in Eq. (6.58) is determined by

$$h^2 = 1 - \alpha^2 = 1 - \left(\frac{3\delta - 1}{2\delta}\right)^2.$$

With these equations in place the conditional distribution is simplified to

$$\theta_t \mid \theta_{t-1} \sim N(\alpha\theta_{t-1} + (1-\alpha)\bar{\theta}, h^2\mathbf{V}_{t-1}), \tag{6.60}$$

which is used to approximate the mixture (6.58). Assuming that at time $t-1$ we have a sample $\theta_{t-1}^{(j)}$, then $\theta_t \mid \theta_{t-1}^{(j)} \sim N(m_{t-1}^{(j)}, h^2\mathbf{V}_{t-1})$ is the Gaussian density $f_\theta(m_{t-1}^{(j)}, h^2\mathbf{V}_{t-1})$ in the kernel density (6.58). Hence, we have used the artificial evolution (6.57) employed with the covariance structure described above in order to obtain the conditional distribution (6.60) and update the sample of θ from time $t-1$ to time t. To the following we give a summary of Liu and West algorithm.

Particle Filter Algorithm II (PF-II)
In the state space model (6.3) for each $t = 1, 2, \ldots, n$ the following apply:

1. Simulate N particles $(\beta_0^{(1)}, \theta_0^{(1)}, \ldots, \beta_0^{(N)}, \theta_0^{(N)})$ from the prior $p(\beta_0, \theta)$. This might be facilitated by simulating $\theta_0^{(j)} = \theta^{(j)}$ from the prior $p(\theta)$ and $\beta_0^{(j)}$ from $p(\beta_0 \mid \theta_0^{(j)})$.

2. a. For each $i = 1, 2, \ldots, N$ calculate prior point estimates $\mu_t^{(j)}$ of β_t and $m_t^{(j)}$ of θ as

$$\mu_t^{(j)} = \mathrm{E}(\beta_t \mid \beta_{t-1}^{(j)}, \theta_{t-1}^{(j)}) \quad \text{and} \quad m_t^{(j)} = \alpha \theta_{t-1}^{(j)} + (1 - \alpha)\bar{\theta}_{t-1},$$

   where α is the smoothing parameter and $\bar{\theta}_{t-1}$ is the Monte Carlo mean of $\theta_{t-1}^{(1)}, \ldots, \theta_{t-1}^{(N)}$.
   b. Sample an auxiliary index variable k from the set $\{1, 2, \ldots, N\}$, with probability proportional to

$$g_t^{(k)} \propto w_{t-1}^{(k)} p(y_t \mid \mu_t^{(k)}, m_{t-1}^{(k)}).$$

   c. Sample a new parameter vector $\theta_t^{(k)}$ from the k-th component of the mixture,

$$\theta_t^{(k)} \sim N(m_{t-1}^{(k)}, h^2 \mathbf{V}_{t-1})$$

   where $h^2 = 1 - \alpha^2$.
   d. Sample a single state vector $\beta_t^{(k)}$ from the state distribution

$$p(\beta_t \mid \beta_{t-1}^{(k)}, \theta_t^{(k)}).$$

   e. Compute the corresponding weight

$$w_t^{(k)} = \frac{p(y_t \mid \beta_t^{(k)}, \theta_t^{(k)})}{p(y_t \mid \mu_t^{(k)}, m_t^{(k)})}.$$

   f. Repeat Steps (b)-(e) to obtain a set of posterior approximations $(\beta_t^{(j)}, \theta_t^{(j)})$, for $i = 1, 2, \ldots, N$.

6.7.9   Case Study: Analysis of Asthma Data

In this section we discuss in some detail a case study, reported in Triantafyllopoulos et al. (2019), which illustrates the use and utility of the models and methods described in the previous sections. We consider data consisting of daily medical contacts for schoolchildren aged between 5 and 16 years old who suffered from asthma over a seven-year period between 1999 and 2005 in England. This data, reported in Julious et al. (2011), are depicted in Fig. 6.4 (top panel). The lower panel of this figure shows weekly counts of medical contacts and this is the primary data set we consider in this section. The main reason for this aggregation is to account for the weekend effect. A primary interest related to these data involves short-term forecasting of the count of asthma patients. This can provide vital input in hospital bed availability and requirements as well as hospital staff availability and planning of resources.

Figure 6.4 suggests that the weekly data appear to be a non-stationary time series. There appears to be some evidence of seasonality, but this is not persistent and modelling it in the dynamic model did not provide an improvement. We consider

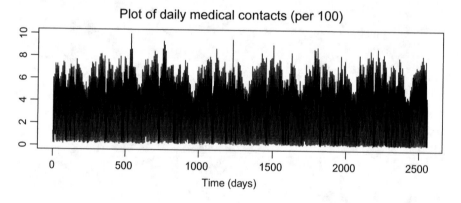

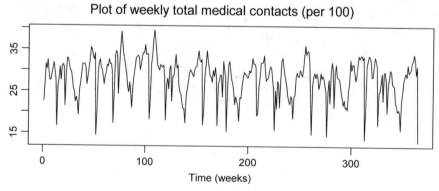

Fig. 6.4 Daily and weekly total medical contacts for asthmatic children

the Poisson and the negative binomial dynamic models, discussed in Sect. 6.2.2. The Poisson model consists of observation (6.8) and transition (6.10), i.e.

$$y_t \mid \beta_t \sim \text{Poisson}[\exp(\beta_t)] \quad \text{and} \quad \beta_t = \beta_{t-1} + \zeta_t,$$

where $\zeta_t \sim N(0, Z)$. Here the rate of the Poisson distribution is $\lambda_t = \exp(\beta_t)$ (the canonical logarithmic link is used) and the static hyperparameter of the model is the state variance Z. The random walk evolution of the states is motivated by weak evidence of stationarity, supported from Fig. 6.4 and from autocorrelation plots, not reported here.

The negative binomial dynamic model consists of observation (6.11) and transition (6.13), i.e.

$$y_t \mid \beta_t \sim \text{NegBinomial}\left[\lambda_t = \lambda, \frac{\exp(\beta_t)}{1 + \exp(\beta_t)}\right] \quad \text{and} \quad \beta_t = \beta_{t-1} + \zeta_t,$$

where the logarithmic link is used, the probability of success is $\pi_t = \exp(\beta_t)[1 + \exp(\beta_t)]^{-1}$ and the static parameters are λ and Z. More details about these models are provided in Sect. 6.2.2.

The Liu and West filter (hereinafter LW) discussed in Sect. 6.7.8 is applied to the asthma weekly time series data with the proposed models. We have used throughout $N = 1000$ particles and a high discount factor $\delta = 0.995$, which corresponds to $\alpha \approx 0.997$ and $h \approx 0.071$. Liu and West (2001) discuss Gaussian mixtures for the prior of each of the hyperparameters (see also Sect. 6.7.8). For parameters where support is not the real line these authors make use of transformations to map the support of these parameters to the real line, e.g. for the variance Z, one can work with $\log Z$. In this section we consider a gamma prior for Z (for both Poisson and negative binomial), i.e. $Z \sim G(2, 0.1)$. For the size λ of the negative binomial we consider three possibilities (a) a gamma prior $\lambda \sim G(2, 0.1)$, (b) a uniform prior $\lambda \sim U(0, 50)$ and (c) a uniform prior for λ^{-1}, $\lambda^{-1} \sim U(0, 1)$. In (a) the gamma prior is unbounded from above to allow large values of λ; moreover, this gamma prior is a weakly informative prior with prior mode 10 and prior variance 200. In (b) the non-informative prior is bounded above, but a large value 50 is chosen; here the prior mean is 25 and the variance is 208.33. In (c) the non-informative prior for λ^{-1} gives the Poisson model when $\lambda^{-1} = 0$ in the boundary.

Figure 6.5 exhibits the final 105 observations of the real data together with one-week-ahead forecasting by using the LW algorithm for the Poisson and the negative binomial models. The LW with Poisson model gives a number of forecasts closer to the real data than does the negative binomial, but for some observations the negative binomial model outperforms the Poisson model.

For the three negative binomial models, Fig. 6.6 shows the estimates of the parameters λ and Z (here all data points considered 1–365). We see that all three priors for λ considered here produce estimates of λ in the bound $(0, 2.5)$, for all t. This indicates that forecasts generated by the negative binomial model have left skewed distributions with the variance being larger than the mean. We observe from

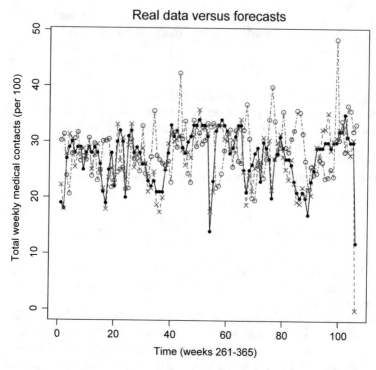

Fig. 6.5 Real data (black solid line with solid points) against one step ahead forecasts for the Poisson model (red dashed line with cross), and negative binomial model (blue dotted line with circle)

Fig. 6.6 that the state variance Z (gamma prior model) is more stable after about 200 time points compared to the other two models. Moreover, the credible bounds of Z (gamma prior model) are the most narrow with their values not exceeding the value of 5 and be consistently less than 4 after 200 time points. Figure 6.7 shows posterior mode and credible bounds of the state variance Z under the Poisson model. We observe that after 200 time points the mode is quite stable and the credible bounds do not exceed the value of 2. The low estimated values of the dispersion parameter λ in Fig. 6.6 put forward the negative binomial model and provide evidence against the Poisson (we would expect λ to be large to favour the Poisson). A close look at Fig. 6.5 reveals that at the start of the series some of the negative binomial forecasts are poor, while towards the end the negative binomial provides some impressive forecasts ($t = 105$).

Finally, Fig. 6.8 shows histograms of the count for both the Poisson and the negative binomial models. The histograms are picked at three points of time ($t = 8, 70, 105$) to reflect on the performance of the two models at different times; plotted are the true observations (vertical lines). We remark that for some points of time the Poisson model is better (e.g. for $t = 8$, corresponding to week 267) and at some

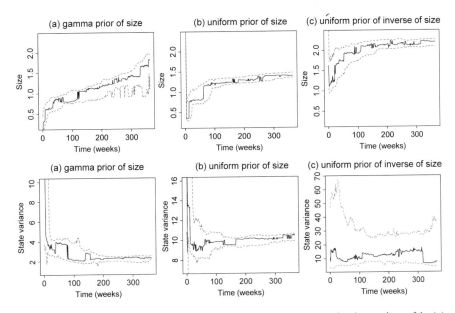

Fig. 6.6 Posterior estimates of the size λ and state variance Z under the three priors of λ: **(a)** gamma prior, **(b)** uniform prior and **(c)** uniform prior for λ^{-1}. Shown are posterior modes with 95% credible bounds

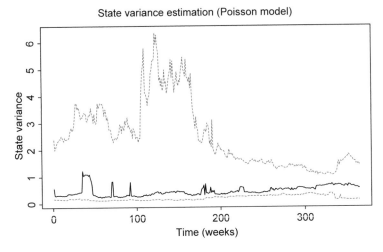

Fig. 6.7 Posterior estimates of the state variance Z using the Poisson model; shown are the posterior mode and 95% credible bounds

points the negative binomial model is better (e.g. for $t = 105$, corresponding to week 365). Both models are reasonable and provide good forecast performance, but there is little support for symmetric histograms for the data (we split the data in time-intervals of length 40 and we found that they were skewed).

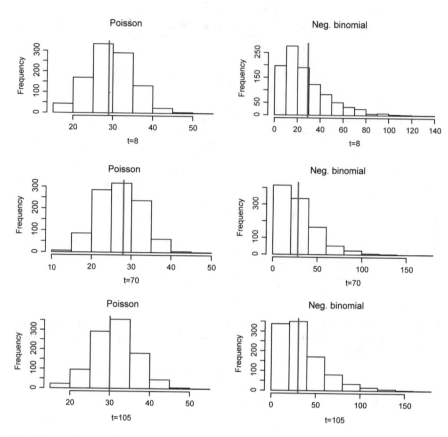

Fig. 6.8 Empirical predictive densities of the Poisson model (left panel) and the negative binomial model (right panel). Densities are plotted at three points of time $t = 8, 70, 105$ and the observed counts are depicted by the vertical lines

6.8   Markov Chain Monte Carlo Inference

There are a number of Markov chain Monte Carlo (MCMC) procedures aimed at Bayesian inference of non-linear and non-Gaussian state space models. These procedures tend to be model-specific and there is not a general procedure. For example, Carlin et al. (1992) propose a Gibbs sampling algorithm for the model of Sect. 6.7.7. For the class of dynamic generalised linear models, which includes a wide number of popular models (Sect. 6.2), Gamerman (1998) develops MCMC procedure based on Metropolis-Hastings and Gibbs sampling; this procedure is discussed in Gamerman and Lopes (2006) and is described below.

6.8.1 *Metropolis-Hastings Algorithm*

In this section we describe the basic notion of Metropolis-Hastings MCMC estimation procedure. Consider the problem of sampling from a target distribution with density $\pi(\theta)$ of some random vector θ; in Bayesian inference this typically will be a posterior distribution. Assuming this is a complicated distribution to sample from MCMC proposes sampling from a Marko chain whose stationary distribution is $\pi(\cdot)$; see Sect. 5.7 for a discussion of the basic notion of MCMC and for a discussion of the Gibbs sampler.

The basic idea of Metropolis-Hastings algorithms is to use a kernel density to simulate states of a Markov chain, instead of sampling from the target density $\pi(\cdot)$, which is complicated. To the following we discuss how this kernel may be chosen. Let $p(\theta, \phi)$ be a kernel so that it defines a Markov chain whose stationary distribution is $\pi(\theta)$. One way to achieve this is by adopting the condition

$$\pi(\theta)p(\theta, \phi) = \pi(\phi)p(\phi, \theta),$$

for all θ and ϕ. This condition defines a reversible chain whose stationary distribution is $\pi(\theta)$; for details see Gamerman and Lopes (2006, Section 4.6).

The density kernel $p(\theta, \phi)$ consists of a transition kernel $q(\theta, \phi)$ and of a probability $\alpha(\theta, \phi)$ so that

$$p(\theta, \phi) = q(\theta, \phi)\alpha(\theta, \phi). \tag{6.61}$$

There is a positive probability for the chain to remain at the current value θ

$$p(\theta, \theta) = P(\text{chain remains at current value } \theta)$$
$$= 1 - P(\text{chain moves from } \theta \text{ to } \phi, \quad \phi \neq \theta)$$
$$= 1 - \int_A q(\theta, \phi)\, d\phi > 0,$$

where A is the domain of θ.

Equation (6.61) defines a density kernel, which describes the distribution of moving the chain from the current value θ to ϕ. Specifically, for any subset B of A we have

$$p(\theta, \phi \in B) = \begin{cases} 1 - \int_A q(\theta, \phi)\, d\phi + \int_B q(\theta, \phi)\, d\phi, & \theta \in B \\ \int_B q(\theta, \phi)\, d\phi, & \theta \notin B \end{cases} \tag{6.62}$$

Hastings proposed to define $\alpha(\theta, \phi)$ as

$$\alpha(\theta, \phi) = \min\left\{1, \frac{\pi(\phi)q(\phi, \theta)}{\pi(\theta)q(\theta, \phi)}\right\} \tag{6.63}$$

so that the transition kernel $q(\theta, \phi)$ defines a reversible chain. Algorithms adopting Eqs. (6.62)–(6.63) are known as Metropolis-Hastings algorithms, hereinafter referred to as M-H. The basic M-H algorithm is described below:

1. Set an initial value for $\theta^{(0)}$. Set the counter $j = 1$.
2. For each $j = 1, 2, \ldots, N$

   a. Generate a proposal state ϕ^* from the density kernel $q(\theta^{(j-1)}, \phi)$.
   b. Evaluate the acceptance probability $\alpha(\theta^{(j-1)}, \phi^*)$, using (6.63).
   c. Draw a single value u from a Uniform distribution $U(0, 1)$. If $u < \alpha(\theta^{(j-1)}, \phi^*)$, then the proposal ϕ^* is accepted and we set $\theta^{(j)} = \phi^*$. If $u \geq \alpha(\theta^{(j-1)}, \phi^*)$, the proposal ϕ^* is rejected and the chain remains at state $\theta^{(j-1)}$; in this case we set $\theta^{(j)} = \theta^{(j-1)}$.

It can be shown that as $N \to \infty$ the chain is simulated from the stationary distribution $\pi(\theta)$. The algorithm was first proposed by Metropolis et al. (1953) and further extended by Hastings (1970), hence its name as Metropolis-Hastings algorithm.

For the application of this algorithm the kernel $q(\theta, \phi)$ has to be chosen. A typical choice suggests that q is symmetric, so that $q(\theta, \phi) = q(\phi, \theta)$. This choice, initially proposed by Metropolis et al. (1953), (6.63), simplifies to

$$\alpha(\theta, \phi) = \min\left\{1, \frac{\pi(\phi)}{\pi(\theta)}\right\}, \qquad (6.64)$$

which does not depend on q. Equation (6.64) offers a simple interpretation. Suppose that the chain is at state $\theta^{(j-1)}$, for some iteration $j - 1$. A new proposal ϕ^* is generated from q, according to the algorithm above. If $\pi(\phi^*)$ is small in comparison to $\pi(\theta^{(j-1)})$, the move is rejected and the acceptance probability α, which is the ratio $\pi(\phi^*)/\pi(\theta^{(j-1)})$, should be small. This indicates that $\phi*$ comes from an area of low probability in $\pi(\cdot)$ and $\theta^{(j-1)}$ is more plausible value for the chain. If the ratio $\pi(\phi^*)/\pi(\theta^{(j-1)})$ is large, then ϕ^* is accepted and the chain moves from $\theta^{(j-1)}$ to ϕ^* (the proposal ϕ^* is accepted).

In order to apply the algorithm a kernel $q(\theta, \phi)$ has to be chosen. This has to be a density which is easy to simulate from. The most common choice, which was originally proposed in Metropolis et al. (1953), is the random walk choice. According to this ϕ^* is generated from $\phi^* = \theta^{(j)} = \theta^{(j-1)} + w_j$, where w_j is a random variable following a normal distribution $N(0, V)$, for some variance V (other symmetric distributions such a Student t may be used). In other words ϕ^* is generated from an $N(\theta^{(j-1)}, V)$. The variance V is crucial in the application of the algorithm. A large value of V will result in large transition proposals from $\theta^{(j-1)}$ to ϕ^* and these are likely to generate very low acceptance rates. This can cause delays of the algorithm and even convergence problems of the chain. If the variance V is small, then the value of ϕ^* is close to $\theta^{(j-1)}$, which is likely to result in high acceptance rates. This in turn may cause delays as the chain moves very slowly and will require a large number of iterations to achieve convergence. There is not an optimal acceptance rate, but several authors have reported that rates in the range 10–15% work best. Hence after some experimentation the variance V can be chosen to achieve a desirable acceptance rate. So far we have assumed that θ is scalar. If θ is a vector, the above discussed are valid with small modifications.

Next we give a toy example to illustrate the M-H algorithm. Suppose we wish to simulate a sample from a gamma distribution $\theta \sim G(2, 3)$, or from target density $\pi(\theta) = 4.5\theta \exp(-3\theta)$, for $\theta > 0$. We remark that M-H algorithm should only be used for distributions that direct sampling is not available, but in this case we use this example for illustration.

We apply the M-H algorithm with arbitrary initial value $\theta^{(0)} = 3$, using a random walk kernel, with $V = 1$. We have used $N = 10{,}000$ iterations and the first 1000 iterations are used for training or burn in. Figure 6.9 shows the trace plot (the plot of the simulated values of $\theta^{(j)}$, for $j = 1, 2, \ldots, 10{,}000$) and the histogram of $\theta^{(j)}$, $j = 1001, 1002, , \ldots, 10{,}000$. The first 1000 values $\theta^{(j)}$, depicted in the top panel of the figure by the vertical line, are used for training the chain and removed from the histogram in the lower panel. The gamma density is plotted and we remark that

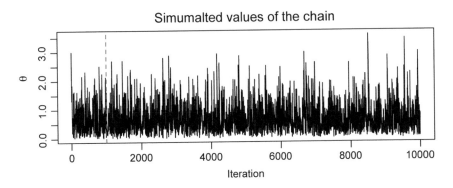

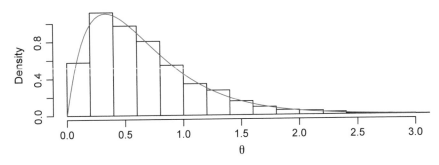

Fig. 6.9 Trace plot (top panel) and histogram (lower panel) of simulations using the M-H algorithm. The target distribution is a gamma $G(2, 3)$ and is depicted in the lower panel with red line. The vertical line in the top panel depicts the burn in period at 1000 iterations

the simulated values of the chain approximate well the true density function of the gamma distribution.

6.8.2 MCMC for Dynamic Generalised Linear Models

Consider the dynamic generalised linear model (DGLM) defined by Eqs. (6.4)–(6.6) together with the prior (6.7) placed on β_0. Suppose that the covariance variance $\mathbf{Z}_t = \mathbf{Z}$ is time-invariant and an inverse Wishart prior is placed on $\mathbf{Z}$, i.e. $\mathbf{Z} \sim IW(v, \mathbf{S})$, for some degrees of freedom v and some scale matrix $\mathbf{S}$. For more details about the DGLM see the discussion of Sect. 6.2. The discussion below is based on Gamerman (1998), but other approaches are available, see e.g. Shephard and Pitt (1997). Book-length coverage of MCMC for DGLMs can be found in Gamerman and Lopes (2006, Section 6.5.3) and Fahrmeir and Tutz (2001, Section 8.3).

Suppose we have a collection of observations $y = (y_1, y_2, \ldots, y_n)^\top$ and we wish to obtain approximations of the posterior distribution of $\beta_t \mid y$ and $\mathbf{Z} \mid y$, for each $t = 1, 2, \ldots, n$. Let $\beta^\top = (\beta_1^\top, \ldots, \beta_n^\top)$ be a vector, which includes all state vectors from $t = 1$ to $t = n$. The joint posterior distribution of β and $\mathbf{Z}$ is

$$p(\beta, \mathbf{Z} \mid y) \propto p(y \mid \beta, \mathbf{Z}) p(\beta \mid \mathbf{Z}) p(\mathbf{Z})$$

$$= \prod_{t=1}^{n} p(y_t \mid \beta_t) \prod_{t=1}^{n} p(\beta_t \mid \beta_{t-1}, \mathbf{Z}) p(\mathbf{Z}) \qquad (6.65)$$

Isolating block β we have

$$p(\beta \mid y) \propto \prod_{t=1}^{n} p(y_t \mid \beta_t) \prod_{t=1}^{n} p(\beta_t \mid \beta_{t-1}, \mathbf{Z}) p(\mathbf{Z})$$

$$\propto \exp\left\{ \sum_{t=1}^{n} \frac{1}{a_t} \left[\gamma_t^\top z(y_t) + b(\gamma_t) \right] \right.$$

$$\left. - \frac{1}{2} \sum_{t=1}^{n} (\beta_t - \mathbf{F}_t \beta_{t-1})^\top \mathbf{Z}^{-1} (\beta_t - \mathbf{F}_t \beta_{t-1}) \right\}, \qquad (6.66)$$

where $p(y_t \mid \beta_t)$ is the exponential family density (6.4) and $p(\beta_t \mid \beta_{t-1}, \mathbf{Z})$ is a multivariate normal density implied by the state evolution, $\beta_t \mid \beta_{t-1}, \mathbf{Z} \sim N(\mathbf{F}\beta_{t-1}, \mathbf{Z})$.

Isolating block β_t, for $t = 1, \ldots, n - 1$ we have

$$p(\beta_t \mid y) \propto p(y_t \mid \beta_t) p(\beta_t \mid \beta_{t-1}, \mathbf{Z}) p(\beta_{t+1} \mid \beta_t, \mathbf{Z})$$

$$\propto \exp\left\{ \frac{1}{a_t}\left[\gamma_t^\top z(y_t) + b(\gamma_t) \right] - \frac{1}{2}(\beta_t - \mathbf{F}_t\beta_{t-1})^\top \mathbf{Z}^{-1}(\beta_t - \mathbf{F}_t\beta_{t-1}) \right.$$

$$\left. - \frac{1}{2}(\beta_{t+1} - \mathbf{F}_{t+1}\beta_t)^\top \mathbf{Z}^{-1}(\beta_{t+1} - \mathbf{F}_{t+1}\beta_t) \right\} \tag{6.67}$$

while for $t = n$ the posterior of β_n is

$$p(\beta_n \mid y) \propto p(y_n\beta_n) p(\beta_n \mid \beta_{n-1}, \mathbf{Z}) \propto \exp\left\{ \frac{1}{a_n}\left[\gamma_n^\top z(y_n) + b(\gamma_n) \right] \right.$$

$$\left. - \frac{1}{2}(\beta_n - \mathbf{F}_n\beta_{n-1})^\top \mathbf{Z}^{-1}(\beta_n - \mathbf{F}_n\beta_{n-1}) \right\}. \tag{6.68}$$

Finally, from (6.65) the posterior of $\mathbf{Z}$ is

$$p(\mathbf{Z} \mid y) \propto \prod_{t=1}^{n} p(\beta_t \mid \beta_{t-1}, \mathbf{Z}) p(\mathbf{Z})$$

$$\propto \prod_{t=1}^{n} |\mathbf{Z}|^{-1/2} \exp\left\{ -\frac{1}{2}\text{trace}\left[(\beta_t - \mathbf{F}_t\beta_{t-1})(\beta_t - \mathbf{F}_t\beta_{t-1})^\top \mathbf{Z}^{-1} \right] \right\}$$

$$\times |\mathbf{Z}|^{-(\nu+p+1)/2} \exp\left\{ -\frac{1}{2}\text{trace}\left(\mathbf{S}\mathbf{Z}^{-1} \right) \right\}$$

$$= |\mathbf{Z}|^{-(\nu+n+p+1)/2}$$

$$\times \exp\left\{ -\frac{1}{2}\text{trace}\left[\sum_{t=1}^{n}(\beta_t - \mathbf{F}_t\beta_{t-1})(\beta_t - \mathbf{F}_t\beta_{t-1})^\top + \mathbf{S} \right]\mathbf{Z}^{-1} \right\},$$

which is proportional to the inverse Wishart distribution

$$\mathbf{Z} \mid y \sim IW\left[\nu + n, \sum_{t=1}^{n}(\beta_t - \mathbf{F}_t\beta_{t-1})(\beta_t - \mathbf{F}_t\beta_{t-1})^\top + \mathbf{S} \right]. \tag{6.69}$$

The states may be sampled all together in the block β, using posterior (6.66) or using the individual posteriors (6.67) and (6.68), for each time t. None of these posteriors are known distributions, from which we can sample and hence for the sampling of the states we have to resort to a Metropolis-Hastings algorithm. Assuming we have sampled the states β_t, the covariance matrix $\mathbf{Z}$ can be sampled in a Gibbs step, since we can easily sample from the inverse Wishart distribution (6.69). Hence the proposed algorithm is a so-called *hybrid* algorithm, which combines Gibbs and Metropolis-Hastings sampling. Specifically, conditionally on an iteration of a Gibbs step for $\mathbf{Z}$ we can use a M-H step and sample the states β_t (either all together in the block β or one by one for each time t). The random walk chain is not proven to have good performance as a proposal distribution, see e.g. the discussion in Gamerman and Lopes (2006, Section 6.5.3). Instead, the proposal distribution of β (or β_t if sampling β_t individually) can be based on the smoothed distribution of $\beta \mid y$ (or $\beta_t \mid y$, if we sample β_t individually) assuming the response y_t is Gaussian, with matching mean and variance as the true distribution of $p(y_t \mid \gamma_t)$. These distributions provided by the fixed-interval smoothing (see Theorem 3.4 at Sect. 3.3.1) are multivariate normal and are easy to sample from. An alternative approach is to use as proposal distribution a multivariate normal distribution, with moments provided by the approximations of West et al. (1985); see also the review of Triantafyllopoulos (2009).

The algorithm, which is summarised below, is essentially a Metropolis within Gibbs algorithm. The convergence of a hybrid algorithm and related aspects of its performance are discussed in Gamerman and Lopes (2006, Chapter 6) and in references therein.

MCMC Algorithm for the DGLM

In the dynamic generalised linear model (6.4)–(6.6), with the priors on β_0 and $\mathbf{Z}$ as above, the following apply:

1. Set initial values of the states $\beta^{(0)\top} = [\beta_1^{(0)\top}, \dots, \beta_n^{(0)\top}]$ and covariance matrix $\mathbf{Z}^{(0)}$. Set the iteration counter to $j = 1$.
2. For each iteration $j = 1, \dots, N$, draw β^* from the proposal density provided by an application of the fixed-interval smoothing (see Theorem 3.4 at Sect. 3.3.1), assuming that $y \mid \beta$ is Gaussian with matching moments the moments of $p(y \mid \beta)$ of (6.4).
3. Calculate the acceptance probability $\alpha(\beta^{(j-1)}, \beta^*)$ and draw a single u from a uniform distribution $U(0, 1)$. If $u < \alpha(\beta^{(j-1)}, \beta^*)$, the proposal β^* is accepted and we set $\beta^{(j)} = \beta^*$, otherwise the move is rejected and we set $\beta^{(j)} = \beta^{(j-1)}$.
4. Draw $\mathbf{Z}^{(j)}$ from the inverse Wishart distribution (6.69) where $\beta_t = \beta_t^{(j)}$ obtained from Step 3.

If the states are updated individually, then steps 2 and 3 are replaced by

2′ For each iteration $j = 1, \dots, N$, draw β_t^* from the proposal density provided by an application of the fixed-interval smoothing (see Theorem 3.4 at Sect. 3.3.1), assuming that $y_t \mid \beta_t$ is Gaussian with matching moments the moments of $p(y_t \mid \beta_t)$ of (6.4).

(continued)

3′ Calculate the acceptance probability $\alpha(\beta_t^{(j-1)}, \beta_t^*)$ and draw a single u_t from a uniform distribution $U(0, 1)$. If $u_t < \alpha(\beta^{(j-1)}, \beta_t^*)$, the proposal β_t^* is accepted and we set $\beta_t^{(j)} = \beta_t^*$, otherwise the move is rejected and we set $\beta_t^{(j)} = \beta_t^{(j-1)}$.

Some comments are in order. First of all the algorithm provides in-sample estimation for the states β_t, i.e. the approximation of the densities $p(\beta_t \mid y)$ are the smoothed densities, while $p(\mathbf{Z} \mid y)$ is the posterior distribution of $\mathbf{Z}$. If approximations of the posterior distribution of $p(\beta_t \mid y_{1:t})$ are required, then the algorithm has to be applied repeatedly over time. Likewise a single application of the algorithm can provide approximation of the forecast distribution $p(y_{n+k} \mid y)$, for some integer k. If forecasts are required sequentially over time, the algorithm has to be applied repeatedly and this can cause the algorithm to be slow.

It might be desirable to update the block β at once at each iteration, hence adopt steps 2 and 3 in the above algorithm. This can be effective as at each iteration we compute the acceptance probability once. However, if the dimension of the state vector β_t is medium or high or if the length of the data n is medium or large, then β will be high dimensional and the M-H step is likely to experience convergence problems. Random walk chains are not advisable for the proposal distribution, as they create highly correlated states, which result in slow convergence. A second difficulty related to calibrating the chain in order to obtain optimal acceptance rates. The approximate fixed interval smoothing for the proposal distribution is a good option. Another possibility, proposed in Shephard and Pitt (1997), is to combine independent chains and chains from the prior distribution of the states.

We conclude this section by considering a simple simulation study in illustrate estimation based on the MCMC scheme of this section. We simulate 70 observations from the dynamic Poisson model (6.8)–(6.10), i.e.

$$y_t \mid \beta_t \sim \text{Poisson}[\exp(\beta_t)] \quad \text{and} \quad \beta_t = \beta_{t-1} + \zeta_t,$$

where $\zeta_t \sim N(0, Z)$. The states β_t, which here are the log rates of the Poisson, are simulated by a random walk with initial state simulated by $\beta_0 \sim N(0, 10)$ and a state variance $Z = 2.5$. The top left panel of Fig. 6.10 shows the simulated states (solid lines). For the estimation of the states and the state variance, we have used the block hybrid MCMC scheme described above. A random walk chain and a Gaussian proposal are used: experimentation has led to set the variance of the proposal distribution equal to 0.6 in order to achieve an acceptance probability equal to 28.53% (the small variance allows for small moves of the chain, which works here as β_t is univariate and is unimodal). The chain is run for 10,000 iterations, and the first 1000 are used as burn-in and the last 1000 are used for illustration purposes.

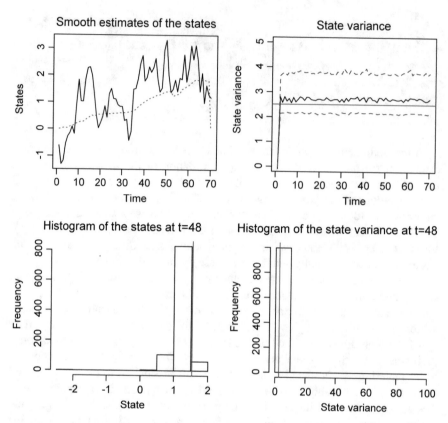

Fig. 6.10 MCMC estimation for the dynamic Poisson model. Top left panel (estimation of β_t): smoothed mode (dotted line) and true states (solid line). Top right panel (estimation of Z): posterior mode (solid line), 95% credible bounds (dashed line) and true value (horizontal line). Bottom panels: histogram of smooth estimates of β_{48} (shown is the true value of $\beta_{48} = 1.5284$, depicted by the vertical line, and histogram of posterior estimate of Z at $t = 48$; the vertical line indicate the true value $Z = 2.5$

Trace plots and their correlograms (not shown here) indicate convergence of the chain. Figure 6.10 shows smoothed estimates of the states (top left panel), posterior mode of the state variance Z, together with 95% credible intervals (top right panel), histogram of the estimated state β_{48} at time $t = 48$ and histogram of the state variance Z at $t = 48$. We remark that the smooth estimates of the states seems to follow well the simulated states, the posterior mode of the Z is close to the true value $Z = 2.5$, although it does seem to slightly overestimate the variance.

6.9 Dynamic Survival Models

6.9.1 Proportional Hazards Model

Over the past 50 years survival modelling has been developed extensively in medicine (Martinussen & Scheike, 2006; van Houwelingen & Putter, 2012), economics (Gamerman & West, 1987; Djeundje & Crook, 2019) and other disciplines (Cox, 1972; Cox & Oakes, 1984; Collett, 2003). Survival analysis is focused around the survival function

$$S(t) = P(T > t),$$

for some random variable T denoting survival time. Hence the survival function at time t is the probability T exceeding t. We consider that T is generated by a density function $f(\cdot)$, with cumulative distribution function $F(\cdot)$, so that $S(t) = 1 - F(t)$.

Associated with the survival function is the hazard function $\lambda(t)$ defined as

$$\lambda(t) = \lim_{\Delta t \to 0} \frac{P(t < T \le t + \Delta t \mid T > t)}{\Delta t}.$$

It follows that

$$\lambda(t) = \frac{f(t)}{S(t)},$$

$$S(t) = \exp\left(-\int_0^t \lambda(u)\, du\right),$$

$$f(t) = \lambda(t) \exp\left(-\int_0^t \lambda(u)\, du\right).$$

Of particular interest is the case of T following an exponential distribution, which is discussed next.

Assuming that T follows an exponential distribution $T \sim \text{Exp}(\lambda)$, with density function

$$f(t) = \lambda \exp(-\lambda t),$$

then it is easy to show that the associated hazard function is constant, i.e.

$$\lambda(t) = \lambda$$

and the survival function is

$$S(t) = \exp(-\lambda t).$$

Often observations of survival times $T = t$ are observed in discrete time. Usually, discrete time points $t_0, t_1, \ldots, t_{s-1}$ are picked and time $t \geq 0$ is partitioned into s intervals of time $I_1, I_2, \ldots, I_s$, defined as $I_1 = [t_0, t_1), I_2 = [t_1, t_2), \ldots, I_s = [t_{s-1}, \infty)$, for some $s > 1$. Survival times are then observed within these intervals of time.

The basic model of Cox (1972) considers a hazard function

$$\lambda(t) = \lambda_0(t) \exp(x^\top \beta), \qquad (6.70)$$

where $\lambda_0(t)$ is a baseline hazard function, x is a vector of p covariates or $x = [x_1, \ldots, x_p]^\top$ and β is a p-dimensional coefficient vector subject to estimation. This model is known as *proportional hazards model*, as the covariates x affect the hazard only by a proportionality factor (Cox, 1972). Other approaches for the modelling of the hazard function include the use of generalised linear models such as logistic regression and Poisson regression; see e.g. the review of Kearns et al. (2019).

6.9.2 Dynamic Survival Model

Time-varying covariates are discussed in Cox and Oakes (1984) who develop partial-likelihood estimation introduced in Cox (1975) to deal with time-dependent covariates; see also Kedem and Fokianos (2002) for a discussion of partial likelihood in the context of generalised linear modelling. Gamerman (1991) extended the above models to incorporate time-varying effects of the coefficients β, by replacing this parameter vector by a time-varying vector of coefficients $\beta(t)$. Hence (6.70) is replaced by

$$\lambda(t) = \exp\left[\beta_0(t) + \sum_{k=1}^{p} x_k^\top \beta_k(t) \right] = \exp[z^\top \beta(t)], \qquad (6.71)$$

where $z^\top = [1, x^\top]$ and $\beta_0(t)$ denotes the log-baseline. A piecewise exponential distribution is considered, where hazard function $\lambda(t)$ is a step-function, or

$$\lambda(t) = \begin{cases} \lambda_1, & t \in I_1 = [t_0, t_1], \\ \vdots & \vdots \\ \lambda_i, & t \in I_i = (t_{i-1}, t_i], \\ \vdots & \vdots \\ \lambda_s, & t \in I_s = (t_{s-1}, \infty), \end{cases}$$

with corresponding coefficients $\beta(t)$, which are allowed to vary according to a random walk within each interval I_i, or

$$\beta_0(t) = \beta_{0i}, \quad t \in I_i, \tag{6.72}$$

$$\beta_k(t) = \beta_{ki}, \quad t \in I_i, \tag{6.73}$$

for $i = 1, \ldots, s$ and $k = 1, \ldots, p$. Gamerman (1991) and Hemming and Shaw (2002) consider a random walk evolution for β_{0t} and β_{ki}, i.e.

$$\beta_{0i} = \beta_{0,i-1} + \zeta_{0i}, \quad \zeta_{it} \sim N(0, Z_0), \tag{6.74}$$

$$\beta_{ki} = \beta_{k,i-1} + \zeta_{ki}, \quad \zeta_{ki} \sim N(0, Z_k), \tag{6.75}$$

for some variances Z_0 and Z_k. This is supported by considering that locally we would expect $E(\beta_{ki}) = E(\beta_{k,i-1})$, but with increased uncertainty, hence the random walk evolution. It is assumed that β_{ki} and β_{lj} are independent for any $k \neq l$ and for any i, j.

A Gaussian prior distribution is set for β_{0t} and β_{ki}, i.e.

$$\beta_{0i} \sim N(\hat{\beta}_{0i}, P_{0i}) \quad \text{and} \quad \beta_{ki} \sim N(\hat{\beta}_{ki}, P_{ki}), \tag{6.76}$$

for some known means $\hat{\beta}_{0i}, \hat{\beta}_{ki}$ and variances P_{0i}, P_{ki}, for each $i = 1, \ldots, N$.

The dynamic survival model adopts hazard function $\lambda(t)$ of (6.72) (together with the piece-wise exponential structure), which is equivalent to T following an exponential distribution $T \sim \text{Exp}(\lambda(t))$. The states in (6.72)–(6.73) attain evolution equations (6.74)–(6.75) and the model is completed by specifying the priors (6.76).

In survival analysis it is very common to experience *censoring*. If for some units the event we are measuring has occurred then we know the exact waiting time and there is no censoring. If, however, the event has not occurred all we know is that the waiting time exceeds the observation time. This is the case of censoring. For example, imagine that we collect time to death in a cancer study. If we know that patient i died at time t_i^* in the interval $[t_i, t_{i+1})$, we have no censoring (as we know the time up to death). If, however, the patient has not died in the interval $[t_i, t_{i+1})$, all we know is that death time $t^* \geq t_{i+1}$. There are various mechanisms of censoring, the most common being Type I, Type II and random censoring; here we briefly discuss random censoring, but for a more in-depth discussion the reader is referred to Fahrmeir and Tutz (2001, Chapter 9) and in Collett (2003). In random censoring each unit is labelled either a lifetime T_i or a censoring time C_i, both of which are assumed to be random variables. We then observe $Y_i = \min(T_i, C_i)$ and an indicator variable δ_i, which tells us whether an observation is terminated by death ($Y_i = T_i$) or by censoring ($Y_i = C_i$).

Suppose that we observe n units, with survival function $S(t)$ and associated density $p(t)$ and hazard $\lambda(t)$. If we observe unit i for time t_i. If the unit died at t_i, then its contribution to the likelihood function is

$$L_i = f(t_i) = S(t_i)\lambda(t_i).$$

On the other hand, if the unit survives at t_i, then its contribution to the likelihood is just its survival function, or

$$L_i = S(t_i).$$

We can combine the two events in the likelihood function, which is given by

$$L = \prod_{i=1}^{n} L_i = \prod_{i=1}^{n} \lambda(t_i)^{\delta_i} S(t_i), \tag{6.77}$$

where $\delta_i = 1$, if T_i is observed (lifetime is observed) and $\delta_i = 0$, if C_i is observed (censoring is observed).

For the piece-wise exponential model described above the hazard is a piece-wise constant function which changes at each interval I_i. Let the interval at which censoring occurred for observation i is denoted by $I_{m_{i+1}}$. The contribution of observation i to the likelihood is broken into the probability of surviving each interval (up to interval $I_{m_{i+1}}$) conditionally on surviving at the previous interval. Hence

$$S(t_i) = \prod_{j=1}^{m_i} P(T > t_j \mid T > t_{j-1}) P(T > t_i \mid > t_{m_i})$$

and from (6.77) the likelihood function is

$$L = \prod_{i=1}^{n} \prod_{j=1}^{m_i} P(T > t_j \mid T > t_{j-1}) P(T > t_i > t_{m_i}) \lambda(t_i)^{\delta_i}. \tag{6.78}$$

Fix $t > t_{j-1}$ ($j = 1, \ldots, m_i$). Then the survival function in interval I_j is

$$P(T > t \mid T > t_{j-1}) = \exp\left[-\int_{t_{j-1}}^{t} \lambda(u)\, du\right]$$

$$= \exp\left[-\int_{t_{j-1}}^{t} \exp(z^\top \beta_j)\, du\right]$$

$$= \exp[-\exp(z^\top \beta_j)(t - t_{j-1})],$$

where $\beta_j = [\beta_{0j}, \beta_{1j}, \ldots, \beta_{pj}]^\top$. Hence likelihood (6.78) is

$$
L = \prod_{i=1}^{n} \exp\left[-\sum_{j=1}^{m_i} \exp(z_i^\top \beta_j)(t_j - t_{j-1}) \right] \exp[-\exp(z_i^\top \beta_{m_{i+1}})(t_i - t_{m_i}]
$$

$$
\times \exp(z_i^\top \beta_{m_{i+1}} \delta_i)
$$

$$
= \exp\left[-\sum_{i=1}^{n}\sum_{j=1}^{m_i} \exp(z_i^\top \beta_j)(t_j - t_{j-1}) - \sum_{i=1}^{n} \exp(z_i^\top \beta_{m_{i+1}})(t_i - t_{m_i}) \right.
$$

$$
\left. + \sum_{i=1}^{n} z_i^\top \beta_{m_{i+1}} \delta_i \right].
$$

Suppose data $\{t_0, t_1, \ldots, t_{s-1}\}$ on s survival times are available. We wish to estimate the coefficients $\beta(t)$ so as to provide an estimate of the hazard function $\lambda(t)$. Maximum likelihood estimation is possible, but will need to resort to numerical methods; for a discussion see Fahrmeir and Tutz (2001, Chapter 9). Most of the literature seems to focus on Bayesian estimation instead.

Gamerman and West (1987) and Gamerman (1991) propose approximate Bayesian inference based on the approach of West et al. (1985). Wilson and Farrow (2017) adopt a Bayes linear kinematics approach for inference of dynamic survival modelling and apply their model to survival times of leukaemia patients. Fahrmeir (1994) propose a penalised likelihood estimation approach, leading to Kalman-type smoothing algorithms; see also Fahrmeir and Tutz (2001, Chapter 9) for this approach. Hemming and Shaw (2002, 2005) and Wagner (2011) propose inference based on Markov chain Monte Carlo (MCMC) methods. A comparison between the Bayes linear estimation of Gamerman (1991) and MCMC inference with the focus on medical application is given in He et al. (2010). A dynamic cured fraction model is developed in Kearns et al. (2021) and is shown to improve extrapolations in the hazard function of curative treatments, whilst being robust to model misspecification.

6.10 Exercises

1. Consider that the time series $\{y_t\}$ is generated by the gamma distribution $y_t \mid \alpha_t, \beta_t \sim G(\alpha_t, \beta_t)$, for some parameters α_t and β_t. Show that y_t belongs to the exponential family of distributions (6.4), with $\gamma_t = -\beta_t$, $a_t = 1$, $z(y_t) = y_t$, $b(\gamma_t) = -\alpha_t \log \beta_t$ and $c(y_t) = (\alpha_t - 1) \log y_t - \log \Gamma(\alpha_t)$, where $\Gamma(\cdot)$ denotes the gamma function. Observe that for $\alpha_t = 1$ the exponential distribution is obtained, i.e. $y_t \mid \beta_t \sim \mathrm{Exp}(\beta_t)$.

2. Extend the exponential family (6.4) to accommodate for a matrix-variate time series as follows. Let $\mathbf{y}_t$ be a matrix-variate time series. We shall say that the distribution of $p(\mathbf{y}_t \mid \boldsymbol{\gamma}_t)$ defines the natural matrix-variate exponential family of distribution if $\mathrm{vec}(\mathbf{y}_t)$ attains the exponential form (6.4) with natural parameter vector $\mathrm{vec}(\boldsymbol{\gamma}_t)$ and where, for simplicity, $z(\cdot)$ is assumed to be the identity function. Show that the matrix-variate exponential family can be written as

$$p(\mathbf{y}_t \mid \boldsymbol{\gamma}_t) = \exp\left\{\frac{1}{a_t}[\mathrm{trace}(\boldsymbol{\gamma}_t^\top \mathbf{y}_t) - b(\boldsymbol{\gamma}_t)] + c(\mathbf{y}_t)\right\},$$

where the functions $b(\cdot)$ and $c(\cdot)$ map the matrices $\boldsymbol{\gamma}_t$ and $\mathbf{y}_t$, respectively, to the real line.

3. Consider that the $p \times p$ matrix-variate time series $\{\mathbf{y}_t\}$ is generated by a Wishart distribution, with density function

$$p(\mathbf{y}_t \mid n, \boldsymbol{\Sigma}_t) = \frac{2^{-np/2}|\boldsymbol{\Sigma}_t|^{-n/2}}{\Gamma_p(n/2)}|\mathbf{y}_t|^{(n-p-1)/2}\exp\left[-\frac{1}{2}\mathrm{trace}(\boldsymbol{\Sigma}_t^{-1}\mathbf{y}_t)\right],$$

where $n > p - 1$ are the degrees of freedom, $\boldsymbol{\Sigma}_t$ is the scale covariance parameter matrix and $\Gamma_p(\cdot)$ denotes the multivariate gamma function. The Wishart distribution is discussed in Sect. 5.5.2.

   Show that the above Wishart distribution belongs to the matrix-variate exponential family of Exercise 2, with $\boldsymbol{\gamma}_t = -2^{-1}\boldsymbol{\Sigma}_t^{-1}$, $b(\boldsymbol{\gamma}_t) = 2^{-1}n\log|\boldsymbol{\Sigma}_t|$ and $c(\mathbf{y}_t) = 2^{-1}(n - p - 1)\log|\mathbf{y}_t| - 2^{-1}np\log 2 - \log\Gamma_p(n/2)$.

4. In the context of the power local level model of Sect. 6.5 show that if $\delta = 1$ prior (6.36) implies that $\beta_t = \psi_t(\beta_0)$, for some deterministic function $\psi_t(\cdot)$, i.e. the state β_t depends stochastically only on β_0.

   As a special case of the above consider the Gaussian local level model $y_t = \beta_t + \epsilon_t$ and $\beta_t = \beta_{t-1} + \zeta_t$, where $\epsilon_t \sim N(0, \sigma^2)$ and $\zeta_t \sim N(0, Z_t)$, for some variances σ^2 and Z_t and remaining assumptions as in the state space model (3.10a)–(3.10b). Adopting the discounting approach of Eq. (6.35) for $\delta = 1$, establish that $y_t = \beta_0 + \epsilon_t$, i.e. this model is reduced to a static simple linear regression model.

5. Consider the Poisson-gamma local level model of Sect. 6.5.2. Define the transition

$$\lambda_t = \frac{\lambda_{t-1}\xi_t}{\delta}, \tag{6.79}$$

where $\lambda_{t-1} \mid y_{1:t-1} \sim G(\alpha_{t-1}, \beta_{t-1})$, ξ_t is a random variable, which is independent of λ_{t-1} and follows the beta distribution $\xi_t \sim \mathrm{Beta}[\delta\alpha_{t-1}, (1 - \delta)\beta_{t-1}]$, for some discount factor δ. Show that

$$\lambda_t \mid y_{1:t-1} \sim G(\delta\alpha_{t-1}, \delta\beta_{t-1}).$$

Hence establish that transition (6.79) yields exactly the same prior distribution $\lambda_t \mid y_{1:t-1} \sim G(\delta\alpha_{t-1}, \delta\beta_{t-1})$, which is the result of the power law (6.36).

6. Consider that the categorical time series y_t is modelled with a binomial model

$$p(y_t \mid \pi_t) = \binom{n_t}{y_t} \pi_t^{y_t} (1 - \pi_t)^{n_t - y_t}, \quad y_t = 0, 1, 2, \ldots, n_t$$

for some known integer $n_t > 0$. Let the posterior distribution of π_{t-1} given $y_{1:t-1}$ be a beta distribution, $\pi_{t-1} \mid y_{1:t-1} \sim \text{Beta}(\alpha_{t-1}, \beta_{t-1})$, for some known $\alpha_{t-1}, \beta_{t-1} > 0$. In the context of power local level models adopt a discount factor δ.

a. Show that the prior distribution of π_t, given $y_{1:t-1}$ is a beta distribution $\pi_t = \pi \mid y_{1:t-1} \sim \text{Beta}(\delta\alpha_{t-1} - \delta + 1, \delta\beta_{t-1} - \delta + 1)$.

b. Upon y_t being observed, show that the posterior distribution of π_t, given $y_{1:t} = (y_t, y_{1:t-1})$ is a beta distribution $\pi_t \mid y_{1:t} \sim \text{Beta}(\alpha_t, \beta_t)$, where $\alpha_t = \delta\alpha_{t-1} - \delta + 1 + y_t$ and $\beta_t = \delta\beta_{t-1} - \delta + 1 + n_t - y_t$.

c. Given information $y_{1:t}$, show that the one-step forecast distribution of y_{t+1} is

$$p(y_{t+1} \mid y_{1:t}) = \frac{(\delta\beta_t + y_{t+1})^{\delta\alpha_t + 2 - \delta}}{\Gamma(\delta\alpha_t + 2 - \delta)},$$

where α_t and β_t are computed as above.

7. Suppose that the time series y_t representing continuous proportion follows a beta distribution $y_t \mid \beta_t \sim \text{Beta}(1, \beta_t)$, so that

$$p(y_t \mid \beta_t) = \beta_t (1 - y_t)^{\beta_t - 1}, \quad 0 \le y_t \le \beta_t,$$

for some $\beta_t > 0$. Assume that at time $t - 1$ the posterior distribution of β_{t-1} is a gamma distribution with known parameters d_{t-1} and g_{t-1}, i.e. $\beta_{t-1} = \beta \sim G(d_{t-1}, g_{t-1})$. Adopt power local level models for a known discount factor δ.

a. Show that the prior distribution of $\beta_t = \beta$ given $y_{1:t-1}$ is a gamma distribution $\beta_t = \beta \mid y_{1:t-1} \sim G(\delta d_{t-1} - \delta + 1, \delta g_{t-1})$.

b. Given observation y_t, show that the posterior distribution of β_t a gamma distribution, i.e. $\beta_t = \beta \mid y_{1:t} \sim G(d_t, g_t)$, where $d_t = \delta d_{t-1} - \delta + 2$ and $g_t = \delta g_{t-1} - \log(1 - y_t)$.

8. Let y_t be a time series defined on an interval $[0, \beta_t]$, for some time-varying parameter $\beta_t > 0$. Suppose that, given β_t, y_t follows a uniform distribution, so that

$$p(y_t \mid \beta_t) = \frac{1}{\beta_t}, \quad 0 \le y_t \le \beta_t.$$

Assume that at time $t - 1$, the posterior distribution of $\beta_{t-1} = \beta$ is a Pareto distribution (see also Exercise 10) with density function

$$p(\beta_{t-1} = \beta \mid y_{1:t-1}) = \frac{d_{t-1}g_{t-1}^{d_{t-1}}}{\beta^{d_{t-1}+1}}, \quad \beta \geq g_{t-1},$$

for some known parameters $d_{t-1}, g_{t-1} > 0$. We shall write $\beta_{t-1} = \beta \mid y_{1:t-1} \sim$ Pareto(d_{t-1}, g_{t-1}) to denote this distribution. Within the context of power local level modelling adopt a discount factor δ.

a. Show that the prior distribution of $\beta_t = \beta$ given information $y_{1:t-1}$ is a pareto distribution, $\beta_t = \beta \mid y_{1:t-1} \sim$ Pareto($\delta d_{t-1} + \delta - 1, g_{t-1}$).
b. Upon observing y_t update the information as $y_{1:t} = (y_{1:t-1}, y_t)$. Show that, given $y_{1:t}$, the posterior distribution of $\beta_t = \beta$ is a Pareto distribution $\beta_t \mid y_{1:t} \sim$ Pareto(d_t, g_t), where $d_t = \delta d_{t-1} + \delta - 1$ and $g_t = \max\{y_t, g_{t-1}\}$.

9. Consider the dynamic Poisson model

$$y_t \mid \lambda_t \sim \text{Poisson}(\lambda_t),$$

so that the log rate follows a random walk model

$$\log \lambda_t = \beta_t = \beta_{t-1} + \zeta_t,$$

where $\zeta_t \sim N(0, Z)$, for some variance Z.

Show that given λ_{t-1}, the rate λ_t follows a log-normal distribution, with density function

$$p(\lambda_t \mid \lambda_{t-1}) = \frac{\lambda_{t-1}}{\lambda_t Z \sqrt{2\pi}} \exp\left\{ -\frac{(\log \lambda_t - \log \lambda_{t-1})^2}{2Z} \right\}.$$

Show that, given λ_{t-1}, the mean and the variance of λ_t are

$$E(\lambda_t \mid \lambda_{t-1}) = \lambda_{t-1} \exp\left(\frac{Z}{2}\right) \quad \text{and}$$

$$\text{Var}(\lambda_t \mid \lambda_{t-1}) = \lambda_{t-1}^2 [\exp(Z) - 1] \exp(Z).$$

10. Consider that the random vector $X = [X_1, \ldots, X_k]^\top$ follows the Dirichlet distribution with parameter vector $\alpha = [\alpha_1, \ldots, \alpha_k]^\top$. The density function of X is

$$p(x) = \frac{1}{D(\alpha)} \prod_{i=1}^{k} x_i^{\alpha_i - 1},$$

where $D(\alpha)$ is the Dirichlet function and $\sum_{i=1}^{k} x_i = 1$; this distribution is discussed in Sect. 6.2.4.

a. Show that for $k = 2$ (beta distribution), the mode of X_1 is $(\alpha_1 - 1)/(\alpha_1 + \alpha_2 - 2)$.

b. Show that for $k = 3$, the modes of X_1, X_2 are $(\alpha_1 - 1)/(\alpha_1 + \alpha_2 + \alpha_3 - 3)$ and $(\alpha_2 - 1)/(\alpha_1 + \alpha_2 + \alpha_3 - 3)$, respectively.

c. Deduce that for any $k \geq 2$, the mode of X_i is

$$\frac{\alpha_i - 1}{\sum_{j=1}^{k} \alpha_j - k},$$

for $i = 1, \ldots, k - 1$.

11. Consider that the time series $\{y_t\}$ is generated from a multinomial distribution with a probability vector $\pi_t = [\pi_{1t}, \ldots, \pi_{kt}]^{\top}$ on k categories, with joint probability mass function

$$p(y_t \mid \pi_t) = \frac{\lambda_t!}{\prod_{i=1}^{k} y_{it}!} \prod_{i=1}^{k} \pi_{it}^{y_{it}},$$

where $\lambda_t = \sum_{i=1}^{k} y_{it}$ and $\sum_{i=1}^{k} \pi_{it} = 1$. This distribution is described in Sects. 2.3.2 and 6.2.3.

Suppose that at time $t - 1$ the posterior distribution of π_{t-1} is a Dirichlet distribution, written as $\pi_{t-1} = \pi \sim \mathrm{Dir}(\alpha_{t-1})$, so that its density is

$$p(\pi_{t-1} = \pi \mid y_{1:t-1}) = \frac{1}{D(\alpha_{t-1})} \prod_{i=1}^{k} \pi_j^{\alpha_{i,t-1} - 1},$$

where $\pi = [\pi_1, \ldots, \pi_k]^{\top}$, $\alpha_{t-1} = [\alpha_{1,t-1}, \ldots, \alpha_{k,t-1}]^{\top}$ and $D(\cdot)$ denotes the Dirichlet function. This distribution is discussed in more detail in Sect. 6.2.4.

a. Upon observing y_t, show that the posterior distribution of $\pi_t = \pi$, given $y_{1:t}$ is $\pi_t = \pi \mid y_{1:t} \sim \mathrm{Dir}(\alpha_t)$, where

$$\alpha_{it} = \delta \alpha_{i,t-1} + 1 - \delta + y_{it},$$

with $\alpha_t = [\alpha_{1t}, \ldots, \alpha_{kt}]^{\top}$.

b. Using the posterior distribution of $\pi_t = \pi$ and the prior distribution of $\pi_{t+1} = \pi$, given information $y_{1:t}$, verify that the mode of $\pi_{t+1} = \pi$ is equal to the mode of $\pi_t = \pi$.

HINT: you may use the result of Exercise 10.
c. Show that the one-step forecast distribution of y_{t+1} is

$$p(y_{t+1} \mid y_{1:t}) = \frac{\lambda_{t+1}! D(\alpha_{t+1}^* + y_{t+1})}{\prod_{i=1}^{k} y_{i,t+1}! D(\alpha_{t+1}^*)},$$

where $\alpha_{t+1}^* = [\alpha_{1,t+1}^*, \ldots, \alpha_{k,t+1}^*]^\top$ and $\alpha_{i,t+1}^* = \delta \alpha_{it} + 1 - \delta$.

12. Consider the binomial model (6.14)–(6.16) for the categorical time series y_t with size λ_t (see model (6.14)–(6.16) for details). Suppose that at time t the prior mean vector of β_t, $\hat{\beta}_{t|t-1}$, is known. Show that this model can be approximated by the linear and Gaussian model with observation equation

$$y_t \approx \frac{\lambda_t \exp(x_t^\top \hat{\beta}_{t|t-1})}{[1 + \exp(x_t^\top \hat{\beta}_{t|t-1})]^2} x_t^\top \beta_t + \varepsilon_t,$$

where x_t is the design vector of the linear predictor $\eta_t = x_t^\top \beta_t$ (see model (6.14)–(6.16) for a full description of the binomial state space model), $\varepsilon_t \sim N(\mu_t, V_t)$, with

$$\mu_t = \frac{\lambda_t \exp(x_t^\top \hat{\beta}_{t|t-1})}{[1 + \exp(x_t^\top \hat{\beta}_{t|t-1})]^2} \left[1 + \exp(x_t^\top \hat{\beta}_{t|t-1}) - x_t^\top \hat{\beta}_{t|t-1} \right]$$

and

$$V_t = \frac{\lambda_t \exp(x_t^\top \hat{\beta}_{t|t-1})}{[1 + \exp(x_t^\top \hat{\beta}_{t|t-1})]^2}.$$

13. Consider the conditionally Gaussian state space model of Sect. 6.7.7, given by equations

$$y_t = \frac{\beta_t^2}{20} + \epsilon_t \quad \text{and} \quad \beta_t = \alpha \beta_{t-1} + \frac{\gamma \beta_{t-1}}{1 + \beta_{t-1}^2} + \delta \cos[1.2(t-1)] + \zeta_t,$$

where ϵ_t and ζ_t independently follow Gaussian distributions, i.e. $\epsilon_t \sim N(0, \sigma^2)$ and $\zeta_t \sim N(0, Z)$; it is also assumed that ϵ_t, ζ_t are independent of the initial state β_0, which is assumed to follow a Gaussian distribution too, i.e. $\beta_0 \sim N(0, 10)$.

a. Show that conditionally on β_{t-1} the mean vector and covariance matrix of $[\beta_t, y_t]^\top$ are

$$\mu_t = \begin{bmatrix} c_t \\ d_t \end{bmatrix}, \quad \mathbf{M}_t = \begin{bmatrix} Z & c_t^3 + 3c_t Z \\ c_t^3 + 3c_t Z & V_t \end{bmatrix},$$

where

$$c_t = \alpha\beta_{t-1} + \frac{\gamma\beta_{t-1}}{1 + \beta_{t-1}^2} + \delta\cos(1.2t - 1.2),$$

$$d_t = \frac{c_t^2 + Z}{20} \quad \text{and} \quad V_t = \frac{3Z^2 + 2c_t Z}{400} + \sigma^2.$$

b. Assuming that given β_{t-1}, the joint distribution of β_t and y_t is Gaussian, show the conditional distribution of $\beta_t \mid \beta_{t-1}, y_t$ is $N(d_t^*, V_t^*)$, with mean and variance

$$d_t^* = c_t + \frac{c_t^3 + 3c_t Z}{V_t}(y_t - d_t),$$

$$V_t^* = Z - \frac{(c_t^3 + 3c_t Z)^2}{V_t}.$$

c. Hence propose a particle filter with states $\beta_t^{(i)}$ are sampled from the importance function $q(\beta_t \mid \beta_{t-1}, y_t) \equiv N(d_t^*, V_t^*)$.

14. Consider the conditionally Gaussian model of Exercises 13.

a. Simulate 100 states $\beta_1, \ldots, \beta_{100}$ and 100 observations $y_1, \ldots, y_{100}$ from this model using $\alpha = 0.5, \gamma = 25, \delta = 15, \sigma^2 = 8$ and $Z = 10$.

b. Define the vector of static parameters $\theta = [\alpha, \gamma, \delta, \sigma^2, Z]^\top$ and suppose you do not know these values from (1) above. Set the following priors for the elements of θ:

$$\alpha \sim U(0, 1) \quad \gamma \sim G(2, 0.1), \quad \delta \sim G(1, 0.1),$$

$$\sigma^2 \sim IG(3, 10), \quad \text{and} \quad Z \sim IG(3, 50).$$

Fit the Liu and West particle filter (Algorithm PF-II on p. 311) and provide credible bounds for the states β_t and for the elements of θ.

15. Kitagawa (1987) proposes a binary process in order to model the number of daily occurrences of rainfall over 1 mm in Tokyo (1983–1984). The following data is a weekly aggregation of this data.

	Week 1	Week 2	Week 3	Week 4
January	4	0	4	2
February	1	2	3	3
March	2	4	6	8
April	5	5	7	2
May	4	2	5	4
June	2	7	9	11
July	4	5	10	3
August	0	2	7	2
September	3	5	10	7
October	1	5	7	3
November	2	2	4	1
December	0	1	3	3

Make an analysis of this data using a dynamic Poisson or a dynamic negative binomial model. You will need to suggest a transition model for the states of your state space model. You will also need to specify priors for the initial state and the hyperparameters of your model.

16. The data `aids` available in the package `dobson` consist of 20 observations on counts of cases of AIDS patients in Australia, sampled over quarters from 1984 to 1988. The data can be accessed in R using the commands:

```
> library(dobson)
> data(aids)
> attach(aids)
```

a. Consider the dynamic Poisson model

$$y_t \mid \lambda_t \sim \text{Poisson}(\lambda_t), \tag{6.80}$$

$$\log \lambda_t = \log \lambda_{t-1} + \zeta_t, \quad \zeta_t \sim N(0, Z), \tag{6.81}$$

for some variance Z. Fit this model to the data using the Liu and West filter (Algorithm PF-II on p. 311) with a static parameter $\theta = Z$. Use a gamma prior for Z, i.e. $Z \sim G(0.1, 0.1)$ (giving mean $E(Z) = 1$ and variance $\text{Var}(Z) = 10$).

b. A more careful analysis reveals that this data is over-dispersed. A simple analysis shows that the sample mean of the counts is 15.643, while the sample variance is 224.555. This casts doubt over the choice of the above Poisson model. Extend this model by considering that

$$\frac{y_t}{c} \sim \text{Poisson}(\lambda_t), \tag{6.82}$$

where $c > 0$ is some constant and λ_t is driven by evolution (6.81) of part (a). Show that

$$E(y_t) = c\lambda_t \quad \text{and} \quad \text{Var}(y_t) = c^2\lambda_t$$

and show for $c > 1$ this model is suitable for over-dispersed data.

i. Show that by choosing $c = 14.3511$ we can deal with over-dispersion and use a Liu and West filter (Algorithm PF-II on p. 311) with a static parameter $\theta = Z$ (see the same gamma prior for Z as in part (a)). Comment on its performance compared to the model of part (a).

ii. Consider model (6.82) together with evolution (6.81), where now c is subject to estimation. Fit this model using the Liu and West filter (Algorithm PF-II on p. 311), with a static parameter vector $\theta = (c, Z)^\top$. Choose a gamma prior for Z as in part (a) and experiment on prior distribution c from (1) $c \sim G(0.1, 0.1)$ and (2) $c - 1 \sim G(0.1, 0.1)$. In prior (1) we allow c to be smaller than one (leading to under-dispersion), $c = 1$ (leading to no over/under dispersion) and $c > 1$ (leading to over-dispersion). In prior (2) $c > 1$ is sampled from a gamma distribution plus 1 (hence referring to over-dispersion). Experiment with these priors and comment on their performance with relevance to part (b-i) and part (a).

17. The data `aihrio` available in the package `pgam` in R consist of counts of hospital admissions outcomes in Universidade do Estado do Rio de Janeiro, Rio de Janeiro, Brazil. The data set gives hospital admissions over 11 different outcomes collected over time length of 365 days. Available are also air quality covariates, including temperature, humidity and rainfall. This data set is a reduced data set from the source: Secretary for the Environment of the Rio de Janeiro City, Brazilian Ministry of Defence and Brazilian Ministry of Health. The data can be accessed in R using the commands:

```
> library(pgam)
> data(aihrio)
> attach(aihrio)
```

Consider a dynamic Poisson model

$$y_t \mid \lambda_t \sim \text{Poisson}(\lambda_t),$$

$$\log \lambda_t = \beta_{0t} + x_{1t}\beta_{1t} + x_{2t}\beta_{2t} + x_{3t}\beta_{3t}$$

$$\beta_{it} = \beta_{i,t-1} + \zeta_{1i}, \quad \zeta_{it} \sim N(0, Z_i), \quad i = 0, 1, 2, 3,$$

where y_t denotes the count of admission ITRESP65 (this is the first variable of admissions) x_{1t} is the maximum temperature at day t, x_{2t} is the humidity of day t and x_{3t} is the total rainfall at time t. Here we may assume that ζ_{it} and ζ_{jt} are independent for $i \neq j$ and a normal prior is set for $\beta_{i,0} \sim N(0, 100)$. Fit this model using the Liu and West particle filter PF-II (see p. 311). You will have to set suitable priors for the estimation of the hyperparameters Z_1, Z_2, Z_3.

18. In Exercise 15 observations consisting of annual number of telephone calls made in Belgium were considered. Make the histogram or otherwise of this data to demonstrate that the data are highly skewed, suggesting that an exponential distribution might describe the data better than the Gaussian assumption, considered in the Gaussian state space model of Exercise 15. Suggest a suitable non-Gaussian state space model with the exponential distribution as a response and re-analyse the data using this model.

19. In Exercises 21 data consisting of monthly totals of car drivers killed or seriously injured were considered. Re-analyse the data suggesting a non-Gaussian state space model. Compare the performance of this state space model with your analysis in Exercise 21.

20. Hundred observations of annual total flow from the river Nile are collected from 1871 to 1970 (source: Durbin and Koopman (2012)). The data are available in R via the library MASS:

```
> library(MASS)
> data(Nile)
> y <- Nile
```

This data set was considered in Exercise 16 in Chap. 4. Suggest a suitable non-Gaussian state space model to analyse this data.

21. Hundred observations are collected on internet usage per minute (source: Makridakis et al. (1998)). The data are available in R via the library MASS:

```
> library(MASS)
> data(WWWusage)
> y <- WWWusage
```

Make the histogram of this data or otherwise establish that a Gaussian state space model is not appropriate to describe this data. Suggest a non-Gaussian state space model and use it to analyse this data.

22. One hundred and fifty-three daily observations of ozone from May to September in New York are collected (source Rousseeuw and Leroy (1987, p. 86, Table 6)). Ozone reflects on air-quality, see e.g. Bersimis and Triantafyllopoulos (2020). Collected are also daily measurements on solar radiation, wind speed and temperature. The data are available in R via the library MASS:

```
> library(MASS)
> data(airquality)
```

Note that this data set was considered in Exercise 7 of Chap. 5. Suggest a suitable non-Gaussian state space model with Ozone as the response and variables Solar.R, Wind and Temp as time-varying covariates. See ?airquality for a description of the data. Fit this model to the data and provide credible intervals for the time-varying coefficients of the covariates.

23. In an experiment in Sweden the effect of speed limit is studied. Drivers were given the option of speed limit and the count of accidents were recorded (source

Svenson (1981); the data are also discussed in Venables and Ripley (2002)). The data consists of 184 daily data, with recorded count of accidents and a binary variable whether or not speed limit was applied. The data are available in R via the library MASS:

```
> library(MASS)
> data(Traffic)
```

Suggest a suitable non-Gaussian state space model for this data set. Fit the model to the data and provide an estimate of the probability of the speed limit being applied. Based on your analysis do you think the speed limit has a positive effect in the drivers' performance with the view to reduce accidents?

Chapter 7
The State Space Model in Finance

The application of the state space model to economics and finance has been in the forefront of development of the Kalman filter and related algorithms. From the local level model, introduced in 1960 by John Muth (1960), to the textbooks of Andrew Harvey (1989), the state space model has played a key role in financial econometrics. This chapter aims to give some of the applications of the Kalman filter to finance in order to illustrate its contribution to finance.

We begin in Sect. 7.1 by considering the problem of regression with autocorrelated error structure. This problem is known from the 1960s, with the work of Zellner and Tiao (1964). It is shown how a state space model can accommodate the inclusion of autocorrelated errors in a regression model. Section 7.2 discusses stationarity and causality in autoregressive models. This section does not relate directly to state space models, but it is necessary for the development of Sects. 7.3 and 7.5, which follow. In Exercise 10, stationarity conditions are used in order to determine tradable periods, within the context of statistical arbitrage. In Sects. 7.2.2 and 7.2.3, stationarity conditions are derived in the space of autoregressive coefficients. Stochastic volatility models are discussed in Sect. 7.3; the state space model has been very successful in describing volatility as a stochastic process using conditionally Gaussian state space models (see also Sects. 1.3.3 and 6.3). Section 7.3 discusses Bayesian inference of univariate stochastic volatility models, consisting of MCMC inference (Sect. 7.3.3) and particle filter-based inference (Sects. 7.3.4 and 7.3.5). In particular, Sect. 7.3.4 considers sequential Monte Carlo estimation for the same volatility model for which MCMC is discussed in Sect. 7.3.3. Section 7.3.5 considers a stochastic volatility model with returns exhibiting skewness and heavy tails and discusses sequential Monte Carlo inference for that model. Multivariate stochastic volatility models are discussed in Sect. 7.4. We start by extending some of the univariate models, and in Sect. 7.4.2 we discuss in some detail Wishart autoregressive processes that are used to describe the stochastic process of the volatility. The problem of asset allocation and optimal portfolio selection is discussed in Sect. 7.4.3; this includes unconstrained and constrained portfolio strategies, and data consisting of the common constituents of the Dow Jones Average Industrial index are used for illustration purposes. Statistical arbitrage, and in particular a strategy known as *pairs trading*, is discussed in Sect. 7.5: the basic idea is to take advantage of relative mispricing of two assets and aid decision-making for profitable trades. This is facilitated by detecting mean-reversion over time using state space modelling and is discussed in some detail in that section.

7.1   Regression with Autocorrelated Errors

In regression modelling, the well known Gauss–Markov assumptions state that the errors are independent or at least uncorrelated, see e.g. Bingham and Fry (2010). Early work in the statistics literature involves the relaxation of this assumption, by assuming that the errors are inter-dependent and estimating their time-invariant correlation by maximum likelihood techniques, see e.g. McGilchrist and Sandland (1979). In economic data, it is very common that after a regression model is fitted, the residuals are in fact autocorrelated. This phenomenon is important because failure to deal with such autocorrelations may lead to poor model fit. This problem is known to economists as the *time series problem* in regression, and there are many studies devoted to it, see e.g. Zellner and Tiao (1964) and Beach and MacKinnon (1978). Most work in the aforementioned literature is considering that the error terms in regression are following an AR(1) time series model. In this section we show that a general class of regression models with autocorrelated errors can be put in state space form.

Consider first the simple linear regression model with autocorrelated errors, defined by

$$y_t = \beta_0 + \beta_1 x_{1t} + \varepsilon_t, \tag{7.1}$$

where ε_t follows an autoregressive model of order one, abbreviated as AR(1), i.e.

$$\varepsilon_t = \phi \varepsilon_{t-1} + v_t, \quad v_t \sim N(0, \sigma_v^2). \tag{7.2}$$

Here, y_t is the response variable, x_{1t} is a time-varying covariate, β_0 and β_1 are static coefficients (intercept and slope) and ϕ_1 is the AR coefficient. The sequence $\{v_t\}$ is a white noise and so if $\phi = 0$, the above model reduces to the usual simple regression with independent errors. If, on the other hand $\phi \neq 0$, then ε_t and ε_s are dependent or correlated, for $t \neq s$. For a complete treatise of autoregressive time series models, the reader is referred to Box et al. (2008) and Brockwell and Davis (1991).

Model (7.1)–(7.2) can be put in state space form if we write

$$y_t = [1, x_{1t}, 1] \begin{bmatrix} \beta_0 \\ \beta_1 \\ \varepsilon_t \end{bmatrix} = x_t^\top \beta_t \tag{7.3}$$

and

$$\beta_t = \begin{bmatrix} \beta_0 \\ \beta_1 \\ \varepsilon_t \end{bmatrix} = \begin{bmatrix} 1 & 0 & 0 \\ 0 & 1 & 0 \\ 0 & 0 & \phi_1 \end{bmatrix} \begin{bmatrix} \beta_0 \\ \beta_1 \\ \varepsilon_{t-1} \end{bmatrix} + \begin{bmatrix} 0 \\ 0 \\ v_t \end{bmatrix}$$

$$= F\beta_{t-1} + \zeta_t.$$

We note that, with the definition of the state space model (3.10a)–(3.10b), $\epsilon_t = 0$ (with probability 1) or $\sigma^2 = 0$.

The above discussion motivates a more general model (time-varying regression with autocorrelated errors), defined by time-varying regression equation

$$y_t = \beta_{0t} + \beta_{1t} x_{1t} + \cdots + \beta_{pt} x_{pt} + \varepsilon_t, \tag{7.4}$$

together with the autocorrelated error equation

$$\varepsilon_t = \phi_1 \varepsilon_{t-1} + \cdots + \phi_d \varepsilon_{t-d} + v_t, \quad v_t \sim N(0, \sigma_v^2),$$

where $x_{1t}, \ldots, x_{pt}$ are p time-varying covariates, $\phi_1, \ldots, \phi_d$ are d static AR coefficients and $\{v_t\}$ is a white noise sequence. We assume that β_{it} follows a random walk, for $i = 0, 1, \ldots, p$. This model can be written in state space form with transition equation

$$\begin{bmatrix} \beta_{0t} \\ \beta_{1t} \\ \vdots \\ \beta_{pt} \\ \hline \varepsilon_t \\ \varepsilon_{t-1} \\ \vdots \\ \varepsilon_{t-d+1} \end{bmatrix} = \begin{bmatrix} 1 & 0 & \cdots & 0 & 0 & 0 & \cdots & 0 \\ 0 & 1 & \cdots & 0 & 0 & 0 & \cdots & 0 \\ \vdots & \vdots & \ddots & \vdots & \vdots & \vdots & \ddots & \vdots \\ 0 & 0 & \cdots & 1 & 0 & 0 & \cdots & 0 \\ \hline 0 & 0 & \cdots & 0 & \phi_1 & \phi_2 & \cdots & \phi_d \\ 0 & 0 & \cdots & 0 & 1 & 0 & \cdots & 0 \\ \vdots & \vdots & \ddots & \vdots & \vdots & \vdots & \ddots & \vdots \\ 0 & 0 & \cdots & 0 & 0 & 0 & \cdots & 0 \end{bmatrix} \begin{bmatrix} \beta_{0,t-1} \\ \beta_{1,t-1} \\ \vdots \\ \beta_{p,t-1} \\ \hline \varepsilon_{t-1} \\ \varepsilon_{t-2} \\ \vdots \\ \varepsilon_{t-d} \end{bmatrix} + \begin{bmatrix} \zeta_{0t} \\ \zeta_{1t} \\ \vdots \\ \zeta_{pt} \\ \hline v_t \\ 0 \\ \vdots \\ 0 \end{bmatrix}$$

or $\beta_t = F\beta_{t-1} + \zeta_t$ and with observation equation

$$y_t = [1, x_{1t}, \ldots, x_{pt}, 1, 0, \ldots, 0]\beta_t = x_t^\top \beta_t.$$

The covariance matrix of ζ_t is given by

$$
\mathbf{Z}_t = \left[\begin{array}{cccc|cccc}
Z_{0t} & 0 & \cdots & 0 & 0 & 0 & \cdots & 0 \\
0 & Z_{1t} & \cdots & 0 & 0 & 0 & \cdots & 0 \\
\vdots & \vdots & \ddots & \vdots & \vdots & \vdots & \ddots & \vdots \\
0 & 0 & \cdots & Z_{pt} & 0 & 0 & \cdots & 0 \\
0 & 0 & \cdots & 0 & \sigma_v^2 & 0 & \cdots & 0 \\
0 & 0 & \cdots & 0 & 0 & 0 & \cdots & 0 \\
\vdots & \vdots & \ddots & \vdots & \vdots & \vdots & \ddots & \vdots \\
0 & 0 & \cdots & 0 & 0 & 0 & \cdots & 0
\end{array}\right].
$$

We can see model (7.3) is a special case of the above model, if we set $p = 1$ and $d = 1$. If we set $\phi_1 = \phi_2 = \cdots = \phi_d = 0$, then $\varepsilon_t = v_t$ is an i.i.d. sequence, and this reduces the model to a time-varying regression model with independent errors, discussed in Sect. 4.1.5 above. If we set $Z_{0t} = Z_{1t} = \cdots = Z_{pt} = 0$, then $\beta_{it} = \beta_{i,t-1}$, for all i, and the model is reduced to a static regression model with autocorrelated errors. Some coefficients β_{jt} may be static (by setting $Z_{jt} = 0$), but other coefficients β_{kt} may be time-varying (by setting $Z_{kt} > 0$), e.g. this may be the case, if static covariates may be included.

Example 7.1 (IBM and Intel Share Prices)   In this example we consider historical prices of IBM and Intel Corporation share prices (in US$). These shares trade at the Dow Jones Industrial Index and are provided by http://finance.yahoo.com/. The data are recorded in daily frequency (excluding weekends) from 18 June 2001 to 17 June 2003 and they are depicted in Fig. 7.1. We observe the two share prices are evolving, following a similar pattern, but Intel prices appear to have larger variance throughout time. This similar pattern is closely linked to cointegration, for a discussion of which see Engle and Granger (1987).

It is believed that since both IBM and Intel operate in similar markets, their prices will be related in some way. The first model we consider is the static regression model of Intel share price at time t (covariate x_{1t}) on the IBM share price at t (response y_t), given by

$$
y_t = \beta_0 + \beta_1 x_{1t} + \varepsilon_t, \quad \varepsilon_t \sim N(0, 1).
$$

This model assumes that $\{\varepsilon_t\}$ is a white noise (independent errors). Using ordinary linear regression, we obtain the estimates of β_0 and β_1 as $\hat{\beta}_0 = 39.667$ and $\hat{\beta}_1 = 2.159$. The validity of the independence assumption of $\{\varepsilon_t\}$ can be measured by the residuals, which, if the above mentioned assumption is valid and the model fit is good (see e.g. Bingham and Fry (2010)), should form an independent or at least uncorrelated sequence as well.

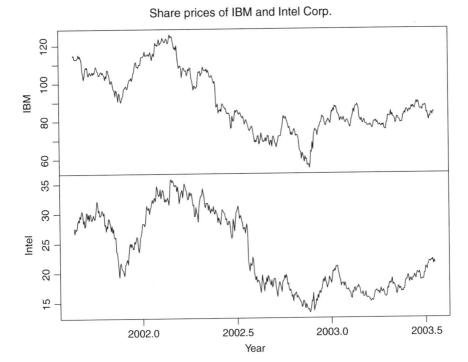

Fig. 7.1 IBM and Intel share prices

Motivated from this observation, we consider a static model with autocorrelated errors, that is, model (7.1)–(7.2). For this model, we have used $\phi_1 = 0.8$ and $\sigma_v^2 = 10$. For the initial state β_1, we have used $\beta_1 \sim N(0, 0.001\mathbf{I})$. Figure 7.2 plots the one-step forecasts using the above model with autocorrelated errors, together with the one-step forecasts from the previous model with independent errors and the actual prices of IBM. Clearly, the model with autocorrelated errors produces forecasts much closer to the observed, while the other model fails to provide reasonable forecasts after 2002.

7.2   Stationarity and Autoregressive Models

7.2.1   Stationarity and Causality

The notion of stationarity has been central to time series analysis and econometrics. Stationarity is a characteristic of some stochastic processes, which studied how

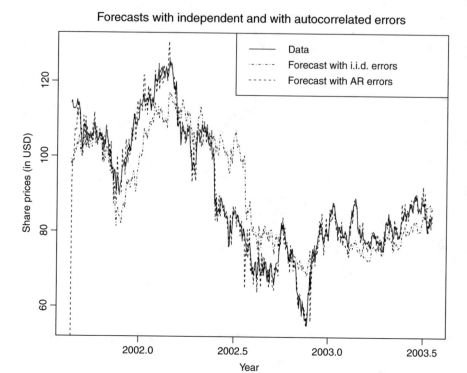

Fig. 7.2 Forecasts using linear regression with independent (i.i.d.) errors and with autocorrelated (AR) errors for the IBM–Intel data

stable is a stochastic process. In particular, a stochastic process $\{y_t\}$ is said to be *strictly stationary*, if the joint distribution of $(y_{t_1}, y_{t_2}, \ldots, y_{t_k})^\top$ is the same as the joint distribution of $(y_{t_1+h}, y_{t_2+h}, \ldots, y_{t_k+h})^\top$, for any selection of points of time $t_1, \ldots, t_k$ and some shift h. This suggests that the stochastic process is *stable* as the distribution of the stochastic process over any subset of time is the same, i.e. y_1, y_2 has the same distribution with y_{1001}, y_{1002}. An important implication of this is that the mean $\mathrm{E}(y_t)$, the variance $\mathrm{Var}(y_t)$ and the covariance $\mathrm{Cov}(y_t, y_{t+h})$ do not depend on time t. As strict stationarity is very restrictive to be attained by real-life processes, a weaker form of stationarity is often employed. According to this, a process $\{y_t\}$ is said to be *weakly stationary* or *second-order stationary* if $\mathrm{E}(y_t)$, $\mathrm{Var}(y_t)$ and $\mathrm{Cov}(y_t, y_{t+h})$ do not depend on time t, see e.g. Brockwell and Davis (2016, Chapter 2).

Example 7.2 (AR(1) Model) We consider the autoregressive model of order one, abbreviated as AR(1), defined as

$$y_t = \phi y_{t-1} + \epsilon_t,$$

where $\{\epsilon_t\}$ is a white noise process (independently and identically distributed, with zero mean and some variance σ^2). For $|\phi| < 1$, this process can be written as

$$y_t = \sum_{j=0}^{\infty} \phi^j \epsilon_{t-j}. \tag{7.5}$$

Using this form, it is easy to verify that the mean of y_t is $E(y_t) = 0$, the variance is

$$\text{Var}(y_t) = \sum_{j=0}^{\infty} \phi^2 \text{Var}(\epsilon_{t-j}) = \frac{\sigma^2}{1 - \phi^2}$$

and the covariance is

$$\text{Cov}(y_t, y_{t+h}) = E(y_t y_{t+h}) = \sigma^2 \sum_{j=0}^{\infty} \phi^j \phi^{j+h} = \frac{\sigma^2 \phi^h}{1 - \phi^2},$$

for $h > 0$. Hence, the process $\{y_t\}$ is weakly stationary for $|\phi| < 1$. If $|\phi| > 1$, the representation (7.5) is not valid. In this case it is possible to express y_t as a series of ϵ_{t+s}, for $s > 0$ and to show that $\{y_t\}$ is still stationary (for a proof see Brockwell and Davis (2016, Chapter 2)). If $\phi = \pm 1$, then $\{y_t\}$ is non-stationary (note that for $\phi = 1$ the model is reduced to a random walk model). However, as it is not desirable to express y_t as a function of ϵ_{t+s}, for $s > 0$, we shall restrict our focus to $|\phi| < 1$. This guarantees stationarity, and it ensures that from (7.5) y_t is written as a linear combination of ϵ_{t-s}, for $s \geq 0$.

As it is desirable that y_t is expressed as a linear combination of present and past elements of ϵ_{t-s},

The above discussion of the AR(1) model motivates the notion of *causality*, which is closely related to stationarity. Consider the general autoregressive model of order p, abbreviated as AR(p), defined as

$$y_t = \phi_1 y_{t-1} + \cdots + \phi_p y_{t-p} + \epsilon_t, \tag{7.6}$$

where ϕ_i are p AR coefficients and $\{\epsilon_t\}$ is a white noise sequence with variance σ^2. If there is an infinite linear representation of y_t in terms of ϵ_{t-s}, for $s \geq 0$ as such

$$y_t = \sum_{j=0}^{\infty} \psi_j \epsilon_{t-j},$$

for some known coefficients ψ_j, then it is said that $\{y_t\}$ is *causal*. In other words, this means that y_t can be determined by past and present values of ϵ_t. Hence for the AR(1) model discussed above, if $|\phi| < 1$, the model is causal and stationary. A central question in time series analysis is to give conditions on the space of the AR parameters to ensure stationarity and causality of model (7.6). The answer to this question is provided by the zeroes of the so-called characteristic polynomial

$$\phi(z) = 1 - \phi_1 z - \phi_2 z^2 - \cdots - \phi_p z^p, \qquad (7.7)$$

where z is a complex-valued argument.

Indeed, model (7.6) is stationary if the roots ρ_j of $\phi(z)$ satisfy $\rho_j \neq \pm 1$, and it is causal if $|\rho_j| > 1$, for any $j = 1, \ldots, p$ (perhaps including multiple roots). For a proof of this fundamental result, the reader is referred to Brockwell and Davis (1991), Brockwell and Davis (2016, Chapter 3). It follows that the condition $|\rho_j| > 1$ ensures that model (7.6) is causal and stationary. For example (7.2), $\phi(z) = 1 - \phi z$ and the single root is $\rho = 1/\phi$. Hence the condition $|\rho| > 1$ of causality coincides with the condition $|\phi| < 1$, which was established in Eq. (7.5). From the above discussion, it is clear that in order to establish causality and stationarity of the AR(p) model, the roots of the polynomial (7.7) need to be computed for specific values of $\phi_1, \ldots, \phi_p$.

There has been a significant amount of literature dealing with the problem of deriving causality conditions for model (7.6) or conditions to establish that the roots of $\phi(z)$ lie outside the unit circle in the complex plain. According to Chipman (1950, pp. 370–371), who provides a historical account of this topic, Schur and Cohn have developed a set of conditions for all roots of $\phi(z)$ to lie outside the unit circle in the complex plain. The algorithm, which is known as the Schur–Cohn algorithm, avoids the direct calculation of the roots of $\phi(z)$, which for large p may not be possible. Instead, it proposes a number of inequalities based on the calculation of determinants of matrices with dimensions which increase at each step. Samuelson (1941) and Wise (1956) have independently developed effectively identical conditions for causality and stationarity of model (7.6), which avoid the computation proposed by Schur and Cohn. Their conditions consist of a number of iterative inequalities involving the coefficients $\phi_1, \ldots, \phi_p$. However, the iterative nature of these inequalities is criticised by Barndorff-Nielsen and Schou (1973). For cases of AR(2) and AR(3), it is possible to derive simple conditions to ensure the roots of $\phi(z)$ lie outside the unit circle. Other than for education purposes, the advantage of these conditions is that they give us a vision of the causality and stationarity region for these popular models. In the next sections we discuss these conditions and provide in detail their derivations.

7.2.2   Stationarity Conditions for AR(2)

Consider the autoregressive model of order two (AR(2)), given by (7.6). Causality and stationarity are implied for this model, if the roots of the characteristic polynomial

$$\phi(x) = 1 - \phi_1 x - \phi_2 x^2 \tag{7.8}$$

lie outside the unit circle. For any pair of ϕ_1 and ϕ_2, one can find the roots of $\phi(x)$ and see whether they lie outside the unit circle in the complex plane. However, a more elaborate way is to find a causality and stationarity region implied from $|x| > 1$ and involving only ϕ_1 and ϕ_2 so that there will be no need to compute the two roots of $\phi(x)$ each time. We shall prove that the necessary and sufficient conditions for the causality and stationarity of AR(2) are

$$\phi_1 + \phi_2 < 1 \tag{7.9}$$

$$-\phi_1 + \phi_2 < 1 \tag{7.10}$$

$$|\phi_2| < 1. \tag{7.11}$$

These conditions are proven in Priestley (1981), Box et al. (2008) and Shumway and Stoffer (2017, pp. 89–90), and we shall follow a similar but slightly more detailed proof here.

Necessity   First we shall show that if the model is causal and stationary (or that the roots of (7.8) lie outside the unit circle), then conditions (7.9)–(7.11) are satisfied. Set $z = 1/x$ so that (7.8) becomes $\phi(z)^* = z^2 - \phi_1 z - \phi_2$, and from the causality assumption we have $|\rho_{1,2}| < 1$, where ρ_i are the two roots of $\phi(z)^*$. From the binomial, the two roots are $\rho_{1,2} = (\phi_1 \pm \sqrt{\phi^2 + 4\phi_2})/2$ and so we have $\rho_1 \rho_2 = -\phi_2$. Hence, $|\phi_2| = |\rho_1||\rho_2| < 1$, and hence (7.11) is satisfied. Also,

$$\phi_1 + \phi_2 = \rho_1 + \rho_2 - \rho_1 \rho_2 = \rho_1(1 - \rho_2) + \rho_2 < 1,$$

as $\rho_1 < 1$ and $1 - \rho_2 > 0$. Hence, (7.9) is satisfied. Likewise,

$$-\phi_1 + \phi_2 = -\rho_1 - \rho_2 - \rho_1 \rho_2 = -\rho_1(1 + \rho_2) - \rho_2 < 1,$$

as $-\rho_1 < 1$ and $1 + \rho_2 > 0$. Hence, (7.10) is satisfied too.

Sufficiency Now we shall show that conditions (7.9)–(7.11) are sufficient for the stationarity and causality of AR(2) or that if they are satisfied, the roots of (7.8) lie outside the unit circle.

First assume that $\phi_1^2 + 4\phi_2 \geq 0$ (so that the roots ρ_1 and ρ_2 are real). We shall show $|\rho_i| < 1$, for $i = 1, 2$. If $\phi_1 \geq -\sqrt{\phi_1^2 + 4\phi_2}$, then $|\rho_1| = (\phi_1 + \sqrt{\phi_1^2 + 4\phi_2})/2$. The proof is by contradiction. Suppose that $|\rho_1| \geq 1$, then $\phi_1 + \sqrt{\phi_1^2 + 4\phi_2} \geq 2$ or $\phi_1^2 + 4\phi_2 \geq (2 - \phi_1)^2$, which implies $\phi_1 + \phi_2 \geq 1$. This contradicts condition (7.9), hence $\rho_1| < 1$. If $\phi_1 < -\sqrt{\phi_1^2 + 4\phi_2}$, then $|\rho_1| = -(\phi_1 + \sqrt{\phi_1^2 + 4\phi_2})/2$, and assuming as before $|\rho_1| \geq 1$, we obtain $\sqrt{\phi_1^2 + 4\phi_2} \leq -\phi_1 - 2 \leq 0$, which is again a contradiction, as $\phi_1 > -2$. Hence, in any case, $|\rho_1| < 1$. The proof of $|\rho_2| < 1$ is similar and is left to the reader as an exercise.

Suppose now that $\phi_1^2 + 4\phi_2 < 0$ (the roots ρ_1 and ρ_2 are complex and conjugate). Since ρ_1 and ρ_2 are conjugate, they have the same modulus and so it suffices to show $|\rho_1| < 1$ only. We notice that $\rho_1 = \phi_1/2 + i\sqrt{-\phi_1^2 - 4\phi_2}/2$. Then $|\rho_1| = \sqrt{-\phi_2}$, with $\phi_2 < 0$. Thus, $|\rho_1|^2 = -\phi_2 = |\phi_2| < 1$, from condition (7.11). This completes the proof. Figure 7.3 shows the stationarity region of the AR(2) model.

7.2.3 Stationarity Conditions for AR(3)

Consider the autoregressive model of order two (AR(3)), given by (7.6). Causality and stationarity are implied for this model, if the roots of the characteristic polynomial

$$\phi_1 + \phi_2 + \phi_3 < 1, \tag{7.12}$$

$$-\phi_1 + \phi_2 - \phi_3 < 1, \tag{7.13}$$

$$\phi_3(\phi_3 - \phi_1) - \phi_2 < 1, \tag{7.14}$$

$$|\phi_3| < 1. \tag{7.15}$$

These conditions are stated without proof in Barndorff-Nielsen and Schou (1973, p. 409). A proof of this result may be derived directly from the Schur–Cohn criterion or by using the Samuelson conditions (Samuelson, 1941) and is provided in Okuguchi and Irie (1990). Below we provide an alternative proof, which is motivated by the proof for the AR(2) model in Sect. 7.2.2.

We show that conditions (7.12)–(7.15) are necessary and sufficient for the stationarity of $\{y_t\}$ generated by an AR(3) model. We first give some preliminary material used in the proof.

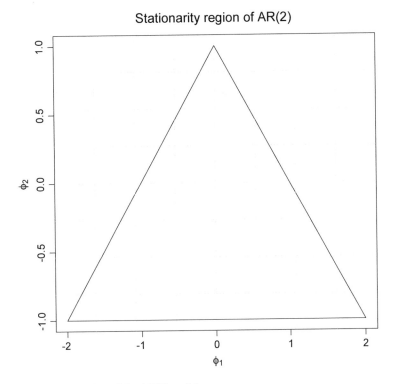

Fig. 7.3   Stationarity region of the AR(2) model

Let $\phi(x) = 1 - \phi_1 x - \phi_2 x^2 - \phi_3 x^3$ be the characteristic polynomial (in the complex-valued x). The time series $\{y_t\}$ is stationary if and only if the roots of $\phi(x)$ lie outside the unit circle or equivalently if the roots of

$$z^3 - \phi_1 z^2 - \phi_2 z - \phi_3 = 0 \tag{7.16}$$

are within the unit circle, where $z = x^{-1}$.

We give the correspondence of the roots ρ_1, ρ_2 and ρ_3 of (7.16) and the coefficients ϕ_1, ϕ_2 and ϕ_3. We write (7.16) as

$$(z - \rho_1)(z - \rho_2)(z - \rho_3) = 0$$

and expand it to get

$$z^2 - (\rho_1 + \rho_2 + \rho_3)z^2 + (\rho_1\rho_2 + \rho_1\rho_3 + \rho_2\rho_3)z - \rho_1\rho_2\rho_3 = 0. \tag{7.17}$$

If we compare Eqs. (7.16) and (7.17), we obtain

$$\phi_1 = \rho_1 + \rho_2 + \rho_3, \tag{7.18}$$

$$\phi_2 = -\rho_1\rho_2 - \rho_1\rho_3 - \rho_2\rho_3, \tag{7.19}$$

$$\phi_3 = \rho_1\rho_2\rho_3. \tag{7.20}$$

Necessity We show that if $\{y_t\}$ is stationary, then (7.12)–(7.15) are satisfied. Under the assumption of stationarity, we have $|\rho_i| < 1$ for all $i = 1, 2, 3$, which from (7.20) immediately implies condition (7.15).

Since $p = 3$, either ρ_1, ρ_2 and ρ_3 are all real, or one of them is real and the other two are conjugate complex roots. First consider ρ_1, ρ_2 and ρ_3 are real.

$$
\begin{aligned}
\phi_1 + \phi_2 + \phi_3 &= \rho_1 + \rho_2 + \rho_3 - \rho_1\rho_2 - \rho_1\rho_3 + \rho_1\rho_2\rho_3 \\
&= \rho_1(1 - \rho_2) + \rho_3(1 - \rho_2) - \rho_1\rho_3(1 - \rho_2) + \rho_2 \\
&= (1 - \rho_2)(\rho_1 + \rho_3 - \rho_1\rho_3) + \rho_2 \\
&< 1 - \rho_2 + \rho_2 = 1,
\end{aligned} \tag{7.21}
$$

since $|\rho_2| < 1$ and

$$\rho_1 + \rho_3 - \rho_1\rho_3 = \rho_1(1 - \rho_3) + \rho_3 < 1, \quad \text{as} \quad |\rho_1| < 1.$$

Similarly for (7.13), we have

$$
\begin{aligned}
-\phi_1 + \phi_2 - \phi_3 &= -\rho_1 - \rho_2 - \rho_3 - \rho_1\rho_2 - \rho_2\rho_3 - \rho_1\rho_3 - \rho_1\rho_2\rho_3 \\
&= -\rho_1(1 + \rho_2) - \rho_3(1 + \rho_2) - \rho_1\rho_3(1 + \rho_2) - \rho_2 \\
&= (1 + \rho_2)(-\rho_1 - \rho_3 - \rho_1\rho_3) - \rho_2 \\
&< 1 + \rho_2 - \rho_2 = 1,
\end{aligned} \tag{7.22}
$$

since $|\rho_2| < 1$ and

$$-\rho_1 - \rho_3 - \rho_1\rho_3 = -\rho_1(1 + \rho_3) - \rho_3 < 1, \quad \text{as} \quad |\rho_1| < 1.$$

Finally, for (7.13), we have

$$
\begin{aligned}
\phi_3(\phi_3 - \phi_1) - \phi_2 &= \rho_1\rho_2\rho_3(\rho_1\rho_2\rho_3 - \rho_1 - \rho_2 - \rho_3) + \rho_1\rho_2 + \rho_1\rho_3 + \rho_2\rho_3 \\
&= (1 - \rho_1\rho_3)(\rho_1\rho_2 - \rho_1\rho_2^2\rho_3 + rho_2\rho_3) + \rho_1\rho_3 \\
&= (1 - \rho_1\rho_3)[\rho_1\rho_2(1 - \rho_2\rho_3) + \rho_2\rho_3] + \rho_1\rho_3 \\
&< 1 - \rho_1\rho_3 + \rho_1\rho_3 = 1,
\end{aligned} \tag{7.23}
$$

since $|\rho_1\rho_3| < 1$ and

$$\rho_1\rho_2(1 - \rho_2\rho_3) + \rho_2\rho_3 < 1, \quad \text{as} \quad |\rho_1\rho_2| < 1.$$

Now suppose that there are two conjugate complex roots and one real root. Without loss of generality, suppose that ρ_2 is real and ρ_1 and ρ_3 are the two complex roots. Write $\rho_1 = a + bi$ and $\rho_3 = a - bi$, for some real $a.b$, where i denotes the imaginary unit. We have $\rho_1 + \rho_3 - \rho_1\rho_3 = 2a - a^2 - b^2 < 1$, since $|\rho_1| = \sqrt{a^2 + b^2} < 1$; hence, from (7.21), we have $\phi_1 + \phi_2 + \phi_3 < 1$. Similarly, from $-\rho_1 - \rho_3 - \rho_1\rho_3 = -2a - a^2 - b^2 < 1$ and (7.22), we obtain $-\phi_1 + \phi_2 - \phi_3 < 1$. Also,

$$\rho_1\rho_2(1 - \rho_2\rho_3) + \rho_2\rho_3 = \rho_2(a + bi)[(1 - \rho_2(a - bi)] + \rho_2(a - bi)$$
$$= 2\rho_2 a - (\rho_2 a)^2 - (\rho_2 b)^2 < 1,$$

since ρ_2 is real and $(\rho_2 a)^2 + (\rho_2 b)^2 < 1$. Thus from (7.23), it is $\phi_3(\phi_3 - \phi_1) - \phi_2 < 1$. This proves (7.12)–(7.15) for complex roots.

Sufficiency   Now we prove that if conditions (7.12)–(7.15) are satisfied, then $\{y_t\}$ is stationary. From condition (7.15) and Eq. (7.20), at least one of $|\rho_1|$, $|\rho_2|$ and $|\rho_3|$ must be strictly less than one. Without loss of generality, suppose $|\rho_2| < 1$.

Consider first the case of real roots ρ_1, ρ_2 and ρ_3. Assume that $\{y_t\}$ were not stationary, that is, $|\rho_1| \geq 1$ or $|\rho_3| \geq 1$. If $\rho_3 > 1$, then $\rho_1 + \rho_3(1 - \rho_1) \geq \rho_1 + 1 - \rho_1 = 1$, and hence from (7.21), we have $\phi_1 + \phi_2 + \phi_3 \geq 1$, which contradicts condition (7.12). If $\rho_3 < -1$, then $-\rho_1 - \rho_3(1 + \rho_1) \geq -\rho_1 + 1 + \rho_1 = 1$, and hence from (7.22), we have $-\phi_1 + \phi_2 - \phi_3 \geq 1$, which contradicts condition (7.13). By interchanging the roles of ρ_1 and ρ_3, we obtain that $|\rho_1| \geq 1$ contracts either (7.12) or (7.13). Hence, it is necessarily $|\rho_1| < 1$, $|\rho_2| < 1$ and $|\rho_3| < 1$, i.e. $\{y_t\}$ is stationary.

Consider now the case of two conjugate complex roots. As before and without loss of generality, we assume that ρ_1 and ρ_3 are complex, while ρ_2 is real. Write as before $\rho_1 = a + bi$ and $\rho_3 = a - bi$. As before, from condition (7.15) and $|\phi_3| = |\rho_1||\rho_2||\rho_3| < 1$, we have that at least one of ρ_1, ρ_2 and ρ_3 has modulus less than one; without loss of generality assume $|\rho_2| < 1$. Suppose that it was $|\rho_1| = |\rho_3| = \sqrt{a^2 + b^2} \geq 1$. From (7.23), we have

$$\phi_3(\phi_3 - \phi_1) - \phi_2 = (1 - a^2 - b^2)[2\rho_2 a - (\rho_2 a)^2 - (\rho_2 b)^2] + a^2 + b^2. \quad (7.24)$$

Put $u = a^2 + b^2$, $A = 2\rho_2 a - (\rho_2 a)^2 - (\rho_2 b)^2$ and $B = (1 - u)A + u$. Since $u \geq 1$, if $A \leq 1$, we have $B \geq 1$. We note that for $A \leq 0$, $(1 - u)A \geq 0$ and so $B = (1 - u)A + u \geq 1$, since $u \geq 1$. If $0 < A \leq 1$, then $(1 - A)u \leq 1 - A$ or $B = (1 - u)A + u \geq 1$. We can see that $A > 1$ is not possible. Indeed with the definition of A as above, we have

$$(\rho_2 a - 1)^2 = 1 + \rho_2^2 a^2 - 2\rho_2 a \geq -\rho_2^2 b^2 \quad \text{or} \quad A = 2\rho_2 a - \rho_2^2 a^2 - \rho_2^2 b^2 \leq 1.$$

Thus, from (7.24), we have $\phi_3(\phi_3 - \phi_1) - \phi_2 \geq 1$, which contradicts (7.14). Hence $|\rho_1| < 1$, $|\rho_2| < 1$ and $|\rho_3| < 1$, i.e. $\{y_t\}$ is stationary.

7.3 Univariate Stochastic Volatility Models

7.3.1 Returns and Volatility

In finance interest frequently lies on the valuation and price of financial assets. An asset can be a share or stock trading in the stock market, or a commodity, or some other financial instrument. An investor might be interested in a particular asset or a number of assets and wishing to hold a portfolio of the most profitable assets. Focusing on a single asset, it is long claimed that the prices of assets follow a random walk process, with variance which is erratically increasing; this is the so-called *random walk hypothesis*, see e.g. Cootner (1964) and Fama (1965). The large variance of the random walk over time has the implication that the forecast variance of a future price of the asset is very high and so forecasts are totally unreliable; see e.g. Example 3.3 and Exercise 6. Instead it is much more common to work on the *return* of an asset. Let p_t denote the price of an asset at time t, and let us assume we sample p_t at equally spaced intervals (usually at daily frequency). The *simple return* of the asset is

$$y_t^{(s)} = p_t - p_{t-1},$$

suggesting that y_t is the shock needed to be added to p_{t-1} to give p_t. An alternative to the simple returns is the so-called *geometric* returns defined as

$$y_t^{(g)} = \frac{p_t - p_{t-1}}{p_{t-1}}.$$

Finally, widely used are the logarithmic returns (or log-returns) defined as

$$y_t^{(l)} = \log p_t - \log p_{t-1}.$$

Using the well known identity, $\log(x + 1) \approx x$, for small x. If $p_t/p_{t-1} - 1$ is small, then $y_t^{(l)} \approx y_t^{(g)}$. Hence statistical analysis using log-returns will provide similar results to analysis using geometric returns. The above definition of the log-returns suggests the model $p_t = p_{t-1} \exp[y_t^{(l)}]$, while that of the geometric returns suggests the model $p_t = (y_t + 1)p_{t-1}$.

The returns using any of the definitions above are expected to have mean close to zero, as it is expected that historically p_t is close to p_{t-1}; this is more pronounced when we sample the returns at daily frequency. The variance of the returns, which is known as *volatility*, plays an important role in financial decisions and has been on the centre of financial econometrics over the past 50 years. For the purposes of this chapter, we shall define the volatility at time t, as the conditional variance of y_t, i.e.

$$\sigma_t^2 = \text{Var}(y_t \mid \sigma_t^2),$$

where y_t denotes the return at time t. Properties of returns and volatility are the subject of many studies, sometimes referred to as *stylised facts* of the returns and volatility. The following is a short summary, and for more information the reader is referred to Tsay (2002, Chapter 1).

1. the historical mean of the returns is very close to zero;
2. the returns have heavy tails, typically heaver than the Gaussian and Student t distributions;
3. the distribution of the returns is asymmetric, typically following a left skew distribution;
4. the volatility is time-varying;
5. the volatility is observed in clusters, i.e. there are periods of time with a certain level of volatility.

Skewness in the returns suggests that positive and negative returns are not equally likely. In periods of market decline (as for example in the credit crunch of 2008), investors lose confidence and negative returns exhibit low frequency; instead, in periods of growth, investments increase and the positive returns are more likely. The characteristics of skewness are studied by many econometricians, see e.g. Bakshi et al. (2003) and the references therein. The tails of the returns are heavier than the Gaussian and the Student t distributions. Hence there is significant mass under negative and positive returns of considerable magnitude, in either side. This is demonstrated empirically in many studies and theoretically in others, see e.g. Bingham and Kiesel (2002). For a detailed discussion of the characteristics of returns and their distribution the reader is referred to Tsay (2002, Chapter 1).

7.3.2 Stochastic Volatility Model

Engle (1982) and Bollerslev (1986) introduced the generalised autoregressive conditional heteroskedastic (GARCH) models to estimate the volatility σ_t^2. GARCH and their numerous generalisations assume that the volatility is a function of the lagged squared returns. Let y_t denote the log-returns and σ_t^2 the volatility at time t, and let $\{\epsilon_t\}$ be a sequence of independent innovations. The GARCH specification sets

$$y_t = \sigma_t \epsilon_t, \quad \epsilon_t \sim N(0, 1)$$

and

$$\sigma_t^2 = \alpha_0 + \alpha_1 y_{t-1}^2 + \cdots + \alpha_p y_{t-p}^2 + \beta_1 \sigma_{t-1}^2 + \cdots + \beta_q \sigma_{t-q}^2, \qquad (7.25)$$

for some positive integers p and q and some parameters α_i and β_j with $\alpha_0 > 0$ and $\sum_{i=1}^{p} \alpha_i + \sum_{j=1}^{q} \beta_j < 1$. Given a set of observed returns $y_1, \ldots, y_n$, estimates of α_i and β_j may be obtained by maximising the log-likelihood function

$$\log p(y_1, \ldots, y_n \mid \alpha_i, \beta_j) = -\frac{n}{2} \log(2\pi) - \frac{1}{2} \sum_{t=1}^{n} \log \sigma_t^2 - \frac{1}{2} \sum_{t=1}^{n} \frac{y_t^2}{\sigma_t^2},$$

where σ_t^2 is given in (7.25). Since σ_t^2 is a deterministic function of α_i and β_j, we can readily obtain the maximum likelihood estimate of σ_t^2 by plugging in (7.25) the maximum likelihood estimates of α_i and β_j. In this specification the values of p, q are assumed known, but it may be possible to be estimated from the data. Other specifications of the GARCH model involve a number of improvements including replacing the Gaussian distribution of the innovations ϵ_t by a Student t distribution. For a good review of GARCH-type volatility models, the reader is referred to Tsay (2002).

As pointed out in Harvey et al. (1994), the GARCH family of volatility models suffers from three drawbacks: (1) the volatility is given as a deterministic function of past squared returns, (2) it is not parsimonious as it includes many parameters and the likelihood maximisation might suffer from local maxima and (3) it does not offer good generalisations to multivariate models, without the compromise of the curse of dimensionality (too many parameters). As a result in the 90s efforts were devoted to developing alternative models, which would overcome the above drawbacks. The idea of treating the volatility σ_t^2 as a stochastic process was advocated by a number of authors, see e.g. Harvey et al. (1994) and Jacquier et al. (1994). Both of these studies adopt the stochastic volatility model described in Sects. 1.3.3 and 6.3, but they develop different inference. In the sequel we shall discuss the basic model adopted by many authors, see e.g. Jacquier et al. (1994) and Kim et al. (1998).

Consider that log-returns $\{y_t\}$ are generated from the model

$$y_t = \exp(h_t/2)\epsilon_t, \quad \epsilon_t \sim N(0, 1), \qquad (7.26a)$$

$$h_t - \mu = \phi(h_{t-1} - \mu) + \omega_t. \quad \omega_t \sim N(0, \sigma_\omega^2), \qquad (7.26b)$$

where $\{\epsilon_t\}$ is an independent sequence of innovations, $\{\omega_t\}$ is an independent sequence of innovations and ϵ_t is independent of ω_s, for any t, s. The model can be cast in state space form as in Sect. 1.3.3. From the observation model (7.26a), the returns are distributed as

$$y_t \mid h_t \sim N\left[0, \exp(h_t)\right];$$

hence, h_t is the logarithm of the volatility at time t. From the evolution (7.26b), the log-volatility h_t follows an autoregressive model of order one, for a discussion of which see Example 7.2. Following the discussion of this example, the unconditional distribution of this model is

$$h_t \mid \mu, \phi, \sigma_\omega \sim N\left[\mu, \frac{\sigma_\omega^2}{1-\phi^2}\right],$$

for $-1 < \phi < 1$. The hyperparameters of this model are μ (the mean of h_t), ϕ (the autoregressive coefficient) and σ_ω^2 (the variance of the innovations ω_t). The model is completed by setting a prior distribution for h_0, i.e. $h_0 \sim N(m, C)$, for some parameters m and C, which may depend on the hyperparameters.

By taking logarithms in the square returns y_t^2, Eq. (7.26a) can be written as

$$\log y_t^2 = h_t + \log \epsilon_t^2, \quad \epsilon_t^2 \sim \chi_1^2. \tag{7.27}$$

This process linearises the non-Gaussian state space model (7.26a)–(7.26b). The squared innovations ϵ_t^2 follow a chi square distribution with one degree of freedom. As it is pointed out in Harvey et al. (1994, page 250), the mean and variance of the random variable $\log \epsilon_t^2$ are approximately equal to -1.27 and $\pi^2/2$, respectively. The derivation of these moments makes use of approximations of the digamma and trigamma functions, see Exercise 3. Hence, Harvey et al. (1994) propose the following state space model

$$\log y_t^2 = -1.27 + h_t + \xi_t,$$

$$h_t = \mu(1 - \phi) + \phi h_{t-1} + \omega_t, \quad \omega_t \sim N(0, \sigma_\omega^2),$$

where ξ_t follows a shifted log-gamma distribution, with zero mean and variance $\pi^2/2$. Assuming that μ, ϕ and σ_ω^2 are known, the extended Kalman filter may be applied and is used to calculate the likelihood function conditional on these hyperparameters; for a detailed discussion of the extended Kalman filter see Sect. 6.6. The hyperparameters μ, ϕ and σ_ω^2 are estimated by quasi-maximum likelihood estimation. One of the most attractive properties of this model is that it can be easily generalised to the multivariate case when y_t forms a vector of log-returns of several assets; for more details, the reader is referred to Harvey et al. (1994). In the next section we discuss Markov chain Monte Carlo estimation (MCMC) inference essentially proposed in Jacquier et al. (1994).

7.3.3   MCMC Inference of Stochastic Volatility Models

Consider that log-returns are generated from model (7.26a)–(7.26b), and let $\theta = [\mu, \phi, \sigma_\omega^2]^\top$ be the vector of hyperparameters. Given a set of observed returns

$y_1, \ldots, y_n$, we want to estimate the log-volatility process $\{h_t\}$ and the hyperparameters μ, ϕ and σ_ω^2.

Write $y = [y_1, \ldots, y_n]^\top$ the vector of all observations (returns) and $h = [h_1, \ldots, h_n]^\top$ the vector of all log-volatilities. Following a Gibbs sampling scheme, we wish to sample from the conditional distributions

1. h, conditionally on y and μ, ϕ and σ_ω^2;
2. μ, ϕ, conditionally on y, h and σ_ω^2, and σ_ω^2, conditionally on y, h, μ and ϕ.

Step (2) is simpler and we start with that. First observe that from Eq. (7.26b), conditionally on h and y and μ, ϕ and σ_ω^2 are independent; hence, we drop y from step (2). Conditionally on h, (7.26b) can be seen as a linear model, with observation h_t, intercept $\mu(1 - \phi)$, slope ϕ and innovation variance σ_ω^2. This model can be written compactly as $h = \mathbf{X}\beta + \omega$, where $\beta = [\mu(1 - \phi), \phi]^\top$ and

$$
\mathbf{X} = \begin{bmatrix} 1 & h_0 \\ 1 & h_1 \\ \vdots & \vdots \\ 1 & h_{n-1} \end{bmatrix}, \quad \omega = \begin{bmatrix} \omega_1 \\ \omega_2 \\ \vdots \\ \omega_n \end{bmatrix}
$$

so that $\omega \sim N(0, \sigma_\omega^2 \mathbf{I})$. Conditionally on σ_ω^2, the posterior distribution of β is $\beta \mid h, \sigma_\omega^2 \sim N(\hat{\beta}, \mathbf{P})$, where $\hat{\beta}$ and $\mathbf{P}$ are provided by Eq. (2.22).

For the estimation of σ_ω^2, we consider an inverse gamma prior, i.e.

$$
\sigma_\omega^2 \sim IG\left(\frac{\nu}{2}, \frac{S}{2}\right),
$$

for some known ν and S. Then the posterior distribution of $1/\sigma_\omega^2$ is

$$
p\left(\frac{1}{\sigma_\omega^2} \mid h, \mu, \phi\right) \propto p(h \mid \mu, \phi, \sigma_\omega^2) p\left(\frac{1}{\sigma_\omega^2}\right)
$$

$$
\propto \left(\frac{1}{\sigma_\omega^2}\right)^{(\nu+1)/2} \exp\left\{-\frac{1}{2}\left[(h - \mathbf{X}\beta)^\top (h - \mathbf{X}\beta) + S\right]\right\}
$$

so that

$$
\sigma_\omega^2 \mid h, \mu, \phi \sim IG\left(\frac{\nu^*}{2}, \frac{S^*}{2}\right), \tag{7.28}
$$

with

$$
\nu^* = \nu + 1 \quad \text{and} \quad S^* = S + (h - \mathbf{X}\beta)^\top (h - \mathbf{X}\beta).
$$

Step (2) suggests that conditionally on a sample of h and σ_ω^2, we can sample μ and ϕ using the bivariate Gaussian distribution above, and then conditionally on h, μ and ϕ, we can sample σ_ω^2 using the inverse gamma distribution (7.28).

Moving on to step (1), we need to provide the conditional distribution of the log-volatilities h, given y and μ, ϕ and σ_ω^2. This is more involved, because this conditional distribution is not easy to sample from. In this step we need to resort to a Metropolis move. We start by looking at the above conditional distribution.

For any $1 \leq t \leq n - 1$, we have

$$p(h_t \mid h_{t-1}, h_{t+1}, y_t, \theta) \propto p(y_t \mid h_t) p(h_t \mid h_{t-1}, \theta) p(h_{t+1} \mid h_t, \theta)$$

$$= \frac{1}{\sqrt{2\pi \exp(h_t)}} \exp\left[-\frac{y_t^2}{2\exp(h_t)}\right] \frac{1}{\sqrt{2\pi}\sigma_\omega}$$

$$\times \exp\left\{-\frac{1}{2\sigma_\omega^2}\left[(h_t - \mu(1-\phi) - \phi h_{t-1})^2\right.\right.$$

$$\left.\left. + (h_{t+1} - \mu(1-\phi) - \phi h_t)^2\right]\right\} \qquad (7.29)$$

$$\propto \exp\left[-\frac{1}{2}\left(h_t + \frac{y_t^2}{\exp(h_t)}\right) - \frac{(1+\phi^2)}{2\sigma_\omega^2}(h_t - \lambda_t)^2\right],$$

$$(7.30)$$

where we have completed the square in (7.29), resulting in

$$\lambda_t = \frac{\mu(1-\phi)^2 + (1-\phi)(h_{t+1} + h_{t-1})}{1 + (1-\phi)^2}.$$

For $t = n$, we have

$$p(h_n \mid h_{n-1}, y_n, \theta) \propto p(y_n \mid h_n) p(h_n \mid h_{n-1}, \theta)$$

$$\propto \exp\left\{-\frac{1}{2}\left[h_n + \frac{y_n^2}{\exp(h_n)}\right.\right.$$

$$\left.\left. + (h_n - \mu(1-\phi) - \phi h_{n-1})^2\right]\right\}. \qquad (7.31)$$

It is not easy to sample from distributions (7.30) and (7.31) because of the denominator of the fractions $y_t^2 / \exp(h_t)$ and $y_n^2 / \exp(h_n)$. Hence, in order to sample from these distributions, we need to resort to a Metropolis step. The general Metropolis–Hastings algorithm is described in Sect. 6.8.1 and is discussed in Gamerman and Lopes (2006, Section 4.6). In order to apply the Metropolis step here, we need to choose a proposal kernel, from which we can sample a draw $h_t^{(j)}$, for each time t. There are two ways this can be achieved: (a) we can simulate $h_t^{(j)}$ from the prior distribution $h_t \mid h_{t-1}, \mid \mu, \phi, \sigma_\omega^2$ and deploy rejection sampling to select $h_t^{(j)}$ or to keep $h_t^{(j-1)}$ and (b) we can use a random walk chain $h_t^{(j)} = h_t^{(j-1)} + w_j$ to sample $h_t^{(j)}$ and then adopt an accept/reject step as in Sect. 6.8.1 (for a definition of w_j see p. 318). In scheme (a), we create an independent chain, and the Markov property is guaranteed by the simple accept/reject step; general discussion on independent chains in Metropolis is given in Gamerman and Lopes (2006, Section 6.3.3) and discussion in the context of stochastic volatility is given in Jacquier et al. (1994).

The proposed scheme (with (a) or (b)) proposes a hybrid MCMC algorithm, which cycles through between the Gibbs step (1) and the Metropolis step (2), and is outlined below.

MCMC Algorithm for the Stochastic Volatility Model
In the stochastic volatility model (7.26a)–(7.26b), with the priors on μ, ϕ and σ_ω^2 as above, the following apply:

1. Set initial values of $\mu^{(0)}$, $\phi^{(0)}$ and $\sigma_\omega^{2(0)}$.
   Set initial values of log-volatilities $h_0^{(0)}, h_1^{(0)}, \ldots, h_n^{(0)}$. For each time $t = 1, 2, \ldots, n$ and for $i, j = 1, 2, \ldots, N$:

   a. Generate h^* from the proposal $N(h_t^{(j-1)}, V)$, for some variance V.
   b. Calculate the acceptance probability $\alpha(h_t^{(j-1)}, h^*) = \min(1, p(h^*)/p(h_t^{(j-1)})$, where $p(h_t^{(j-1)})$ is calculated using (7.30) and (7.31), conditionally on $\mu = \mu^{(i-1)}$, $\phi = \phi^{(i-1)}$ and $\sigma_\omega^{2(i-1)}$. Draw a single u from a uniform distribution $U(0, 1)$. If $u < \alpha(h_t^{(j-1)}, h^*)$, the proposal h^* is accepted and we set $h_t^{(j)} = h^*$, otherwise the move is rejected and we set $h_t^{(j)} = h_t^{(j-1)}$.

2. Conditionally on $\sigma_\omega^{2(i-1)}$, draw $\beta^{(i)} = [\mu^{(i)}, \phi^{(i)}]^\top$ from $N(\hat{\beta}^{(i-1)}, \mathbf{P}^{(i-1)})$, where $\hat{\beta}^{(i-1)}$ and $\mathbf{P}^{(i-1)}$ are provided by (2.22) if we set $h = h^{(i-1)}$ using the linear model of p. 358.
3. Conditionally on $\mu^{(i)}$ and $\phi^{(i)}$, draw $\sigma_\omega^{2(i)}$ from the inverted gamma distribution (7.28).

Some comments are in order. Step 1 above (the Metropolis step) is indicated by (a) and (b) in the above algorithm and generate the log-volatilities for given hyperparameters μ, ϕ and σ_ω^2. Step 2 (Gibbs sampling) is indicated by (1)–(3) in the algorithm above. Hence, this is a hybrid MCMC algorithm, for a discussion of which the reader is referred to Gamerman and Lopes (2006).

7.3.4   *Particle Filter Inference of Stochastic Volatility Models*

In this section we shall present two stochastic volatility models and we shall discuss
sequential Monte Carlo inference for both of them.

   Consider model (7.26a)–(7.26b), and let $\theta = [\mu, \phi, \sigma_\omega^2]^\top$ be the vector of
hyperparameters as before. The prior of h_t, given h_{t-1} and θ, follows from (7.26b)
as

$$h_t \mid h_{t-1}, \theta \sim N[\mu(1 - \phi) + \phi h_{t-1}, \sigma_\omega^2]. \qquad (7.32)$$

Assume initial value for the log-volatilities $h_0^{(i)}$ (these may be set initially or
simulated from a Gaussian distribution with zero mean and some variance). Given
θ, we can simulate $h_t^{(i)}$ from prior (7.32) (Bootstrap filter) and $y_t \mid h_t^{(i)}, \theta \sim$
$N[0, \exp(h_t)]$. We can then apply the Bootstrap filter (see Sect. 6.7.4).

   However, in practice θ will not be known and subject to estimation. Two possible
approaches are the approaches of Storvik (2002) and Liu and West (2001). The
resulting Storvik filter is similar to step (2) of the MCMC stochastic volatility
algorithm described in the previous section and is briefly described in Sect. 6.7.8.
The Liu and West filter is described in detail in Sect. 6.7.8.2. The Liu and West filter
(Particle Filter Algorithm II) is summaried on p. 311.

   For the application of the Liu and West filter, we have chosen the following
priors:

$$h_0 \sim N(0, 10), \quad \mu \sim N(0, 10), \quad \phi \sim U(-1, 1) \quad \text{and} \quad \sigma_\omega^2 \sim G(2, 2).$$

Some comments on these priors are in order. The prior of the log-volatility h_0 at
$t = 0$ is set to be Gaussian with zero mean and variance 10. The zero mean reflects
weak belief on whether the log-volatility is positive or negative and the variance of
10 reflects moderate uncertainty around this. Similar comments apply to the choice
of the prior distribution of μ, the mean of h_0. From the condition of stationarity
$-1 < \phi < 1$ of the AR coefficient ϕ, a uniform distribution $U(-1, 1)$ is chosen.
This simple consideration is a non-informative prior specification for ϕ. Finally, the
gamma prior distribution of σ_ω^2 is chosen so that the prior mean of σ_ω^2 is equal to
one (which seems a moderate/small value) with associated variance $1/2$.

   In the application of the Liu and West filter, we consider Gaussian mixtures for
the estimation of μ, ϕ and σ_ω^2. Since the support of σ_ω^2 is $[0, \infty)$ and the support of
ϕ is $(-1, 1)$, in the mixtures, we use $\log \sigma_\omega^2$ and $\log[(1 - \phi)/(1 + \phi)]$, which both
are $\mathbb{R}$.

   We fit the above model to the log-returns of the IBM closing of Sect. 1.3.3.
The data, plotted in Fig. 1.7, consist of 1776 log-return observations sampled at
daily frequency (trading days). The Liu and West filter is fitted using the above
priors and 1000 particles at each point of time. Following some experimentation,
we use $\delta = 0.995$ the value of the discount factor (used to compute the smoothing
factor $\alpha = (3\delta - 1)/(2\delta)$ of mean of each mixture, see p. 311). This corresponds

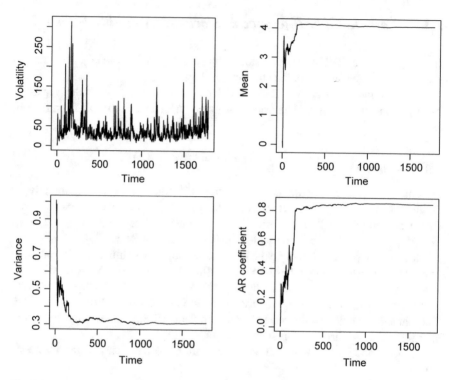

Fig. 7.4 Posterior mode of volatilities $\exp(h_t)$ and hyperparameters μ (mean), σ_ω^2 (variance) and ϕ (AR coefficient)

to a value of the smoothing parameter $\alpha = 0.997$. Figure 7.4 shows the modes of the posterior samples of the log-volatilities h_t (top left panel), the mean μ (top right panel), the variance σ_ω^2 (bottom left panel) and the AR coefficient ϕ (bottom right). The posterior modes of μ, σ_ω^2 and ϕ converge to stable values; the average of μ is 4.016, of σ_ω^2 is 0.331 and of ϕ is 0.8 (rounded to 3 decimal points). These stable values validate the choice we have made of these hyperparameters to be time-invariant. The value of ϕ confirms that the process $\{h_t\}$ is stationary, but with quite high autocorrelation structure. At each point of time, we can calculate histograms or empirical distributions of the posterior sample of h_t, μ, σ_ω^2 and ϕ and so we can compute posterior credible intervals from these samples. Figure 7.5 shows the histograms of h_{1776}, μ, σ_ω^2 and ϕ at time $t = 1776$. Hence we can use Fig. 7.6 to zoom out and see the big picture (estimation over time) and Fig. 7.5 to zoom in and study the posterior distribution of the parameters at a particular point of time.

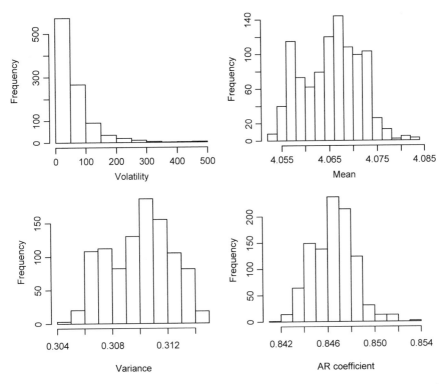

Fig. 7.5 Histograms at $t = 1776$ of volatilitiy $\exp(h_{1776})$ and hyperparameters μ (mean), σ_ω^2 (variance) and ϕ (AR coefficient)

7.3.5   Particle Filter Inference of Stochastic Volatility Models with Asymmetric Returns

One disadvantage of the models discussed above is the Gaussian distribution assumed for the log-returns in (7.26a). As is discussed in Sect. 7.3.1, financial returns have heavy tails (typically heavier than the normal distribution) and are skewed and not symmetric. The asymmetry of the returns reflects upon the fact that negative returns have different probability mass than positive returns. Hence, the normal or Gaussian distribution assumed for the returns is not appropriate. The need for considering asymmetric and heavy tailed distribution for returns is discussed in the finance literature, see e.g. Bingham and Kiesel (2002), Bingham and Kiesel (2004) and Ass and Haff (2006) among others.

A suitable asymmetric distribution is the skew Student t distribution of Fernandez and Steel (1998); this distribution is suitable for modelling financial returns, because it incorporates fat tails and asymmetry. In the sequel we shall briefly describe it. Suppose that a random variable ϵ following a symmetric distribution,

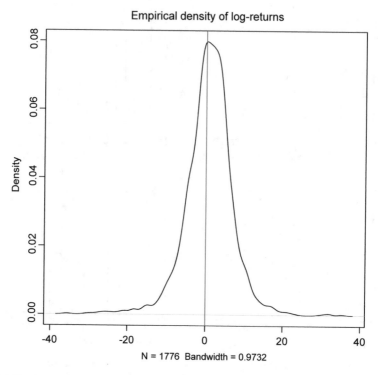

Fig. 7.6 Empirical density of the IBM log-returns: there appears to be slight positive skewness

e.g. the standard Student t distribution with $\nu > 0$ degrees of freedom, with density function

$$f(\epsilon) = \frac{\Gamma\left(\frac{\nu+1}{2}\right)}{\sqrt{\nu\pi}\,\Gamma\left(\frac{\nu}{2}\right)}\left(1 + \frac{\epsilon^2}{\nu}\right)^{-(\nu+1)/2}.$$

Fernandez and Steel (1998) introduce a distribution with density $p(x)$ as the asymmetric or skew version of $f(\cdot)$

$$p(\epsilon \mid \gamma) = \frac{2}{\gamma + \frac{1}{\gamma}}\left[f\left(\frac{\epsilon}{\gamma}\right)I_{[0,\infty]}(\epsilon) + f(\gamma\epsilon)I_{(-\infty,0)}(\epsilon)\right], \qquad (7.33)$$

where $\gamma \neq 0$ is a parameter which controls the skewness of the distribution and $I_A(\epsilon)$ denotes the indicator function, so that $I_A(\epsilon) = 1$, if $\epsilon \in A$ and is zero otherwise, where A is a subset of the domain of $f(\cdot)$. Equation (7.33) suggests that if $\epsilon \geq 0$, the $p(\epsilon \mid \gamma)$ is the symmetric density $f(\cdot)$ scaled by γ, while if $\epsilon < 0$, $p(\epsilon \mid \gamma)$ is the symmetric density $f(\cdot)$ scaled by $1/\gamma$. In particular, (7.33) implies that

$$p(\epsilon \mid \gamma) = p\left(-\epsilon \mid \frac{1}{\gamma}\right),$$

so that by inverting γ we gain the mirror image of $p(\epsilon)$ around zero.

We observe that by setting $\gamma = 1$, Eq. (7.33) implies $p(\epsilon \mid \gamma = 1) = f(\epsilon)$. Hence, $\gamma = 1$ returns the symmetric Student t distribution.

We shall show that

$$\frac{P(\epsilon \geq 0 \mid \gamma)}{P(\epsilon < 0 \mid \gamma)} = \gamma^2. \tag{7.34}$$

This clearly indicates that the mass allocated left of $\epsilon = 0$ and the mass allocated on the right of $\epsilon = 0$ are unequal (they are equal if and only if $\gamma = 1$).

To prove (7.34), first note that

$$P(\epsilon \geq 0 \mid \gamma) = \int_0^\infty p(u)\, du = \frac{2}{\gamma + \frac{1}{\gamma}} \int_0^\infty f\left(\frac{u}{\gamma}\right) du. \tag{7.35}$$

On the other hand,

$$P(\epsilon < 0 \mid \gamma) = \int_{-\infty}^0 p(u)\, du = \frac{2}{\gamma + \frac{1}{\gamma}} \int_{-\infty}^0 f(\gamma u)\, du.$$

We apply the following change of variable $\gamma u = s/\gamma$, so that

$$P(\epsilon < 0 \mid \gamma) = \frac{2}{\gamma + \frac{1}{\gamma}} \frac{1}{\gamma^2} \int_0^\infty f\left(\frac{s}{\gamma}\right) ds, \tag{7.36}$$

since $f(\cdot)$ is a symmetric distribution. Combining (7.35) and (7.36), we obtain (7.34) as required.

Let k be a positive integer. The k-th (raw) moment of ϵ, given γ, is

$$E(\epsilon^k \mid \gamma) = \frac{2}{\gamma + \frac{1}{\gamma}} \left(\gamma^{k+1} + \frac{(-1)^k}{\gamma^{k+1}}\right) 2 \int_0^\infty u^k f(u)\, du. \tag{7.37}$$

To prove this, first we write

$$E(\epsilon^k \mid \gamma) = \int_{-\infty}^\infty p(\epsilon \mid \gamma)\, d\gamma$$

$$= \frac{2}{\gamma + \frac{1}{\gamma}} \left[\int_0^\infty \epsilon^k f\left(\frac{\epsilon}{\gamma}\right) d\epsilon + \int_{-\infty}^0 \epsilon^k f(\gamma\epsilon)\, d\epsilon\right]. \tag{7.38}$$

With the change of variable $\epsilon/\gamma = u$, the first integral of (7.38) becomes

$$\int_0^\infty \epsilon^k f\left(\frac{\epsilon}{\gamma}\right) d\epsilon = \gamma^{k+1} \int_0^\infty u^k f(u)\, du. \tag{7.39}$$

With the change of variable $\gamma\epsilon = s$, the second integral of (7.38) is

$$\int_{-\infty}^0 \epsilon^k f(\gamma\epsilon)\, d\epsilon = \frac{1}{\gamma^{k+1}} \int_{-\infty}^0 s^k f(s)\, ds = \frac{(-1)^k}{\gamma^{k+1}} \int_0^\infty s^k f(s)\, ds. \tag{7.40}$$

Substituting (7.39) and (7.40) into (7.38), we obtain (7.37) .

Using the well known moments of the Student t distribution, we have that $E(\epsilon^k \mid \gamma) = 0$, if k is odd. If k is even, using (7.37), the k-th (raw) moment of ϵ is

$$E(\epsilon^k \mid \gamma) = \frac{2}{\gamma + \frac{1}{\gamma}}\left(\gamma^{k+1} + \frac{1}{\gamma^{k+1}}\right) \frac{v^{k/2}}{\sqrt{p}\,\Gamma\left(\frac{v}{2}\right)} \Gamma\left(\frac{k+1}{2}\right) \Gamma\left(\frac{v-k}{2}\right),$$

where $0 < k < v$. If $k \geq v$, the moment does not exist. It follows that $E(\epsilon \mid \gamma) = 0$ and

$$\mathrm{Var}(\epsilon \mid \gamma) = E(\epsilon^2 \mid \gamma) = \frac{2(\gamma^6 + 1)v}{\gamma^2(\gamma^2 + 1)(v - 2)}, \tag{7.41}$$

for $v > 2$.

A measure of skewness γ_M based on the mode, due to Arnold and Groeneveld (1995), is defined as

$$\gamma_M = 1 - 2F(M),$$

where $F(\cdot)$ is the cumulative distribution function and M is the mode of a unimodal distribution. Arnold and Groeneveld (1995) propose this measure to order continuous distributions according to their skewness. The mode of the skew t distribution of (7.33) is $M = 0$ (since $f(\epsilon)$ has mode 0). We observe that from $\int_{-\infty}^0 f(u)\, du = 1/2$, if we apply the change of variable $\gamma\epsilon = u$, we get

$$1 - 4\gamma \int_{-\infty}^0 f(\gamma\epsilon)\, d\epsilon = -1.$$

Now from this equation, we obtain

$$\gamma_M = 1 - 2F(0) = 1 - \frac{4\gamma}{\gamma^2 + 1}\int_{-\infty}^0 f(\gamma\epsilon)\, d\epsilon = \frac{\gamma^2 - 1}{\gamma^2 + 1}. \tag{7.42}$$

For $\gamma \in (0, \infty)$, the function $\gamma_M(\gamma)$ is strictly increasing (in γ) and $\gamma_M \in (-1, 1)$. We note that when $\gamma \approx 0$, then $\gamma_M \approx -1$; when $\gamma \to \infty$, then $\gamma_M \approx 1$. For $\gamma = 1$ (symmetric distribution), we have $\gamma_M = 0$; positive skewness is implied if $\gamma_M > 0$ and negative skewness is implied if $\gamma_M < 0$. For more details and properties of γ_M, the reader is referred to Arnold and Groeneveld (1995) and Fernandez and Steel (1998).

We turn our attention to the stochastic volatility model (7.26a)–(7.26b). As before, let y_t be the log-return at time t. We shall reconsider the Gaussian assumption of Eq. (7.26a) and we shall replace it by a skew t distribution. Hence, we redefine the stochastic volatility model as

$$y_t = \exp(h_t/2)\epsilon_t, \quad \epsilon_t \sim ST_v, \tag{7.43a}$$

$$h_t - \mu = \phi(h_{t-1} - \mu) + \omega_t. \quad \omega_t \sim N(0, \sigma_\omega^2), \tag{7.43b}$$

where ST_v denotes a skew t distribution with v degrees of freedom and with density (7.33). The assumptions of process $\{h_t\}$ are as in (7.26b), but we note that h_t is not the log-volatility in model (7.43a)–(7.43b). Indeed, for $v > 2$, from Eq. (7.41), the volatility of y_t (the conditional variance of the log-returns) is

$$\sigma_t^2 = \text{Var}(y_t \mid \gamma, h_t, v) = \frac{2(\gamma^6 + 1)v \exp(h_t)}{\gamma^2(\gamma^2 + 1)(v - 2)}. \tag{7.44}$$

The vector of hyperparameters is now $\theta = (\mu, \phi, \sigma_\omega^2, v, \gamma)^\top$. In the application of the Liu and West particle filter (Sect. 6.7.8.2), the priors of μ, ϕ and σ_ω^2 are the same as in Sect. 7.3.4. For the degrees of freedom v, we can set a gamma prior. However, as commented below, estimating the degrees of freedom this way due to the fat tails of the log-returns may result in an estimate $\hat{v} < 2$. In this case the volatility (7.44) does not exist. Although some other measure of the volatility can be considered (e.g. replacing the ratio $v/(v - 2)$ in (7.44) with some 'suitable' constant, it is generally recommended to avoid such an approach. In the implementation of this model for the IBM log-returns below, we specify $v = 3$, so that it is relatively low in order to capture fat tails and allow $\text{Var}(y_t \mid \gamma, h_t, v)$ to exist. It remains then to set the prior of γ. We set a gamma prior on γ, i.e. $\gamma \sim G(0.1, 0.1)$, so that $\text{E}(\gamma) = 1$ and $\text{Var}(\gamma) = 10$. This prior suggests that our prior belief of the returns is centred on a symmetric distribution ($\gamma = 1$), with variance 10.

Figure 7.6 plots the empirical density of the IBM log-returns. This figure indicates slight positive skewness (the mass of the positive values of the density is larger than that of the negative values, indicating asymmetry). We fit the stochastic volatility model (7.43a)–(7.43b) for this data. Figure 7.7 plots the posterior mode of volatility, together with modes of hyperparameters μ (mean), σ_ω^2 (variance), ϕ (AR coefficient) and γ (skewness parameter). These hyperparameters converge to stable values: the average of μ is 2.355, of σ_ω^2 is 0.514, of ϕ is 0.759 and of γ is 2.886. The degrees of freedom are set to $v = 3$, so that the variance of y_t (hence the volatility) exists.

The estimate volatility plotted on the top right panel of Fig. 7.7 is

$$\hat{\sigma}_t^2 = \frac{3(\hat{\gamma}^6 + 1)\exp(\hat{h}_t)}{\hat{\gamma}^2(\hat{\gamma}^2 + 1)},$$

where $\hat{\gamma} = 2.886$, $\nu = 3$ and $\hat{h}_t$ is the mode of $h_t^{(1)}, \ldots, h_t^{(1000)}$, for each point of time t. The mean (taken over time) of the posterior modes of the skewness parameter γ (2.886) suggests that the log-returns exhibit small positive skewness, which agrees with the shape of the distribution of the log-returns in Fig. 7.6. The skewness measure based on the mode of (7.42) is equal to 0.786, which again agrees with positive skewness. A formal comparison between the skew t and the Gaussian model (7.26a)–(7.26b) is not performed; however, we expect that the skew t model to perform better, as it incorporates a certain degree of fatness in the tails and asymmetry of the returns distribution. Figure 7.8 shows the histograms of volatility $\hat{\sigma}_{1776}^2$ and hyperparameters μ (mean), σ_ω^2 (variance), ϕ (AR coefficient) and γ (skewness parameter).

Figure 7.7 shows that the estimated volatility seems to pick higher peaks of the log-returns in the model incorporating asymmetry. We observe that in high volatile

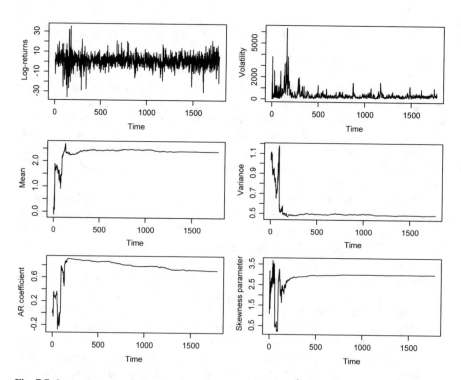

Fig. 7.7 Log-returns (top left), posterior mode of volatilities $\hat{\sigma}_t^2$ (top right) and posterior modes of hyperparameters μ (mean), σ_ω^2 (variance), ϕ (AR coefficient) and γ (skewness parameter)

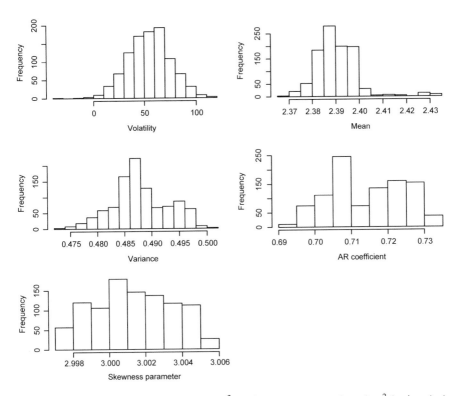

Fig. 7.8 Histograms at $t = 1776$ of volatility σ_t^2 and hyperparameters μ (mean), σ_ω^2 (variance), ϕ (AR coefficient) and γ (skewness parameter)

periods, such as in the beginning of the series (see top left panel of Fig. 7.7), the estimated volatility from the skew t model is considerably larger than that of the Gaussian model (compare Figs. 7.7 and 7.6). In periods of low volatility, the two methods are more comparable, see e.g. the histograms of the estimated volatility at $t = 1776$ (Figs. 7.5 and 7.8). Although a formal comparison between the two models is not performed, our analysis for this data set favours strongly the model with the incorporation of skew t distribution for the log-returns.

7.4   Multivariate Stochastic Volatility Models

7.4.1   Motivation and General Overview

Section 7.3 discussed inference for univariate stochastic volatility models. These models consider a scalar returns time series (or the returns of a single asset). However, in practice we are usually interested in the volatility and inter-dependence of several assets, for example, to enable asset allocation and optimise portfolio performance (Markowitz, 1959; Aguilar & West, 2000; Han, 2006; Brandt & Santa-Clara, 2006). The recognition that assets and other financial instruments populate financial markets and are therefore subject to market restrictions, co-dependences and socio-economic and political change. Hence, it is important to study returns as vector time series and their conditional covariance matrix as their volatility. To set our notation, consider that we hold $p > 1$ assets and at time t the i-th constituent return (log-return or geometric return, see Sect. 7.3.1) is denoted by y_{it}. We shall define the vector return at time t as

$$y_t = \begin{bmatrix} y_{1t} \\ \vdots \\ y_{pt} \end{bmatrix}. \tag{7.45}$$

The conditional covariance matrix of y_t

$$\Sigma_t = \text{Var}(y_t \mid \Sigma_t)$$

is known as the volatility matrix.

The volatility has been the centre of a large body of research over the past 40 years (Liesenfeld & Richard, 2003; Asai et al., 2006; Yu & Meyer, 2006). This is because Σ_t holds important information about the uncertainty of assets over time (the diagonal element of Σ_t is the marginal volatility) and the cross-correlation of the returns, which is important in the construction of portfolio allocation (Markowitz, 1959; Han, 2006; Brandt & Santa-Clara, 2006).

As in univariate volatility models, there are two main classes of volatility models, the multivariate generalised autoregressive conditional heteroskedasticity (MGARCH) and the multivariate stochastic volatility models. MGARCH models, which are reviewed in Bauwens et al. (2006), usually adopt a maximum likelihood inference. One of the challenges they face is the so-called *curse of dimensionality*, because there are too many parameters to estimate and likelihood maximisation might prove to be challenging and in risk to be affected by local maxima. Another challenge affecting the models is the restriction on the parameter space so that to make sure that the volatility matrix is a non-negative definite symmetric matrix (i.e. covariance matrix). In order to overcome these challenges, clever model specifications have been suggested, which trade-off model complexity and dimensionality reduction. The most notable studies are the conditional correlation specification of Engle (2002) and the improvement of this model, the orthogonal GARCH specification of Van Der Weide (2002). We shall briefly describe the above GO-GARCH to get an idea of what is the modelling set-up for these models.

Consider that a p-dimensional log-return vector y_t, defined in (7.45), is observed. It is assumed that y_t follows a multivariate normal distribution, with zero mean vector and covariance matrix, the volatility Σ_t, or $y_t \sim N(0, \Sigma_t)$. These authors make the assumption that y_t is governed by a linear combination of uncorrelated economic components x_t, or $y_t = \mathbf{Z}x_t$, where $\mathbf{Z}$ is a $p \times q$ parameter matrix and x_t is a q-dimensional column vector consisting of q independent component time series. If $\mathbf{H}_t = \mathrm{diag}(h_{1,t}, \ldots, h_{q,t})$ denotes the volatility matrix of $x_t = [x_{1t}, \ldots, x_{qt}]^\top$, then Van Der Weide (2002) considers a univariate GARCH(r, s) for each of the components x_{it}, $i = 1, \ldots, q$. For example, for $r = s = 1$, the GARCH specification for x_{it} is

$$h_{it} = (1 - \alpha_i - \beta_i) + \alpha_i y_{i,t-1}^2 + \beta_i h_{i,t-1},$$

where α_i and β_i are the parameters for the i-th GARCH model of x_{it}. This specification defines the volatility model for $\mathbf{H}_t$, and hence the volatility matrix $\boldsymbol{\Sigma}_t$ of y_t is $\boldsymbol{\Sigma}_t = \mathbf{Z}\mathbf{H}_t\mathbf{Z}^\top$. Estimation of $\mathbf{Z}$ and $h_{1t}, \ldots, h_{qt}$ is carried out by maximum likelihood estimation, for details of which the reader is referred to Van Der Weide (2002).

Other specifications of multivariate GARCH are available, including Bayesian inference of GARCH models, see e.g. Vrontos et al. (2003), Virbickaite et al. (2015) and Iqbal and Triantafyllopoulos (2019) among others.

Multivariate stochastic volatility models consider that the volatility (covariance) matrix $\boldsymbol{\Sigma}_t$ of y_t is a stochastic process. There are generally two main classes of model specifications: (a) extensions of the univariate stochastic volatility model (Sect. 7.3) as reported in Chib et al. (2006) and (b) models that use a suitable stochastic process for $\boldsymbol{\Sigma}_t$.

Below we shall briefly describe a simplified version of the model specification of Chib et al. (2006). In our specification we do not include jumps considered by Chib et al. (2006). Let us denote $y_t = [y_{1t}, y_{pt}]^\top$ be the log-returns vector and consider the model

$$y_t = \mathbf{B} f_t + u_t,$$

where f_t is a q-dimensional vector of factors ($q < p$), and $\mathbf{B} = (b_{ij})_{i,j=1,\ldots,q}$ is a parameter matrix, with $b_{ij} = 0$, for $j > i$ and $b_{ii} = 1$. The vector of innovations $u_t = [u_{1t}, \ldots, u_{pt}]^\top$ has the following specification. Each u_{it} follows a Student t distribution and u_{it} is independent of u_{jt}, for $i \neq j$, with the hierarchical structure

$$u_{it} = \frac{\epsilon_{it}}{\sqrt{\lambda_{it}}}, \quad \lambda_{it} \sim G\left(\frac{\nu_i}{2}, \frac{\nu_i}{2}\right),$$

where $\epsilon_t = [\epsilon_{1t}, \ldots, \epsilon_{pt}]^\top$ is independent of $f_t = [f_{1t}, \ldots, f_{qt}]^\top$, each of which follows multivariate Gaussian distributions with zero mean vectors and covariance matrices $\mathbf{V}_t$ and $\mathbf{D}_t$, or $\epsilon_t \sim N(0, \mathbf{V}_t)$ and $f_t \sim N(0, \mathbf{Q}_t)$ and ϵ_t is independent of f_t. The covariance matrices $\mathbf{V}_t$ and $\mathbf{D}_t$ are specified by

$$\mathrm{diag}(\mathbf{V}_t, \mathbf{D}_t) = \mathrm{diag}[\exp(h_{1t}), \ldots, \exp(h_{p+q,t})],$$

where h_{jt} is an autoregressive process

$$h_{jt} - \mu_j = \phi_j(h_{j,t-1} - \mu_j) + \zeta_{jt}, \quad \zeta_{jt} \sim N(0, \sigma_j^2) \qquad (7.46)$$

and ζ_{it} is independent of ζ_{jt}, for $i \neq j$ and $j = 1, \ldots, p+q$. The parameters of this model are $\mathbf{B}$, v_i, h_{jt}, μ_j, ϕ_j and σ_j^2, for $i = 1, \ldots, p$ and $j = 1, \ldots, p+q$.

Let β be the number of parameters of $\mathbf{B}$ after imposing the restriction on its elements (see above). Chib et al. (2006) show that sampling β and f_t in one block, conditionally on the other parameters and then sampling these parameters conditioned on $\mathbf{B}$ and f_t is ineffective and computationally slow for large values of p and q. Conditionally on $\mathbf{B}$, λ_j, y, the model can be decomposed into p univariate conditionally Gaussian state space models, with observation model

$$y_{it} \mid \mathbf{B}, \lambda_j \sim N(b_i \mathbf{D}_t b_i^\top + \lambda_{it} \exp(h_{it})), \qquad (7.47)$$

where $b_i = [b_{i1}, \ldots, b_{iq}]$ is the i-th row vector of B and h_{it} are specified as in (7.46), for $i = 1, \ldots, p$.

Hence we can draw samples of h_{jt}, μ_j, ϕ_j and σ_j^2 from each of the above univariate conditionally Gaussian state space models; note that model (7.46)–(7.47) is very similar to the univariate model (7.26a)–(7.26b), for which MCMC was discussed in Sect. 7.3.3. The quantity λ_{jt} can be sampled easily by using the gamma prior once h_{it} is sampled. Chib et al. (2006) discuss how $\mathbf{B}$ can be sampled once we have sampled other parameters, and for the full details the reader is referred to that reference. One of the advantages of this MCMC algorithm is that it devolves covariance sampling to a number of univariate sampling. This makes it scalable to both dimensions p and q. Hence it has been successful dealing with high dimensional returns.

In the next section we describe two stochastic volatility models, which model the volatility matrix $\mathbf{\Sigma}_t$ of the returns as a stochastic process directly (and not via log-volatilities as in the model above).

7.4.2   Wishart Autoregressive Stochastic Volatility Models

Considering the univariate stochastic volatility model (7.26a)–(7.26b) is natural, since we can always define $\sigma_t^2 = \text{Var}(y_t) = \exp(h_t)$. In the multivariate case, however, when y_t is a vector, it is more challenging to define $\mathbf{\Sigma}_t$ in a similar way. While Chib et al. (2006) define the volatilities (diagonal elements of $\mathbf{\Sigma}_t$) separately, their overall hierarchical model structure may not be always desirable to adopt. Ishihara et al. (2016) propose a different generalisation of model (7.26a)–(7.26b) using the matrix exponential, see also Chiu et al. (1996). All these approaches are based on the log-volatility being modelled with a Gaussian or other symmetric distribution. This is basically adopted so as to address the issue of the restrictions of $\mathbf{\Sigma}_t$, being a symmetric and positive definite (or more generally a non-negative definite) matrix. It has been therefore of interest to propose a variance–covariance stochastic process for Σ_t.

Shephard (1994a) proposes variance stochastic process to describe the evolution of univariate volatility. His *local scale* models assume a Gaussian model for the returns

$$y_t \mid \phi_t \sim N(0, \phi_t^{-1}), \tag{7.48}$$

where the volatility $\sigma_t^2 = \mathrm{Var}(y_t \mid \phi_t) = \phi_t^{-1}$ is specified in terms of the precision of the returns ϕ_t.

The precision ϕ_t follows a variance (local scale) law

$$\phi_t = \delta^{-1}\phi_{t-1}\eta_t, \tag{7.49}$$

where δ is a discount factor and η_t follows independently of ϕ_{t-1} a beta distribution

$$\eta_t \sim B[\delta v_{t-1}, (1-\delta)v_{t-1}], \tag{7.50}$$

with v_{t-1} being some degrees of freedom. Placing a prior gamma distribution on ϕ_{t-1}, it follows that the posterior distribution of ϕ_t is a gamma distribution, and from (7.48) it follows that, unconditionally of ϕ_t, the returns follow a Student t distribution. Also, it follows that the posterior distribution of the volatility σ_t^2 is an inverse gamma distribution. This model borrows the gamma variance law (7.49) first proposed in Smith and Miller (1986) and then used in Harvey and Fernandes (1989); in this book this variance law is detailed in Sect. 6.5.2 and in Exercise 5 (p. 330). The variance law (7.49)–(7.50) defines a random walk type evolution, in the sense that

$$E(\phi_t \mid \phi_{t-1}) = \phi_{t-1} \quad \text{and} \quad \mathrm{Var}(\phi_t \mid \phi_{t-1}) = \frac{(1-\delta)\phi_{t-1}^2}{\delta(v_{t-1}+1)}$$

and hence the name *local scale* model. Unlike model (7.26a)–(7.26b), which proposes an autoregressive type evolution of the volatility, model (7.48)–(7.50) proposes a local level or random walk type evolution for the volatility. This might be unacceptable in the long run (as volatility is expected to be mean stationary), but in the short term (locally), this setting can work. An attractive feature of model (7.48)–(7.50) is that it is analytically tractable and hence no approximation or simulation steps involved.

In the mid 1990s, there were efforts led by Harald Uhlig to generalise the gamma variance law of Smith and Miller (1986) and Shephard (1994a) to the multivariate case. The Wishart distribution was the obvious candidate to replace the gamma distribution of the precision matrix $\Phi_t = \Sigma_t^{-1}$, where Σ_t is the volatility matrix of a returns vector y_t. The multivariate beta distribution was already known and some results between the Wishart and multivariate beta are reported in Khatri and Pillai (1965) and Muirhead (1982). However, Uhlig (1994) observed that the Wishart-beta conjugacy of Muirhead (1982) implied that the degrees of freedom increased with no bound over time with the serious consequence that the mean of the precision to go to infinity. Uhlig (1994) realised that in order to preserve the degrees of freedom similarly as in (7.49)–(7.50), one needs to replace the multivariate beta by the singular multivariate beta distribution. Hence he re-established the Wishart-beta covariance law for a singular multivariate beta distribution and he proved a number of important results in Uhlig (1994). This work, of significant theoretical and practical consideration, has led to a number of papers that have improved the arguments put forward by Uhlig (1994) and solved a conjecture by Uhlig (Díaz-García & Gutiérrez, 1997, 1998); see also Konno (1988). In the context of stochastic volatility, Uhlig (1997) made use of the Wishart–singular beta conjugacy; his method involved a simulation step in order to estimate autoregressive parameters of the volatility stochastic process, but his approach is fast due to the conjugacy mentioned above.

This body of research has led to the so-called *Wishart autoregressive processes* (WAR), see e.g. Bru (1991), Philipov and Glickman (2006), Gourieroux et al. (2009), Triantafyllopoulos (2008b, 2011b, 2012, 2014), Hata and Sekine (2013), Bäuerle and Li (2013) and Yu et al. (2017) among others.

In the sequel we describe the model of Triantafyllopoulos (2012). Consider that, at time t, the log-return vector y_t follows a p-variate Gaussian distribution with mean vector μ and covariance matrix Σ_t, i.e.

$$y_t = \mu + \Sigma_t^{1/2}\epsilon_t, \quad \epsilon_t \sim N(0, \mathbf{I}), \tag{7.51}$$

where $\Sigma_t^{1/2}$ denotes the square root matrix of Σ_t, and the sequence of $\{\epsilon_t\}$ follows a p-dimensional Gaussian white noise process with unit diagonal variances (here $\mathbf{I}$ denotes the $p \times p$ identity matrix).

Define the precision matrix $\Phi_t = \Sigma_t^{-1}$ (assuming that matrix Σ_t is positive definite). Φ_t is assumed to follow a Wishart autoregressive process based on Uhlig's random walk representation (Uhlig, 1994). Initially it is assumed that Φ_0 follows a Wishart distribution with some known degrees of freedom $n_0 > p - 1$ and scale matrix $\mathbf{F}_0$, written as $\Phi_0 \sim W(n_0, \mathbf{F}_0)$ (see Sect. 5.5.2 for a discussion of the Wishart and inverse Wishart distributions). The transition model for Φ_t is defined by

$$\Phi_t = k\mathbf{A}\mathcal{U}(\Phi_{t-1})^\top \beta_t \mathcal{U}(\Phi_{t-1})\mathbf{A}^\top + \Lambda_t, \tag{7.52}$$

where k is a constant to be determined, $\mathbf{A}$ is a $p \times p$ autoregressive parameter matrix, Λ_t is a $p \times p$ symmetric non-negative definite matrix and $\mathcal{U}(\Phi_{t-1})$ denotes the upper triangular matrix of the Choleski decomposition of the matrix Φ_{t-1}. Matrix β_t follows a singular multivariate beta distribution with parameters $a/2$ and $b/2$ to be specified.

Triantafyllopoulos (2012) sets $\mathbf{\Lambda}_t = \mathbf{0}$, which is also used in Uhlig (1994, 1997), but it might be not appropriate in the long run as $t \to \infty$, because the above autoregressive structure means that $\mathbf{\Phi}_t$ will concentrate around the zero matrix. However, for the local scale models discussed here, this issue is not a problem. The autoregressive (AR) feature (or characterisation) of model (7.52) is depicted by noticing

$$E(\mathbf{\Phi}_t \mid \mathbf{\Phi}_{t-1}) = k A \mathcal{U}(\mathbf{\Phi}_{t-1})^\top E(\boldsymbol{\beta}_t) \mathcal{U}(\mathbf{\Phi}_{t-1}) A^\top$$
$$= A \mathbf{\Phi}_{t-1} A^\top,$$

where $E(\boldsymbol{\beta}_t) = k^{-1}\mathbf{I}$. The parameters a an b of the bets distribution are conveniently chosen (a is a function of δ and $b = 1$) so that $E(\boldsymbol{\beta}_t) = k^{-1}\mathbf{I}$. For the random walk model Uhlig (1994) shows that $\boldsymbol{\beta}_t$ has to follow a singular beta distribution for this to be possible, in order to have $b < p - 1$ (because for a non-singular distribution b is greater than $p - 1$ (Muirhead, 1982)).

Given data $y_{1:t} = \{y_1, \ldots, y_t\}$, we wish to provide the posterior distribution of the volatility matrix $\mathbf{\Sigma}_t$. Triantafyllopoulos (2012) adopts a two-step approach for inference. In Step 1 the posterior distribution $p(\mathbf{\Phi}_t \mid A, y_{1:t})$ of the precision matrix $\mathbf{\Phi}_t$ is provided. This is facilitated using the conjugacy of the beta and Wishart distributions, under the evolution (7.52). In Step 2 the AR parameter matrix A is estimated by maximising the log-posterior function; for this to end, Triantafyllopoulos (2012) proposes a Newton–Raphson method; for a discussion on posterior mode estimation, the reader is referred to Fahrmeir and Tutz (2001, Section 8.3.1). Details of both steps are provided in the sequel.

Conditionally on A, assume that $\mathbf{\Phi}_{t-1}$ has the posterior distribution $\mathbf{\Phi}_{t-1} \mid A, y_{1:t-1} \sim W(n + p - 1, \mathbf{F}_{t-1})$, where $\mathbf{F}_{t-1}$ implicitly depends on A and $n = (1 - \delta)^{-1}$, for a discount or forgetting factor $0 < \delta < 1$. Starting at $t = 1$, this is consistent with the prior of $\mathbf{\Phi}_0$, if we set $n_0 = n + p - 1$. If we specify $a = \delta(1 - \delta)^{-1} + p - 1$ and $b = 1$, we see from Uhlig (1994) that $k^{-1}A^{-1}\mathbf{\Phi}_t \mid A, y_{1:t-1} \sim W(\delta n + p - 1, \mathbf{F}_{t-1})$, or $\mathbf{\Phi}_t \mid A, y_{1:t-1} \sim W(\delta n + p - 1, kA\mathbf{F}_{t-1}A^\top)$. From the above, it is $E(\mathbf{\Phi}_{t-1} \mid A, y_{1:t-1}) = (n + p - 1)\mathbf{F}_{t-1}$ and $E(\mathbf{\Phi}_t \mid A, y_{1:t-1}) = (\delta n + p - 1)kA\mathbf{F}_{t-1}A^\top$, and so by equalising these two expectations, we obtain

$$k = \frac{n + p - 1}{\delta n + p - 1} = \frac{\delta(1 - p) + p}{\delta(2 - p) + p - 1}.$$

Under the above setting, this value of k guarantees the autoregressive property of the model, expressed by $E(\mathbf{\Phi}_t \mid A, y_{1:t-1}) = A E(\mathbf{\Phi}_{t-1} \mid A, y_{1:t-1})A^\top$.

We note that $a > p - 1$, but $1 = b < p - 1$, the latter of which being responsible for the singularity of the beta distribution. The singular beta density, being defined on the Stiefel manifold, replaces the determinant of $\mathbf{I} - \boldsymbol{\beta}_t$ (which is zero) by the only positive eigenvalue of that matrix (due to $b = 1$). On the other hand, the determinant of $\boldsymbol{\beta}_t$ remains positive as $a > p - 1$, and thus all p eigenvalues of $\boldsymbol{\beta}_t$ are positive;

this beta distribution is discussed in Uhlig (1994) and Díaz-García and Gutiérrez (1997, 1998).

Having established the prior $\Phi_t \mid \mathbf{A}, y_{1:t-1} \sim W(\delta n + p - 1, k\mathbf{A}\mathbf{F}_{t-1}\mathbf{A}^\top)$, the posterior distribution follows by a similar argument as in Triantafyllopoulos (2008b)

$$\Phi_t \mid \mathbf{A}, y_{1:t} \sim W(n + p - 1, \mathbf{F}_t), \tag{7.53}$$

where $e_t = y_t - \mu$ is the residual vector and $\mathbf{F}_t = (e_t e_t^\top + (k\mathbf{A}\mathbf{F}_{t-1}\mathbf{A}^\top))^{-1}$. From the above reference, the one-step forecast distribution of y_t is a p-variate Student t distribution with δn degrees of freedom and spread matrix $\delta^{-1}n^{-1}(k\mathbf{A}\mathbf{F}_{t-1}\mathbf{A}^\top)^{-1}$, i.e. $y_t \mid \mathbf{A}, y_{1:t-1} \sim t_p(\delta n, \mu, \delta^{-1}n^{-1}(k\mathbf{A}\mathbf{F}_{t-1}\mathbf{A}^\top)^{-1})$. This completes step 1 (conditional inference on the AR matrix $\mathbf{A}$).

Moving on to step 2, from the joint prior density $f(\Phi_t, \mathbf{A} \mid y_{1:t-1}) = f(\Phi_t \mid \mathbf{A}, y_{1:t-1}) f(\mathbf{A} \mid y_{1:t-1})$ and from an application of Bayes theorem for $(\Phi_t, \mathbf{A})$, we have $f(\Phi_t, \mathbf{A} \mid y_{1:t}) \propto f(y_t \mid \Phi_t) f(\Phi_t \mid \mathbf{A}, y_{1:t-1}) f(\mathbf{A} \mid y_{1:t-1})$, so that

$$f(\mathbf{A} \mid y_{1:t}) \propto f(\mathbf{A} \mid y_{1:t-1}) \int f(y_t \mid \Phi_t) f(\Phi_t \mid \mathbf{A}, y_{1:t-1}) \, d\Phi_t. \tag{7.54}$$

From the forecast distribution of y_t, the integral of (7.54) is

$$\int f(y_t \mid \Phi_t) f(\Phi_t \mid \mathbf{A}, y_{1:t-1}) \, d\Phi_t \propto |e_t e_t^\top + (k\mathbf{A}\mathbf{F}_{t-1}\mathbf{A}^\top)^{-1}|^{-(\delta n + p)/2},$$

and so

$$f(\mathbf{A} \mid y_{1:t}) \propto f(\mathbf{A}) \prod_{j=1}^{t} |e_j e_j^\top + (k\mathbf{A}\mathbf{F}_{j-1}\mathbf{A}^\top)^{-1}|^{-(\delta n + p)/2},$$

where $f(\mathbf{A})$ is the prior density of $\mathbf{A}$. The prior density of $\mathbf{A}$ is assumed to be a matrix-variate Gaussian distribution, i.e.

$$\mathbf{A} \sim N(\mathbf{M}_A, \mathbf{V}_A, \mathbf{W}_A), \tag{7.55}$$

where $\mathbf{M}_A = E(\mathbf{A})$ is the prior mean matrix and $\mathbf{V}_A$ and $\mathbf{W}_A$ are $p \times p$ left and right covariance matrices; see Gupta and Nagar (1999) for a detailed account on the matrix normal distribution, and Sect. 5.5.2 discusses matrix-variate Gaussian distributions. Given the above references, (7.55) implies that $\text{vec}(\mathbf{A})$ follows a p^2-dimensional multivariate distribution with mean vector $\text{vec}(\mathbf{A})$ and covariance matrix $\mathbf{W}_A \otimes \mathbf{V}_A$, where $\text{vec}(\cdot)$ denotes the column stacking operator of an unrestricted matrix and $\otimes$ denotes the Kronecker product of two matrices.

In order to find the mode $\hat{\mathbf{A}}$ of the posterior $f(\mathbf{A} \mid y_{1:t})$, we note that the matrix equation $\partial f(\mathbf{A} \mid y_{1:t})/\partial \mathbf{A} = \mathbf{0}$ (with respect to $\mathbf{A}$) does not appear to admit an analytical solution. Thus, we approximate the true mode $\hat{\mathbf{A}}$, by employing the Newton–Raphson method, according to which at each time t, for iteration $i = 1, 2, \ldots$, we compute $\hat{\mathbf{A}}^{(i)}$ using the formula

$$\text{vec}(\hat{\mathbf{A}}^{(i)}) = \text{vec}(\hat{\mathbf{A}}^{(i-1)}) + \left(\frac{\partial^2 \log f(\mathbf{A} \mid y_{1:t})}{\partial \text{vec}(\mathbf{A}) \partial \text{vec}(\mathbf{A})^\top} \right)^{-1} \Bigg|_{\mathbf{A}=\hat{\mathbf{A}}^{(i-1)}} \frac{\partial \log f(\mathbf{A} \mid y_{1:t})}{\partial \text{vec}(\mathbf{A})} \Bigg|_{\mathbf{A}=\hat{\mathbf{A}}^{(i-1)}},$$
(7.56)

where $\hat{\mathbf{A}}^{(0)}$ is initially given and $\text{vec}(\cdot)$ denotes the column stacking operator of an unrestricted matrix. Under some regulatory conditions (Shumway & Stoffer, 2017), the algorithm converges to the true mode $\hat{\mathbf{A}}$.

The log-posterior of $\mathbf{A}$ is

$$\log f(\mathbf{A} \mid y_{1:t}) = \log c + \log f(\mathbf{A}) - \frac{\delta n + p}{2} \sum_{j=1}^{t} \log |e_j e_j^\top + (k\mathbf{A}\mathbf{F}_{j-1}\mathbf{A}^\top)^{-1}|,$$

where c is the proportionality constant of $f(\mathbf{A} \mid y_{1:t})$.

From the prior density (7.55) of $\mathbf{A}$, we have

$$\frac{\partial \log f(\mathbf{A} \mid y_{1:t})}{\partial \text{vec}(\mathbf{A})} = -(\mathbf{W}_A^{-1} \otimes \mathbf{V}_A^{-1})(\text{vec}(\mathbf{A}) - \text{vec}(\mathbf{M}_A))$$

$$-k(\delta n + p) \sum_{j=1}^{t} \left((\mathbf{F}_{j-1} \otimes e_j e_j')\text{vec}(k\mathbf{A}\mathbf{F}_{j-1}\mathbf{A}^\top e_j e_j^\top + \mathbf{I})^{-1})\mathbf{A} \right.$$

$$\left. -(\mathbf{F}_{j-1} \otimes \mathbf{I})\text{vec}(k\mathbf{A}\mathbf{F}_{j-1}\mathbf{A}^\top)^{-1}\mathbf{A} \right).$$
(7.57)

To obtain the Hessian matrix of (7.56), we differentiate (7.57), i.e.

$$\frac{\partial^2 \log f(\mathbf{A} \mid y_{1:t})}{\partial \text{vec}(\mathbf{A}) \partial \text{vec}(\mathbf{A})^\top} = -\mathbf{W}_A^{-1} \otimes \mathbf{V}_A^{-1} + k(\delta n + p) \sum_{j=1}^{t} \left((\mathbf{F}_{j-1} \otimes e_j e_j^\top) \right.$$

$$\times (k\mathbf{F}_{j-1}\mathbf{A}^\top e_j e_j^\top + \mathbf{A}^{-1})^{-1} \otimes (k\mathbf{F}_{j-1}\mathbf{A}^\top e_j e_j^\top + \mathbf{A}^{-1})^{-1}$$

$$\times ((e_j e_j^\top \otimes k\mathbf{F}_{j-1})\mathbf{K}_p - \mathbf{A}^{-1} \otimes \mathbf{A}^{-1}) - (\mathbf{F}_{j-1} \otimes \mathbf{I})(k^{-1}\mathbf{A}^{\top-1}\mathbf{F}_{j-1})$$

$$\left. \otimes (k^{-1}\mathbf{A}^{\top-1}\mathbf{F}_{j-1})((\mathbf{I} \otimes k\mathbf{F}_{j-1})\mathbf{K}_p) \right),$$
(7.58)

where $\mathbf{K}_p$ is the $p^2 \times p^2$ vec-permutation matrix, i.e. $\text{vec}(\mathbf{A}^\top) = \mathbf{K}_p\text{vec}(\mathbf{A})$. For the proof of (7.57) and (7.58), the reader is referred to Triantafyllopoulos (2012).

Hence in step 2 the posterior mode $\hat{\mathbf{A}}^{(i)}$ is obtained by the iterative procedure of the Newton algorithm (7.56). Convergence is achieved when the Frobenius matrix norm between two successive iterations is less than a given tolerance threshold, or

$\| \hat{\mathbf{A}}^{(i)} - \hat{\mathbf{A}}^{(i-1)} \|_F \le$ Tol, where $\| \mathbf{X} \|_F = \sqrt{\sum_{i=1}^{n} \sum_{j=1}^{m} x_{ij}^2}$ is the Frobenius norm of an $n \times m$ matrix $\mathbf{X} = (x_{ij})$ and Tol is a pre-specified tolerance threshold.

We illustrate the above methodology with data consisting of five foreign exchange rates *vis-à-vis* the US dollar. The exchange rates are the Canadian dollar (CAD), Euro (EUR), Japanese Yen (JPY), British pound (GBP) and Australian dollar (AUD), all expressed as a number of units of the foreign currency per US dollar. The sample period runs from 4 January 1999 until 31 December 2009 and corresponds to 2760 observations, sampled at daily frequencies. This data set was obtained from the Pacific Exchange Rate Service of the University of British Columbia (http://fx.sauder.ubc.ca/). The data is transformed to log returns. In the first two years (4 January 1999 to 31 December 2001), we use the data for pre-processing purposes, in order to obtain sample estimates for μ and Σ_0. Then, starting at 2 January 2002, we run the volatility algorithm, in order to obtain forecasts of the volatility matrix.

Figure 7.9 shows the absolute returns together with the out of sample predicted marginal volatilities (the diagonal elements of the predicted volatility matrix $\hat{\Sigma}_t$, conditioned upon information $y_{1:t-1}$ sequentially for $t = 1, \ldots, N$ starting at 2 January 2002), and Fig. 7.10 shows the out of sample predicted correlations. Figure 7.9 indicates the good out of sample forecasting performance of the volatility, while Fig. 7.10 shows the dynamics of the correlation. Figure 7.11 shows the estimates of the diagonal elements of $\mathbf{A} = (A_{ij})_{i,j=1,\ldots,5}$. We note that A_{11} and A_{55} indicate a structural change after 2008, which highlights the abrupt increase in the volatility at that period, being evident by the left panel of Fig. 7.9 for CAD (relevant to A_{11}) and AUD (relevant to A_{55}). We also note that initially, the values of A_{ii} are centred around one ($\mathbf{A}$ indicates the autocorrelation of the precision process $\{\Phi_t\}$). In Fig. 7.11, we see that the A_{ii}'s gradually increase, and after 2003 the estimated values of A_{ii} are centred around 16.4, although for more conclusive comments one needs to look at the off-diagonal elements of $\mathbf{A}$ too. For the Newton–Raphson algorithm, we have used a stoppage tolerance Tol $= 0.0001$, and this was achieved for a minimum of 4 iterations and a maximum of 10 iterations.

One of the advantages of this model is that step 1 is completed using the conjugacy between the Wishart and the singular beta distribution and step 2 may require only a small number of iterations for the Newton algorithm to converge (here only 10 iterations were needed). Hence, there is no simulation step involved, and as a result the model is scalable. It can handle a large number of dimensions p, although it is expected that with the number of dimensions the number of iterations of the Newton algorithm will increase. As a rule of thumb experimentation shows that if the dimension of the vector of the returns is p, the number of iterations required is about $2p$.

A disadvantage of the above algorithms is that they make the assumption of the multivariate normal distribution for the returns. As discussed in Sects. 7.3.1 and 7.3.4, this assumption may not hold true in practice and should be replaced by a suitable multivariate skew t distribution or some other asymmetric distribution with fat tails. As the skew t distribution of Fernandez and Steel (1998) does not

Absolute returns and predicted marginal volatilities

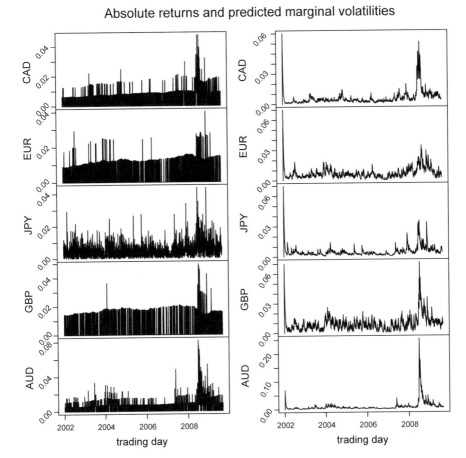

Fig. 7.9 Absolute returns and standard deviations of the out of sample predicted volatility, with $\delta = 0.7$

generalise in the multivariate case, a candidate is the skew t distributions of Azzalini and Capitanio (2003); see also Parisi and Liseo (2018).

7.4.3 *Portfolio Optimisation and Asset Allocation*

7.4.3.1 Problem Statement

In this section we discuss the classical mean–variance portfolio construction (Markowitz, 1952, 1959; Lintner, 1965). Following the seminal work of Markowitz (1952), a large amount of research is devoted to the construction, selection and assessment of optimal portfolio and risk management, see e.g. Han (2006); Soyer

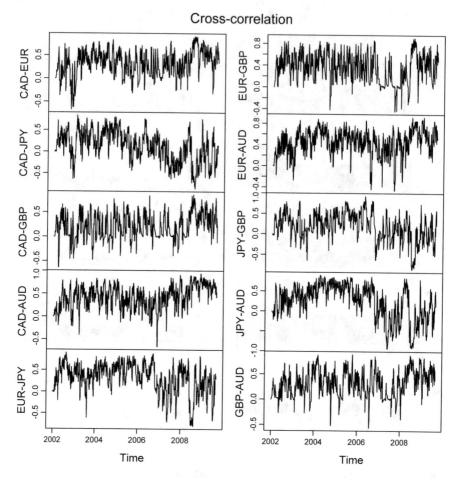

Fig. 7.10 Out of sample predictions of the cross-correlations between the five exchange rates, with $\delta = 0.7$

and Tanyeri (2006); Adcock (2007, 2010) for some recent contributions. Book-length coverage of asset allocation techniques and risk management can be found in Meucci (2005) and Schulmerich et al. (2015). Below we discuss the two basic portfolio selection strategies due to Markowitz (1952; 1959), but we apply them sequentially over time as in Aguilar and West (2000) and Soyer and Tanyeri (2006).

Suppose that we hold p assets, so that asset i is associated with log-returns y_{it}, $i = 1, \ldots, p$. For example, these may be the 30 constituents of the Dow Jones Industrial Average index or the subset of the S&P500 index. We define the vector of log-returns $y_t = [y_{1t}, \ldots, y_{tp}]^{\top}$, as discussed in Sect. 7.4.1 above. Adopting a multivariate volatility model, we obtain an estimate of $\mu = \mathrm{E}(y_t)$ and of the

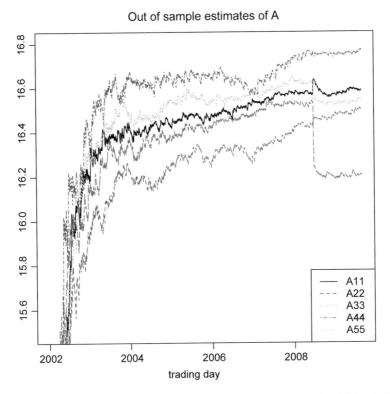

Fig. 7.11 Out of sample estimates of the diagonal elements A_{ii} of $A = \{A_{ij}\}$, with $\delta = 0.7$

volatility matrix $\boldsymbol{\Sigma}_t$, as discussed in the previous sections. We define the portfolio returns

$$r_t = \sum_{i=1}^{p} w_i y_{it} = w^\top y_t,$$

where $w = [w_1, \ldots, w_p]^\top$ is a vector of p weights associated with the p assets. It is assumed that $\sum_{i=1}^{p} w_i = w^\top 1_p = 1$, where $1_p = [1, \ldots, 1]^\top$. Sometimes we may assume that all weights are non-negative, but the optimisation procedures described below do allow for negative weights. A negative weight is usually associated with short-selling an asset. The portfolio selection problem consists of choosing weights w in an optimal way; this usually means either minimising the portfolio variance $\mathrm{Var}(r_t)$ or maximising the portfolio returns $\mathrm{E}(r_t)$. Below we discuss two possible strategies, the unconstrained portfolio (UP) and the constrained portfolio (CP).

7.4.3.2 Unconstrained Portfolio Selection

In the UP problem, we minimise the portfolio variance

$$\sigma_t^2 = \text{Var}(r_t) = \text{Var}(w^\top y_t) = w^\top \boldsymbol{\Sigma}_t w, \tag{7.59}$$

subject to the constraint that $w^\top \mu = m$, where m is a target portfolio mean. This portfolio selection minimises the portfolio variance or volatility given a certain value for the expected return. It is also equivalent of maximising the expected portfolio return $E(r_t)$, given a fixed target value for the portfolio variance.

We will use the optimisation technique using Lagrange multipliers. Let λ be a Lagrange multiplier and define the function

$$L(w, \lambda) = \frac{1}{2} w^\top \boldsymbol{\Sigma}_t w + \lambda(m - w^\top \mu).$$

The partial derivatives of L with respect to w and λ are

$$\frac{\partial L}{\partial w} = \boldsymbol{\Sigma}_t w - \lambda \mu \quad \text{and} \quad \frac{\partial L}{\partial \lambda} = m - w^\top \mu.$$

Equalising the partial derivatives to zero, we obtain

$$w = \lambda \boldsymbol{\Sigma}_t^{-1} \mu, \tag{7.60}$$

$$w^\top \mu = m. \tag{7.61}$$

Substituting w of (7.60) into (7.61), we solve for λ as

$$\lambda = \frac{m}{\mu^\top \boldsymbol{\Sigma}_t^{-1} \mu},$$

and putting this back in (7.60), we obtain the optimal weight as

$$w = \frac{m \boldsymbol{\Sigma}_t^{-1} \mu}{\mu^\top \boldsymbol{\Sigma}_t^{-1} \mu}. \tag{7.62}$$

However, one of the disadvantages of this approach is that it does not guarantee that the weights sum to 1, since

$$w^\top 1_p = \frac{m \mu^\top \boldsymbol{\Sigma}_t 1_p}{\mu^\top \boldsymbol{\Sigma}_t^{-1} \mu},$$

which is equal to 1, if $\mu = m 1_p$, but in general $\sum_{i=1}^p w_i \neq 1$. One possible solution is to set $\mu = m 1_p$; this is not a good choice as it sets the expectation of the return

$E(y_{it})$ equal to m, for each $i = 1, \ldots, p$. Another solution would be to calculate $w = [w_1, \ldots, w_p]^\top$ and then to recalculate (adjust) $w_p = 1 - \sum_{i=1}^{p-1} w_i$. However, this might be problematic too, as there is no guarantee that w_p will be non-negative. The minimum portfolio variance is

$$\sigma_t^2 = w^\top \mathbf{\Sigma}_t^{-1} w = \frac{m^2}{\mu^\top \mathbf{\Sigma}_t^{-1} \mu},$$

and if $\mu = m 1_p$, then $\sigma_t^2 = (1_p^\top \mathbf{\Sigma}_t^{-1} 1_p)^{-1}$.

7.4.3.3 Constrained Portfolio Selection

Moving on to the constraint portfolio (CP), we minimise the portfolio variance (7.59) subject to the constraints $w^\top \mu = m$ and $w^\top 1_p = 1$. We define the function

$$L(w, \lambda_1, \lambda_2) = \frac{1}{2} w^\top \mathbf{\Sigma}_t w + \lambda_1 (m - w^\top \mu) + \lambda_2 (1 - w^\top 1_p),$$

where λ_1 and λ_2 are Lagrange multipliers. The partial derivatives set to zero lead to

$$\mathbf{\Sigma}_t w - \lambda_1 \mu - \lambda_2 1_p = 0, \tag{7.63}$$

$$m - w^\top \mu = 0, \tag{7.64}$$

$$1 - w^\top = 0. \tag{7.65}$$

From (7.63), we solve for w,

$$w = \lambda_1 \mathbf{\Sigma}_t^{-1} \mu + \lambda_2 \mathbf{\Sigma}_t^{-1} 1_p. \tag{7.66}$$

We solve the system of equations (7.64)–(7.65) for λ_1 and λ_2 and then put them back to (7.66).

From (7.64), we have

$$w^\top \mu = \lambda_1 \mu^\top \mathbf{\Sigma}_t^{-1} \mu + \lambda_2 \mu \mathbf{\Sigma}_t^{-1} = m$$

and

$$w^\top 1_p = \lambda_1 \mu^\top \mathbf{\Sigma}_t^{-1} 1_p + \lambda_2 1_p^\top \mathbf{\Sigma}_t^{-1} 1_p.$$

We can write these two simultaneous equations as a linear system

$$\begin{bmatrix} a & b \\ b & c \end{bmatrix} \begin{bmatrix} \lambda_1 \\ \lambda_2 \end{bmatrix} = \begin{bmatrix} m \\ 1 \end{bmatrix},$$

where $a = \mu^\top \Sigma_t^{-1} \mu$, $b = \mu^\top \Sigma_t^{-1} 1_p$ and $c = 1_p^\top \Sigma_t^{-1} 1_p$. The solution of λ_1 and λ_2 is

$$\lambda_1 = \frac{cm - b}{ac - b^2} = \frac{m 1_p^\top \Sigma_t^{-1} 1_p - \mu^\top \Sigma_t^{-1} 1_p}{(\mu^\top \Sigma_t^{-1} \mu)(1_p^\top \Sigma_t^{-1} 1_p) - (\mu^\top \Sigma_t^{-1} 1_p)^2},$$

$$\lambda_2 = \frac{a - bm}{ac - b^2} = \frac{\mu^\top \Sigma_t^{-1} \mu - m\mu^\top \Sigma_t^{-1} 1_p}{(\mu^\top \Sigma_t^{-1} \mu)(1_p^\top \Sigma_t^{-1} 1_p) - (\mu^\top \Sigma_t^{-1} 1_p)^2}.$$

Hence from (7.66), the optimal weight vector under CP is

$$w = \Sigma_t^{-1} \frac{m(1_p^\top \Sigma_t^{-1} 1_p)\mu - (\mu^\top \Sigma_t^{-1} 1_p)\mu + (\mu^\top \Sigma_t^{-1} \mu)1_p - m(\mu^\top \Sigma_t^{-1} 1_p)1_p}{(\mu^\top \Sigma_t^{-1} \mu)(1_p^\top \Sigma_t^{-1} 1_p) - (\mu^\top \Sigma_t^{-1} 1_p)^2}.$$

$$(7.67)$$

In the application of the above portfolio selection strategies, first we adopt a model for the returns y_t, such as those described in Sects. 7.4.1 and 7.4.2. Hence we obtain estimates $\hat{\mu}$ and $\hat{\Sigma}_t$ of the mean vector μ and of the volatility matrix Σ_t, for any $t = 1, 2, \ldots, n$. In a first reading, $\hat{\mu}$ might be obtained as the historical mean of the returns and $\hat{\Sigma}_t$ might be the mode of the posterior distribution of Σ. These quantities are loaded into Eqs. (7.62) and (7.67) to get the UP and CP weights and hence to determine the proportion of asset to be allocated to the portfolio (assuming equal transaction costs along the range of the assets).

For illustration purposes we consider a data set consisting of the common constituents of the Dow Jones Industrial Average index over the period of 18 June 2001 to 4 September 2009. There are 30 assets included in the index and their closing prices are observed at each trading day (excluding weekends and bank holidays). The data is transformed into log-returns y_t and the first 637 observation returns (corresponding to a period of 18 June 2001 to 31 December 2003) are used in order to obtain the historical value $\hat{\mu}$ as the estimate of μ. The average (over the 30 assets) of this mean is -0.0001912098. We then fit the Wishart autoregressive volatility model of Sect. 7.4.2 for times $t = 638 - 2066$ (corresponding to 1 January 2004 to 4 September 2009). The reader will notice that this period of time includes the start of the 2008 global financial crises triggered by the sub-prime mortgage crises in US. Hence, a portfolio performance is likely to be affected by high volatile and uncertain times. The estimated volatility $\hat{\Sigma}_t$ is obtained using the Wishart volatility model: this estimate is the mode of the posterior distribution $\Sigma_t \mid A = \hat{A}^{(i)}, y_{1:t} \sim IW(n+29, F_t)$ (the inverse Wishart distribution with degrees of freedom $n + 29$ and scale matrix F_t), where $n = (1 - \delta)^{-1}$ and i is the iteration point of the Newton at convergence; for more details see Eq. (7.53).

Given the mean vector $\hat{\mu}$ and the covariance matrices $\hat{\Sigma}_t$, we fit the unconstrained portfolio (UP) and the constrained portfolio (CP), for two values of the discount factor $\delta = 0.7$ and $\delta = 0.95$. A measure of the performance of the chosen allocation is the average risk, defined as $\bar{R} = 1700^{-1} \sum_{t=1}^{n} w^{(t)\top} \hat{\Sigma}_t w^{(t)}$, where

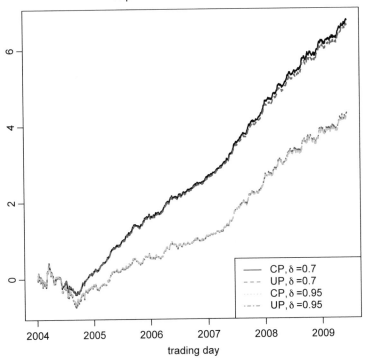

Fig. 7.12 Cumulative returns for the Dow Jones data; shown are constrained portfolio (CP) and unconstrained portfolio (UP) using discount factors $\delta = 0.7$ and $\delta = 0.95$

$w^{(t)}$ is vector of weights at time $t = 1, \ldots, 1700$ (corresponding to times from 638 to 2067 or from 1 January 2004 to 4 September 2009). A visual way to appreciate the portfolio performance is the plot of the cumulative portfolio returns, defined by $r_t^{(c)} = \sum_{i=1}^{t} r_j = \sum_{j=1}^{t} w^{\top} y_j$, for $t = 1, \ldots, 1700$. Figure 7.12 plots the cumulative returns for the two portfolio selection (UP and CP) for the two values of the discount factor δ. The first observation is that the UP and the CP in each case provide very similar asset allocation, with the cumulative returns being very close to each other (under a fixed value of δ). The portfolio under the low discount factor ($\delta = 0.7$) produces consistently higher cumulative returns and has therefore superior performance compared to the other portfolio. We conclude that the constrained portfolio where the volatility is estimated using $\delta = 0.7$ is the best performer for this data set. The average risks using the model with $\delta = 0.7$ are 0.02449 (CP) and 0.02297 (UP) and using the model with $\delta = 0.95$ are 0.039 (CP) and 0.0365 (UP); included are also the average risks for the equal weight portfolio – also known as naive portfolio – 0.552 (model with $\delta = 0.7$) and 0.769 (model with $\delta = 0.95$). These results confirm that the volatility model with $\delta = 0.7$ provides lower average risk compared to the model using $\delta = 0.95$. There is little difference between the

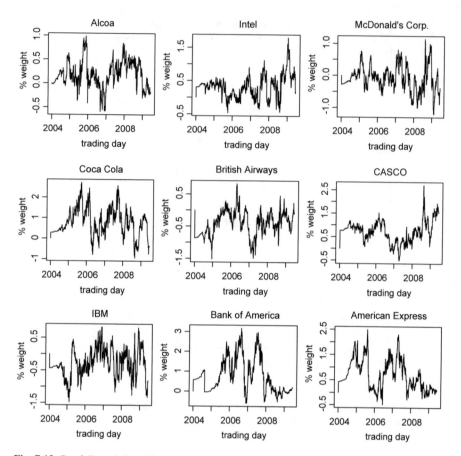

Fig. 7.13 Portfolio weights of 9 assets under the CP portfolio ($\delta = 0.7$)

performance of the CP and UP in the risk, but here the UP has slightly lower average risk. As expected the naive portfolio allocation provides higher risk.

Figure 7.13 shows the estimated portfolio weights over time for 9 assets, for the CP where the volatility is estimated using $\delta = 0.7$. The unconstrained and constrained optimisation procedures described above do not impose the assumption that $w_i \geq 0$. Hence, negative weights can appear, e.g. in the UP, a negative weight is likely to occur either due to negative covariances or due to negative returns. Usually, a negative portfolio corresponds to investing less at the given proportion, i.e. short-selling the given asset. If such a strategy is not desired, then the optimisation of the portfolio should include the additional constraint that $w_i \geq 0$. There are practical strategies that allow more delicate constraints to be included such that $w_i \geq d_i$, where d_i is some margin and $d_i > 0$. For this and more strategies, the reader is referred to Meucci (2005) and Schulmerich et al. (2015).

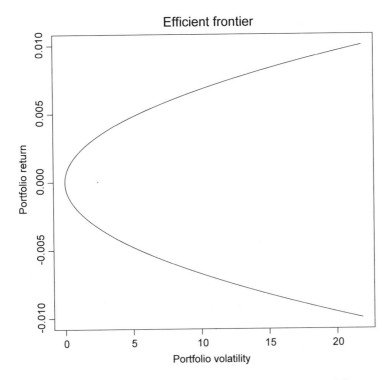

Fig. 7.14 Efficient frontier plot for the Dow Jones data. Shown are the portfolio returns as a function to the portfolio volatility

The efficient frontier shows a number of optimal portfolios that maximise the expected portfolio return for a given level or risk (or square root of the portfolio volatility) or minimise the risk for a given portfolio return (Markowitz, 1952, 1959; Lintner, 1965). Figure 7.14 plots the efficient frontier for the Dow Jones data, using a CP strategy, where the volatility is estimated using a discount factor $\delta = 0.7$. Shown is the average of the volatility portfolio $1700^{-1} \sum_{t=1}^{1700} w^{(t)\top} \hat{\Sigma}_t w^{(t)}$ for a given target portfolio return m. This figure shows the range of optimal portfolios (on the curve) for a range of target returns. As we can see when the average risk tends to zero, the return is also close to zero. While as the target return increases the risk increases. The plot is useful in identifying maximum return for a given risk. In practice a risk level may be specified (usually a range of risk which the investor will be comfortable to accept) and a maximum return should be identified. More details of similar portfolio strategies can be found in Meucci (2005) and Schulmerich et al. (2015) and the references therein.

7.5 Pairs Trading

7.5.1 Introduction and Basic Concept

Statistical arbitrage consists of a collection of trading strategies based on statistics and econometrics with the aim to return profit to an investor. Pairs trading is a market-neutral trading philosophy, which exploits a very basic trading rule in the stock market: buy low and short-sell high. Market-neutral strategies consist of trading rules that do not depend on the overall performance of the stock market or other financial markets. Such strategies are generally focusing on the relative movements of selected assets and therefore are not dependent on the overall performance of the markets.

The idea behind pairs trading can be traced back in the 1930s Cowles 3rd and Jones (1937), but it appeared formally in the 1980s with the work of Nunzio Tartaglia and his quantitative group at Morgan Stanley investment bank. Algorithmic pairs trading deploys trading strategies and related decision-making that can be implemented in the computer without human intervention. Recently, there has been a growing interest in pairs trading and in related market-neutral trading approaches, see Elliott et al. (2005), Gatev et al. (2006), Zhang and Zhang (2008), Triantafyllopoulos and Montana (2011) and Montana et al. (2009). Book-length discussion of pairs trading is also available Vidyamurthy (2004), Pole (2007). A recent chapter devoted to the implementation of pairs trading can be found in Nolan and Lang (2015, Chapter 6).

Considering the spread of two assets A and B, defined as the difference of the prices of A and B, pairs trading assumes that the spread attains an equilibrium or that the spread in the long run reverts to its historical mean. The main idea behind pairs trading is to propose trades based upon the relative temporary mispricing of the two assets. For example, suppose that the equilibrium of the spread is $10 (US dollars) and today the two assets trade at $40 and $10, or with spread 40−10=30. Then, pairs trading suggests to *go short* (or short-sell) asset A (as this is likely to be overpriced at $40) and to *go long* (or buy) asset B (as this is likely to be underpriced at $10). If the spread reverts to its mean, the price of asset A will decrease and/or the price of asset B will increase, either of which can return a healthy profit.

7.5.2 State Space Models for Mean-Reverted Spreads

In this section we briefly describe the model-based approach inference and prediction of pairs trading following Elliott et al. (2005) and Triantafyllopoulos and Montana (2011). Consider two assets A and B, and write p_t^A the price of asset A at time t and p_t^B the price of asset B at time t. We shall consider these prices are observed at a daily frequency, but lower frequency may be considered with few modifications. We form the *spread* of these two assets as

$$y_t = p_t^A - p_t^B. \tag{7.68}$$

The concept of pairs trading described in Sect. 7.5.1 is based on the assumption of mean-reversion of the spread y_t. Mean-reversion suggests that the mean of y_t is not time-dependent, for any $t > t_0$, for some time point t_0. That means that after some time t_0, the spread reaches some equilibrium level. The interpretation of mean-reversion in this context is that if asset A is overpriced compared to asset B (positive spread) at time t, then at some later time $t + k$ the positions will be reversed (asset B will be overpriced compared to A, negative spread), for some $k > 0$. In this set-up at t, we can open a trading position to go short on asset A and go long on B, so that at time $t + k$ we can close the position and realise profit. Mean-reversion is closely related to stationarity (see Sect. 7.2), but mean-reversion is a weaker condition, because it does not require that the variance be time-invariant. For example, the stochastic volatility model (7.26a)–(7.26b) is mean-reverted, as its mean is zero, but it is not weakly stationary, as its variance $\exp(h_t)$ is time-dependent.

The principle idea of Elliott et al. (2005) is to deploy a time series model for the spread y_t, which estimation will enable discovery of mean-reversion. In particular, these authors consider a state process x_t, which is generated by mean-reversion, so that

$$x_t - x_{t-1} = a - bx_{t-1} + \omega_t, \tag{7.69}$$

where $0 < b < 2$ and a is an unrestricted real number and the initial state is x_0. The restriction on b is imposed, to ensure stationarity of x_t. The innovation series $\{\omega_t\}$ is assumed to be a white noise (i.i.d. with zero mean and some variance σ_ω^2), and it is further assumed to be Gaussian. We can rearrange (7.69) to

$$x_t = a + (1 - b)x_{t-1} + \omega_t, \tag{7.70}$$

so that x_t is an autoregressive model of order one, with some non-zero mean. This can also be written $x_t - \mu = \phi(x_{t-1} - \mu) + \omega_t$, where $\phi = (1 - b)$ and $a = \mu(1 - \phi) = \mu b$, so that $x_t - \mu$ is the zero-mean AR(1) model of section (7.5). It follows that $\mu = a/b$, which is the equilibrium mean of x_t. Together with the state process x_t, Elliott et al. (2005) consider that the spread y_t is a noisy version of x_t or

$$y_t = x_t + \epsilon_t, \tag{7.71}$$

where ϵ_t is a Gaussian white noise with variance σ_ϵ^2, which is assumed to be independent of ω_t, for all t. Model (7.70)–(7.71) is a state space model, with state x_t and observation y_t. Assuming a Gaussian prior for x_0 and based on a collection of observations $\{y_1, \ldots, y_n\}$, we can estimate a and b and hence establish whether $0 < b < 2$, so as to justify mean-reversion for this model.

A simple trading rule based on model (7.70)–(7.71) is outlined next. If the spread y_{t+1} was known at time t, it would be easy to make a decision as to which asset to short-sell and which to buy. If say, we knew that $y_t < y_{t+1}$, we could anticipate that the price of A was likely to increase at $t+1$ and/or the price of B was likely to decrease at $t+1$ (relative to their prices at t). As a result at time t, we could buy asset A and short-sell B. At $t+1$, with $y_{t+1} > y_t$, we would realise a profit when we close the position, if asset A was sold and B was bought (minus transaction costs).

However, at time t, the spread y_{t+1} is not known, and hence we resort to forecasting it. Let $\hat{y}_{t+1} = \hat{p}_{t+1}^A - \hat{p}_{t+1}^B$ denote the one-step ahead forecast mean of the spread y_{t+1} at time t.

Suppose that we wish to open a trading position at time t. We first check whether the spread is expected to be mean-reverted at $t+1$, i.e. we see whether $0 < \hat{b} < 2$. If $\hat{b} < 0$ or $\hat{b} \geq 2$, we decide not to trade and so we do not open a position at t. If $0 < \hat{b} < 2$, we open a trading position according to the rule: buy a unit of A and short-sell a unit of B, if $\hat{y}_{t+1} - h > y_t$, and short-sell a unit of A and buy a unit of B, if $\hat{y}_{t+1} + h < y_t$. Here $h > 0$ is a margin that allows some uncertainty to guarantee that the unknown y_{t+1} at time t falls in the range $[\hat{y}_{t+1} - h, \hat{y}_{t+1} + h]$. For example, suppose that at time t, the spread is equal to $y_t = 20$ and that we project that at time $t+1$ the spread prediction goes up to $\hat{y}_{t+1} = 21$. As there is uncertainty around this prediction, it is equally likely that the true value of y_{t+1} be 23 (2 units higher than y_t) or 19 (one unit lower than y_t), each of which returns a different trading rule (buy/sell or sell/buy); in particular the latter ($y_{t+1} = 19 < y_t$) can result in a loss, if we implement the rule $y_t > \hat{y}_{t+1}$. For this reason, introducing h prevents this happening. In this simple example, if we operate with h as 10% of $\hat{y}_{t+1}$ or 2.1, then $\hat{y}_{t+1} - h = 18.9 < 20 = y_t$ and so we will not open the position: buy A and short-sell B. Likewise $\hat{y}_{t+1} + h = 22.1 > 20 = y_t$ and so we do not open the position: short-sell A and buy B. In such a case, we make the decision not to open a trading position at t, because the predicted $\hat{y}_{t+1}$ does not create a safe margin in the spread to allow for a probable profit. We can see that the lower the value of h, the more transactions we operate (we are more exposed to risk), and the higher the value of h, the less transactions we operate (we are more conservative). For more information on this strategy, the reader is referred to Triantafyllopoulos and Han (2013); other strategy procedures are described in Nolan and Lang (2015, Chapter 6).

7.5.3 Time-Varying Autoregressive Models for Trading-Spreads

The above model can be extended in various ways. First, the definitions of the spread (7.68) may not serve a number of practical situations. For example, the data set in Fig. 7.15 shows historical values of SouthWest Airlines and Exxon Mobile share prices. We notice that the spread $y_t = p_t^A - p_t^B$ exhibits high variability, which can

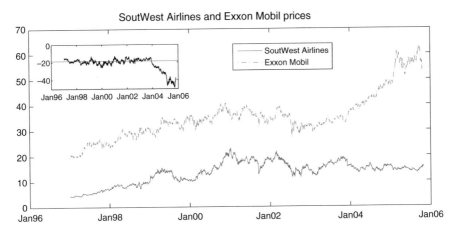

Fig. 7.15 SouthWest Airlines and Exxon Mobile share prices

cause problems in modelling with a Gaussian state space model described above. This may be overcome by considering the log-spread y_t as $y_t = \log p_t^A - \log p_t^B$, which is the logarithm of the ratio of the prices of the two assets. However, still the state space model (7.69)–(7.71) might be inadequate if the spread exhibits clear lack of mean-reversion as well as mean-reversion periods, as in Fig. 7.15. This is caused firstly by the constancy of the parameters a and b and secondly because the spread is too complex to be described by a simple state space model for all times t. In Triantafyllopoulos and Montana (2011), this issue is overcome by considering the spread as

$$y_t = p_t^A - \alpha - \beta p_t^B, \qquad (7.72)$$

where α and β are the intercept and the slope of the linear model if we regress p_t^A against p_t^B; here, A is the SouthWest Airlines and B the Exxon Mobile. *Cointegration* is the term used when a linear combination of two non-stationary time series (here p_t^A and p_t^B) is weakly stationary. In its most simple form, a linear model is fitted as

$$p_t^A = \alpha + \beta p_t^B + y_t \qquad (7.73)$$

and p_t^A and p_t^B are cointegrated, if the residual y_t is a weakly stationary time series. In the classical analysis of cointegration, a hypothesis test is set to test whether $\{y_t\}$ is weakly stationary or not. For example, the Dickey–Fuller distribution (Dickey & Fuller, 1979) is used to construct the cointegration test following Engle and Granger (1987) and Phillips and Perron (1988). Since then there is a host of publications related to cointegration; from a frequentist standpoint, we signal out the Johansen test Johansen (1991), and from a Bayesian standpoint we signal out Strachan and Inder (2004). In the context of pairs trading, the residual y_t of the linear model in (7.73) is the spread of the two assets. Instead of testing whether $\{y_t\}$ is weakly stationary, we aim to establish for which periods of time $\{y_t\}$ is mean-reverted and this is achieved by fitting the state space models (7.70)–(7.71) or the one described below.

Coming back to Fig. 7.15, we observe that up to January 2004, the spread seems to be stable (the two share prices appear to be parallel up to 2004). However, after that point of time, SouthWest Airlines see their prices increase, while Exxon Mobile share prices remain relatively stable. This is clearly depicted in the inset plot, which shows the difference $x_t^A - x_t^B$; we see that up to January 2004, spread (7.68) fluctuates around its equilibrium level, but after that date the spread decreases dramatically. This observation suggests that mean-reversion is lost after January 2004, and hence we can speak of periods of mean-reversion. In the search of detecting such periods, the parameters a and b in the state space model above (see e.g. (7.69)) should be time-varying, hence able to capture the local performance of the spread. Moving towards this direction, Triantafyllopoulos and Montana (2011) consider the time-varying autoregressive model

$$y_t = A_t + B_t y_{t-1} + \epsilon_t, \tag{7.74}$$
$$A_t = \phi_1 A_{t-1} + \zeta_{1t}, \quad B_t = \phi_2 B_{t-1} + \zeta_{2t},$$

where ϕ_1 and ϕ_2 are the AR coefficients, usually being assumed to satisfy $|\phi_i| < 1$ so that A_t and B_t may be weakly stationary processes.

Setting $\beta_t = [A_t, B_t]^\top$ and $c_t = [1, y_{t-1}]^\top$, the model can be expressed in state space form,

$$y_t = c_t^\top \beta_t + \epsilon_t, \tag{7.75a}$$
$$\beta_t = \mathbf{F}\beta_{t-1} + \zeta_t, \tag{7.75b}$$

with $\mathbf{F} = \text{diag}(\phi_1, \phi_2)$ and error structure governed by the observation error $\epsilon_t \sim N(0, \sigma_\epsilon^2)$ and the evolution error vector $\zeta_t = [\zeta_{1t}, \zeta_{2t}]' \sim N(0, \sigma_\epsilon^2 \mathbf{V}_t)$, for some covariance matrix $\mathbf{V}_t = \{V_{ij,t}\}_{i,j=1,2}$. It is assumed that the innovation series $\{\epsilon_t\}$ and $\{\zeta_t\}$ are individually and mutually independent and they are also independent of the initial state vector β_0.

As noted earlier mean-reversion for model (7.69)–(7.71) is ensured if $0 < b < 2$. For model (7.75a)–(7.75b), the following theorem establishes conditions of mean-reversion.

Theorem 7.1 *If $\{y_t\}$ is generated from model (7.75a)–(7.75b), then, conditionally on a realised sequence $B_1, \ldots, B_t$, $\{y_t\}$ is mean-reverting if one of the two conditions applies:*

(a) $\phi_1 = \phi_2 = 1$, $\mathbf{V}_t = \mathbf{O}$ and $|B_0| < 1$;
(b) $|\phi_1| < 1$ and $|\phi_2| < 1$, $\mathbf{V}_t$ is bounded and $|B_t| < 1$, for all $t \geq t_0$ and for some integer $t_0 > 0$.

Before we prove this theorem we make some comments. If $A_t = A$ and $B_t = B$, which is achieved by setting $\phi_1 = \phi_2 = 1$ and $V_t = O$, model (7.75a)–(7.75b) reduces to the standard AR(1) model and Theorem 7.1 reduces to the standard stationarity condition $|B| < 1$ (Example 7.2). The theorem allows $A_t = A$ to be time-invariant and B_t to be time-varying; this is obtained if we set $\phi_1 = 1$, $V_{11,t} = V_{12,t} = 0$, $|\phi_2| < 1$ and $V_{22,t} > 0$ (similarly we can have A_t time-varying and $B_t = B$ to be time-invariant). Moreover, following the approach of Elliott et al. (2005), estimates $\hat{B}_t$ of B_t can be used to check the mean-reversion or $|\hat{B}_t| < 1$, for all $t > t_0$ for case (b) of Theorem 7.1. However, this fails to take into account the uncertainty around the estimate of B_t and so in Triantafyllopoulos and Montana (2011) it is proposed that posterior bounds of B_t are first calculated and are required to be less than one in absolute value. Hence this procedure allows for online checking of mean-reversion.

Proof of Theorem 7.1 With $\phi_1 = \phi_2 = 1$ and $V_t = O$, the state space model (7.75a) reduces to the AR model $y_t = A + B y_{t-1} + \epsilon_t$, where $A_t = A$ and $B_t = B$ and it is trivial to verify that $\{y_t\}$ is mean-reverting if $|B_0| < 1$, see also Example 7.2. This completes (a).

Proceeding now to (b), from the AR model for A_t, we note that $E(A_t) = 0$. From (7.75a), write y_t recursively as

$$y_t = A_t + B_t y_{t-1} + \epsilon_t = A_t + B_t A_{t-1} + B_t B_{t-1} y_{t-2} + B_t \epsilon_{t-1} + \epsilon_t = \cdots$$

$$= y_1 \prod_{i=2}^{t} B_i + \sum_{j=0}^{t-3} \prod_{i=0}^{j} B_{t-i} A_{t-j-1} + A_t + \sum_{j=0}^{t-3} \prod_{i=0}^{j} B_{t-i} \epsilon_{t-j-1} + \epsilon_t.$$

We write $B_{1:t} = \{B_1, \ldots, B_t\}$, for $t = 1, \ldots, n$. Since $\{\epsilon_t\}$ is a white noise, we have

$$E(y_t \mid B_{1:t}) = y_1 \prod_{i=2}^{t} B_i. \tag{7.76}$$

This is a convergent series if $|B_t| < 1$, for all $t > t_0$, for some positive integer t_0. To see this first write $x_t^{(1)} = \prod_{i=2}^{t} B_i$, which is a decreasing series as $|x_{t+1}^{(1)}/x_t^{(1)}| = |B_{t+1}| < 1$. Also $\{x_t^{(1)}\}$ is bounded as $|x_t^{(1)}| = \prod_{i=2}^{t} |B_i| < 1$ and so $\{x_t^{(1)}\}$ is convergent.

For the variance of y_t, we have

$$\mathrm{Var}(y_t \mid B_{1:t}) = \mathrm{Var}(A_t) + \sum_{j=0}^{t-3} \prod_{i=0}^{j} B_{t-i}^2 \mathrm{Var}(A_{t-j-1}) + \sum_{j=0}^{t-3} \prod_{i=0}^{j} B_{t-i}^2 \mathrm{Var}(\epsilon_{t-j-1})$$

$$+ \mathrm{Var}(\epsilon_t) + \sum_{j=0}^{t-3} \prod_{i=0}^{j} B_{t-i} \mathrm{Cov}(A_t, A_{t-j-1})$$

$$\leq \sigma_\epsilon^2 + \frac{\sigma_\epsilon^2 V_{11}}{1 - \phi_1^2} + \left(\frac{\sigma_\epsilon^2 V_{11}}{1 - \phi_1^2} + \sigma_\epsilon^2\right) \sum_{j=0}^{t-3} \prod_{i=0}^{j} B_{t-i}^2 \qquad (7.77)$$

$$+ \frac{\sigma_\epsilon^2 V_{11}}{1 - \phi_1^2} \sum_{j=0}^{t-3} \phi_1^{j+1} \prod_{i=0}^{j} B_{t-i},$$

where it is used that

$$\mathrm{Var}(A_t) \leq \frac{\sigma_\epsilon^2 V_{11}}{1 - \phi_1^2} \quad \text{and} \quad \mathrm{Cov}(A_t, A_{t-j-1}) \leq \frac{\sigma_\epsilon^2 V_{11}}{1 - \phi_1^2},$$

for $V_{11,t} \leq V_{11}$, since from the hypothesis $\mathbf{V}_t$ is bounded, and so there exists some $V_{11} > 0$ so that $V_{11,t} \leq V_{11}$.

Now we show that the series $x_t^{(2)} = \sum_{j=0}^{t-3} \prod_{i=0}^{j} B_{t-i}^2$ and $x_t^{(3)} = \sum_{j=0}^{t-3} \phi_1^{j+1} \prod_{i=0}^{j} B_{t-i}$ are both convergent. For the former series we note that given $|B_t| < 1$, we can find some B so that $|B_t| < |B| < 1$, from which it follows that

$$|x_t^{(2)}| \leq \sum_{j=0}^{t-3} \prod_{i=0}^{j} |B_{t-i}| \leq \sum_{j=0}^{t-3} \prod_{i=0}^{j} |B| = \sum_{j=0}^{t-3} |B|^{j+1},$$

which is proportional to a geometric series that converges for $|B| < 1$, and since $x_t^{(2)}$ is a positive series, it follows that $\{x_t^{(2)}\}$ is convergent.

For the series $\{x_t^{(3)}\}$, we follow an analogous argument, i.e. for B satisfying $|B_t| < |B| < 1$, we obtain

$$|x_t^{(3)}| \leq \sum_{j=0}^{t-3} |\phi_1 B|^{j+1},$$

which shows that $x_t^{(3)}$ is convergent as $\sum_{j=0}^{t-3} |\phi_1 B|^{j+1}$ is a geometric series with $|\phi_1 B| < 1$ and $x_t^{(3)}$ is a positive series.

With these convergence results in place, the convergence of $\mathrm{Var}(y_t \mid B_{1:t})$ is obvious. Given, $B_1, \ldots, B_t$, we have shown that the mean and the variance of $\{y_t\}$ are convergent and so $\{y_t\}$ is mean-reverting. $\square$

We fit the state space model (7.75a)–(7.75b) to the SouthWest Airlines and Exxon Mobile data set of Fig. 7.15. Spread (7.72) is computed at each point of time t using the recursive least squares algorithm (see e.g. Sect. 3.1.2). For the application of the state space model (7.75a)–(7.75b), we have used the SOP model of Sect. 4.3, where matrix $\mathbf{V}_t$ is specified with two discount factors δ_1 and δ_2. After some experimentation with log-likelihood optimisation, we have selected values

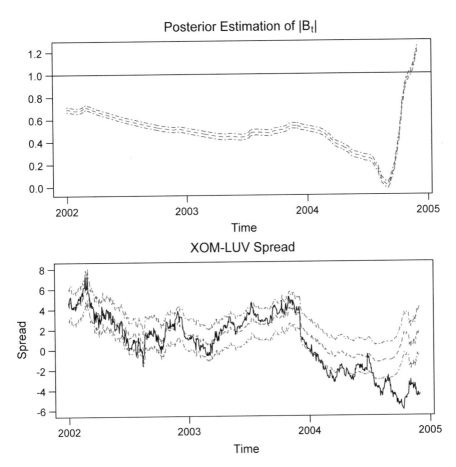

Fig. 7.16 Exxon Mobile (XOM) and SouthWest Airlines (LUV) spread. The top panel shows posterior mean (dashed line) and absolute credible bounds of B_t (dashed/dotted lines), and the lower panel shows observed spread (solid line) and forecast mean (dashed line) with 95% forecast interval (dotted/dashed lines)

$\phi_1 = 0.1, \phi_2 = 0.99839, \delta_1 = 0.992$ and $\delta_2 = 0.995$; details of this are provided in Triantafyllopoulos and Montana (2011). Figure 7.16 plots absolute posterior means together with 95% credible bounds of B_t (top panel) and one-step ahead forecast mean and 95% forecast interval of the spread y_t (lower panel); for illustration purposes only the last 731 points of time are shown, covering the period of 1 January 2002 to 30 November 2004. The horizontal line on the top panel indicates the threshold of one, to which $|\hat{B}_t|$ and its bound are compared. We can see that towards the end of 2004, $\hat{B}_t$ exceeds the threshold, and hence the model detects lack of mean-reversion at this point of time. This is depicted in the lower panel as the observed spread is clearly out of the forecast interval in that period of time, suggesting that the spread moves away from the forecast mean. The forecast performance is reasonable for up to January 2004 with most observed values of the spread lying within the forecast interval.

7.6   Exercises

1. Consider the autoregressive model

$$y_t = \alpha y_{t-1} + \left(\frac{1}{3} - \alpha\right) y_{t-2} + \epsilon_t,$$

where ϵ_t is a white noise process. Show that if $-1/3 < \alpha < 4/3$, this model is weakly stationary and causal.

2. Show that the autoregressive model

$$y_t = 0.2 y_{t-1} - 0.5 y_{t-2} + 0.4 y_{t-3} + \epsilon_t$$

is weakly stationary and causal model, where ϵ_t is a white noise process.

3. Consider a random variable X, which follows the gamma distribution with parameters α and β, with density function

$$p(x) = \frac{\beta^\alpha}{\Gamma(\alpha)} x^{\alpha-1} \exp(-\beta x),$$

where $\Gamma(\alpha)$ is the gamma function with argument α.

a. Show that the density function of $Y = \log X$ is

$$p(y) = \frac{\beta^\alpha}{\Gamma(\alpha)} \exp\left[\alpha y - \beta \exp(y)\right].$$

b. Show that the moment generating function of Y is

$$M_Y(z) = \frac{\Gamma(\alpha + z)}{\Gamma(\alpha)\beta^\alpha}.$$

c. Use (b) to verify that the mean and the variance of Y are

$$E(Y) = \psi(\alpha) - \log \beta \quad \text{and} \quad \text{Var}(Y) = \frac{d\psi(\alpha)}{d\alpha},$$

where $\psi(\alpha)$ is the digamma function with argument α and $d\psi(\alpha)/d\alpha$ is the trigamma function with argument α.

d. Use these results in the context of the stochastic volatility model of p. 357 (see also Eq. (7.27)) in order to show that

$$E(\log \epsilon_t^2) = -1.270363 \quad \text{and} \quad \text{Var}(\log \epsilon_t^2) = \frac{\pi^2}{2}.$$

You may use the functions `digamma` and `trigamma` in R to evaluate the digamma and the trigamma functions at specific arguments.

4. Consider the stochastic volatility model

$$y_t = \exp(h_t/2)\epsilon_t, \quad \epsilon_t \sim N(0, 1),$$
$$h_t - 0.2 = 0.9(h_{t-1} - 0.2) + \omega_t. \quad \omega_t \sim N(0, 1),$$

where ϵ_t is a white noise process, which is independent of ω_t, for all t.

a. Write down the steps of the Bootstrap particle filter algorithm (see Sect. 6.7.4) for this model.
b. Simulate 200 log-volatilities $h_1, \ldots, h_{200}$ and 200 log-returns $y_1, \ldots, y_{200}$ from this model.
c. Fit the Bootstrap filter algorithm and provide histograms of the approximate posterior distribution of the log-volatilities at times $t = 50, 100, 150, 200$. Plot one-step ahead forecast modes with 95% forecast intervals of y_t against the observed values of y_t. Hence, comment on the performance of the Bootstrap filter for this model.

5. Consider the stochastic volatility model of Exercise 4.

a. Apply the MCMC algorithm (7.3.3) to the simulated data of Exercise 4. Assuming that $\mu = 0.2$, $\phi = 0.9$ and $\sigma_\omega^2 = 1$ were unknown, draw samples of $h_t^{(j)}$, $\mu^{(j)}$, $\phi^{(j)}$ and $\sigma_\omega^{2(j)}$, for $t = 1, \ldots, 200$; $j = 1, \ldots, 10,000$.
b. Determine how $h_t^{(j)}$, $\mu^{(j)}$, $\phi^{(j)}$ and $\sigma_\omega^{2(j)}$ are compared to the true values of h_t, $\mu = 0.2$, $\phi = 0.9$ and $\sigma_\omega^2 = 1$.
c. Compare the approximate posteriors at times $t = 50, 100, 150, 200$, and hence provide a comparison between the MCMC algorithm and the Bootstrap filter.

6. The data in the table below show closing prices (in US dollar) of IBM shares, covering a period from 2 January 2002 to 28 March 2002.

January	Closing price	February	Closing price	March	Closing price
02/01	121.5	01/02	108	01/03	103.02
03/01	123.66	04/02	106.8	04/03	105.9
04/01	125.6	05/02	106.3	05/03	105.67
07/01	124.05	06/02	106.63	06/03	106.3
08/01	124.7	07/02	103.91	07/03	103.71
09/01	124.49	08/02	104.99	08/03	105.09
10/01	122.14	11/02	107.38	11/03	105.24
11/01	120.31	12/02	106.57	12/03	108.5
14/01	118.05	13/02	108.07	13/03	107.18
15/01	118.85	14/02	107.89	. 14/03	106.6
16/01	117.4	15/02	102.89	15/03	106.79
17/01	119.9	19/02	101.3	18/03	106.35
18/01	114.25	20/02	99.31	20/03	105.5
22/01	110.5	21/02	96.38	21/03	106.78
23/01	107.9	22/02	98.45	22/03	105.6
24/01	108.72	25/02	98.3	25/03	103.56
25/01	109.28	26/02	97.15	26/03	102.9
28/01	108.15	27/02	97.83	27/03	103.39
29/01	103	28/02	98.12	28/03	104
30/01	105.55				
31/01	107.89				

   a. Convert the data to log-returns and plot a histogram of the log-returns.
   b. Fit the stochastic volatility model (7.26a)–(7.26b) using the MCMC algorithm (7.3.3).

7. A collection of 2780 daily observations of returns of the Standards and Poors
500 Index is available in R by executing the commands:

```
> library(MASS)
> data("SP500")
> y <- SP500
```

   a. Perform an exploratory data analysis of this data set. Produce summary
statistics and a histogram of this data set.
   b. Fit the stochastic volatility model (7.26a)–(7.26b) using the MCMC algorithm (7.3.3), and draw samples for the volatility $\exp(h_t)$ and the hyperparameters μ, ϕ and σ_ω^2 of this model.

8. The stochastic volatility model (7.26a)–(7.26b) assumes that the mean of the
returns y_t is zero. The mean of the returns is expected to be very close to
zero, and in some financial applications it may play an important role, e.g.
in asset allocation and portfolio optimisation. In many volatility models, the
mean is calculated using historical returns and subtracted from the returns as
necessary. Model (7.26a)–(7.26b) can be extended to allow for a drift in the

mean. Consider the model

$$y_t = m_t + \exp(h_t/2)\epsilon_t, \quad \epsilon_t \sim N(0, 1), \tag{7.78}$$

$$m_t = x_t^\top \beta_t \quad \text{and} \quad \beta_t = \mathbf{F}\beta_{t-1} + \zeta_t, \tag{7.79}$$

$$h_t - \mu = \phi(h_{t-1} - \mu) + \omega_t, \quad \omega_t \sim N(0, \sigma_\omega^2), \tag{7.80}$$

where m_t, the mean of y_t, follows a state space model, with design vector x_t, state β_t, transition matrix $\mathbf{F}$ and innovation vector ζ_t. It may be assumed that ζ_t is independent of ω_t and ω_t, for all t. We signal out three specifications for the above state space model for m_t:

a. $m_t = m$ is a historical mean of returns.
b. m_t follows the autoregressive specification

$$m_t = \gamma_0 + \sum_{i=0}^{p} \gamma_i y_{t-i},$$

where γ_0 is an intercept and $\gamma_1, \ldots, \gamma_p$ are AR coefficients.
c. m_t follows a time-varying autoregressive specification

$$m_t = \gamma_{0,t} + \sum_{i=0}^{p} \gamma_{it} y_{t-i},$$

where γ_{it} follows random walk evolution, or $\gamma_{it} = \gamma_{i,t-1} + \zeta_{it}$, for some innovations ζ_{it} such that ζ_{it} is independent of ζ_{jt}, for $i \neq j$ and $i, j = 1, \ldots, p$. We may further assume that $\zeta_{it} \sim N(0, Z_i)$, for some variances Z_i.

For each of the specifications (a)–(c) above, propose an MCMC algorithm for the estimation of m_t, h_t, μ, ϕ and σ_ω^2 in model (7.78)–(7.80). This should extend the hybrid MCMC algorithm 7.3.3.

9. Extend the stochastic volatility model (7.26a)–(7.26b) to the multivariate case (Harvey et al., 1994, Section 3). Suppose you observe a vector of returns $y_t = [y_{1t}, \ldots, y_{pt}]^\top$, where y_{it} is the log-return of the i-th asset at time t. The univariate model (7.26a)–(7.26b) can be extended by considering

$$y_{it} = \exp(h_{it}/2)\epsilon_{it},$$

$$h_{it} - \mu_i = \phi_i(h_{i,t-1} - \mu_i) + \omega_{it}. \quad \omega_{it} \sim N(0, \sigma_{\omega,i}^2),$$

where $\omega_t = [\omega_{1t}, \ldots, \omega_{pt}]^\top$, $\epsilon_t = [\epsilon_{it}, \ldots, \epsilon_{pt}]^\top$. In this specification ω_{it} and ω_{jt} are independent, for $i \neq j$, and ϵ_t follows a multivariate normal distribution with zero mean vector and covariance matrix $\mathbf{\Sigma}$.

  a. Show that

$$\text{Var}(y_{it} \mid h_{it}) = \exp(h_{it})\sigma_{ii}$$
$$\text{Cov}(y_{it}, y_{jt} \mid h_{it}, h_{jt}) = \exp(h_{it}/2)\exp(h_{jt}/2)\sigma_{ij},$$

so that the correlation of y_{it} and y_{jt} is

$$\text{Cor}(y_{it}, y_{jt} \mid h_{it}, h_{jt}) = \frac{\sigma_{ij}}{\sqrt{\sigma_{ii}\sigma_{jj}}},$$

where σ_{ij} is the (ij)-th element of the covariance matrix $\boldsymbol{\Sigma}$. Hence, this model relates to the constant conditional correlation multivariate GARCH model, introduced in Bollerslev (1990) and further discussed in Bauwens et al. (2006).
  b. Write down the steps of an MCMC algorithm similar to algorithm (7.3.3) for the estimation of h_t, μ_i, ϕ_i, σ_{ω_i} and $\boldsymbol{\Sigma}$.

10. Extend the state space model (7.70)–(7.71) for modelling the spread y_t as follows. Consider that the state process x_t of (7.70) follows an autoregressive model of order 2, defined by

$$x_t = \phi_0 + \phi_1 y_{t-1} + \phi_2 y_{t-2} + \omega_t, \tag{7.81}$$

where ϕ_1 and ϕ_2 are the AR coefficients, ϕ_0 is the intercept and ω_t is a white noise as defined in (7.70). We shall assume that y_t follows the observation equation (7.71).

  a. Cast model (7.81)–(7.71) in state space form. Define the state vector $\beta_t = [1, x_t, x_{t-1}]^\top$, and write the model in the form

$$y_t = c_t^\top \beta_t + \epsilon_t \quad \text{and} \quad \beta_t = F\beta_{t-1} + \zeta_t,$$

   and hence determine the design vector c_t, the transition matrix F and the innovations vector ζ_t.
  b. Suppose that for a set of observed spreads $y_1, \ldots, y_n$, we have estimates of β_t and of F, the latter may depend on the parameters ϕ_i, for $i = 0, 1, 2$. Explain how $\hat{F}$ the estimate of F may be computed.
  c. Using the results of Sect. 7.2.2, explain how $\hat{F}$ may be used in order to identify periods of mean-reversion, and hence determine tradable periods.

11. Historical daily closing prices of Pepsi Corporation Inc (PEP) and Coca-Cola Company (KO) are available from the website of Yahoo Finance!: https://uk.finance.yahoo.com (this is the UK website; for people outside the UK, a Google search of 'Yahoo Finance!' will find the correct website. In order to download the data, visit the above website and search for the abbreviations 'PEP' and 'KO' and then look at historical data and set 5-year period on daily

prices. This will download daily prices over a 5-year window for each company separately. We recommend to work with closing prices, but other options are available. The closing prices of the two companies should be brought together on a single spreadsheet, from which one may obtain their spread. Follow the above procedure to download closing prices of the two assets.

a. Compute the simple spread $y_t = p_t^A - p_t^B$, where A and B denote the two companies.
b. Compute the log-spread $y_t^l = \log p_t^A - \log p_t^B$.
c. Compute the cointegration spread $y_t^c = p_t^A - \alpha - \beta p_t^B$, where α and β are determined by recursive least squares estimation.

Perform an exploratory data analysis for each of (a)-(c) by providing summary statistics and histograms of the spreads. Which spread do you think will be more suitable to fit the time-varying model of Sect. 7.5.3?

12. In the context of Exercise 11, fit the state space model (7.75a)–(7.75b) using the spread you have identified in Exercise 11 as more suitable. Identify tradable periods, and provide the one-step ahead forecasts of the spread.

13. Consider two assets (1) and (2) and denote their prices at time t by $p_t^{(1)}$ and $p_t^{(2)}$, respectively. Define their spread as $s_t = p_t^{(1)} - p_t^{(2)}$, for $t = 1, 2, \dots$. Consider the trading rule; if $s_t > 0$, then go long 2 and go short 1, if $s_t < 0$, then go long 1 and go short 2 and if $s_t = 0$, then take no action.

a. Define the returns $r_t^{(i)} = p_t^{(i)} - p_{t-1}^{(i)}$, for $i = 1, 2$. Define the spread on the returns

$$s_t^* = r_t^{(1)} - r_t^{(2)} = s_t - s_{t-1}.$$

Show that the trading rule described above is equivalent to

- If $s_t^* + s_{t-1} > 0$, go short 1 / long 2.
- If $s_t^* + s_{t-1} < 0$, go long 1 / short 2.
- If $s_t^* + s_{t-1} = 0$, no action.

b. Consider now the log-returns

$$r_t^{(i)} = \log p_t^{(i)} - \log p_{t-1}^{(i)}, \quad i = 1, 2,$$

and as before define the spread on the returns

$$s_t^* = r_t^{(1)} - r_t^{(2)}.$$

Show that the trading rule is now equivalent to

- If $s_t^* > \log p_{t-1}^{(2)} - \log p_{t-1}^{(1)}$, go short 1 / long 2.
- If $s_t^* < \log p_{t-1}^{(2)} - \log p_{t-1}^{(1)}$, go long 1 / short 2.
- If $s_t^* = \log p_{t-1}^{(2)} - \log p_{t-1}^{(1)}$, no action.

c. The geometric returns are defined as

$$r_t^{(i)} = \frac{p_t^{(i)} - p_{t-1}^{(i)}}{p_{t-1}^{(i)}} = \frac{p_t^{(i)}}{p_{t-1}^{(i)}} - 1, \quad i = 1, 2.$$

Using again the return spread

$$s_t^* = r_t^{(1)} - r_t^{(2)},$$

show that the trading rule is now equivalent to

- if $(s_t^* + 1 + r_t^{(2)})(1 + r_t^{(2)}) > \frac{p_{t-1}^{(1)}}{p_{t-1}^{(2)}}$, go short 1 / long 2.

- if $(s_t^* + 1 + r_t^{(2)})(1 + r_t^{(2)}) < \frac{p_{t-1}^{(1)}}{p_{t-1}^{(2)}}$, go long 1 / short 2.

- if $(s_t^* + 1 + r_t^{(2)})(1 + r_t^{(2)}) = \frac{p_{t-1}^{(1)}}{p_{t-1}^{(2)}}$, no action.

Chapter 8
Dynamic Systems and Control

State space models have played a significant role in the development of dynamic systems. Dynamic systems are usually driven by a system of differential equations and the state space framework has been used to represent and identify a dynamic system. Indeed the state space representation of a dynamic system reduces higher order linear differential equations to a system of first-order differential equations, with which solution is simple. Central to the study of a dynamic system is the notion of stability, which effectively studies the dynamic behaviour of the system for small perturbation of states and inputs of this system. The compact form of state space systems is used to study the stability of dynamic systems. More importantly checking the stability of linear systems is reduced to obtaining the eigenstructure of components of this system. For non-linear systems indirect and direct Lyapunov criteria are used to study the stability of these systems. In 1960 R. E. Kalman published his seminal paper on the Kalman filter, (Kalman, 1960) and in 1961 R. E. Kalman and R. S. Bucy published a continuous-time version of the filter, known as *Kalman–Bucy* filter, (Kalman & Bucy, 1961). The Kalman–Bucy filter can be regarded as a Wiener filter with time-varying parameters, but it benefits by not making the assumption of stationarity. Hence the Kalman–Bucy filter has provided a good estimation approach to describe continuous-time systems, which might be non-stationary. More importantly the filter is able to describe complex systems which are subject to uncertainty and noise, see e.g. Schweppe (1973) and Yedavalli (2016).

We begin by describing dynamic systems in Sect. 8.1. For linear systems driven by linear differential equations, their solution can be obtained by using Laplace transforms as described in Sect. 8.1.3. Section 8.2 introduces the *state* of a system and develops state representation of a system. Focusing on linear systems central to this is the solution of the state differential equation in Sect. 8.2.3. We discuss discrete-time and continuous-time systems and we discuss how a continuous-time system can be discretised. System stability is discussed in Sect. 8.3. Stability of linear systems is discussed via the eigenstructure of the state matrix and stability of non-linear systems is discussed via Lyapunov's criteria (indirect and direct methods). The Kalman–Bucy filter is discussed in Sect. 8.4. We start with the discrete-time Kalman filter (also discussed throughout in this book) and then we move on to describe the continuous-time Kalman–Bucy filter. The convergence of the error covariance matrix and the resulting steady state of continuous-time systems are discussed in Sect. 8.4.3. Section 8.4.4 considers an extension of the Kalman–Bucy filter, known as *extended Kalman filter* which aims to apply the filter to non-linear systems. Very closely connected to the description of dynamic systems is the concept of control. Section 8.5 discusses feedback control and in particular the popular proportional, integral, derivative (PID) controller. Finally, Sect. 8.5.2 considers a case study of control of a twin rotor experimental rig system.

8.1 Dynamic Systems

8.1.1 Basic Principles

A *system* is a collection of components, which interact with each other in order to perform some purposeful operation. This operation is known as the *behaviour* of the system. A system is usually evolving over time, hence it is sometimes referred to as *dynamic* system. In most practical situations the components of the system consist of m input variables, denoted by $x_i(t)$ ($t \in T_1$) and d output variables $y_j(t)$ ($t \in T_2$), for $i = 1, \ldots, m$ and $j = 1, \ldots, d$ and sets T_1 and T_2 subsets of the real field. In many situation we shall have $T_1 = T_2$. If T_1 and T_2 are discrete sets, the system is referred to as *discrete-time system*; if the input and output signals are continuous-time signals, then the system is known as *continuous-time system*. In this chapter we shall discuss both discrete-time and continuous-time modelling together, unless stated otherwise. We shall write $x(t) = [x_1(t), \ldots, x_m(t)]^\top$ the input vector for $t \in T_1$ and $y(t) = [y_1(t), \ldots, y_p(t)]^\top$ the output vector, for $t \in T_2$. The *operation* or *transformation* of the system is a mathematical mapping $\mathcal{F}(\cdot)$, which transforms $x(t)$ to $y(t)$, i.e.

$$y(t) = \mathcal{F}[x(t), t \in T_1](t), \quad t \in T_2,$$

hence it is a mapping from the m-dimensional vector space of inputs to the p-dimensional vector space of outputs. For further information on dynamic systems the reader is referred to Ogata (1970); linear systems are covered in detail in Zadeh and Desoer (1979).

Example 8.1 (Position of a Moving Vehicle) Consider a moving vehicle, with speed $v(t)$ at time t, for some time $t \in T_1 = [t_0, t_1]$, where t_0 is a starting time and t_1 is the finish-time. In this interval of time, the vehicle's position $y(t)$ is

$$y(t) = \int_{t_0}^{t} v(\tau) \, d\tau + y(t_0), \quad t_0 < t < t_1,$$

where $y(t_0)$ is the initial position of the vehicle. This is a continuous-time dynamic system with input $v(t)$, output $y(t)$ and transformation

$$\mathcal{F}[v(t), \quad t \in T_1](t) = \int_{t_0}^{t} v(\tau)\, d\tau + y(t_0).$$

Given the input function $v(t)$ and the initial position $y(t_0)$, calculation of the above integral can provide the output function $y(t)$.

8.1.2 Linear Systems

As it is evident from Example 8.1 the transformation operator $\mathcal{F}(\cdot)$ plays an important role in the behaviour of a system. A wide class of systems, known as *Linear Systems* are described below.

A system is said to be linear on T_2 if for all $t \in T_2$ it is

$$\mathcal{F}[ax](t) = a\mathcal{F}[x](t) \quad \text{and} \quad \mathcal{F}[x_1 + x_2](x) = \mathcal{F}[x_1](t) + \mathcal{F}[x_2](t),$$

for inputs x_1 and x_2 and for a constant a.

It is trivial to see that with $y(t_0) = 0$, the system of Example 8.1 is linear. However, for $y(t_0) \neq 0$, the system is not linear.

The following theorem provides an important property of linear systems. First we introduce the notion of the *impulse response* of a system. The impulse signal or impulse response is the output of the system, if a certain input pulse is applied. The impulse response provides the reaction of the system for a short-pulse input. The impulse response $h(t, t_0)$ is the value of the output, for an input equal to the Dirac delta function $\delta(t - t_0)$ (see Sect. 6.7.2), for a discrete-time system the Dirac delta is replaced by the Kronecker delta, defined as $\delta(t, t_0) = 1$, for $t = t_0$ and $\delta(t, t_0) = 0$, for $t \neq t_0$.

Theorem 8.1 *A single-input, single-output system is linear if and only if for any input $x(t)$ the output $y(t)$ has the following expressions*

$$y(t) = \int_{T_1} x(\tau)h(t, \tau)\, d\tau, \quad \text{for continuous-time system,} \tag{8.1}$$

$$y(t) = \sum_{\tau_k \in T_1} x(\tau_k)h(t, \tau_k), \quad \text{for discrete-time system,} \tag{8.2}$$

where $h(t, t_0)$ is the impulse response function defined above and $\mathcal{F}(\cdot)$ in the continuous-time system is assumed to be a continuous function in x.

Proof The proof mimics the proof of Minkler and Minkler (1993, Chapter 3). Following these authors we shall give the proof for the continuous-time case. The discrete-time case follows readily by replacing integration by finite sums.

First we shall prove sufficiency, so that if Eq. (8.1) holds true (continuous-time case), then the system is linear. This is trivial to show by the definition of the linear system above and basic integration properties, and is left to the reader as an exercise.

Moving on, we shall prove necessity, so that if the system is linear, then it has the representation (8.1) (continuous-time case). Since the system is linear we have

$$\mathcal{F}[x(t_0)\delta(\tau - t_0)](t) = x(t_0)\mathcal{F}[\delta(\tau - t_0)](t) = x(t_0)h(t, t_0),$$

using the definition of the impulse response. It then follows that

$$\int_{T_1} x(t_0)h(t, t_0)\, dt_0 = \int_{T_1} \mathcal{F}[x(t_0)\delta(\tau - t_0)](t)\, dt_0$$

$$= \lim_{\max_i |\Delta t_{0,i}| \to 0} \sum_i \mathcal{F}[x(t_{0,i})\delta(\tau - t_{0,i}), \tau \in T_1](t)\Delta t_{0,i}$$

$$= \mathcal{F}\left[\lim_{\max_i |\Delta t_{0,i}| \to 0} \sum_i x(t_{0,i})\delta(\tau - t_{0,i})\Delta t_{0,i}, \tau \in T_1\right](t)$$

$$= \mathcal{F}\left[\int_{T_1} x(t_0)\delta(t - t_0)\, dt_0, \tau \in T_1\right](t)$$

$$= \mathcal{F}[x(\tau), \tau \in T_1](t) = y(t).$$

□

The theorem above is presented for a single input $x(t)$ and a single output $y(t)$. The theorem holds true in the vector case when $x(t)$ is a $m \times 1$ vector of inputs and $y(t)$ a $d \times 1$ vector of outputs; the modifications are relatively straightforward and for a detailed discussion the reader is referred to Minkler and Minkler (1993, pp. 74–75). Below we discuss two classes of dynamic systems, namely incrementally linear and time-invariant linear systems.

We have noted above that Example 8.1 is a linear system if the starting position of the vehicle is $y(t_0) = 0$. The definition of linear systems above does not allow for $\mathcal{F}$ being a linear function of the form $\mathcal{F}[ax + b](t)$, for $b \neq 0$. A dynamic system, with $\mathcal{F}(\cdot)$, which satisfies

$$\mathcal{F}[ax + b](t) = a\mathcal{F}[x](t) + b \quad \text{and} \quad \mathcal{F}[x_1 + x_2](x) = \mathcal{F}[x_1](t) + \mathcal{F}[x_2](t),$$

for any inputs $x_1, x_2 \in T_1$ and any constants a, b, is known as *incrementally linear system*. In order to explain the name of this definition, consider an incrementally linear system with $\mathcal{F}[ax + b](t) = a\mathcal{F}[x](t) + b$. We can decompose the output signal $y(t) = \mathcal{F}[x(t), t \in T_1](t)$, as $y(t) = z(t) + b$, where $z(t)$ is an output with transformation operator $\mathcal{F}'[ax(t), t \in T_1](t) = a\mathcal{F}'[x(t), t \in T_1](t)$, for $t \in T_2$. Hence $y(t)$ is equal to the output from a linear system plus the constant b, hence the name incrementally linear. We can see that the system of Example 8.1 is incrementally linear as we can see that $b = y(t_0)$. Many of the properties of linear systems are shared by incrementally linear systems. For example, Theorem 8.1 can be extended to accommodate an incrementally linear system. In this case for a single-input single-output system, Eq. (8.1) is replaced by $y(t) = g(t) + \int_{T_1} x(\tau)h(t, \tau)\, d\tau$ (continuous-time case), where $g(t)$ is a function reflecting on the constant b of the system. A similar equation applies for the case of discrete-time systems and as in liner systems these are extended to accommodate systems with vector input and vector output. Because incrementally linear systems are met in practice more often than linear systems and they still share many properties, some authors use the term *linear system* to actually mean *incrementally linear system* and we shall adopt this convention in this chapter.

We complete this section by briefly discussing time-invariant systems. A system, with transformation operator $\mathcal{F}[x(t), t \in T_1](t)$ is said to be time-invariant (or stationary) if $y(t) = \mathcal{F}[x(t), t \in T_1](t)$ is equal to $y(t - t_0) = \mathcal{F}[x(t - t_0), t - t_0 \in T_1](t)$, for any $t_0 < t$. In other words, the output signal $y(t)$ is time-invariant and is determined by $y(t_0) = \mathcal{F}[x(t_0)](t_0)$. In a time-invariant system we have $\mathcal{F}[\delta(\tau - t_0), \tau - t_0 \in T_1](t) = h(t - t_0) = h(t, t_0)$, the impulse response of the system. From Theorem 8.1 the system is linear if and only if

$$y(t) = \int_{T_1} x(\tau)h(t - \tau)\, d\tau, \quad \text{for continuous-time system,}$$

$$y(t) = \sum_{\tau_k \in T_1} x(\tau_k)h(t - \tau_k), \quad \text{for discrete-time system.}$$

8.1.3 Laplace Transform

The transformation mapping $\mathcal{F}(\cdot)$ of a dynamic system relates the input of the system to the output of the system. This mapping is usually described by a differential or integral equation (for a continuous-time system) or by a difference equation (for a discrete-time system); see e.g. various examples of Robinson (2012). In Example 8.1 at time t the position of the vehicle $y(t)$ is given by the integral of the velocity of the vehicle plus the initial position $y(0)$. Hence, in order to determine the position $y(t)$ the above integral must be computed or approximated. In general, once $\mathcal{F}$ is determined, a system engineer usually desires to solve a differential equation in order to answer the main questions of the system. These questions may involve to determine the stability of the system over time and provide values of the output $y(t)$ for small changes of the values of the input vector $x(t)$. For linear systems, the differential equation(s) may be solved by employing the Laplace transform (time domain) or the discrete Fourier transform (frequency domain); for more information the reader is referred to Ogata (1970), Oppenheim and Willsky (1983) and Robinson (2012). Below we describe the method of Laplace transforms and we show how it can be used to solve linear differential equations.

Given a continuous function $f(t)$, for $t \geq 0$, the *Laplace transform* of $f(t)$ is a function $F(s)$ with domain the complex plain, defined by

$$F(s) = \mathcal{L}[f](s) = \int_0^\infty f(t)e^{-st}\,dt, \quad s \in \mathbb{C}. \tag{8.3}$$

Some authors call this the one-sided Laplace transform, defined in $[0, \infty)$ and the (two-sided) Laplace transform, if the limits in the integral are $-\infty$ and $+\infty$. In this section we shall consider functions defined for $t \geq 0$ and we shall call (8.3) as the Laplace transform.

Function $f(t)$ is said to be the *inverse Laplace transform* of $F(s)$ and can be defined by the integral

$$f(t) = \frac{1}{2\pi i} \lim_{N \to \infty} \int_{\gamma - iN}^{\gamma + iN} F(s)e^{st}\,ds,$$

with the integration is done along $\text{Re}(s) = \gamma$ (the real part of s) in the complex plain. More details about the definition and conditions of the convergence of these integrals are given in Dyke (1999).

Some basic properties of the Laplace transform are given below. For the functions below it is implicitly assumed that their domain is $[0, \infty)$.

1. *Linearity.* If $f(t)$ and $g(t)$ are functions, and a, b are any real numbers, then

$$\mathcal{L}[af + bg](s) = a\mathcal{L}[f](s) + b\mathcal{L}[g](s).$$

2. *Differentiation.* If $\mathcal{L}[f](s)$ and $\mathcal{L}[f'](s)$ exist, for some function $f(t)$, then

$$\mathcal{L}[f'](s) = s\mathcal{L}[f](s) - f(0).$$

3. *Integration.* If $f(t)$ is a continuous function, then

$$\mathcal{L}\left[\int_0^t f(\tau)\,d\tau\right] = \frac{1}{s}\mathcal{L}[f](s).$$

4. *Convolution.* Let $f_1(t)$, $f_2(t)$ be non-negative valued functions, with convolution

$$g(t) = (f_1(t) * f_2(t)) = \int_0^\infty f_1(\tau)f_2(t - \tau)\,d\tau = \int_0^\infty f_1(t - \tau)f_2(\tau).\,d\tau.$$

Then, the Laplace transform of $g(t)$ is equal to $F_1(s)F_2(s)$. By definition (8.3) it follows that $g(t)$ is the inverse Laplace transform of $F_1(s)F_2(s)$.

Consider now the linear system generated by the differential equation

$$\frac{d^n y}{dt^n} + a_{n-1}\frac{d^{n-1} y}{dt^{n-1}} + \cdots + a_1\frac{dy}{dt} + a_0 y = b_m\frac{d^m u}{dt^m} + \cdots + \frac{du}{dt} + b_0 u, \qquad (8.4)$$

with initial conditions $y(0), y'(0), \ldots, y^{(n-1)}(0), \; u(0), u'(0), \ldots, u^{(m-1)}(0)$, where $y = y(t)$ is the output function, $u = u(t)$ is a known input function, a_i, b_j are real numbers $(i = 1, \ldots, n; \; j = 1, \ldots, m)$ and $m \leq n$.

This differential equation can be solved by taking Laplace transforms in (8.4) and using property (2). Indeed applying (2) repeatedly we have

$$\mathcal{L}\left[\frac{d^k f}{dt^k}\right](s) = s^k \mathcal{L}[f](s) - s^{k-1} f(0) - s^{k-2} f'(0) - \cdots - f^{(k-1)}(0).$$

Gathering terms for all values of $k = 1, 2, \ldots, n$ we obtain $Y(s)$ the Laplace transform of y as a function of $U(s)$ the Laplace transform of u as

$$p(s)Y(s) - r(s) = q(s)U(s) - \ell(s), \qquad (8.5)$$

where $p(s), q(s)$ are polynomial functions on s, defined as

$$p(s) = a_0 + a_1 s + \cdots + a_{n-1} s^{n-1} + s^n, \qquad q(s) = b_0 + b_1 s + \cdots + b_m s^m$$

and the functions $r(s), \ell(s)$ are

$$r(s) = \sum_{i=0}^{n-1} \sum_{k=i+1}^{n} a_k s^{k-1-i} y^{(i)}(0), \qquad a_n = 1$$

$$\ell(s) = \begin{cases} \sum_{i=0}^{m-1} \sum_{k=i+1}^{m} b_k s^{k-1-i} s^{k-1-i} u^{(i)}(0), & m \geq 1 \\ 0, & m = 0 \end{cases}$$

Solving (8.5) for $Y(s)$ we obtain the Laplace transform of y as

$$Y(s) = \frac{r(s) - \ell(s)}{p(s)} + \frac{q(s)}{p(s)} U(s). \qquad (8.6)$$

The first fraction of the right hand side of (8.6) is a proper rational function, which depends only on initial conditions. The second term is the convolution of $h(t)$ and $u(t)$, where $h(t)$ is the inverse Laplace transformation of $q(s)/p(s)$ and $u(t)$ is the inverse Laplace transformation of $U(s)$.

Let $g(t)$ be the inverse Laplace transform of $[r(s) - \ell(s)]/p(s)$, then the solution of (8.4) can be written as

$$y(t) = g(t) + \int_0^\infty h(\tau)u(t-\tau)\,d\tau.$$

This shows that (8.4) is an incrementally linear system, with impulse response function $h(t)$ given by

$$h(t) = \mathcal{L}^{-1}\left[\frac{q(s)}{p(s)}\right].$$

The Laplace transform of $h(t)$

$$H(s) = \frac{q(s)}{p(s)}$$

is called the *system transfer function* of the (incrementally) linear system (8.4). Using these results it follows that the solution of the differential equation (8.4) involves computation of Laplace transforms. Laplace transforms of some useful functions are tabulated, see e.g. Minkler and Minkler (1993, Chapter 4). See Exercise 1 for the calculation of the Laplace transform of two useful functions.

Example 8.2 (Hookean Spring Force Dynamics) Consider the Hookean spring force example of Sect. 1.3.4, describing the motion of an object, which is attached to a wall on a spring. The position of the object $y(t)$ at time t follows the differential equation

$$m\frac{d^2y(t)}{dt^2} + k_1\frac{dy(t)}{dt} + k_2y(t) = u(t), \tag{8.7}$$

where $dy(t)/dt$ is the velocity of the object at t, $u(t)$ is an applied force at t and the constants k_1, k_2 are the viscous friction coefficient and the spring constant, respectively. The system is shown below in Fig. 8.1.

Fig. 8.1 Spring single-mass system, including a spring and damping

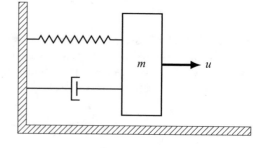

Let $Y(s)$ and $U(s)$ be the Laplace transforms of (8.7). Suppose that the initial conditions are

$$\left.\frac{dy(t)}{dt}\right|_{t=0} = y(0) = u(0) = 0. \tag{8.8}$$

The Laplace transform of (8.7) is

$$Y(s) = \frac{q(s)}{p(s)}U(s) = \frac{1}{ms^2 + k_1 s + k_2}U(s),$$

since $r(s) - \ell(s) = 0$ from the initial conditions. Thus, the system transfer function is

$$H(s) = \frac{1}{ms^2 + k_1 s + k_2}$$

and $g(t) = 0$ (again from the initial conditions).

If we set in (8.7) $m = 1$, $k_1 = 2$, $k_2 = 1$ and $u(t) = e^{-t}$, then we have $U(s) = (s+1)^{-1}$ and $H(s) = (s^2 + 2s + 1)^{-1} = (s+1)^{-2}$. Hence the Laplace transform of y is

$$Y(s) = H(s)U(s) = \frac{1}{(s+1)^3}.$$

From Exercise 1 we see that the inverse Laplace transform of $Y(s)$ is

$$y(t) = \frac{1}{2}e^{-t}t^2,$$

which is the solution of (8.7) under the initial conditions (8.8).

8.2   State Space Representation of Dynamic Systems

8.2.1   State Variables and State of a System

Cambridge dictionary defines *state* as 'a condition or way of being that exists at a particular time'. A *state variable* of a dynamic system is a variable of the system, used to describe the mathematical state of a dynamic system. According to Minkler and Minkler (1993) in dynamic systems a state of a system is the smallest collection of variables so that their knowledge at time $t = t_0$ together with knowledge of input variables at all times $t \geq t_0$ completely determine the system for any $t \geq t_0$, assuming no external force is applied to the system.

Example 8.3 (Hookean Spring Force Revisited) Consider the Hookean spring force example 8.2 above, where the position of the object $y(t)$ is generated by the differential equation (8.7). The state of this system $x(t)$ can be represented by the vector comprising $y(t)$ and $u(t)$, or

$$x(t) = \begin{bmatrix} y(t) \\ u(t) \end{bmatrix},$$

so that knowledge of $x(t)$, for $t \geq t_0$ determines the system.

8.2.2   Continuous-Time State Space Model

In general, in continuous-time systems, we shall write $x(t)$ the state vector, where $x(t)$ implicitly depends on $x_0 = x(t_0)$. In discrete-time systems we shall write x_k, where again x_k depends implicitly on x_{k_0}. The state space representation of a dynamic system is a set of two equations involving states: in the system state the first derivative $\dot{x}$ of the state vector x is given as a function of x, input vector u and time t, so that

$$\dot{x} = f(x, u, t), \tag{8.9}$$

for some function $f(\cdot)$ and $x = x(t)$ is implicit. The second equations, known as observation equation, links observations $y = y(t)$ to the state vector x

$$y = g(x, u, t), \tag{8.10}$$

where $g(\cdot)$ is a suitable function. These functions may incorporate innovations, as discussed below for the linear state space model.

Equations (8.9)–(8.10) define a state space model, which can represent or describe a dynamic system. It is implicitly understood that $u = u(t)$ is known for all $t \geq t_0$ and that $x_0 = x(t_0)$ is known at time $t = t_0$. Usually it is required to solve the differential equation (8.9) for $x(t)$. We shall denote by $\phi(t, x_0)$ the general solution of (8.9), which is a function of time t and also depends on the initial condition $x_0 = x(0)$; in Sect. 8.2.3 we provide this solution for linear systems.

Linear systems are represented by linear state space models, so that the functions $f(\cdot)$ and $g(\cdot)$ are linear functions. In particular the following observation and system equations define a linear state space model or representation of a linear system

$$\dot{x}(t) = \mathbf{F}x(t) + \mathbf{G}u(t) + \zeta(t) \tag{8.11a}$$

$$y(t) = \mathbf{H}x(t) + \epsilon(t), \tag{8.11b}$$

where $\zeta(t)$ and $\epsilon(t)$ are random innovations. In a deterministic system $\zeta(t) = \epsilon(t) = 0$, with probability one. In practical situations these innovations are needed to cater for observations and states exhibiting uncertainty. In the state space model (8.11a)–(8.11b) the matrices $\mathbf{F}$, $\mathbf{G}$ and $\mathbf{H}$ are time-invariant and in this case the model is known as *time-invariant linear state space model*; if at least one of these matrices are time-dependent the model is called *time-varying state linear space model*. For most of what follows we shall work with time-invariant linear state space models. Note that if the output $y(t)$ is a scalar, then $\mathbf{H} = H$ is a row vector. Likewise, if the input u is scalar, then $\mathbf{G} = G$ is a column vector. We shall treat these components as matrices to cover generality and hence use boldface, but the reader should remember that some may be vectors.

Example 8.4 (Hookean Spring Force Dynamics) In the context of Example 8.3 write the state vector $x(t)$ as

$$x(t) = \begin{bmatrix} x_1(t) \\ x_2(t) \end{bmatrix}$$

with $x_1(t) = y(t)$ (the position of the object at time t) and $x_2(t) = \dot{y}(t) = \dot{x}_1(t)$. From the differential equation (8.7) we have

$$\dot{x}(t) = \begin{bmatrix} \dot{x}_1(t) \\ \dot{x}_2(t) \end{bmatrix} = \begin{bmatrix} 0 & 1 \\ -\frac{k_1}{m} & -\frac{k_2}{m} \end{bmatrix} \begin{bmatrix} x_1(t) \\ x_2(t) \end{bmatrix} + \begin{bmatrix} 0 \\ \frac{1}{m} \end{bmatrix} u(t)$$

$$= \mathbf{F}x(t) + \mathbf{G}u(t), \tag{8.12}$$

where $\dot{x}_2(t) = \ddot{x}_1(t) = \ddot{y}(t)$.
   Also

$$y(t) = x_1(t) = [1, 0] \begin{bmatrix} x_1(t) \\ x_2(t) \end{bmatrix} = Hx(t). \tag{8.13}$$

Equations (8.12)–(8.13) define an invariant linear state space model.

Example 8.5 Consider the dynamic system, which is described by the differential equation

$$\frac{d^3 y}{d t^3} - 4\frac{d^2 y}{d t^2} + \frac{dy}{d t} + 2y = u_1 - 2u_2, \tag{8.14}$$

where the time t is implicit in all functions. In order to write this model in state space form we define $y = x_1$, $\dot{y} = x_2$, $\ddot{y} = x_3$ and $\dddot{y} = \dot{x}_3$. From this and the differential equation (8.14) we obtain

$$\dot{x}_3 = 4\dot{x}_2 - \dot{x}_1 - 2x_1 + u_1 - 2u_2 = 4x_3 - x_2 - 2x_1 + u_1 - 2u_2.$$

Hence the system equation is

$$\dot{x} = \begin{bmatrix} \dot{x}_1 \\ \dot{x}_2 \\ \dot{x}_3 \end{bmatrix} = \begin{bmatrix} 0 & 1 & 0 \\ 0 & 0 & 1 \\ -2 & -1 & 4 \end{bmatrix} \begin{bmatrix} x_1 \\ x_2 \\ x_3 \end{bmatrix} + \begin{bmatrix} 0 & 0 \\ 0 & 0 \\ 1 & -2 \end{bmatrix} \begin{bmatrix} u_1 \\ u_2 \end{bmatrix}$$

$$= Fx + Gu$$

and

$$y = x_1 = [1, 0, 0] \begin{bmatrix} x_1 \\ x_2 \\ x_3 \end{bmatrix} = Hx.$$

In general, consider a linear system with input $u = u(t)$ and output $y = y(t)$ is generated by the differential equation

$$\frac{d^n y}{d t^n} + a_1 \frac{d^{n-1} y}{d t^{n-1}} + \cdots + a_{n-1} \frac{dy}{dt} + a_n y = u, \tag{8.15}$$

where a_i are real-valued coefficients, $i = 1, 2, \ldots, n$. A state space representation of this system can be obtained if we define the state vector

$$x = \begin{bmatrix} x_1 \\ x_2 \\ \vdots \\ x_n \end{bmatrix} = \begin{bmatrix} y \\ y^{(1)} \\ \vdots \\ y^{(n-1)} \end{bmatrix},$$

where $y^{(i)}$ denotes the i-th derivative of the function y ($i = 1, 2, \ldots, n$). With this definition in place, the system equation is

$$\dot{x} = \begin{bmatrix} x_1 \\ x_2 \\ \vdots \\ x_n \end{bmatrix} = \begin{bmatrix} 0 & 1 & 0 & \cdots & 0 \\ 0 & 0 & 1 & \cdots & 0 \\ \vdots & \vdots & \vdots & \ddots & \vdots \\ -a_n & -a_{n-1} & -a_{n-2} & \cdots & -a_1 \end{bmatrix} \begin{bmatrix} x_1 \\ x_2 \\ \vdots \\ x_n \end{bmatrix} + \begin{bmatrix} 0 \\ 0 \\ \vdots \\ 1 \end{bmatrix} u$$

$$= Fx + Gu \tag{8.16}$$

and the measurement or observation equation is

$$y = [1, 0, \ldots, 0] \begin{bmatrix} x_1 \\ x_2 \\ \vdots \\ x_n \end{bmatrix} = Hx. \tag{8.17}$$

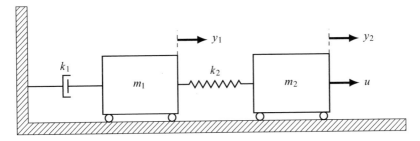

Fig. 8.2 Double-object spring. The two objects are connected with a spring, while the first object is attached to the wall with a damper

Equations (8.16)–(8.17) define the state space representation of the linear system described by the differential equation (8.15). Note that matrix **F** is a companion matrix, which is discussed in some detail in Sect. 4.2.2.

Example 8.6 (Two Objects Connected with a Spring and Damper) Consider two objects connected via a spring, where the first object is connected via a damp on the wall (see Fig. 8.2). Object 1 (with mass m_1) is connected to the wall via a damper, with damping coefficient k_1. On the other end Object 1 is attached to a spring with coefficient k_2, which other end is connected to Object 2 (with mass m_2). A force $u(t)$ is applied on Object 2 and the output of this system is the positions $y_1(t)$ and $y_2(t)$ of the two objects.

This system can be described by the system of differential equations

$$m_1 \frac{d^2 y_1(t)}{dt^2} + k_1 \frac{dy_1(t)}{dt} + k_2[y_1(t) - y_2(t)] = 0$$

$$m_2 \frac{d^2 y_2(t)}{dt^2} + k_2[y_2(t) - y_1(t)] = u(t).$$

This can be written as

$$\ddot{y}_1 = -\frac{k_1}{m_1}\dot{y} - \frac{k_2}{m_1}y_1 + \frac{k_2}{m_1}y_2,$$

$$\ddot{y}_2 = -\frac{k_2}{m_2}y_2 + \frac{k_2}{m_2}y_1 + \frac{u}{m_2}.$$

If we define the state vector

$$x = \begin{bmatrix} \dot{y}_1 \\ y_1 \\ \dot{y}_2 \\ y_2 \end{bmatrix},$$

then the state space representation of this system is

$$
\dot{x} = \begin{bmatrix} \ddot{y}_1 \\ \dot{y}_1 \\ \ddot{y}_2 \\ \dot{y}_2 \end{bmatrix} = \begin{bmatrix} -\frac{k_1}{m_1} & -\frac{k_2}{m_1} & 0 & \frac{k_2}{m_1} \\ 1 & 0 & 0 & 0 \\ 0 & \frac{k_2}{m_2} & 0 & -\frac{k_2}{m_2} \\ 0 & 0 & 1 & 0 \end{bmatrix} \begin{bmatrix} \dot{y}_1 \\ y_1 \\ \dot{y}_2 \\ y_2 \end{bmatrix} + \begin{bmatrix} 0 \\ 0 \\ \frac{1}{m_2} \\ 0 \end{bmatrix} u
$$

$$
= Fx + Gu \tag{8.18}
$$

and

$$
y = \begin{bmatrix} 0 & 1 & 0 & 0 \\ 0 & 0 & 0 & 1 \end{bmatrix} \begin{bmatrix} \dot{y}_1 \\ y_1 \\ \dot{y}_2 \\ y_2 \end{bmatrix} = Hx. \tag{8.19}
$$

8.2.3   Solution of the State Differential Equation

In this section we give the solution of the state differential equation $\dot{x}(t) = Fx(t) + Gu(t) + \zeta(t)$ of the time-invariant state space model (8.11a)–(8.11b).

Theorem 8.2 *Consider the time-invariant state space model* (8.11a)–(8.11b) *with the initial condition* $x(t_0)$, *for some* t_0. *The solution of the state differential equation* (8.11a) *is given by*

$$
x(t) = \exp[F(t - t_0)]x(t_0) + \int_{t_0}^{t} \exp[F(t - \tau)][Gu(\tau) + \zeta(\tau)]\, d\tau. \tag{8.20}
$$

Proof The proof is by direct differentiation with respect to t. With the matrix exponential of a square matrix A, defined by $\exp(A) = \sum_{n=0}^{\infty}(1/n!)A^n$, we obtain

$$
\dot{x}(t) = F \exp[F(t - t_0)]x(t_0) + \frac{d}{dt}\exp(Ft)\int_{t_0}^{t}\exp(-F\tau)C(\tau)\,d\tau
$$

$$
= F\left\{\exp[F(t - t_0)] + \int_{t_0}^{t}\exp[F(t - \tau)]C(\tau)\,d\tau\right\} + C(t)
$$

$$
= Fx(t) + C(\tau),
$$

where $C(t) = Gu(t) + \zeta(t)$.   □

Some comments are in order. The matrix $\Phi(t, t_0) = \exp[F(t - t_0)]$ is called state transition matrix and it satisfies $\dot{\Phi}(t, t_0) = F\Phi(t, t_0)$ and $\Phi(t_0, t_0) = I$. With this in place the solution $x(t)$ of (8.20) can be written as

$$x(t) = \mathbf{\Phi}(t, t_0)x(t_0) + \int_{t_0}^{t} \mathbf{\Phi}(t, \tau)[\mathbf{G}u(\tau) + \zeta(\tau)]\,d\tau. \qquad (8.21)$$

Under some conditions (8.21) is the solution of the time-varying state differential equation (when $\mathbf{F}(t)$ and $\mathbf{G}(t)$ are time-varying matrices) where now $\mathbf{G}$ is replaced by $\mathbf{G}(\tau)$ and $\mathbf{\Phi}(t, t_0)$ is a suitable matrix satisfying certain conditions; for details the reader is referred to Minkler and Minkler (1993, Chapter 4).

As was earlier mentioned we can find the solution the differential equation driving the dynamics of a system using the Laplace or Fourier transforms. The appeal of Theorem 8.2 is that it can provide the solution of high dimensional linear systems or systems driven by high-order differential equations.

8.2.4 Discrete-Time State Space Model

The above discussed can be applied to discrete-time dynamic systems. First we shall define the discrete-time state space representation of a system and then we will provide the version of Theorem 8.2 for discrete-time.

Consider an equally spaced partition of an interval of the real line $[t_0, t_N]$, consisting of time points $t_0, t_1, \ldots, t_k, \ldots, t_N$. There are $N + 1$ points in total, with equal length $\Delta t = t_{k+1} - t_k$, for $k = 0, 1, \ldots, N - 1$. This implies that $t_{k+1} = (k + 1)\Delta t$. The discrete-time equivalent to the continuous-time state space model (8.9)–(8.10) is

$$x[(k + 1)\Delta t] = f(x[k\Delta t], u[k\Delta t]),$$

$$y[k\Delta t] = g(x[k\Delta t], u[k\Delta t]),$$

for suitable functions $f(\cdot)$ and $g(\cdot)$. It is also understood that the initial state $x(t_0)$ is known.

A linear system may be represented by a discrete-time and time-invariant linear state space model

$$x[(k + 1)\Delta t] = \mathbf{F}x(k\Delta t) + \mathbf{G}u(k\Delta t) + \zeta_k, \qquad (8.22a)$$

$$y(k\Delta t) = \mathbf{H}x(k\Delta t) + \epsilon_k, \qquad (8.22b)$$

where matrices $\mathbf{F}$, $\mathbf{G}$ and $\mathbf{H}$ are time-varying and ζ_k, ϵ_t are innovations. If some of $\mathbf{F}$, $\mathbf{G}$ or $\mathbf{H}$ are time-varying the model is called time-varying linear state space model. Here, the innovations are assumed to be white noise and independent of each other. The initial state $x(t_0)$ is assumed to be known. The model specification is completed by specifying the error distributions of ζ_k and ϵ_k; we shall discuss these in Sect. 8.4.1 below.

State space model (8.22a)–(8.22b) can be used to approximate the continuous-time model (8.11a)–(8.11b). Consider the partition of $[t_0, t_N]$ considered above and take $t_{k+1} = (k + 1)\Delta t$. From solution (8.21) for $x(t)$ evaluated at $t = t_k$

$$x[(k + 1)\Delta t] = \mathbf{\Phi}[(k + 1)\Delta t, k\Delta]x(k\Delta t) + \int_{k\Delta t}^{(k+1)\Delta t} \mathbf{\Phi}[(k + 1)\Delta t, \tau][\mathbf{G}u(\tau) + \zeta(\tau)]\,d\tau.$$
$$(8.23)$$

We assume that $u(\tau)$ changes slowly with respect to Δt; this can be supported by considering Δt short enough. Hence, we assume $u(\tau) \approx u(k\Delta)$, for $k\Delta t \leq \tau \leq (k + 1)\Delta t$. Combining this with (8.23) we obtain

$$x[(k + 1)\Delta t] = \mathbf{F}^* x(k\Delta t) + \mathbf{G}^* u(k\Delta t) + \zeta_k, \qquad (8.24)$$

where

$$\mathbf{F}^* = \mathbf{\Phi}[(k + 1)\Delta t, k\Delta t]$$

$$\mathbf{G}^* = \int_{k\Delta t}^{(k+1)\Delta t} \mathbf{\Phi}[(k + 1)\Delta t, \tau]\mathbf{G}\,d\tau,$$

$$\zeta_k = \int_{k\Delta t}^{(k+1)\Delta t} \mathbf{\Phi}[(k + 1)\Delta t, \tau]\zeta(\tau)\,d\tau.$$

The observation equation is given by

$$y(k\Delta t) = \mathbf{H}x(k\Delta t) + \epsilon_k, \qquad (8.25)$$

where $\epsilon_k = \epsilon(k\Delta t)$.

The discrete-time linear state space model (8.24)–(8.25) is a discrete analogue of the continuous-time model (8.11a)–(8.11b) and can be used to approximate it.

Example 8.7 Consider the continuous-time linear system having the state space representation

$$\dot{x} = \begin{bmatrix} \dot{x}_1 \\ \dot{x}_2 \end{bmatrix} = \begin{bmatrix} 1 & \lambda \\ 0 & 1 \end{bmatrix} \begin{bmatrix} x_1 \\ x_2 \end{bmatrix} + \begin{bmatrix} 1 \\ 1 \end{bmatrix} u = Fx + Gu,$$

$$y = [1, -1] \begin{bmatrix} x_1 \\ x_2 \end{bmatrix} = Hx,$$

where u is a scalar input, for some constant λ. In order to discretise this model and write down the discrete-time state space analogue we need to evaluate the matrices $\mathbf{F}^*$ and G^* ($\zeta_k = 0$, since $\zeta(\tau) = 0$).

First we compute $\mathbf{F}^*$. From the definition of $\mathbf{F}$ it is easy to verify that the n-th power of $\mathbf{F}$ is

$$\mathbf{F}^n = \begin{bmatrix} 1 & n\lambda \\ 0 & 1 \end{bmatrix}.$$

This may be proven by induction. Using this and the matrix exponential Taylor expansion we have

$$e^{c\mathbf{F}} = \begin{bmatrix} 1 + c + \frac{c^2}{2!} + \cdots & \lambda c + \frac{2\lambda c^2}{2!} + \cdots \\ 0 & 1 + c + \frac{c^2}{2!} + \cdots \end{bmatrix}$$

$$= \begin{bmatrix} e^c & \lambda c e^c \\ 0 & e^c \end{bmatrix}, \tag{8.26}$$

for some constant c.

Putting $c = \Delta t$ we obtain

$$\mathbf{F}^* = \mathbf{\Phi}[(k+1)\Delta t, k\Delta t] = e^{\Delta t \mathbf{F}} = \begin{bmatrix} e^{\Delta t} & \lambda \Delta t e^{\Delta t} \\ 0 & e^{\Delta t} \end{bmatrix}.$$

From the matrix exponential (8.26) and using $c = (k+1)\Delta t - \tau$ we have

$$G^* = \int_{k\Delta t}^{(k+1)\Delta t} e^{[(k+1)\Delta t - \tau]F}\, d\tau \begin{bmatrix} 1 \\ 1 \end{bmatrix} = \begin{bmatrix} I_1 & I_2 \\ 0 & I_1 \end{bmatrix} \begin{bmatrix} 1 \\ 1 \end{bmatrix}, \tag{8.27}$$

where I_1 and I_2 are the integrals evaluated below

$$I_1 = \int_{k\Delta t}^{(k+1)\Delta t} e^{(k+1)\Delta t - \tau}\, d\tau = \left[-e^{(k+1)\Delta t - \tau} \right]_{k\Delta t}^{(k+1)\Delta t} = e^{\Delta t} - 1$$

and

$$I_2 = \lambda \int_{k\Delta t}^{(k+1)\Delta t} [(k+1)\Delta t - \tau] e^{(k+1)\Delta t - \tau}\, d\tau$$

$$= \lambda(k+1)\Delta t (e^{\Delta t} - 1) + \lambda \int_{k\Delta t}^{(k+1)\Delta t} \tau d e^{(k+1)\Delta t - \tau}$$

$$= \lambda(k+1)\Delta t (e^{\Delta t} - 1) + \lambda \left[\tau e^{(k+1)\Delta t - \tau} \right]_{k\Delta t}^{(k+1)\Delta t} - \lambda \int_{k\Delta t}^{(k+1)\Delta t} e^{(k+1)\Delta t - \tau}\, d\tau$$

$$= \lambda \Delta t e^{\Delta t} - \lambda e^{\Delta t} + 1,$$

where integration by parts was used.

Putting I_1 and I_2 in (8.27) we obtain G^* as

$$G^* = \begin{bmatrix} e^{\Delta t}(1 + \lambda \Delta t - \lambda) \\ e^{\Delta t} - 1 \end{bmatrix}.$$

The discrete-time state space model is

$$x[(k+1)\Delta t] = \mathbf{F}^* x(k\Delta t) + \mathbf{G}^* u = \begin{bmatrix} e^{\Delta t} & \lambda \Delta t e^{\Delta t} \\ 0 & e^{\Delta t} \end{bmatrix} x(k\Delta t) + \begin{bmatrix} e^{\Delta t}(1 + \lambda \Delta t - \lambda) \\ e^{\Delta t} - 1 \end{bmatrix} u$$

$$y(k\Delta t) = [1, -1] x(k\Delta t).$$

Finally, we provide the equivalent of the solution of x_k for discrete-time systems. Consider the time-invariant state space representation (8.22a)–(8.22b) and set $x_k = x(k\Delta t)$, for $\Delta t = 1$. For fixed Δt the discrete set $\{t_0, t_1, t_2, \dots, t_N\}$ is equivalent to $\{k_0, k_0 + 1, \dots, k_0 + N\}$, where $k_0 = t_0$. We shall work with the set $\{k_0, k_0 + 1, \dots, k_0 + N\}$ where Δt is implicit. With this in place, model (8.22a)–(8.22b) can be written as

$$x_{k+1} = \mathbf{F}x_k + \mathbf{G}u_k + \zeta_k, \tag{8.28a}$$

$$y_k = \mathbf{H}x_k + \epsilon_k, \tag{8.28b}$$

where the initial state is $x(k_0)$ is assumed known.

For this state space model the difference state equation (8.28a) has solution

$$x_k = \mathbf{F}^{k-k_0} x_{k_0} + \sum_{j=0}^{k-k_0-1} \mathbf{F}^j (\mathbf{G}u_{k-j-1} + \zeta_{k-j-1})$$

and is the discrete-time equivalent to the solution (8.20) of the differential state equation (8.11a).

8.3 System Stability

8.3.1 Definitions

The concept of *stability* of a system is concerned with the behaviour of the system for a certain amount of input (Bacciotti, 2019). It relates to the basic principle of facilitating bounded energy in the output, for bounded energy in the input. The so-called bounded input bounded output (BIBO) stability requires that if the input is bounded, then the output should be bounded too; details and conditions of BIBO stability are given in Minkler and Minkler (1993, Section 3.4). Introductions to stability in dynamic systems can be found in Sastry (1999), Zinober (2001), Ding (2013), Guo and Han (2018) and Bacciotti (2019).

In this section we discuss stability under the state space representation of a system.

Consider the continuous-time system represented by Eqs. (8.9)–(8.10). State space system stability studies the dynamic behaviour of the state vector over time, for small perturbations of the state and the input functions (sometimes we focus on the states and consider a so-called *free* or *undriven* system, with $u(t) = 0$). Hence we operate with a state differential equation

$$\dot{x} = f(x, u = 0, t). \tag{8.29}$$

A state $x_e = x(t_e)$ of (8.29) is called an equilibrium state if $x(t) = x_e$, for any $t \geq t_e$, or in words: after equilibrium time t_e the state vector is constant and equal to x_e. From (8.29) it follows that $f(x_e, u = 0) = 0$, as

$$\dot{x}_e = \lim_{h \to 0} \frac{x(t_e + h) - x_e}{h} = 0, \tag{8.30}$$

since $x(t_e + h) = x_e$, as $t_e + h > t_e$.

System stability studies the dynamic behaviour of the solution $\phi(t, x_0)$ of (8.29), with initial condition x_0. More specifically, we are interested to know how close to x_e the solution $\phi(t, x_0)$ is, provided x_0 is chosen to be close to x_e. In this case we have

Definition 8.1 An equilibrium state x_e of (8.29) is *stable* at time t_0, if for every $\varepsilon > 0$ there exists $\delta(\varepsilon, t_0) > 0$ such that $\| \phi(t, x_0) - x_e \| \leq \varepsilon$ for all $t \geq t_0$, provided that the initial state x_0 satisfies $\| x_0 - x_e \| \leq \delta(\varepsilon, t_0)$, where $\| \cdot \|$ is a suitable vector norm (usually the Euclidean norm) and $\phi(t, x_0)$ is the general solution of (8.29).

An equilibrium state x_e is stable if it stable at all points t_0. According to this definition, if the initial state is chosen to be in a neighbourhood of the stable equilibrium state x_e, then $\phi(t, x_0)$ will also be in a neighbourhood of x_e.

Note that Definition 8.1 does not imply that $\phi(t, x_0)$ converges to x_e as $t \to \infty$. It simply suggests that $\phi(t, x_0)$ is bounded in a neighbourhood of x_e. The next definition in addition to an equilibrium state being stable it requires it to be the limit of the solution $\phi(t, x_0)$ of the differential equation (8.29).

Definition 8.2 An equilibrium state x_e of (8.29) is *asymptotically stable* if it is stable (Definition 8.1) and if there exists a β-neighbourhood of x_e, with $\| x_0 - x_e \| \leq \beta(t_0)$, so that for any $\kappa > 0$ there exists $t_1(t_0, \kappa)$ with $\| \phi(t, x_0) - x_e \| \leq \kappa$, for all $t \geq t_0 + t_1(t_0, \kappa)$.

In this definition we start again by placing x_0 in a neighbourhood of x_e, but now for all $t \geq t_0 + t_1(t_0, \kappa)$, the solution $\phi(t, x_0)$ converges to x_e as $t \to \infty$.

The dependency of t_0 on δ, β and t_1 in Definitions 8.1 and 8.2 can be dropped in order to provide stronger definitions of stability which are known as *uniform stability* (Definition 8.1 if we drop t_0 dependence from δ) and *uniform asymptotic stability* (if we drop t_0 from the dependence of β and t_1). Finally, we give the definition of stability of a system.

Definition 8.3 The free or undriven system generated by the differential equation (8.29) is *stable, asymptotically stable or uniformly asymptotically stable*, if the zero-equilibrium state $x_e = 0$ is stable, asymptotically stable or uniformly asymptotically stable, respectively.

8.3.2 Stability of Linear Systems

The theory of stability is focused primarily on linear systems and these will be discussed first. A book-length coverage of stability for linear systems can be found in Ding (2013) and Bacciotti (2019) and in the references therein.

Example 8.8 Consider the first-order system with differential equation

$$\dot{x} = Fx(t) + Gu(t), \tag{8.31}$$

where F and G are scalars. For a zero input $u(t) = 0$, let $x_e = 0$ be an equilibrium state. The solution of (8.31) is $x(t) = e^{Ft}x(0)$, for $t_0 = 0$. Hence, the state $x_e = 0$ is unstable if $F > 0$ and stable if $F \leq 0$. In particular, x_e is stable if $F = 0$ and asymptotically stable if $F < 0$.

Consider now the first-order discrete system, with difference equation

$$x_k = Fx_{k-1} + Gu_{k-1}, \tag{8.32}$$

where F and G are scalars as before. For the zero input $u_{k-1} = 0$ let $x_e = 0$ be an equilibrium state. The solution of (8.32) is $x_k = F^k x_0$. Hence, the state $x_e = 0$ is unstable if $|F| > 1$ and stable if $|F| \leq 1$. In particular, x_e is stable if $F = \pm 1$ and asymptotically stable if $|F| < 1$.

Consider now the linear system represented by (8.11a)–(8.11b), with perhaps a time-varying matrix $\mathbf{F}(t)$. The state differential equation is reduced to $\dot{x} = \mathbf{F}(t)x(t)$, given the zero-vector input $u(t) = 0$, so that the system is free or undriven. The general solution (8.21) of the state differential equation satisfy $x(t) = \mathbf{\Phi}(t, t_0)x(t_0)$, where $\mathbf{\Phi}(t, t_0)$ is discussed in Sect. 8.2.3. The following theorem provides conditions of stability for this system.

Theorem 8.3 *The linear system represented by (8.11a)–(8.11b) is uniformly asymptotically stable if and only if*

$$\| \mathbf{\Phi}(t, t_0) \| \leq a_1 e^{-a_2(t-t_0)},$$

for some constants $a_1, a_2 > 0$.

For the proof of this theorem the reader is referred to Kalman and Bertram (1960). To the following we shall use the term *asymptotically stable* to describe a system which is *uniformly asymptotically stable*.

When the linear system is time-invariant, so that $F(t) = F$ is time-invariant, then linear system (8.11a)–(8.11b), then $\Phi(t, t_0) = \exp[F(t - t_0)]$. The following theorem gives necessary and sufficient conditions for the stability of such a system.

Theorem 8.4 *Consider the time-invariant linear system* (8.11a)–(8.11b), *with* $F(t) = F, G(t) = G$ *and* $H(t) = H$. *Assume that the matrix* F *is diagonalisable. The undriven system* $(u(t) = 0)$ *is asymptotically stable if and only if the eigenvalues of* F *have negative real parts.*

Let $\lambda_1, \ldots, \lambda_p$ be the eigenvalues of F, with $\lambda_j = a_j + i b_j$, where a_j and b_j are the real and imaginary parts of λ_j and i is the imaginary unit of the complex plain. Then according to Theorem 8.4

1. If $a_1, \ldots, a_p < 0$, then the system is asymptotically stable.
2. If $a_1, \ldots, a_p \leq 0$, then the system is stable.
3. If there is at least one k such that $a_k > 0$, then the system is unstable.

Proof of Theorem 8.4 Consider the linear and time-invariant system (8.11a)–(8.11b), where F is diagonalisable and $u(t) = 0$ for stability. This means we can write $F = T\Lambda T^{-1}$, where $L = \mathrm{diag}(\lambda_1, \ldots, \lambda_p)$ is the diagonal matrix, with the eigenvalues of F in its diagonal (not necessarily distinct) and T is the $p \times p$ matrix with columns the corresponding standardised eigenvectors of F. Hence we can write

$$
e^{F(t-t_0)} = \sum_{n=0}^{\infty} \frac{(t - t_0)^n}{n!} F^n
$$

$$
= \sum_{n=0}^{\infty} \frac{(t - t_0)^n}{n!} T\Lambda^n T^{-1}
$$

$$
= T e^{(t-t_0)\Lambda} T^{-1}, \tag{8.33}
$$

since T does not depend on n in the infinite sum. Now write $\lambda_j = a_j + i b_j$, where a_j and b_j are the real and imaginary parts of the complex eigenvalue λ_j and i is the imaginary unit. The j-th eigenvalue of $e^{(t-t_0)\Lambda}$ can be written as $c_j e^{d_j i}$, where $c_j = e^{(t-t_0)a_j}$ and $d_j = (t - t_0)b_j$, both implicitly depending on t. With $\| \cdot \|$ the Euclidean matrix norm on the complex plain, from (8.33) we obtain

$$
\| e^{F(t-t_0)} \| \leq \| T \| \| T^{-1} \| \| e^{(t-t_0)\Lambda} \| . \tag{8.34}
$$

Also since $e^{(t-t_0)\Lambda}$ is a diagonal matrix we have

$$
\begin{aligned}
\| e^{(t-t_0)\Lambda} \| &= \| \operatorname{diag}(c_1, \ldots, c_p)\operatorname{diag}(e^{d_1 i}, \ldots, e^{d_p i}) \| \\
&\le \| \operatorname{diag}(c_1, \ldots, c_p) \| \| \operatorname{diag}(e^{d_1 i}, \ldots, e^{d_p i}) \| \\
&= \sqrt{c_1^2 + \cdots + c_p^2} \sqrt{|e^{d_1 i}|^2 + \cdots + |e^{d_p i}|^2} \\
&= \sqrt{p \sum_{j=1}^{p} e^{2(t-t_0)a_j}}.
\end{aligned}
$$

Hence from (8.34) it follows

$$
\| e^{F(t-t_0)} \| \le \sqrt{p \sum_{j=1}^{p} e^{2(t-t_0)a_j}}.
$$

From this equation we have the following conclusion

1. If $a_1, \ldots, a_p < 0$, the system is asymptotically stable, as $\lim_{t \to \infty} e^{F(t-t_0)} = \mathbf{0}$.
2. If $a_1, \ldots, a_p \le 0$, then the system is stable.
3. If there is k such that $a_k > 0$, for at least one $k \in \{1, \ldots, p\}$, then the system is unstable. The limit $\lim_{t \to \infty} e^{F(t-t_0)}$ does not exist.

$\square$

As an illustration of this assertion, consider the time-invariant linear system (8.11a)–(8.11b) and suppose that the system matrix F has one eigenvalue λ, with non-negative real part, i.e. $\lambda = a + ib$, with $a \ge 0$. We will show that the system is not stable, or that

$$
\lim_{t \to \infty} \| e^{F(t-t_0)} x(t_*) \| =
\begin{cases}
\| x_* \|, & \text{if } a = 0, \\
\infty, & \text{if } a > 0,
\end{cases}
\tag{8.35}
$$

where $x(t_*)$ is the eigenvector corresponding to λ.

Since $x(t_*)$ is the eigenvector of $\mathbf{F}$ corresponding to the eigenvalue λ we have $\mathbf{F}x(t_*) = \lambda x(t_*)$. This implies

$$
\mathbf{F}^n x(t_*) = \mathbf{F}^{n-1}[\mathbf{F}x(t_*)] = \lambda \mathbf{F}^{n-1} x(t_*) = \cdots = \lambda^n x(t_*),
$$

for any $n = 0, 1, 2, \ldots$. This shows that the eigenvalue of $\mathbf{F}^n$ is λ^n. Then using the matrix exponential expansion

$$
e^{F(t-t_0)} x(t_*) = \sum_{n=0}^{\infty} \frac{(t-t_0)^n}{n!} \mathbf{F}^n x(t_*) = \sum_{n=0}^{\infty} \frac{\lambda^n (t-t_0)^n}{n!} x(t_*) = e^{\lambda(t-t_0)} x(t_*).
$$

With $\lambda = a + ib$, we get

$$\| e^{\mathbf{F}(t-t_0)} x(t_*) \| = |e^{\lambda(t-t_0)}| \, \| x(t_*) \| = e^{a(t-t_0)} \, \| x(t_*) \|,$$

after using $|e^{ib(t-t_0)}| = 1$, from Euler's formula.

As $t \to \infty$, $e^{a(t-t_0)}$ converges to 1 ($a = 0$) or to ∞ ($a > 0$) and this together with $\| x(t_*) \| > 0$ proves (8.35).

Example 8.9 (Spring Force Dynamics Revisited) We revisit Example 8.4 of the spring force dynamics. In that example a time-invariant state matrix

$$\mathbf{F} = \begin{bmatrix} 0 & 1 \\ -\frac{k_1}{m} & -\frac{k_2}{m} \end{bmatrix}$$

was proposed, where k_1 is the viscus friction coefficient, k_2 is the spring constant and m is the mass of the object. Assume that k_1, k_2 and m satisfy $k_2^2 - 4k_1 m < 0$. In order to see whether this system is asymptotically stable, we shall calculate the eigenvalues of F. The characteristic polynomial $|\mathbf{F} - \lambda I| = 0$ has two complex roots

$$\lambda_{1,2} = -\frac{k_2}{2m} \pm \frac{\sqrt{4mk_1 - k_2^2}}{2m} i.$$

The real parts of $\lambda_{1,2}$ are both negative ($-k_2/(2m)) < 0$ and so the system is asymptotically stable, under the condition $k_2^2 - 4mk_1 < 0$.

Minkler and Minkler (1993, Section 4.3.2) show that BIBO stability is equivalent to state space system stability. More specifically, they show that a time-invariant linear system (8.11a)–(8.11b) is BIBO stable, if and only if it is asymptotically stable. The state space representation of a time-invariant linear system provides an important tool to establish stability of a linear system. For time-invariant model computing the eigenvalues of F is significantly easier than working with integrals involved in BIBO stability. This illustrates the usefulness and versatility of the state space representation of linear systems.

Example 8.10 (RLC Electric Circuit) Consider a LRC electric circuit, consisting of a voltage source or battery (V), a capacitor (C), a resistor (R) and an inductor (L), all serially connected; for a circuit diagram see Fig. 8.3.

Let $u(t)$ be the input voltage of the battery, $x_1(t)$ be the current and $x_2(t)$ the voltage of the capacitor. According to Kirchhoff's voltage law, the total voltage of the system in the battery $u(t)$ is equal to the sum of the amount of voltage in the three units, or

$$u(t) - V_R - V_L - V_C = u(t) - Rx_1(t) - L\dot{x}_1(t) - x_2(t) = 0,$$

where R is the effective resistance of the combined load, source and component, L is the inductance of the inductor and C is the capacitance of the capacitor. Here

Fig. 8.3 RLC electric circuit.
Shown are the voltage source
(V), the capacitor (C), the
resistor (R) and the inductor
(L), all serially connected

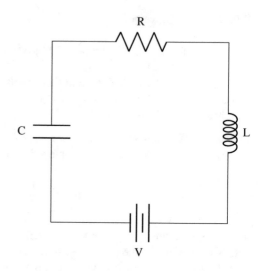

we have used $V_R = Rx_1(t)$ and $V_L = L\dot{x}(t)$ (from Kirchhoff's law). Also from Kirchhoff's law for the current we have

$$x_2(t) = V_C = u(0) + \frac{1}{C} \int_0^t x_1(\tau)\, d\tau,$$

hence

$$x_1(t) = Cx_2(t).$$

If we define the state vector $x(t) = [x_1(t), x_2(t)]^\top$ we have

$$\dot{x}(t) = \begin{bmatrix} -\frac{R}{L} & -\frac{1}{L} \\ -\frac{1}{C} & 0 \end{bmatrix} x(t) + \begin{bmatrix} \frac{1}{L} \\ 0 \end{bmatrix} u(t) = \mathbf{F}x(t) + \mathbf{G}u(t). \tag{8.36}$$

Suppose we are interested in the current $x_1(t) = y(t)$ as output of the system. Hence we write

$$y(t) = [1, 0]x(t) = Hx(t). \tag{8.37}$$

Equations (8.36)–(8.37) define a linear system for the RLC circuit described above.

In order to study the stability of system (8.36)–(8.37) we need to calculate the eigenvalues of $\mathbf{F}$. Solving $|\mathbf{F} - \lambda \mathbf{I}| = 0$ we find the eigenvalues

$$\lambda_{1,2} = \frac{1}{2L} \left(-R \pm \sqrt{R^2 - \frac{4L}{C}} \right).$$

- **Case I.** If $R^2 - 4L/C \geq 0$, then there are two (or one double) real roots

$$\lambda_1 = \frac{1}{2L}\left(-R - \sqrt{R^2 - \frac{4L}{C}}\right) < 0$$

$$\lambda_2 = \frac{1}{2L}\left(-R + \sqrt{R^2 - \frac{4L}{C}}\right).$$

Since $R^2 - 4L/C < R^2$, it follows that $\sqrt{R^2 - 4L/C} < R$, and so $\lambda_2 < 0$ too.
- **Case II.** If $R^2 - aL/C < 0$, there are two conjugate complex roots

$$\lambda_{1,2} = \frac{1}{2L}\left(-R \pm i\sqrt{\frac{4L}{C} - R^2}\right),$$

with real parts $-R/L < 0$.

Hence, in either case the system has eigenvalues with negative real parts and from Theorem 8.4 the system is asymptotically stable.

We remark that the above results hold for discrete-time systems (8.28a)–(8.28b), with minor modifications. Definitions 8.1–8.3 are unchanged, with the only modification that $\phi(t, x_0)$ is the solution of the difference equation (8.28a). Considering the time-invariant linear system (8.28a)–(8.28b) and for zero input $u = 0$ the state difference equation is $x_k = \mathbf{F}^{k-k_0}x_0$. If matrix $\mathbf{F}$ is diagonalisable, then the system is asymptotically stable, if and only if the eigenvalues of $\mathbf{F}$ lie inside the unit circle, or $|\lambda_j| < 1$, for $j = 1, \ldots, p$. This result, which is the district analogue of Theorem 8.4, is schematically proven as follows. Since $\mathbf{F}$ is diagonalisable, we can write $\mathbf{F} = \mathbf{T}\mathbf{\Lambda}\mathbf{T}^{-1}$, where $\mathbf{\Lambda}$ is the diagonal matrix with the eigenvalues of $\mathbf{F}$ in its diagonal and $\mathbf{T}$ is the matrix with columns the standardised eigenvectors of $\mathbf{F}$. This implies $\mathbf{F}^k = \mathbf{T}\mathbf{\Lambda}^k\mathbf{T}^{-1}$. Then

$$\| \mathbf{F}^k \| \leq \| \mathbf{T} \| \| \mathbf{T}^{-1} \| \| \mathbf{\Lambda}^k \|$$

$$= \| \mathbf{T} \| \| \mathbf{T}^{-1} \| \sqrt{|\lambda_1|^{2k} + \cdots + |\lambda_p|^{2k}},$$

which tends to the zero matrix, if $\lim_{k \to \infty} |\lambda_j|^{2k}$, for all $j = 1, \ldots, p$. Hence the system is asymptotically stable if and only if $|\lambda_j| < 1$, for all $j = 1, \ldots, p$. Note that stability for discrete-time time-invariant systems relates to causality and stationarity for time series, see e.g. Sect. 7.2.

8.3.3 Stability of Non-Linear Systems

Stability theory studies the dynamic behaviour of dynamic systems for small perturbations of the states and inputs from equilibrium values. Aleksandr Mikhailovich Lyapunov (6 June 1857–3 November 1918) has made significant contributions to the theory of differential equations and in particular stability theory of dynamical systems, among other fields including mathematical physics and probability theory. For a review of his contributions in mathematics the reader is referred to Smirnov (1992) and Parks (1992) among others. In the context of this chapter, we discuss the *indirect Lyapunov method* (also known as first kind Lyapunov method) and the *direct Lyapunov method* (also known as second kind Lyapunov method). Among the many studies, which highlight the importance of stability theory of non-linear systems, we signal out Zinober (1994), Sastry (1999), Zinober (2001), Ding (2013) and Guo and Han (2018).

8.3.3.1 Lyapunov Indirect Method

Consider the continuous-time system (8.9)–(8.10) or

$$\dot{x} = f(x(t), u(t)), \tag{8.38a}$$

$$y(t) = g(x(t), u(t)), \tag{8.38b}$$

for some known smooth functions $f(\cdot)$ and $g(\cdot)$. Here, unlike system (8.9)–(8.10) we have allowed $g()$ in the measurement equation to depend on the input $u(t)$. The objective of stability theory is to investigate the dynamic behaviour of the system for small perturbations of the input and state functions

$$\Delta u(t) = u(t) - u_e \quad \text{and} \quad \Delta x(t) = x(t) - x_e, \tag{8.39}$$

where u_e is the equilibrium of the input function and x_e is the equilibrium of the state.

We start with Lyapunov's indirect method. The basic idea of this method is to linearise the non-linear model and study the stability of the linearised system in order to make inference for the stability of the non-linear model. Suppose that x_e is an equilibrium state, so that $f(x_e, u_e) = 0$.

From Eq. (8.39) we get

$$
\begin{aligned}
\Delta \dot{x}(t) &= \dot{x}(t) - \dot{x}_e \\
&= f(\Delta x(t) + x_e, \Delta u(t) + u_e) \\
&\approx \frac{\partial f}{\partial x}(x_e, u_e) \Delta x(t) + \frac{\partial f}{\partial u}(x_e, u_e) \Delta u(t) \\
&= \mathbf{F} \Delta x(t) + \mathbf{G} \Delta u(t).
\end{aligned}
\tag{8.40}
$$

Similarly, for the measurement equation (8.38b), we obtain

$$\Delta y(t) \approx \frac{\partial g}{\partial x}(x_e, u_e)\Delta x(t) + \frac{\partial g}{\partial u}(x_e, u_e)\Delta u(t)$$

$$= \mathbf{H}\Delta x(t) + \mathbf{L}\Delta u(t), \qquad (8.41)$$

where $\Delta y(t) = y(t) - g(x_e, u_e)$ is the perturbation of the output from the equilibrium. It follows that the perturbations $\Delta x(t)$ (state), $\Delta y(t)$ (output) and $\Delta u(t)$ (input) are approximately ruled by the linearised system (8.40)–(8.41). Lyapunov's indirect method involves the following steps.

1. Consider the non-linear system (8.38a) with $f(\cdot)$ differentiable, $u(t) = 0$ and $x_e = 0$ (zero-equilibrium state). In this case the system is $\dot{x} = f(x, u = 0)$.
2. Approximate the non-linear system by the linear system (8.40), which can be written as $\dot{x} = \mathbf{F}x$, with $\mathbf{F}$ as in (8.40).
3. If the linear system $\dot{x} = \mathbf{F}x$ is asymptotically stable, then the state $x_e = 0$ is asymptotically stable for the non-linear system $\dot{x} = f(x, u = 0)$ (at a neighbourhood of $x_e = 0$).
4. If the linear system $\dot{x} = \mathbf{F}x$ is unstable, then the non-linear system $\dot{x} = f(x, u = 0)$ is also unstable.
5. If the linear system $\dot{x} = \mathbf{F}x$ is stable, nothing can be said about the stability of the equilibrium state $x_e = 0$ for the non-linear system $\dot{x} = f(x, u = 0)$.

This suggests whether a non-linear system is asymptotically stable or unstable, can be checked by considering the corresponding linear system. If, however, the linear system is stable only (i.e. one or more of the eigenvalues of $\mathbf{F}$ are zero), then the linear system cannot be used to check the stability of the non-linear system. In this case we need to resort to the direct method of Lyapunov (see below).

Example 8.11 Simple gravity pendulum Consider a simple gravity pendulum of Fig. 8.4. An object at the one end of a massless rod is suspended from a pivot and is allowed to swing freely. Due to gravity, the object oscillates towards the vertical (dashed) line at its equilibrium point.

The input of this system is the gravity force $u(t) = mg$, where m is the mass of the object and g is Newton's gravitational constant. The angular displacement $y = y(t)$ is the output of the system, $\dot{y}(t)$ is the angular velocity, $\ddot{y}(t)$ is the angular acceleration and $I = m\ell^2$ is the rational inertia, where ℓ is the length of the pendulum. From Newton's second law for rotation (torque is equal to inertia times angular acceleration) if we assume that there is no friction, it follows that

$$-mg \sin \theta \ell = m\ell^2 \frac{d^2\theta}{dt^2}.$$

For $\theta = y(t)$ and assuming friction is present, this law is written as

$$m\ell^2 \ddot{y}(t) = -\ell mg \sin y(t) - k\dot{y}(t),$$

Fig. 8.4 Simple pendulum

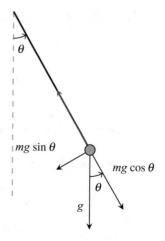

where k is the friction constant. Hence the following non-linear differential equation
is obtained

$$\ddot{y}(t) = -\frac{g}{\ell}\sin y(t) - \frac{k}{m\ell^2}\dot{y}(t). \tag{8.42}$$

We shall put this differential equation in state space form.

Define the state vector $x = [x_1, x_2]^\top$, with $x_1 = y$ and $x_2 = \dot{x}_1 = \dot{y}$. From
Eq. (8.42) we obtain the non-linear system

$$\dot{x}(t) = \begin{bmatrix} \dot{x}_1(t) \\ \dot{x}_2(t) \end{bmatrix} = \begin{bmatrix} f_1(x) \\ f_2(x) \end{bmatrix} = \begin{bmatrix} x_2 \\ -\frac{g}{\ell}\sin x_1 - \frac{k}{m\ell^2}x_2 \end{bmatrix} = f(x, u), \tag{8.43}$$

$$y(t) = [1, 0]x = Hx. \tag{8.44}$$

We remark that this is a non-linear system (the state equation $\dot{x} = f(x, u)$ is non-
linear due to $\sin y$).

In order to check stability of this system we will use Lyapunov's indirect method.
We shall linearise system (8.43)–(8.44) and investigate the stability of the linear
model. The Jacobian matrix of $f(x)$ is

$$J(x_1, x_2) = \begin{bmatrix} 0 & 1 \\ -\frac{g}{\ell}\cos x_1 & -\frac{k}{m\ell^2} \end{bmatrix}.$$

At the equilibrium state $x_e = [0, 0]^\top$, the linearised dynamic model is

$$\Delta\dot{x}(t) = \begin{bmatrix} 0 & 1 \\ -\frac{g}{\ell} & -c \end{bmatrix}\Delta x(t) = F\Delta t, \tag{8.45}$$

$$\Delta y(t) = H\Delta x(t), \tag{8.46}$$

where $c = km^{-1}\ell^{-2}$. The eigenvalues of $\mathbf{F}$ are

$$\lambda_{1,2} = \frac{1}{2}\left(-c \pm \sqrt{c^2 - \frac{4g}{\ell}}\right).$$

If $c^2 - 4g/\ell \geq 0$, there are two (or one double) real roots (root). $\lambda_1 = 2^{-1}(-c - \sqrt{c^2 - 4g/\ell}) < 0$ and $\lambda_1 = 2^{-1}(-c + \sqrt{c^2 - 4g/\ell}) < 0$ too, as $c > \sqrt{c^2 - 4/\ell}$. If $c^2 - 4g/\ell < 0$, there are two conjugate complex eigenvalues $\lambda_{1,2} = 2^{-1}(-c \pm i\sqrt{4g/\ell - c}$. The real part of $\lambda_{1,2}$ is $-c/2 < 0$. Hence, in any case the eigenvalues of $\mathbf{F}$ have negative real parts and from Theorem 8.4 the linear system (8.45)–(8.46) is asymptotically stable. From Lyapunov's indirect method, it follows that the non-linear system (8.43)–(8.44) is asymptotically stable too, or that the zero state $x_e = [0, 0]^\top$ is asymptotically stable.

Consider now the equilibrium state $x_e = [\pi, 0]^\top$. At this state the linearised system is

$$\Delta \dot{x}(t) = \begin{bmatrix} 0 & 1 \\ \frac{g}{\ell} & -c \end{bmatrix} \Delta x(t) = \mathbf{F}\Delta x(t).$$

The eigenvalues of $\mathbf{F}$ are

$$\lambda_{1,2} = \frac{1}{2}\left(-c \pm \sqrt{c^2 + \frac{4g}{\ell}}\right).$$

Here $c^2 + 4g/\ell > 0$ always, hence there are two distinct real eigenvalues $\lambda_1 = 2^{-1}(-c - \sqrt{c^2 + 4g/\ell}) < 0$, but $\lambda_2 = 2^{-1}(-c + \sqrt{c^2 + 4g/\ell}) > 0$, as $c < \sqrt{c^2 + 4g/\ell}$. The equilibrium state $x_e = [\pi, 0]^\top$ of the linear system is unstable (as one eigenvalue is negative and the other one is positive) and so by Lyapunov's first method, x_e is unstable state for the non-linear system too.

8.3.3.2 Lyapunov Direct Method

We are now moving on to discuss Lyapunov's direct method for the stability of non-linear systems. Suppose that a system has a state differential equation

$$\dot{x} = f(x) \tag{8.47}$$

and x_e be an equilibrium state, so that $f(x_e) = 0$.

Consider a scalar function $V(x)$, which is continuously differentiable or $V \in C^1$, where C^1 denotes the set of functions, with continuous first derivatives. V is said to be positive definite (positive semi-definite) in a neighbourhood $D(x_e, r)$ of x_e, if (a) $V(x_e) = 0$ and (b) $V(x) > 0$ ($V(x) \geq 0$) and (c) all sublevel sets of V are bounded, for all $x \in D(x_e, r)$ and $x \neq x_e$, where r denotes the radius of the neighbourhood. Condition (c) is equivalent to $V(x) \to \infty$ as $x \to \infty$. The function V is said to be negative definite (negative semi-definite) if $-V$ is positive definite (positive semi-definite) in $D(x_e, r)$. The function V is said to be indefinite, if it is neither positive definite (positive semi-definite) nor negative definite (negative semi-definite). For example, if A is a symmetric and positive definite matrix, then $V(x) = x^\top A x$ is a positive definite function.

Write $f_i(x)$ the i-th constituent of the vector-valued function $f(x)$, so that

$$f(x) = \begin{bmatrix} f_1(x) \\ f_2(x) \\ \vdots \\ f_p(x) \end{bmatrix},$$

where each $f_i(x)$ is a scalar function with domain a subset of $\mathbb{R}^p$. From the chain rule of differentiation we get

$$\dot{V}(t) = \frac{dV}{dt} = \sum_{i=1}^{p} \frac{\partial V}{\partial x_i} \frac{\partial x_i}{\partial t} = \sum_{i=1}^{p} \frac{\partial V}{\partial x_i} f_i(x(t)). \tag{8.48}$$

If $\dot{V}(x)$ regraded as function of x, $\dot{V}(x)$ is the derivative of V along the system trajectories. The following theorem provides Lyapunov's direct method for the stability of system (8.47).

Theorem 8.5 (Lyapunov's Direct Method) *In the non-linear system* (8.47), *with equilibrium vector* x_e, *the following apply*

1. *If there exists a scalar function* $V(x) \in C^1$ *so that* $V(x)$ *is positive definite in a neighbourhood* $D(x_e, r)$ *and* $\dot{V}(x)$ *is negative semi-definite in* $D(x_e, r)$, *then state* x_e *is stable and if* $x_e = 0$, *system* (8.47) *is stable.*
2. *If there exists a scalar function* $V(x) \in C^1$ *so that* $V(x)$ *is positive definite in a neighbourhood* $D(x_e, r)$ *and* $\dot{V}(x)$ *is negative definite in* $D(x_e, r)$, *then state* x_e *is asymptotically stable and if* $x_e = 0$, *system* (8.47) *is asymptotically stable.*

Proof First we prove (1). From $V(x) > 0$, V is bounded below. Also as $\dot{V} \leq 0$, V is decreasing function with $V(x(0))$ an upper bound. Also

$$V(x(t)) = V(x(0)) + \int_0^t \dot{V}(x(\tau)) \, d\tau \leq V(x(0)).$$

So the trajectory x belongs to the set $x \in \{z \in \mathbb{R}^p : V(z) \leq V(0)\}$, which is bounded. Because V has bounded sublevel sets, it follows that x must be bounded in a neighbourhood $D(x_e, r)$. Hence x_e is stable.

Proceeding now to (2), first note that $V(x_e) = 0$ and from (8.48) it is $\dot{V}(x_e) = 0$, since at equilibrium vector x_e it is $f(x_e) = 0$; see also Eq. (8.30). Since V is bounded below (by zero) and decreasing, it follows that $\lim_{t\to\infty} V(x(t)) = V^* \geq 0$ exists. We will prove that $V^* = 0$. Suppose that $V^* > 0$. V is a continuous function in x and it is zero at x_e ($V(x_e) = 0$) and so with $V^* > 0$ there would exist a neighbourhood $D(x_e, s)$ in which trajectory $x(t)$ never enters ($0 = V(x_e) < V^* \leq V(t)$). Since $\dot{V}$ is continuous function, with $\dot{V}(x) < 0$ and $\dot{V}(x_e) = 0$, there would exist $c > 0$ such that $\dot{V} \leq -c$, for all x. Hence

$$V(x(t)) = V(x(0)) + \int_0^t \dot{V}(x(\tau))\,d\tau \leq V(x(0)) - ct$$

and $V(x(t))$ would become negative at some finite time. This is a contradiction to the assumption that V is positive definite function. Hence $V^* = 0$.

Since V is decreasing with $\lim_{t\to\infty} V(t) = 0$ and $V(x) = 0$, for x_e only, it follows that $\lim_{t\to\infty} x(t) = x_e$. This together with the fact that x_e is stable from (1), prove that x_e is asymptotically stable. $\qquad\square$

Some comments are in order. In the application of Theorem 8.5 a candidate positive definite function V is proposed, usually involving some constants to be specified. Then these constants may be specified in order to satisfy $\dot{V} < 0$ or $\dot{V} \leq 0$ and so establish asymptotic stability or stability. A function V, which satisfies the conditions of Theorem 8.5 is known as *Lyapunov function*. Studying stability for non-linear systems using the direct method involves finding suitable Lyapunov functions. The next two examples illustrate this point.

Example 8.12 Consider the dynamic system

$$\dot{x}_1 = -x_1 + 6x_2,$$
$$\dot{x}_2 = -x_1 - x_2^3,$$

with equilibrium $x_e = [0, 0]^\top$.

Consider the function $V(x) = x_1^2 + cx_2^2$, where $c > 0$ is subject to specification. It is easy to verify that V is a positive definite function, i.e. $V(x) > 0$ and $V(x_e) = 0$. We shall specify c so that $\dot{V}(x) < 0$. Indeed

$$\dot{V}(x) = [2x_1, 2cx_2]\begin{bmatrix} -x_1 + 6x_2 \\ -x_1 - x_2^3 \end{bmatrix}$$
$$= -2x_1^2 + (12 - 2c)x_1 x_2 - 2cx_2^4.$$

If we set $c = 6$, we have

$$\dot{V}(x) = -2x_1^2 - 12x_2^4 < 0,$$

so that $\dot{V}$ is a negative definite function. From Theorem 8.5 it follows that $x_e = [0, 0]^\top$ is asymptotically stable (or that the system is asymptotically stable).

The next example studies the stability of the popular Lorenz system, see Lorenz (1963).

Example 8.13 (Lorenz System) A non-linear system of differential equations was proposed by Lorenz (1963) to approximate fluid heating in the atmosphere. In fact Lorenz considered a case study of a horizontal fluid layer which is heated from below. The warmer fluid from the bottom side of the layer rises and convection currents occur. Let x, y and z be the coordinates of the layer at $\mathbb{R}^3$ Euclidean space. The Lorenz differential equations are

$$\dot{x} = \sigma(y - x), \tag{8.49a}$$

$$\dot{y} = \rho x - y - xz, \tag{8.49b}$$

$$\dot{z} = xy - \beta z, \tag{8.49c}$$

where σ, ρ and β are some positive constants. This model has a complex dynamic behaviour and is extensively used in the literature as an example in chaos theory, see e.g. Hirsch et al. (2012). In this example we study the stability of system (8.49a)–(8.49c), for $0 < \rho < 1$, which results in relatively simple dynamics of the system. For example, setting $\sigma = 10$, $\rho = 28$ and $\beta = 8/3$ results in a chaotic behaviour (see Fig. 8.5 for two realisations of the system). The Lorenz system (also known as Lorenz attractor) is discussed in many textbooks as a non-linear system, which can generate a chaotic dynamic behaviour, see e.g. Ding (2013, pp. 21–24).

We consider a candidate Lyapunov function

$$V = a_1 x^2 + a_2 y^2 + a_3 z^2,$$

where the constants a_1, a_2 and a_3 are to be determined so that V qualifies for a Lyapunov function. The first derivative of V is

$$\dot{V} = [2a_1 x, 2a_2 y, 2a_3 z] \begin{bmatrix} \sigma(y - x) \\ \rho x - y - xz \\ xy - \beta z \end{bmatrix}$$

$$= -2a_1 \sigma x^2 - 2a_2 y^2 - 2a_3 \beta z^2 + 2(a_1 \sigma + a_2 \rho)xy + 2(a_3 - a_2)xyz.$$

If we set $a_2 = a_3 = 1$ and $a_1 = 1/\sigma$ we obtain

$$\dot{V} = -2[x^2 + y^2 + \beta z^2 - (1 + \rho)xy]$$

$$= -2\left[\left(x - \frac{1+\rho}{2}y\right)^2 + \left(1 - \left(\frac{1+\rho}{2}\right)^2\right)y^2 + \beta z^2\right]. \tag{8.50}$$

Since $0 < \rho < 1$, it is $0 < (1 + r)/2 < 1$, which implies $\dot{V} < 0$ in (8.50). That means $\dot{V}$ is a negative definite function and so V is a Lyapunov function. From

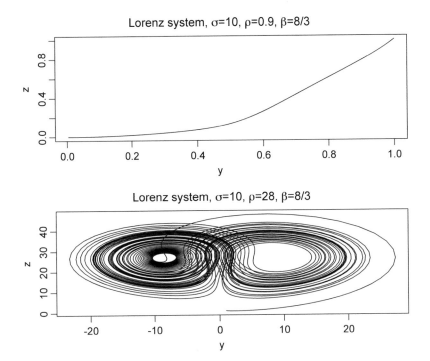

Fig. 8.5 Plot of Lorenz systems: the top panel shows the system with $\sigma = 10$, $\rho = 0.9$, $\beta = 8/3$ discussed in this example and the lower panel plots the same system for $\rho = 28$

Theorem 8.5 the Lorenz system above is asymptotically stable, under the condition $0 < \rho < 1$. Figure 8.5 plots two instances of the Lorenz system (8.49a)–(8.49c). The first instance (top panel) uses $\sigma = 10$, $\rho = 0.9$ and $\beta = 8/3$; we observe that due to ρ being between 0 and 1, the dynamics of this system are simple. This system is shown above to be asymptotically stable. The system in the bottom panel of Fig. 8.5 shows the second instance of the system for $\sigma = 10$, $\rho = 28$ and $\beta = 8/3$. This is a considerably more complex system, which is chaotic. As ρ (and perhaps σ and β) is allowed to increase (starting from lower values) the system can describe fluid heating over time, as described originally in Lorenz (1963).

8.4   Continuous-Time Kalman Filter

In this section we briefly discuss the Kalman filter for discrete-time linear systems (Sect. 8.4.1) and then we use it in order to prove the Kalman filter for continuous-time linear systems, also known as the *Kalman–Bucy* filter (Sect. 8.4.2). Following the seminal papers (Kalman, 1960; Kalman & Bucy, 1961), these filters have been extensively discussed in the literature, as evidenced by influential textbooks

(Anderson & Moore, 1979; Minkler & Minkler, 1993) and (Grewal & Andrews, 2015). We start with the discrete-time Kalman filter.

8.4.1 Discrete-Time Kalman Filter

We consider the discrete-time linear system having the state space representation (8.28a)–(8.28b), which is given below for convenience

$$x_{k+1} = \mathbf{F}_k x_k + \zeta_k, \tag{8.51a}$$

$$y_k = \mathbf{H}_k x_k + \epsilon_k, \tag{8.51b}$$

where the initial state is x_0 is must be specified, see Eq. (8.53) below.. In this representation the input $u_k = 0$ or u_k is absorbed into the state vector x_k, see e.g. Exercise 3. The components $\mathbf{F}_k$ and $\mathbf{H}_k$ are assumed known and the innovation sequences ζ_k and ϵ_k are assumed to be white noises and uncorrelated, i.e. $E(\zeta_k) = 0$, $E(\epsilon_k) = 0$ and

$$E(\zeta_k \zeta_s^\top) = \delta_{ks} \mathbf{Z}_k, \quad E(\epsilon_k \epsilon_s^\top) = \delta_{ks} \mathbf{\Sigma}_k \quad \text{and} \quad E(\zeta_t \epsilon_s^\top) = \mathbf{0}, \tag{8.52}$$

for any k, s, where δ_{ks} denotes the Kronecker delta ($\delta_{ks} = 1$ if $k = s$, and 0 otherwise). In addition we may assume that ζ_k and ϵ_k are Gaussian and so we may write

$$\zeta_k \sim N(0, \mathbf{Z}) \quad \text{and} \quad \epsilon_k \sim N(0, \mathbf{\Sigma}).$$

The initial state is assumed to follow a normal distribution

$$x_0 \sim N(\hat{x}_{0|0}, \mathbf{P}_{0|0}), \tag{8.53}$$

for some mean vector $\hat{x}_{0|0}$ and covariance matrix $\mathbf{P}_{0|0}$. If $x_0 = \hat{x}_{0|0}$ is known with no uncertainty, then $\mathbf{P}_{0|0} = 0$. The prior or initial distribution (8.53) allows for uncertainty around the mean $\hat{x}_{0|0}$ of x_0.

With the output observations $y_{1:k+1} = (y_1, \ldots, y_{k+1})$ the Kalman filter provides estimation of the states x_{k+1} via the conditional distribution

$$x_{k+1} \mid y_{k+1} \sim N(\hat{x}_{k+1|k+1}, \mathbf{P}_{k+1|k+1}), \tag{8.54}$$

where $\hat{x}_{k+1|k+1}$ and $\mathbf{P}_{k+1|k+1}$ are calculated sequentially by the recursions

$$\hat{x}_{k+1|k+1} = \mathbf{F}_k \hat{x}_{k|k} + \mathbf{K}_{k+1}(y_{k+1} - \mathbf{H}_{k+1} \mathbf{F}_k \hat{x}_{k|k}) \tag{8.55}$$

$$\mathbf{K}_{k+1} = \mathbf{P}_{k+1|k} \mathbf{H}_{k+1}^\top (\mathbf{H}_{k+1} \mathbf{P}_{k+1|k} \mathbf{H}_{k+1}^{-1} + \mathbf{\Sigma}_{k+1})^{-1} \tag{8.56}$$

$$\mathbf{P}_{k+1|k} = \mathbf{F}_k \mathbf{P}_{k|k} \mathbf{F}_k^\top + \mathbf{Z}_k \tag{8.57}$$

$$\mathbf{P}_{k+1|k+1} = (\mathbf{I} - \mathbf{K}_{k+1}\mathbf{H}_{k+1})\mathbf{P}_{k+1|k} \tag{8.58}$$

Some comments are in order.

1. Equations (8.54) with recursions (8.55)–(8.58) consist of the Kalman filter for discrete-time systems. The proof of this is given by Theorem 3.2 (see Sect. 3.2) if we replace x_t, β_t and t (Sect. 3.2) by H_k, x_k and k, respectively (assuming that y_k is scalar, univariate Kalman filter). If y_k is a vector, then the reader is referred to the multivariate Kalman filter (Theorem 5.1, in Sect. 5.1).
2. The proof of Theorem 3.2 makes use of normal distribution theory, in order to derive the conditional distribution (8.54). The Kalman filter recursions (8.55)–(8.58) are valid, even if we drop the normality assumption of the innovations but we keep the error structure assumptions (8.52). For the proof of the Kalman filter recursions if the normality assumption is dropped see Theorem 3.3.

8.4.2 Kalman–Bucy Filter

We consider the continuous-time linear system having the state space representation

$$\dot{x}(t) = \mathbf{F}(t)x(t) + \zeta(t), \tag{8.59a}$$

$$y(t) = \mathbf{H}(t)x(t) + \epsilon(t), \tag{8.59b}$$

where $\zeta(t)$ and $\epsilon(t)$ are random innovations and the inputs are absorbed into the states as discussed in the discrete-time system in Sect. 8.4.1 above.

Some discussion should be devoted on the definition of the innovation functions $\zeta(t)$ and $\epsilon(t)$. White noise in continuous-time can be defined by several ways. The most rigorous considers white noise as the derivative of Brownian motion. The Brownian motion is the continuous-time analogue of random walk and hence its derivative (with analogue first-order difference of random walk) can be defined as a white noise process. However, under the Lebesgue-Stieltjes integration (Carter & van Brunt, 2000), the Brownian motion is nowhere differentiable along its path. The definition of its derivative calls upon consideration of Îto integral and stochastic calculus, for a good treatise see Applebaum (2011) and Stroock (2018). As it is out of the scope of this book to cover stochastic calculus, we shall adopt the less-rigorous approach of the definition of $\zeta(t)$ and $\epsilon(t)$, which is adopted by several engineering and signal-processing textbooks, see e.g. Schweppe (1968) and Minkler and Minkler (1993). According to this a formal definition of the distributions of $\zeta(t)$ and $\epsilon(t)$ is left unspecified; instead a continuous-time analogue of assumptions

(8.52) is adopted, or $E(\zeta(t)) = 0$, $E(\epsilon(t)) = 0$ and

$$E(\zeta(t)\zeta^\top(t+\tau)) = \delta(\tau)\mathbf{Z}(t), \quad E(\epsilon(t)\epsilon^\top(t+\tau)) = \delta(\tau)\mathbf{\Sigma}(t) \tag{8.60}$$

$$\text{and} \quad E(\zeta(t)\epsilon^\top(t+\tau)) = \mathbf{0}, \tag{8.61}$$

where $\delta(t)$ is the Dirac function. These assumptions indicate that $\zeta(t)$ is uncorrelated of $\zeta(t+\tau)$, for $\tau \neq 0$, $\epsilon(t)$ is uncorrelated of $\epsilon(t+\tau)$, and that $\zeta(t)$ and $\epsilon(s)$ are uncorrelated for any t, s. It follows that $\text{Var}(\zeta(t)) = \mathbf{Z}(t)$ and $\text{Var}(\epsilon(t)) = \mathbf{\Sigma}(t)$.

For the initial state $x(t_0) = x_0$ we assume it is random and uncorrelated to both $\zeta(t)$ and $\epsilon(t)$ and it has a Gaussian prior distribution

$$x_0 \sim N(\hat{x}(t_0), \mathbf{P}(t_0)), \tag{8.62}$$

for some known mean vector $\hat{x}(t_0)$ and covariance matrix $\mathbf{P}(t_0)$. Moreover, we can accommodate a deterministic x_0, if we set $\mathbf{P}(t_0) = \mathbf{0}$.

We shall assume that $\mathbf{F}(t)$, $\mathbf{H}(t)$, $\mathbf{Z}(t)$ and $\mathbf{\Sigma}(t)$ are known continuous functions. We shall also assume that $\mathbf{\Sigma}(t)$ is positive definite matrix (note that if $y(t)$ is scalar function, then $\mathbf{\Sigma}(t) > 0$). With all these assumptions the following theorem provides the *Kalman–Bucy* filter, see Kalman and Bucy (1961), Anderson (1971), Grewal and Andrews (2015), and Johnson and Núñez (2014).

Theorem 8.6 (Kalman–Bucy Filter) *In the continuous-time linear system* (8.59a)–(8.59b) *with the error structure* (8.60)–(8.61) *and with the prior of x_0 as in* (8.62), *the estimate $\hat{x}(t)$ of the state vector $x(t)$ and the corresponding error covariance matrix $P(t)$ satisfy the following differential equations*

$$\dot{\hat{x}}(t) = \mathbf{F}(t)\hat{x}(t) + \mathbf{K}(t)[y(t) - \mathbf{H}(t)\hat{x}(t)], \tag{8.63}$$

$$\mathbf{K}(t) = \mathbf{P}(t)H^\top(t)\mathbf{\Sigma}(t)^{-1}, \tag{8.64}$$

$$\dot{\mathbf{P}}(t) = \mathbf{F}(t)\mathbf{P}(t) + \mathbf{P}(t)\mathbf{F}^\top(t) - \mathbf{P}(t)\mathbf{H}^\top(t)\mathbf{\Sigma}(t)^{-1}\mathbf{H}(t)\mathbf{P}(t) + \mathbf{Z}(t). \tag{8.65}$$

The proof of Kalman and Bucy (1961) is based on the orthogonality principle (theory of projections) and hence they derived the differential equation (8.63) by generalising the Wiener–Hopf equation, Wiener and Hopf (1931), Crowdy and Luca (2014) and the references therein. The proof below is based on discretising the continuous-time system, using the Kalman-filter for discrete-time and then taking the limit of the time-interval to tend to zero.

Proof The proof is by discretising the linear system using the discrete-time linear system (8.51a)–(8.51b), where $t = k\Delta t$ (see also Sect. 8.2.4 of how we can obtain a discrete approximation of the continuous-time linear system). When Δt tends to 0, the discrete-time system converges to the continuous-time linear system. Hence we define

$$\hat{x}(t) = \lim_{\Delta t \to 0} \hat{x}_k \quad \text{and} \quad \mathbf{P}(t) = \lim_{\Delta t \to 0} \mathbf{P}_{k+1},$$

where $\hat{x}_k = \hat{x}_{k|k}$ and $\mathbf{P}_{k+1} = \mathbf{P}_{k+1|k+1}$ are the mean vector and covariance matrix of x_k, given $y_{1:k}$ in the discrete-time system. From Sect. 8.2.4 we obtain the discrete-time system (8.51a)–(8.51b), with

$$x_k = x(k\Delta t), \quad \mathbf{F}_k = \mathbf{\Phi}[(k+1)\Delta t, k\Delta t], \quad y_k = y(k\Delta t),$$

$$\zeta_k = \int_{k\Delta t}^{(k+1)\Delta t} \mathbf{\Phi}[(k+1)\Delta t, \tau]\zeta(\tau)\, d\tau, \quad \mathbf{H}_k = \mathbf{H}(k\Delta t), \epsilon_k = \epsilon(k\Delta t),$$

where $\mathbf{\Phi}(\cdot)$ is the state transition matrix defined in Sect. 8.2.4.

The next step is to show the following useful results:

1. ζ_k and ϵ_k are uncorrelated and their covariances are approximately equal to

$$\mathbf{Z}_k \approx \Delta t \mathbf{F}_k \mathbf{Z}(k\Delta t)\mathbf{F}_k^\top,$$

$$\mathbf{\Sigma}_k \approx \mathbf{\Sigma}(k\Delta t)/\Delta t.$$

2. For $t = k\Delta t$ we have

$$\mathbf{F}(t) = \lim_{\Delta t \to 0} \frac{\mathbf{F}_k - \mathbf{I}}{\Delta t}. \tag{8.66}$$

To prove (1) we get

$$E(\zeta_k \epsilon_\ell^\top) = E\left\{\int_{k\Delta t}^{(k+1)\Delta t} \mathbf{\Phi}[(k+1)\Delta t, \tau]\zeta(\tau)\, d\tau \epsilon_\ell^\top\right\}$$

$$= \int_{k\Delta t}^{(k+1)\Delta t} \mathbf{\Phi}[(k+1)\Delta t, \tau]E\left\{\zeta(\tau)\epsilon_\ell^\top\right\} d\tau$$

$$= \int_{k\Delta t}^{(k+1)\Delta t} \mathbf{\Phi}[(k+1)\Delta t, \tau]E(\zeta(\tau))E(\epsilon_\ell^\top)\, d\tau$$

$$= 0,$$

since $\zeta(t)$ is uncorrelated of $\epsilon(t)$. This proves that ζ_k and ϵ_k are uncorrelated too.

The covariance matrix of ζ_k is

$$\mathrm{Var}(\zeta_k) = \mathrm{Var}\left\{\int_{k\Delta t}^{(k+1)\Delta t} \mathbf{\Phi}[(k+1)\Delta t, \tau]\zeta(\tau)\, d\tau\right\}$$

$$= \int_{k\Delta t}^{(k+1)\Delta t} \mathbf{\Phi}[(k+1)\Delta t, \tau]\mathbf{Z}(\tau)\mathbf{\Phi}[(k+1)\Delta t, \tau]^\top\, d\tau$$

$$\approx \Delta t\mathbf{\Phi}[(k+1)\Delta t, \tau]\mathbf{Z}(\tau)\mathbf{\Phi}[(k+1)\Delta t, \tau]^\top,$$

using the approximation

$$\int_a^{a+\Delta t} f(x)\,dx \approx \Delta t f(\Delta t),$$

for some a and for small length Δt, where $f(\cdot)$ is a continuous function.
Likewise for ϵ_k we first approximate ϵ_k as

$$\epsilon_k = \epsilon(k\Delta t) \approx \frac{1}{\Delta t} \int_{k\Delta t}^{(k+1)\Delta t} \epsilon(\tau)\,d\tau$$

and then we take the covariance matrix

$$\mathrm{Var}(\epsilon_k) \approx \frac{1}{\Delta t^2} \int_{k\Delta t}^{(k+1)\Delta t} \mathbf{\Sigma}(\tau)\,d\tau \approx \frac{\mathbf{\Sigma}(k\Delta t)}{\Delta t}.$$

This completes the proof of (1).

Moving on to the proof of (2) we first recall that the state transition matrix $\mathbf{\Phi}(t, t_1)$ satisfies the matrix equation $\dot{\mathbf{\Phi}}(t, t_1) = F(t)\mathbf{\Phi}(t, t_1)$, for some t_1; see Sect. 8.2.3. Solving this equation for $\mathbf{F}(t)$ and setting $t_1 = k\Delta t$ we obtain

$$\mathbf{F}(t) = \dot{\mathbf{\Phi}}(t, t_1)\mathbf{\Phi}^{-1}(t, t_1) = \lim_{\Delta t \to 0} \frac{\mathbf{\Phi}(t + \Delta t, t_1) - \mathbf{\Phi}(t, t_1)}{\Delta t}\mathbf{\Phi}^{-1}(t, t_1)$$

$$= \lim_{\Delta t \to 0} \frac{\mathbf{\Phi}(k\Delta t + \Delta t, k\Delta t) - \mathbf{\Phi}(t_1, t_1)}{\Delta t}\mathbf{\Phi}^{-1}(t_1, t_1)$$

$$= \lim_{\Delta t \to 0} \frac{\mathbf{F}_k - \mathbf{I}}{\Delta t},$$

as $\mathbf{\Phi}(t_1, t_1) = I$. This settles the proof of (2).

Proceeding now to the rest of the proof from Eq. (8.66) we have

$$\mathbf{F}_k \approx \Delta t \mathbf{F}(k\Delta t) + \mathbf{I}, \tag{8.67}$$

which is substituted in $\hat{x}_{k+1|k+1} = \hat{x}_k$ of the discrete-time system (8.55)

$$\frac{\hat{x}_{k+1} - \hat{x}_k}{\Delta t} = \mathbf{F}(k\Delta t)\hat{x}_k + \frac{\mathbf{K}_{k+1}}{\Delta t}[y_{k+1} - \mathbf{H}_{k+1}(\Delta t F(k\Delta t) + I)\hat{x}_k]$$

$$= \mathbf{F}(k\Delta t)\hat{x}_k + \frac{\mathbf{K}_{k+1}}{\Delta t}(y_{k+1} - H_{k+1}\mathbf{F}_k\hat{x}_k).$$

Now write this with $k\Delta = t$ and allow k and Δt to vary

$$\frac{\hat{x}_{k+1} - \hat{x}_k}{\Delta t} = \mathbf{F}(t)\hat{x}_k + \frac{\mathbf{K}_{k+1}}{\Delta t}[(y(t + \Delta t) - \mathbf{H}(t + \Delta t)\mathbf{\Phi}(t + \Delta t, t)\hat{x}_k]$$

and by taking limits as Δt tends to zero we obtain

$$\dot{x}(t) = \lim_{\Delta t \to 0} \frac{\hat{x}_{k+1} - \hat{x}_k}{\Delta t} = \mathbf{F}(t)\hat{x}_k + \left(\lim_{\Delta t \to 0} \frac{\mathbf{K}_{k+1}}{\Delta t} \right) [y(t) - \mathbf{H}(t)\hat{x}_k]. \qquad (8.68)$$

Now we need to deal with the limit of $\mathbf{K}_{k+1}/\Delta t$. From the definition of the Kalman gain $\mathbf{K}_{k+1}$ (8.56) in the discrete-time system we get

$$\mathbf{K}_{k+1} = \mathbf{P}_{k+1|k}\mathbf{H}^\top(t + \Delta t) \left[H(t + \Delta t)\mathbf{P}_{k+1|k}\mathbf{H}^\top(t + \Delta t) + \frac{\mathbf{\Sigma}(t + \Delta t)}{\Delta t} \right]^{-1}$$

$$= \Delta t \mathbf{P}_{k+1|k}\mathbf{H}^\top(t + \delta t)[\Delta t H(t + \Delta t)\mathbf{P}_{k+1|k}\mathbf{H}^\top(t + \Delta t) + \mathbf{\Sigma}(t + \Delta t)]^{-1}.$$

We can observe that

$$\lim_{\Delta t \to 0} \mathbf{P}_{k+1|k} = \lim_{\Delta t \to 0} \mathbf{P}_{k|k} = \lim_{\Delta t \to 0} E[(x(t) - \hat{x}(t))(x(t) - \hat{x}(t))^\top] = \mathbf{P}(t).$$

Hence from $\mathbf{K}_{k+1}$ above we have

$$\mathbf{K}(t) = \lim_{\Delta t \to 0} \frac{\mathbf{K}_{k+1}}{\Delta t} = \mathbf{P}(t)\mathbf{H}^\top(t)\mathbf{\Sigma}(t)^{-1}.$$

If we substitute $\mathbf{K}(t)$ into (8.68) we obtain (8.63) as required.

Finally, we will prove Eq. (8.65). From the recursion of $\mathbf{P}_{k+1|k+1}$ for the discrete-time system (see Eq. (8.58)) and using $\mathbf{F}_k$ as in (8.67) we obtain

$$\mathbf{P}_{k+1} = \mathbf{P}_{k+1|k+1} = \mathbf{P}_{k+1|k} - \mathbf{K}_{k+1}\mathbf{H}_{k+1}\mathbf{P}_{k+1|k}$$

$$= \mathbf{F}_k\mathbf{P}_{k|k}\mathbf{F}_k^\top + \mathbf{Z}_k - \mathbf{K}_{k+1}\mathbf{H}_{k+1}\mathbf{P}_{k+1|k}$$

$$= (\mathbf{I} + \Delta t\mathbf{F}(t))\mathbf{P}_k(I + \Delta t\mathbf{F}(t))^\top + \Delta t\mathbf{F}_k\mathbf{Z}(t)\mathbf{F}_k^\top - \mathbf{K}_{k+1}\mathbf{H}_{k+1}\mathbf{P}_{k+1|k}.$$

Rearranging this equation we get

$$\frac{\mathbf{P}_{k+1} - \mathbf{P}_k}{\Delta t} = \mathbf{F}(t)\mathbf{P}_k + \mathbf{P}_k\mathbf{F}^\top(t) + \Delta t\mathbf{F}(t)\mathbf{P}_k\mathbf{F}^\top(t) + \mathbf{F}_k\mathbf{Z}(t)\mathbf{F}_k^\top - \frac{\mathbf{K}_{k+1}}{\Delta t}\mathbf{H}(t + \Delta t)\mathbf{P}_{k+1|k}.$$

By taking the limit as $\Delta t \to 0$ we obtain

$$\dot{\mathbf{P}}(t) = \lim_{\Delta t \to 0} \frac{\mathbf{P}_{k+1} - \mathbf{P}_k}{\Delta t} = \mathbf{F}(t)\mathbf{P}(t) + \mathbf{P}(t)\mathbf{F}^\top(t) - \mathbf{P}(t)\mathbf{H}^\top(t)\mathbf{\Sigma}(t)^{-1}\mathbf{H}(t)\mathbf{P}(t) + \mathbf{Z}(t)$$

and the proof is completed.   □

Some comments are in order. By close observation of Eqs. (8.63) and (8.65) (continuous-time) and recursions (8.55) and (8.58) (discrete-time) we see that the latter can be expressed as difference (discrete time) equations while the former are differential equations (continuous time). In both systems if we have a single output (i.e. $y(t)$ or y_k are scalars), then there is no inversion involved in the computation of the Kalman filter. The stability of the Kalman–Bucy filter has attracted some considerable attraction. An early study on this topic is conducted in Anderson (1971) and a recent one in Kulikov and Kulikova (2018).

The differential equation (8.65) is a Riccati matrix equation and can be solved in closed form only in special cases. In particular, it can be solved in closed form when there are no quadratic terms on the right hand side of the equation. We can see that this is achieved when (a) $\mathbf{H}(t) = \mathbf{0}$ (noise-only measurement equation: $y(t) = \epsilon(t)$) or (b) when $\mathbf{Z}(t) = 0$ (noise-free system equation: $\dot{x}(t) = \mathbf{F}(t)x(t)$). We discuss first case (b).

Consider the continuous-time time-invariant linear system

$$\dot{x}(t) = \mathbf{F}x(t),$$

$$y(t) = \mathbf{H}x(t) + \epsilon(t),$$

where the covariance matrix $\mathbf{\Sigma}(t) = \mathbf{\Sigma}$ is time-invariant and $\mathbf{P}(t_0)$ is known and positive definite. The differential equation (8.65) can be written as

$$\mathbf{P}(t)^{-1}\dot{\mathbf{P}}(t)\mathbf{P}(t)^{-1} = \mathbf{P}(t)^{-1}\mathbf{F} + \mathbf{F}^{\top}\mathbf{P}(t)^{-1} - \mathbf{H}^{\top}\mathbf{\Sigma}^{-1}\mathbf{H}. \qquad (8.69)$$

If we differentiate the equality $\mathbf{P}(t)^{-1}\mathbf{P}(t) = \mathbf{I}$ we have

$$\mathbf{0} = \frac{d\mathbf{P}(t)^{-1}\mathbf{P}(t)}{dt} = \frac{d\mathbf{P}(t)^{-1}}{dt}\mathbf{P}(t) + \mathbf{P}(t)^{-1}\dot{\mathbf{P}}(t)$$

$$\text{or} \quad \frac{d\mathbf{P}(t)^{-1}}{dt} = -\mathbf{P}(t)^{-1}\dot{\mathbf{P}}(t)\mathbf{P}(t)^{-1}. \qquad (8.70)$$

Substitute now (8.70) into (8.69)

$$\frac{d\mathbf{P}(t)^{-1}}{dt} = -\mathbf{P}(t)^{-1}\mathbf{F} - \mathbf{F}^{\top}\mathbf{P}(t)^{-1} + \mathbf{H}^{\top}\mathbf{\Sigma}^{-1}\mathbf{H}.$$

The solution of this differential equation is

$$\mathbf{P}(t)^{-1} = \int_{t_0}^{t} e^{-\mathbf{F}^{\top}(t-w)}\mathbf{H}^{\top}\mathbf{\Sigma}^{-1}\mathbf{H}e^{-\mathbf{F}(t-w)}\,dw + e^{-\mathbf{F}^{\top}(t-t_0)}\mathbf{P}(t_0)^{-1}e^{-\mathbf{F}(t-t_0)}$$

$$= e^{\mathbf{F}^{\top}t}\int_{t_0}^{t} e^{\mathbf{F}^{\top}w}\mathbf{H}^{\top}\mathbf{\Sigma}^{-1}\mathbf{H}e^{\mathbf{F}w}\,dw\,e^{-\mathbf{F}t} + e^{-\mathbf{F}^{\top}(t-t_0)}\mathbf{P}(t_0)^{-1}e^{-\mathbf{F}(t-t_0)}.$$

$$(8.71)$$

To prove this result, we differentiate (8.71) with respect to t, or

$$\frac{d\mathbf{P}(t)^{-1}}{dt} = -\mathbf{F}^\top e^{-\mathbf{F}^\top t} \int_{t_0}^t e^{\mathbf{F}^\top w} \mathbf{H}^\top \mathbf{\Sigma}^{-1} \mathbf{H} e^{\mathbf{F} w} \, dw e^{-\mathbf{F} t} + e^{-\mathbf{F}^\top t} e^{\mathbf{F}^\top t} \mathbf{H}^\top \mathbf{\Sigma}^{-1} \mathbf{H} e^{\mathbf{F} t} e^{-\mathbf{F} t}$$

$$- e^{-\mathbf{F}^\top t} \int_{t_0}^t e^{\mathbf{F}^\top w} \mathbf{H}^\top \mathbf{\Sigma}^{-1} \mathbf{H} e^{\mathbf{F} w} \, dw e^{-\mathbf{F} t} \mathbf{F} - \mathbf{F}^\top e^{-\mathbf{F}^\top (t-t_0)} \mathbf{P}(t_0)^{-1} e^{-\mathbf{F}(t-t_0)}$$

$$- e^{-\mathbf{F}^\top (t-t_0)} \mathbf{P}(t_0)^{-1} e^{\mathbf{F}(t-t_0)} \mathbf{F}$$

$$= -\mathbf{F}^\top \left[\int_{t_0}^t e^{-\mathbf{F}^\top (t-w)} \mathbf{H}^\top \mathbf{\Sigma}^{-1} \mathbf{H} e^{-\mathbf{F}(t-w)} \, dw + e^{-\mathbf{F}^\top (t-t_0)} \mathbf{P}(t_0)^{-1} e^{\mathbf{F}(t-t_0)} \right]$$

$$- \left[\int_{t_0}^t e^{-\mathbf{F}^\top (t-w)} \mathbf{H}^\top \mathbf{\Sigma}^{-1} \mathbf{H} e^{-\mathbf{F}(t-w)} \, dw + e^{-\mathbf{F}^\top (t-t_0)} \mathbf{P}(t_0)^{-1} e^{\mathbf{F}(t-t_0)} \right] \mathbf{F}$$

$$+ \mathbf{H}^\top \mathbf{\Sigma}^{-1} \mathbf{H}$$

$$= -\mathbf{F}^\top \mathbf{P}(t)^{-1} - \mathbf{P}(t)^{-1} \mathbf{F} + \mathbf{H}^\top \mathbf{\Sigma}^{-1} \mathbf{H}.$$

Equation (8.71) can be written as

$$\mathbf{P}(t)^{-1} = \int_0^{t-t_0} e^{-\mathbf{F}^\top w} \mathbf{H}^\top \mathbf{\Sigma}^{-1} \mathbf{H} e^{-\mathbf{F} w} \, dw + e^{-\mathbf{F}^\top (t-t_0)} \mathbf{P}(t_0)^{-1} e^{-\mathbf{F}(t-t_0)}. \qquad (8.72)$$

The solution of $\mathbf{P}(t)$ may be obtained by inverting $\mathbf{P}(t)^{-1}$ from (8.72).

If all eigenvalues of $\mathbf{F}$ have positive real parts, then

$$\mathbf{P} = \lim_{t \to \infty} \mathbf{P}(t) = \left[\int_0^\infty e^{-\mathbf{F}^\top w} \mathbf{H}^\top \mathbf{\Sigma}^{-1} \mathbf{H} e^{-\mathbf{F} w} \, dw \right]^{-1}, \qquad (8.73)$$

since

$$\lim_{t \to \infty} e^{-\mathbf{F}^\top (t-t_0)} = \lim_{t \to \infty} e^{-\mathbf{F}(t-t_0)} = \mathbf{0}.$$

If the system is stable (all eigenvalues of $\mathbf{F}$ have negative real parts) we have

$$\bar{\mathbf{P}}(t) = \left[\int_0^{t-t_0} e^{-\mathbf{F}^\top w} \mathbf{H}^\top \mathbf{\Sigma}^{-1} \mathbf{H} e^{-\mathbf{F} w} \, dw + e^{-\mathbf{F}^\top (t-t_0)} \bar{\mathbf{P}}(t_0)^{-1} e^{-\mathbf{F}(t-t_0)} \right]^{-1}$$

$$= \left[e^{-\mathbf{F}^\top (t-t_0)} \left\{ \int_0^{t-t_0} e^{\mathbf{F}^\top (t-t_0-w)} \mathbf{H}^\top \mathbf{\Sigma}^{-1} \mathbf{H} e^{\mathbf{F}(t-t_0-w)} \, dw + \mathbf{P}(t_0)^{-1} \right\} e^{-\mathbf{F}(t-t_0)} \right]^{-1}$$

$$= e^{\mathbf{F}(t-t_0)} \left[\int_0^{t-t_0} e^{\mathbf{F}^\top w} \mathbf{H}^\top \mathbf{\Sigma}^{-1} \mathbf{H} e^{\mathbf{F} w} \, dw + \mathbf{P}(t_0)^{-1} \right]^{-1} e^{\mathbf{F}^\top (t-t_0)}.$$

Since $\lim_{t\to\infty} e^{F(t-t_0)} = 0$ it follows that

$$\mathbf{P} = \lim_{t\to\infty} \mathbf{P}(t) = \mathbf{0}.$$

Hence we have proven that if the system is stable, the error covariance matrix $\mathbf{P}(t)$ converges to the zero matrix as $t \to \infty$.

Consider now case (a) where $H = 0$, so that the system is

$$\hat{x}(t) = \mathbf{F}x(t) + \zeta(t), \tag{8.74a}$$

$$y(t) = \epsilon(t), \tag{8.74b}$$

where $\zeta(t)$ and $\epsilon(t)$ satisfy assumptions (8.60)–(8.61) and the innovation covariance matrices $\mathbf{Z}(t) = \mathbf{Z}$ and $\boldsymbol{\Sigma}(t) = \boldsymbol{\Sigma}$ are time-invariant. In this case the error covariance differential equation is

$$\frac{d\mathbf{P}(t)}{dt} = \mathbf{F}\mathbf{P}(t) + \mathbf{P}(t)\mathbf{F}^{\top} + \mathbf{Z},$$

which has solution

$$\mathbf{P}(t) = \int_{t_0}^{t} e^{\mathbf{F}(t-w)} \mathbf{Z} e^{\mathbf{F}^{\top}(t-w)} \, dw + e^{\mathbf{F}(t-t_0)} \mathbf{P}(t_0) e^{\mathbf{F}^{\top}(t-t_0)}$$

$$= \int_{0}^{t-t_0} e^{\mathbf{F}w} \mathbf{Z} e^{\mathbf{F}^{T} w} \, dw + e^{\mathbf{F}(t-t_0)} \mathbf{P}(t_0) e^{\mathbf{F}^{\top}(t-t_0)}. \tag{8.75}$$

The proof of this is very similar to the proof of (8.71) and is left to the reader as an exercise. Notice that in the system (8.74a)–(8.74b) measurements $y(t)$ do not provide any information about the states $x(t)$. Another way we can see this case is when H is non-zero, but now $\boldsymbol{\Sigma}$ is equal to infinity. In such a case the measurements $y(t)$ vary erratically around the state $x(t)$ and as before (in the case studied above when $H = 0$) there is no meaningful information the measurements can provide to the states. In practice we can consider a situation where $\boldsymbol{\Sigma}$ is very large instead of infinity and this case might be of interest if the state differential equation is subject to noise, hence $\mathbf{Z}$ must be included. On the other hand, if the state differential equation has small noise (can be assumed as equal to zero) and focus is placed on the measurement or sensor noise (via covariance matrix $\boldsymbol{\Sigma}$), then case (b) should be considered as described above. Exercise 21 studies a linear system which is in case (a). Below we give an example illustrating case (b), $\mathbf{Z} = \mathbf{0}$ and $\boldsymbol{\Sigma} > \mathbf{0}$.

Example 8.14 (Example 8.7 Revisited) Consider the linear system of Example 8.7 where now the measurement equation includes an error term, i.e.

$$y(t) = Hx(k) + \epsilon(k), \quad \epsilon(t) \sim N(0, \sigma^2),$$

with $\lambda = 1$, $u = 0$, $\sigma^2 = 1$, $t_0 = 0$ and $\mathbf{P}(0) = 2^{-1}\mathbf{I}$. From Eq. (8.72) the inverse of $P(t)$ is

$$\mathbf{P}(t)^{-1} = \int_0^t \exp\left\{-w\begin{bmatrix} 1 & 0 \\ 1 & 1 \end{bmatrix}\right\}\begin{bmatrix} 1 & -1 \\ -1 & 1 \end{bmatrix}\exp\left\{-w\begin{bmatrix} 1 & 1 \\ 0 & 1 \end{bmatrix}\right\} dw$$

$$+ \frac{1}{2}\exp\left\{-t\begin{bmatrix} 1 & 0 \\ 1 & 1 \end{bmatrix}\right\}\exp\left\{-t\begin{bmatrix} 1 & 1 \\ 0 & 1 \end{bmatrix}\right\}. \qquad (8.76)$$

Using the equalities of the matrix exponential (see also Example 8.7)

$$e^{c\mathbf{F}} = e^c\begin{bmatrix} 1 & c \\ 0 & 1 \end{bmatrix},$$

for some constant c and $\exp(\mathbf{F})^\top = \exp(\mathbf{F}^\top)$, we have that the integrated part of (8.76) is

$$I = \int_0^t e^{-2w}\begin{bmatrix} 1 & 0 \\ -w & 1 \end{bmatrix}\begin{bmatrix} 1 & -1 \\ -1 & 1 \end{bmatrix}\begin{bmatrix} 1 & -w \\ 0 & 1 \end{bmatrix} dw$$

$$= \int_0^t \begin{bmatrix} e^{-2w} & -(w+1)e^{-2w} \\ -(w+1)e^{-2w} & e^{-2w}(w^2 + w + 1) \end{bmatrix} dw$$

$$= \begin{bmatrix} I_1 & I_2 \\ I_2 & I_3 \end{bmatrix}.$$

These three integrals I_1, I_2 and I_3 are calculated below

$$I_1 = \int_0^t e^{-2w} dw = \left[\frac{1}{2}e^{-2w}\right]_0^t = \frac{1}{2}(1 - e^{-2t}).$$

I_2 and I_3 are calculated by applying successive integration by parts as

$$I_2 = -\int_0^t (w+1)e^{-w} dw = \frac{1}{2}te^{-2t} - \frac{3}{2}(1 - e^{-2t})$$

and

$$I_3 = \int_0^t (w^2 + w + 1)e^{-2w} dw = 1 - e^{-2t} - \frac{t(t+2)}{2}e^{-2t}.$$

The details of the derivations of integrals I_2 and I_3 are omitted and are left as an exercise for the reader.

The non-integrated part of (8.76) is

$$\exp\left\{-t\begin{bmatrix}1&0\\-t&1\end{bmatrix}\right\}\exp\left\{-t\begin{bmatrix}1&-t\\0&1\end{bmatrix}\right\}=e^{-2t}\begin{bmatrix}1&-t\\-t&t^2+1\end{bmatrix}.$$

Substituting this and I into (8.76) we get

$$\mathbf{P}(t)^{-1}=\begin{bmatrix}\frac{1}{2}&\frac{1}{2}(3e^{-2t}-1)\\\frac{1}{2}(3e^{-2t}-1)&1-\frac{1}{2}e^{-2t}(2-t^2).\end{bmatrix}\tag{8.77}$$

Hence matrix $\mathbf{P}(t)$ is computed by inverting $\mathbf{P}(t)^{-1}$ as

$$\mathbf{P}(t)=\frac{2}{1+(4+t^2)e^{-2t}-9e^{-4t}}\begin{bmatrix}2-(2-t^2)e^{-2t}&1-3e^{-2t}\\1-3e^{-2t}&1\end{bmatrix}.\tag{8.78}$$

Note that the limit $\mathbf{P}^{-1}=\lim_{t\to\infty}\mathbf{P}(t)^{-1}$ is obtained by (8.77) as

$$\mathbf{P}^{-1}=\lim_{t\to\infty}\mathbf{P}(t)^{-1}=\begin{bmatrix}\frac{1}{2}&-\frac{1}{2}\\-\frac{1}{2}&1\end{bmatrix}.$$

This can be used to obtain the limit $\mathbf{P}$ of $\mathbf{P}(t)$ as

$$\mathbf{P}=\lim_{t\to\infty}\mathbf{P}(t)=(\mathbf{P}^{-1})^{-1}=\begin{bmatrix}4&2\\2&2\end{bmatrix}.$$

This can be calculated by taking the limit of $\mathbf{P}(t)$ from (8.78), but it is simpler to work out the limit via $\mathbf{P}^{-1}$ as above. Note that as $\mathbf{F}$ has a single eigenvalue $\lambda=1$, with positive real part, the system is not stable. The limit $\mathbf{P}$ may be directly obtained by (8.73). Here we chose to do the full calculations in order to obtain $\mathbf{P}(t)$, which is of interest in its own right.

The steady state of $\hat{x}_t$ is obtained when the Kalman gain $K(t)$ is replaced by its limit K, in the state differential equation

$$\dot{\hat{x}}(t)=(\mathbf{F}-KH)\hat{x}(t)+Ky(t),\tag{8.79}$$

where $K=\lim_{t\to\infty}K(t)=PH^\top/\sigma^2$ is the limit of the Kalman gain. In Exercise 19 we explore the steady state $x(t)$ for this example. The steady state of time-invariant linear systems is further discussed in Sect. 8.4.3 below.

Figure 8.6 plots the variances $P_{11,t}$, $P_{22,t}$ (diagonal elements of matrix $\mathbf{P}(t)$ as in Eq. (8.78), and the covariance $P_{12,t}$ (off-diagonal elements of $\mathbf{P}(t)$), for time $t\in[0,5]$. Plotted are also the limit values of $\mathbf{P}=\{P_{12}\}$, with $P_{11}=4$, $P_{12}=P_{22}=2$ (indicated by the horizontal lines in Fig. 8.6). We remark that convergence is exponentially fast and in principle from about $t=4$, the error covariance matrix

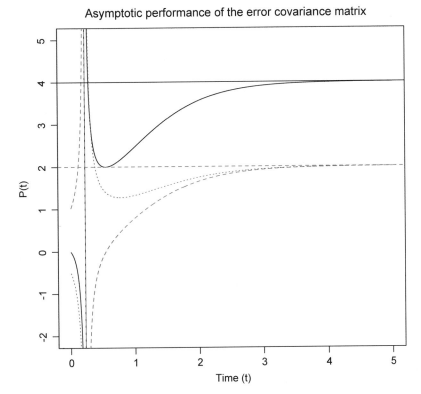

Fig. 8.6 Asymptotic performance of the error covariance matrix $P(t) = \{P_{ij,t}\}_{i,j=1,2}$. The plot shows $P_{11,t}$ (solid line), $P_{12,t}$ (dashed line), $P_{22,t}$ (dotted line), together with their limits $P_{11} = 4$, $P_{12} = P_{22} = 2$

has reached its limit. The next section explores the limit of the error covariance matrix $\mathbf{P}(t)$ for time-invariant continuous-time state space systems.

8.4.3   Observability and Convergence

Kalman introduced the notions of *observability* and *controllability*, in order to provide the asymptotic performance of the Kalman filter (for discrete-time systems) and the Kalman–Bucy filter (for continuous-time systems). Several textbooks discuss in detail observability and controllability of systems, see e.g. Minkler and Minkler (1993),

A system is observable when its states $x(t)$ can be estimated by the output or measurements $y(t)$. In other words, in an observable systems if we have perfect knowledge of the measurements we should be able to compute the states. If the system is not observable there will be some states, which are not possible to estimate from the measurements.

Consider the time-invariant linear system

$$\dot{x}(t) = Fx(t), \tag{8.80a}$$

$$y(t) = Hx(t), \tag{8.80b}$$

where $x(t)$ is a $p \times 1$ state vector, F is a $p \times p$ state matrix, $y(t)$ is a $d \times 1$ measurements vector and H is a $1 \times p$ row vector (here the output $y(t)$ is a scalar function). Note that noise terms may be included, but for the notions of observability we shall not consider them.

Example 8.15 Consider system (8.80a)–(8.80b), with

$$F = \begin{bmatrix} -2 & 0 \\ 0 & -1 \end{bmatrix} \quad \text{and} \quad H = [0, 1].$$

The solution of the state differential is

$$x(t) = \begin{bmatrix} x_1(t) \\ x_2(t) \end{bmatrix} = \begin{bmatrix} x_1(t_0)e^{-2(t-t_0)} \\ x_2(t_0)e^{-(t-t_0)} \end{bmatrix}$$

and the measurement equation implies

$$y(t) = Hx(t) = x_2(t).$$

Hence knowledge of measurements $y(t)$ gives no information about the state $x_1(t)$. We cannot make inference for $x_1(t)$ based on measurements $y(t)$. We can only learn about $x_2(t)$ from $y(t)$. Hence, the state vector $x(t)$ is not observable in this case.

It follows we can reduce the dimension of the state vector, so that the system be observable. Consider the linear system

$$\dot{x}_2(t) = -x_2(t),$$

$$y(t) = x_2(t).$$

This system has output $y(t) = x_2(t)$ as above and $x_2(t) = x_2(t_0)e^{-(t-t_0)}$. In this system knowledge of $y(t)$ provides knowledge of the state $x_2(t)$, hence this system is observable.

Theorem 8.7 *Consider the linear system (8.80a)–(8.80b). This system is observable if and only if the $p \times p$ observability matrix*

$$O = \begin{bmatrix} H \\ HF \\ \vdots \\ HF^{p-1} \end{bmatrix}$$

has full rank p.

For Example 8.15, the observability matrix is

$$O = \begin{bmatrix} H \\ HF \end{bmatrix} = \begin{bmatrix} 0 & 1 \\ 0 & -1 \end{bmatrix},$$

which has rank 1 and so system is not observable. The reduced system has observability matrix (here is reduced to a scalar) $O = H = 1$, which has rank 1 and the system is observable.

Proof of Theorem 8.7 The continuous-time linear system can be approximated by the discrete-time linear system (8.24)–(8.25), with $\mathbf{G} = \mathbf{0}$. The state matrix of this system is

$$\mathbf{F}^* = \Phi((k+1)\Delta t, k\Delta t) = e^{\mathbf{F}\Delta t}. \tag{8.81}$$

We shall prove that system (8.24)–(8.25) is observable if and only if the observability matrix

$$O^* = \begin{bmatrix} H \\ H\mathbf{F}^* \\ \vdots \\ H\mathbf{F}^{*p-1} \end{bmatrix}$$

has rank p. From equation

$$\begin{bmatrix} y(k\Delta t) \\ y[(k+1)\Delta t] \\ \vdots \\ y[(k+p-1)\Delta t] \end{bmatrix} = \begin{bmatrix} H \\ H\mathbf{F}^* \\ \vdots \\ H\mathbf{F}^{*p-1} \end{bmatrix} x(k\Delta t),$$

it follows that $x(k\Delta t)$ can be determined from the outputs $y[(k+j)\Delta t]$, if and only if matrix O^* has full rank p. If this is the case the states are obtained by

$$x(k\Delta t) = O^{*-1}y^*,$$

where $y^* = [y(k\Delta t)], y[(k+1)\Delta t], \ldots, y[(k+p-1)\Delta t]^\top$.

It remains to show that when O^* has rank p, then O has rank p too. With the definition of $\mathbf{F}^*$ as in (8.81) we show that the row vectors $H, HF, \ldots, HF^{p-1}$ are linearly independent. Since O^* is of full rank p, it follows that the row vectors $H, He^{\mathbf{F}\Delta t}, \ldots, He^{\mathbf{F}\Delta t(p-1)}$ are linearly independent. Hence equation

$$\lambda_1 H + \lambda_2 He^{\mathbf{F}\Delta t} + \cdots + \lambda_p He^{\mathbf{F}\Delta t(p-1)} = 0 \tag{8.82}$$

implies $\lambda_1 = \cdots = \lambda_p = 0$. If we differentiate (8.82) with respect to Δt and take the limit as $\Delta t \to 0$, it follows that

$$\lambda_1^* H + \lambda_2^* H\mathbf{F} + \cdots + \lambda_p^* H\mathbf{F}^{p-1} = 0,$$

implies $\lambda_1^* = \cdots = \lambda_p^* = 0$, for $\lambda_j^* = j\lambda_j$ and $j = 1, \ldots, p$. This shows that O is of full rank p and the proof is completed. $\square$

Some comments are in order.

1. Suppose that $\mathbf{F}$ is diagonalisable, so that it has the representation $\mathbf{F} = \mathbf{T}\mathbf{\Lambda}\mathbf{T}^{-1}$, where $\mathbf{\Lambda}$ is the diagonal matrix with diagonal elements the eigenvalues of $\mathbf{F}$ and $\mathbf{T}$ be the matrix with columns the standardised eigenvectors of $\mathbf{F}$. Consider the set of row vectors $H, H\mathbf{F}, \ldots, H\mathbf{F}^{p-1}$ (the rows of the observ-ability matrix O. For constants $\mu_1, \ldots, \mu_p$, equality $\sum_{i=0}^{p-1} \mu_{i+1} H\mathbf{F}^i = 0$ implies $HT \sum_{i=0}^{p-1} \mu_{i+1} \mathbf{\Lambda}^i = 0$. With $HT = [a_1, \ldots, a_p]$, the vectors $H, H\mathbf{F}, \ldots, H\mathbf{F}^{p-1}$ are linearly independent if and only if the determinant

$$D = \begin{vmatrix} a_1 & a_1\lambda_1 & \cdots & a_1\lambda_1^{p-1} \\ a_2 & a_2\lambda_2 & \cdots & a_2\lambda_2^{p-1} \\ \vdots & \vdots & \ddots & \vdots \\ a_p & a_p\lambda_p & \cdots & a_p\lambda_p^{p-1} \end{vmatrix}$$

is non-zero. This determinant is equal to $D = (a_1 a_2 \cdots a_p)^p V$, where V is the Vandermonde determinant. Hence

$$D = \prod_{i=1}^{p} a_i^p \prod_{1 \leq i, j \leq p} (\lambda_j - \lambda_i),$$

where $\lambda_1, \ldots, \lambda_p$ are the eigenvalues of $\mathbf{F}$. For a discussion of this determinant and details on its evaluation see Horn and Johnson (2013).

   It follows that $D \neq 0$, if and only if $a_1, \ldots, a_p \neq 0$ and $\lambda_i \neq \lambda_j$, for $i \neq j$ ($\mathbf{F}$ has distinct eigenvalues). In this case system (8.80a)–(8.80b) is observable, since linear independence of the row vectors $H, H\mathbf{F}, \ldots, H\mathbf{F}^{p-1}$ implies that the observability matrix O is of full rank p.

2. From the proof of Theorem 8.7 if follows that the discrete-time time-invariant system

$$x_{k+1} = \mathbf{F}x_k,$$

$$y_k = Hx_k,$$

is observable if and only if the observability matrix

$$O = \begin{bmatrix} H \\ HF \\ \vdots \\ HF^{p-1} \end{bmatrix}$$

has full rank p. The proof of this is already done in the proof of Theorem 8.7, if we equate $\mathbf{F}^*$ with $\mathbf{F}$ and write $x_k = x(k\Delta t)$ and $y_k = y(k\Delta t)$. A further discussion on the observability for discrete-time state space models is discussed in Sect. 3.5.1.

As we have seen in Sect. 8.4.2 if a linear time-invariant system is asymptotically stable and it has a zero state covariance matrix $\mathbf{Z} = \mathbf{0}$, then the limit of the error covariance matrix $\mathbf{P}(t)$ is the zero matrix $\mathbf{P} = \mathbf{0}$. Additionally, in Sect. 8.4.2 it was shown that if $\mathbf{Z} = \mathbf{0}$ and the state matrix $\mathbf{F}$ has eigenvalues with positive real parts (unstable system), then the error covariance matrix $\mathbf{P}(t)$ convergences to a non-zero limit matrix $\mathbf{P} \neq \mathbf{0}$.

Example 8.14 shows that in an unstable, linear time-invariant system, the limit of $\mathbf{P}(t)$ exists and is equal to $\mathbf{P} > \mathbf{0}$ (strictly positive definite symmetric matrix). For this example, the convergence of the error covariance matrix is illustrated in Fig. 8.6. For linear systems, since the Kalman–Bucy filter inception, the convergence of $\mathbf{P}(t)$ is studied extensively, see e.g. Kalman and Bucy (1961), Jazwinski (1969), Jazwinski (1970), Fitzgerald (1971) and Anderson and Moore (1979) among others. Restricting our attention to linear time-invariant systems, the following theorem gives the main result; more extensive convergence results for time-varying systems are provided in the references above.

Theorem 8.8 *Consider the single output ($y(t)$ being a scalar output) linear time-invariant system (8.59a)–(8.59b), where the components $\mathbf{F}(t) = \mathbf{F}$, $H(t) = H$, $\mathbf{Z}(t) = \mathbf{Z}$ and $\Sigma(t) = \sigma^2$ are time-invariant (hence the system is a time-invariant linear system). Consider the error structure and prior information as in Theorem 8.6 and assume that the system is observable. With these assumptions in place the limit of the error covariance matrix $\mathbf{P}(t)$ exists, i.e.*

$$\lim_{t \to \infty} \mathbf{P}(t) = \mathbf{P}$$

and is independent of the prior covariance matrix $\mathbf{P}(t_0)$.

Proof We shall discretise the continuous-time system (8.59a)–(8.59b) as in the proof of Theorem 8.6. The discrete-time system has measurements $y_k = y(k\Delta t)$ and states $x_k = x(k\Delta t)$, where $\Delta t = t_{k+1} - t_k$ and $k = 0, 1, 2, \ldots$. The discrete system is given by (8.22a)–(8.22b), see Sect. 8.2.4 for a full description of the discrete system.

We establish that for the discrete-time system above, the limit of its error covariance matrix $\mathbf{P}_k$ exists $\lim_{k \to \infty} \mathbf{P}_k = \mathbf{P}$ and is independent of $\mathbf{P}_0 = \mathbf{P}(t_0)$. To prove this first note that since the continuous-time system is observable, it follows that the discrete-time system is observable too (see point (2) on p. 450). Secondly the discrete-time system is a time-invariant system too and as such $\lim_{k \to \infty} \mathbf{P}_k = \mathbf{P}$; for the proof of this see Theorem 3.7 in Sect. 3.5.3.

Returning now to the continuous-time system (8.59a)–(8.59b) we have

$$\lim_{t \to \infty} \mathbf{P}(t) = \lim_{k \to \infty} \lim_{\Delta t \to 0} \mathbf{P}_k = \lim_{\Delta t \to 0} \lim_{k \to \infty} \mathbf{P}_k = \mathbf{P},$$

which does not depend on $\mathbf{P}(t_0)$.   □

Some comments are in order. Under the assumptions of Theorem 8.8, the Kalman gain $K(t)$ also converges to a limit K, i.e. $\lim_{t \to \infty} K(t) = \lim_{t \to \infty} P(t) H^\top / \sigma^2 = PH^\top / \sigma^2 = K$. From the Kalman–Bucy filter (Theorem 8.6) the limiting value $\mathbf{P}$ may be computed by solving the algebraic Riccati equation

$$\mathbf{FP} + \mathbf{PF}^\top - \mathbf{PH}^\top H\mathbf{P}/\sigma^2 + \mathbf{Z} = \mathbf{0}. \tag{8.83}$$

If we replace $K(t)$ by its limit K In the state differential equation of $\hat{x}(t)$ (Theorem 8.8) we obtain the steady state differential equation as

$$\dot{\hat{x}}(t) = \mathbf{F}x(t) + K[y(t) - H\hat{x}(t)]$$
$$= (\mathbf{F} - KH^\top)\hat{x}(t) + Ky(t).$$

This representation has the important implication that, from Theorem 8.2 the solution of the state differential equation is

$$x(t) = e^{(\mathbf{F} - KH^\top)(t - t_0)} x(t_0) + \int_{t_0}^{t} e^{(\mathbf{F} - KH^\top)(t - \tau)} Ky(\tau)\, \tau.$$

The steady state of a system is very useful consideration as it expedites considerably the computation, as $\mathbf{P}(t)$ and K are only computed once. Furthermore, the task of solving the differential equation of $\mathbf{P}(t)$ is reduced to solving the algebraic Riccati equation (8.83). These considerations might be helpful in particular when both $\mathbf{Z}$ and H are non-zero, in which case the algebraic Riccati equation (8.83) can be solved using numerical methods.

These results complement convergence results and the steady state of discrete-time models discussed in some detail in Sect. 3.5.3. (see also Sect. 3.5.2 for steady state of the local level model). For discrete-time systems convergence of Riccati algebraic equations are studied in Chan et al. (1984), see also Anderson and Moore (1979) and Sect. 3.5.3 of this book.

8.4.4 Extended Kalman–Bucy Filter

Systems considered earlier in Sects. 8.4.1, 8.4.2 and 8.4.3 are linear systems. These systems enable the optimal estimation of the states, using the Kalman-Buce filter and under the assumption of observability the convergence of the error covariance system leads to the steady state of the system. However, many systems are non-linear; some examples are discussed in Sect. 8.3.3, see also Sastry (1999), Zinober (2001), Hirsch et al. (2012) and Ding (2013) among others. The problem of lack of linearity and Gaussianity in the Kalman–Bucy filter was first realised by S. F. Schmidt and his team who were working on applying the Kalman filter for the Apollo project. A historical perspective of the Kalman filter and its application to the Apollo project is provided in Grewal and Andrews (2010). These authors discuss how Schmidt's team realised the limitations of applying the Kalman–Bucy filter in the presence of non-linearities. Schmidt's team proposed linearising the non-linear system and reported excellent tracking performance, which benefitted the Apollo project and helped the widespread application and utility of the Kalman-Busy filter. Schmidt has written an account of events led his team discovering the extended Kalman filter, see Schmidt (1981) and McGee and Schmidt (1985). Since then the so-called *extended Kalman filter* has been used extensively as reported in Kappl (1971), Ljung (1979), Morris and Sterling (1979), Schmidt (1981), McGee and Schmidt (1985), Grewal et al. (1991), Grewal and Andrews (2015) and in the references therein. We remark that in this book, in Sect. 8.3.3.1 we have already described linearising a non-linear system, in the context of Lyapunov's indirect method of stability. The extended Kalman filter for discrete-time systems is discussed in Sect. 6.6. In this section we describe the basic form of the extended Kalman filter for continuous-time systems.

Consider a non-linear system, which states $x(t)$ are generated by a non-linear differential equation inflated with noise $\zeta(t)$, or

$$\dot{x}(t) = f(x(t), \zeta(t)), \tag{8.84}$$

where $\zeta(t)$ is a white noise process, which may be assumed to be Gaussian, with covariance matrix $\mathbf{Z}(t)$.

The measurements $y(t)$ are assumed to follow a linear equation

$$y(t) = \mathbf{H}(t)x(t) + \epsilon(t), \tag{8.85}$$

where $\epsilon(t)$ is a Gaussian white noise process with covariance matrix $\mathbf{\Sigma}(t)$. This system assumes innovations (process and measurement) $\zeta(t)$ and $\epsilon(t)$ to satisfy conditions (8.60)–(8.61). It is also assumed that x_0 follows a Gaussian prior distribution as in (8.62). Equations (8.84)–(8.85) define a non-linear system, which measurements follow a linear relationship. It is possible to extend the system to having measurements, which are linked to the states with a non-linear equation too, but here for simplicity we assume linearity in (8.85).

Motivated from the state differential equation of the Kalman–Bucy filter we consider the non-linear

$$\dot{\hat{x}}(t) = f(\hat{x}(t), \zeta(t) = 0) + \mathbf{K}(t)[y(t) - \mathbf{H}(t)\hat{x}(t)], \tag{8.86}$$

where $\hat{x}(t)$ is the estimator of $x(t)$/ Note that, if $f(x, \zeta)$ is a linear function in x and ζ, then systems (8.84)–(8.85) are reduced to a linear model and in this case the proposed state differential equation (8.86) reduced to (8.63) (Kalman–Bucy filter, see Theorem 8.6).

The error we make when considering (8.86) is

$$e(t) = x(t) - \hat{x}(t)$$

which by (8.86) is

$$\begin{aligned}
\dot{e}(t) &= \dot{x} - \dot{\hat{x}}(t) \\
&= f(x(t), \zeta(t)) - f(\hat{x}(t), 0) - \mathbf{K}(t)[y(t) - \mathbf{H}(t)\hat{x}(t)] \\
&= F(e(t), \hat{x}(t), \zeta(t)) - \mathbf{K}(t)\mathbf{H}(t)e(t), \tag{8.87}
\end{aligned}$$

where $F(e(t), \hat{x}(t), \zeta(t)) = (x(t), \zeta(t)) - f(\hat{x}(t), 0)$.

We linearise (8.87) around current estimate $\hat{x}(t)$

$$\begin{aligned}
\dot{e}(t) &= F(0, \hat{x}(t), 0) + \left. \frac{\partial F}{\partial e} \right|_{(e,x,\zeta)=(0,\hat{x},0)} e(t) + \left. \frac{\partial F}{\partial \zeta} \right|_{(e,x,\zeta)=(0,\hat{x},0)} \zeta(t) - \mathbf{K}(t)\mathbf{H}(t)e(t) \\
&= \bar{\mathbf{F}}(t)e(t) + \tilde{\mathbf{F}}(t)\zeta(t) - \mathbf{K}(t)\mathbf{H}(t)e(t),
\end{aligned}$$

where $F(0, \hat{x}(t), 0) = 0$ and

$$\bar{\mathbf{F}}(t) = \left. \frac{\partial F}{\partial e} \right|_{(e,x,\zeta)=(0,\hat{x},0)} = \left. \frac{\partial f}{\partial x} \right|_{(x,\zeta)=(\hat{x},0)}$$

$$\tilde{\mathbf{F}}(t) = \left. \frac{\partial F}{\partial \zeta} \right|_{(e,x,\zeta)=(0,\hat{x},0)} = \left. \frac{\partial f}{\partial \zeta} \right|_{(x,\zeta)=(\hat{x},0)}.$$

With this approximation in place the extended Kalman filter differential equations are

$$\dot{\hat{x}}(t) = f(\hat{x}(t), 0) + \mathbf{K}(t)[y(t) - \mathbf{H}(t)\hat{x}(t)], \tag{8.88}$$

$$\mathbf{K}(t) = \mathbf{P}(t)\mathbf{H}^{\top}(t)\mathbf{\Sigma}(t)^{-1}, \tag{8.89}$$

$$\dot{\mathbf{P}}(t) = \bar{\mathbf{F}}(t)\mathbf{P}(t) + \mathbf{P}(t)\bar{\mathbf{F}}^{\top}(t) - \mathbf{P}(t)\mathbf{H}^{\top}(t)\mathbf{\Sigma}(t)^{-1}\mathbf{H}(t)\mathbf{P}(t) + \tilde{\mathbf{F}}(t)\mathbf{Z}(t)\tilde{\mathbf{F}}(t)^{\top}. \tag{8.90}$$

Some comments are in order.

1. If $f(x(t), \zeta(t)) = \mathbf{F}(t)x(t) + \zeta(t)$, where here $\mathbf{F}(t)$ is a system matrix, then (8.88)–(8.90) reduce to the Kalman–Bucy filtering equations (see Theorem 8.6).
2. It is possible to incorporate a non-linear function in the measurement equation (8.85) so that

$$y(t) = h(x(t), \epsilon(t)).$$

In this case the resulting filter includes the partial derivatives

$$\left. \frac{\partial h}{\partial x} \right|_{(x,\epsilon)=(\hat{x},0)} \quad \text{and} \quad \left. \frac{\partial h}{\partial \epsilon} \right|_{(x,\epsilon)=(\hat{x},0)}$$

evaluated at the current estimate $\hat{x}(t)$ and $\epsilon = 0$.

3. In this section we considered the continuous-time extended Kalman filter. In the discrete-time extended Kalman filter, discussed in Sect. 6.6, first we compute estimated states $\hat{x}_k$ at point $k\Delta t$ and then $\partial f/\partial x$, $\partial f/\partial \zeta$ are evaluated at $\hat{x}_k$. This means that

$$\left. \frac{\partial f}{\partial x} \right|_{x=\hat{x}_k} \quad \text{and} \quad \left. \frac{\partial f}{\partial \zeta} \right|_{x=\hat{x}_k}$$

are used in order to compute $\hat{x}_{k+1}$ at point $(k+1)\Delta t$. As a result, in the discrete-time version we can compute the sequence $\hat{x}_1, \ldots, \hat{x}_k, \ldots$ sequentially. This is considerably simpler and more efficient compared to the continuous-time system, where $\bar{\mathbf{F}}(t)$, $\tilde{\mathbf{F}}(t)$ are combined in the differential equations (8.88)–(8.90) and cannot be computed separately. This observation suggests the extended Kalman filter for the continuous-time system is not very practical. In applications the continuous-time system is first discretised and then the discrete-time extended Kalman filter is used.

8.5 Feedback Control

In this section we discuss *feedback control*, an essential part of practicing dynamic systems in engineering, see e.g. Biernson (1989), Zinober (2001), Tokhi and Azad (2008), Johnson and Moradi (2010), Stevens et al. (2016), Graf (2016) and Franklin et al. (2018) among others. We start by describing the basic framework of control and in particular of the popular proportional, integral, derivative controller, known as PID-controller (Franklin et al., 2018). Section 8.5.2 gives an illustration of control, for a twin rotor experimental rig, which can be used to test aircraft movement stability.

8.5.1   The PID-Controller

Consider a continuous-time system (8.9)–(8.10), which might be single input single
output (SISO) or multiple input multiple output (MIMO). In this section we shall be
concerned with univariate measurements, or that the output $y(t)$ is scalar. Feedback
control for MIMO systems is more challenging, because output variables may
interact. In many situations the output $y(t)$ is measured subject to considerable noise
or other undesirable effects, e.g. long-term oscillations or instabilities reflecting
large uncertainty. Figure 8.7 plots simulated stable process (panel (a)) and unstable
process (panel (b)) together with the desirable or reference output indicated by the
straight lines in each panel. We see that the stable process in (a) after the first 20
points is rather close to the reference output (fixed at $x = 20$, whereas the process

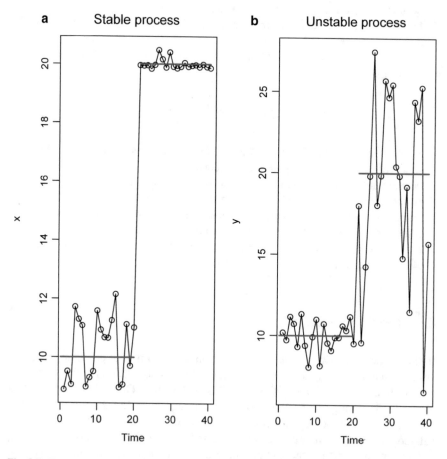

Fig. 8.7 Example of stable process (**a**) and unstable process (**b**). The reference signal is denoted
in both cases with the red line

in (b) exhibits considerable instability with high levels of noise around the reference mark of $y = 20$.

Control theory studies a mathematical device, known as *controller*, which enables the process stabilisation starting from a process that would be unstable if no action is taken. There are effectively two kinds of control under which a controller may operate: *open loop* and *closed loop*. A simple example of control is the house-boiler which provides a household with heating. The boiler may be activated manually whenever heating is needed. In this case there is no control and the temperature can reach high levels or low levels and only manual interventions can be made. If a timer is used, heating is turned on automatically at a specific duration, e.g. in the morning or at night. A system operating with the timer is an example of open-loop control. The boiler turns on / off according to a pre-specified setting, which does not depend on the amount of heating in the building. If there is a thermostat used, then the boiler is automatically turned on when the temperature is below some pre-specified level, say 18°C. In this case, the temperature of the building is fed back to the boiler and the boiler will stop when the temperature of 18°C is reached. This is an example of a closed-loop feedback control, in which the amount of heating in the building is fed into the boiler via the thermostat. Closed-loop feedback control is met in many engineering designs, as the amount of the output signal is fed back into the system. Below we shall briefly describe the PID-controller, which is perhaps one of the most popular controllers used.

The set-point or reference output denoted by $r(t)$ is a desired signal, which the system aim to get close to. Hence we can measure the error $e(t)$ as the difference of the output $y(t)$ from our system from the reference signal, i.e. $e(t) = r(t) - y(t)$. If $e(t) = 0$, then we have achieved the desired signal and there is no need for control. If $e(t)$ is large in absolute value, then a control signal $u(t)$ is deployed. The popular *PID-controller* is defined as a function of $e(t)$ and given by

$$u(t) = k_p e(t) + k_i \int_0^t e(\tau) \, d\tau + k_d \frac{de(t)}{dt}, \qquad (8.91)$$

where k_p, k_i and k_d are constants. This control comprises three elements, the proportional one $k_p e(t)$ (P), the integral one $k_i \int_0^t e(\tau) \, d\tau$ (I) and the derivative one $k_d de/dt$ (D), hence the name PID-controller. Equation (8.91) can be written as

$$u(t) = k_p \left[e(t) + \frac{1}{T_i} \int_0^t e(\tau) \, d\tau + T_d \frac{de(t)}{dt} \right], \qquad (8.92)$$

where T_i is known as the integral time and T_d as the derivative time, with $k_i = k_p/T_i$, $k_p T_d = k_d$.

The proportional term works in a similar way as the term $K(t)[y(t) - H(t)\hat{x}(t)]$ of Eq. (8.63) in the Kalman–Bucy filter (see Theorem 8.6). It aims to move the output towards minimising the error $e(t)$. If we set $k_i = 0$ and $k_d = 0$, then the controller includes only the P term, but this is frequently not enough as the output signal can oscillate retaining always a constant error $e(t)$. The I term improves the controller as it basically sums the errors over time. It has the property that if we use the P and I terms together, the process output agrees with the reference when the system reaches the steady state. This argument can be shown as follows. Assume that the system at time t_s has reached the stead state, so that $u(t) = u_s$ and $e(t) = e_s$, for an $t \geq t_s$. We shall prove this argument by the method of contradiction. Assume that $e_s \neq 0$. Then from (8.91) we have

$$u_s = k_p e_s + k_i \int_0^t e_s \, d\tau = k_p e_s + k_i e_s t.$$

In this equation, the right hand side u_s is time-invariant (as the system is in steady state) and the right hand side depends on t, which is a contradiction. Hence it must be $e_s = 0$. The D part of the controller can improve stability in the transient state. The constants k_p, k_i and k_d can be tuned so as to maximise the efficiency of the algorithm or minimise $e(t)$ for short time path. PID feedback control exhibits notable similarities to the discrete-based feedback adjustment in process monitoring as described in Box et al. (2009) and Triantafyllopoulos et al. (2005).

The following diagram schematically illustrates the feedback control algorithm. At a particular time, y is available and with the reference signal r e is computed. This is then entered into the PID-controller and the control signal u is computed using (8.91). This is then inputted into the system and the signal y is produced. This is then filtered via the measurements and fed back into the node so as the new value of e is computed and so forth.

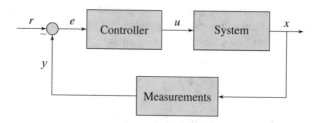

Consider a continuous-time linear system (8.4.2)–(8.4.2) and assume that the system is time-invariant, i.e.

$$\dot{x}(t) = \mathbf{F}x(t) + \mathbf{G}u(t) \quad \text{and} \quad y(t) = Hx(t), \qquad (8.93)$$

where the innovations $\zeta(t)$ and $\epsilon(t)$ are set to be zero with probability one. Here we assume that $y(t)$ is scalar, hence H is a row vector. Suppose we are applying the above PID feedback control. Write $Y(s)$, $X(s)$, $U(s)$, $R(s)$, $E(s)$ the Laplace transforms of $y(t)$, $x(t)$, $u(t)$, $r(t)$, $e(t)$ and $C(s)$, $M(s)$ are the transfer functions of u_t (the controller) and $y(t)$ (measurements), respectively. For a discussion of Laplace transforms and transfer functions see Sect. 8.1.3.

From properties of the Laplace transform (Sect. 8.1.3) we have

$$Y(s) = X(s)U(s) \tag{8.94}$$

$$U(s) = C(s)E(s) \tag{8.95}$$

$$E(s) = R(s) - M(s)Y(s). \tag{8.96}$$

From (8.94) and (8.95) we have

$$E(s) = \frac{Y(s)}{X(s)C(s)},$$

which is put to (8.96) to give the solution

$$Y(s) = \frac{X(s)C(s)}{1 + X(s)C(s)M(s)}R(s). \tag{8.97}$$

Hence the transfer function of the closed-loop feedback control is

$$H(s) = \frac{X(s)C(s)}{1 + X(s)C(s)M(s)}.$$

Fine tuning of the constants k_p, k_i, k_d is required for the feedback control to work properly. The example gives an illustration for the choice of these constants leading to perfect control.

Example 8.16 Consider the first-order linear system (8.93), where F, G and $H = G/F$ are scalars. The transfer function $H^{(X)}(s)$ of $x(t)$ and the transfer function $H^{(Y)}(s)$ of $y(t)$ are

$$H^{(X)}(s) = \frac{G}{s - F} = \frac{A}{1 + sT} \quad \text{and} \quad H^{(Y)}(s) = \frac{1}{1 + sT},$$

where $T = -1/F$, $A = GF^{-1}$ and $F \neq 0$. From (8.91) if we take the Laplace transform we have

$$U(s) = C(s)E(s) = \left(k_p + k_i \frac{1}{s} + k_d s\right) E(s)$$

$$= k\left(1 + \frac{1}{sT_i}\right)(1 + sT_d)E(s),$$

where T_i and T_d are the integral and derivative times, defined in (8.92) and k is a constant. By expanding this it follows that

$$k_p = k\left(1 + \frac{T_d}{T_i}\right), \quad k_i = \frac{k}{T_i} \quad \text{and} \quad k_d = kT_d.$$

We will show that if we choose $k = A^{-1}$, $T_i = T$ and $T_d = T$, then the control is perfect so that $e(t) = 0$. To this end we need to show $y(t) = r(t)$ or $Y(s) = R(s)$. From (8.97) we need to show $1 + C(s)X(c)M(s) = X(s)C(s)$, or

$$M(s) = 1 - \frac{1}{C(s)X(s)}.$$

With the values of k, T_i and T_d stated above, we have

$$1 - \frac{1}{C(s)X(s)} = 1 - \left[\frac{1}{A}\left(1 + \frac{1}{sT}\right)(1 + sT)\frac{A}{1 + sT} \right]^{-1} = \frac{1}{1 + sT} = M(s)$$

and so $y(t) = r(t)$.

8.5.2 Twin Rotor Static Rig for Air-Vehicle Testing

In this section we give an illustration of closed-loop feedback control, based on Triantafyllopoulos et al. (2009). The experimental rig is a twin rotor multi-input multi-output platform, designed by FeedbackInstruments (1996) for experimental use. The platform has many similarities with the motion of the normal helicopter and as a result it can be used as a static-platform for the design and testing of air-vehicles (Seddon & Newman, 2011).

The platform, which is schematically shown in Fig. 8.8, consists of a beam pivoted on its base in such a way that it can rotate freely in both the horizontal and vertical planes. At both ends of the beam there are two rotors, the main and tail rotors, driven by DC motors. The rotation of the main rotor produces a lifting

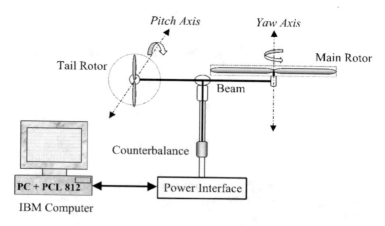

Fig. 8.8 Schematic illustration of the twin rotor MIMO system

force allowing the beam to rise vertically making a rotation around pitch axis. This vertical movement of the beam makes an elevation angle with respect to pitch plane. While, the tail rotor which is smaller than the main rotor is used to make the beam turn left or right around yaw axis producing a rotational angle . A counterbalance arm with a weight at its end is fixed to the beam at the pivot. The state of the beam is described by four process variables: horizontal and vertical angles measured by position sensors fitted at the pivot, and two corresponding angular velocities. Either or both axes of rotation can be locked by means of two locking screws, provided for physically restricting the horizontal or vertical plane TRMS rotation.

In a normal helicopter, changing the angle of attack of the blades controls the aerodynamic force. However, the TRMS is constructed in such a way that the angle of attack is fixed. In this case, the aerodynamic force is controlled by varying the speed of the DC motors. Therefore, the control inputs are supply voltages of the DC motors. A change in the voltage value results in a change of the rotational speed of the propeller, which in turn results in a change of the corresponding position of the beam . The input voltage is limited to ±10 volts.

The experimental designed by FeedbackInstruments (1996), is shown in Fig. 8.9 and was used by our colleagues within the Department of Automatic Control and Systems Engineering at the University of Sheffield. We can see the beam with the two rotors, the DC adaptor and the PC, controlling the voltage that can be inputted in order to activate the system.

A number of studies published for this experimental set-up include Ahmad et al. (2001), Ahmad et al. (2002) and Chalupa et al. (2015) among others. In this study we have focused on a single input $u(t)$ (in voltage) and a single output $y(t)$, which

Fig. 8.9 The twin rotor experimental rig used at Sheffield University

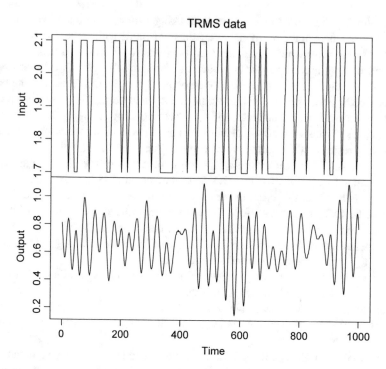

Fig. 8.10 Data from the TRMS system. Top panel shows the input (in volts) and the lower panel shows the output values, sampled every 10 s

is the motion of the tail rotor, translated in voltage too and recorded in the computer using the MATLAB/SIMULINK package. The system has the capability to act as a single rotor device (as we use it here) or as originally constructed as a twin rotor. The tail rotor used is indicated in Fig. 8.9 by the orange indicators (if seen in black and white is the rotor near the wall).

Figure 8.10 plots the data (1000 time points with a sample frequency of 10 s). The top panel of the plot shows the input signal and the bottom panel shows the output. The plot suggests that the output appears to be unstable, with oscillations repeated randomly.

A linear system is adopted and is described in the sequel. The continuous-time system is discretized using $t_k = \Delta k = 10k$, for $k = 0, 1, \ldots$. Following Triantafyllopoulos et al. (2009) we adopt a linear regression model, with time-varying components for $y_k = y(t_k)$

$$y_k = \sum_{i=1}^{8} y_{k-i} x_i + \sum_{j=1}^{8} u_{k-j} x_j + \epsilon_k,$$

where $x_1, \ldots, x_{16}$ are time-invariant coefficients and ϵ_k is a Gaussian white noise with variance σ^2, which is assumed to be uncorrelated of x_i. This model can be put in state space form

$$y_k = H_k x + \epsilon_k, \quad \epsilon_k \sim N(0, \sigma^2), \tag{8.98}$$

where $H_k = [y_{k-1}, \ldots, y_{k-8}, u_{k-1}, \ldots, u_{k-8}]$ and $x = [x_1, \ldots, x_{16}]^\top$ is the state vector.

In the first phase of analysis we fit this model to the data. After some experimentation we chose to include 8 lagged terms y_{k-i} and u_{k-j}. Our criteria of goodness of fit were to minimise the mean squared error (MSE) and maximise the log-likelihood function. The variance of σ^2 was estimated using the Bayesian conjugate approach of Sect. 4.3.3. We have implemented the model with a time-varying state vector $x_k = F x_{k-1} + \zeta_t$, but that introduced more fluctuations on the predictions of y_k, while from Fig. 8.10 the output signal is clearly smooth. Hence we adopt model (8.98). Figure 8.11 shows the one-step ahead forecast errors (left panel) and the

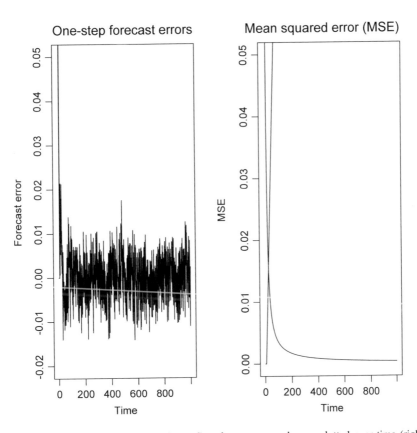

Fig. 8.11 One-step forecast errors (left panel) and mean squared error plotted over time (right panel)

MSE over time. We see that excluding the first few errors, they are small with range between ±0.02 and the MSE is low tending to zero (for example, for $k = 200$, the MSE is 0.0017. The average value of the MSE is 0.0024. The overall mean absolute deviation (MAD) is equal to 0.0089. We conclude that the performance of the model is very good indicating that (8.98) is a good representation of the system.

The next phase of the analysis is to see whether the system will benefit by adopting a feedback control strategy. For this to end we have instructed a step-signal as a reference signal and run the model with no feedback incorporated. Figure 8.12 plots the amplitude of the signal $y(t)$ together with the reference signal, for an open-loop system (no feedback control). We see that the signal oscillates persistently indicating lack of stability and only reaches steady state after considerable time. The vibration of the TRMS dominates the system indicating poor tracking performance of the open-loop approach. Next we apply the closed-loop feedback control using the PID-controller described in Sect. 8.5.1.

The performance of the feedback controller is assessed in terms of time domain performance objectives (overshoot, rise-time, settling-time and steady state error) in comparison to system performance in the open loop without control. The controlled output (elevation angle) is expected to have low overshoot, quick settling time of the residual oscillation and reasonably fast speed of response. Figure 8.13 shows the amplitude of the signal against the reference signal when feedback control is applied. The controller demonstrates superior performance in comparison to the system response without controller (i.e. open-loop configuration) in tracking the

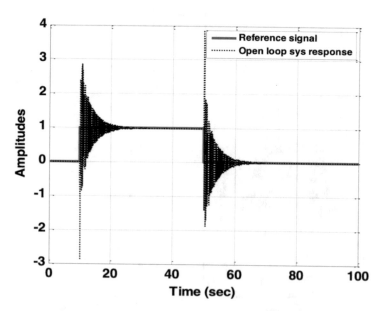

Fig. 8.12 Open-loop system response (no feedback control is applied). The step function indicates the reference signal

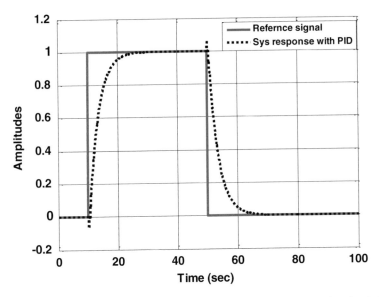

Fig. 8.13 Closed-loop system response (feedback control is applied). The step function indicates the reference signal

reference signal. Its response was characterised by relatively small overshoot, fast rise time and consistent settling time. A smooth system response was observed with no significant overshoot (0.039%) and fast rise time of 6.93 s. The system response with this controller settled within 12.77 s with a steady state error of 2.02% recorded at the settling time. The system paired with the closed-loop feedback controller returns a smooth signal and is clearly a stable system.

8.6   Exercises

1. a. Show that the Laplace transform of $f(t) = e^{-at} t^n$ is

$$\mathcal{L}[f(t)](s) = \frac{n!}{(s+a)^{n+1}},$$

   for $\mathrm{Re}(s) > -a$.
   b. Show that the Laplace transform of $g(t) = e^{-at} \sin(\omega t)$ is

$$\mathcal{L}[g(t)](s) = \frac{\omega}{(s+a)^2 + \omega^2},$$

   where $\mathrm{Re}(s) > -a$.

2. Consider the linear system having the state space representation

$$\dot{x} = \mathbf{F}x + \mathbf{G}u \quad \text{and} \quad y = \mathbf{H}x,$$

for some known components $\mathbf{F}$, $\mathbf{G}$ and $\mathbf{H}$.

Define the state $z = \mathbf{P}x$, where $\mathbf{P}$ is a non-non-singular square matrix. Find a state space representation of y using the state z, i.e.

$$\dot{z} = \mathbf{F}^* z + \mathbf{G}^* u \quad \text{and} \quad y = \mathbf{H}^* z,$$

and identify the components $\mathbf{F}^*$, $\mathbf{G}^*$ and $\mathbf{H}^*$.

3. Consider the continuous-time linear system

$$\dot{x}(t) = \mathbf{F}(t)x(t) + \mathbf{G}(t)u(t) + \zeta(t),$$
$$y(t) = \mathbf{H}(t)x(t) + \mathbf{L}(t)u(t) + \epsilon(t),$$

where $u(t)$ is the input of the system, $y(t)$ is the output of the system and the components $\mathbf{F}(t)$, $\mathbf{G}(t)$, $\mathbf{L}(t)$ and $\mathbf{L}(t)$ are assumed known functions. Suppose that the input functions are generated by the differential equation $\dot{u}(t) = \phi x(t) + \psi$, for some constants ϕ and ψ. Define the state vector

$$x^*(t) = \begin{bmatrix} x(t) \\ u(t) \end{bmatrix}.$$

Find a state space representation of the system using the state vector $x^*(t)$, i.e.

$$\dot{x}^*(t) = \mathbf{F}^*(t)x^*(t) + \zeta^*(t),$$
$$y(t) = \mathbf{H}^*(t)x^*(t) + \epsilon^*(t),$$

and identify the components $\mathbf{F}^*(t)$, $\mathbf{H}^*(t)$ and define the innovation functions $\zeta^*(t)$ and $\epsilon^*(t)$. Hence show that by using the state vector $x^*(t)$ you can incorporate the input $u(t)$ into the states.

4. Consider the linear system

$$\dot{x} = \begin{bmatrix} 1 & 0 & 0 \\ 0 & -1 & 0 \\ 0 & 0 & -4 \end{bmatrix} x + \begin{bmatrix} 1 & -2 \\ -1 & 1 \\ 0 & 1 \end{bmatrix} u,$$
$$y = [0, 1, 1]x,$$

where $x = [x_1, x_2, x_3]^\top$ is the state vector and $u = [u_1, u_2]^\top$ is a vector of inputs.

a. Show that the system is not stable.

b. Define the state

$$z = \begin{bmatrix} x_2 \\ x_3 \end{bmatrix}.$$

Find a state space representation for the system using the state vector z and show that this system is stable.

5. Consider the matrix exponential of some square matrix F defined as

$$e^F = \sum_{n=0}^{\infty} \frac{F^n}{n!}.$$

a. Show that

$$(e^F)^T = e^{F^T}.$$

b. If F and G are $p \times p$ matrices, show that in general

$$e^{F+G} \neq e^F e^G.$$

6. A vehicle experiencing force $f(t) = ma(t)$ in the interval $t_0, T]$, where m is the mass of the vehicle, for some bivariate vector a. If $p(t) = [p_1(t), p_2(t)]^T$ denotes the position of the vehicle at time t with coordinates $p_1(t)$ and $p_2(t)$, then the system is described by the following differential equation

$$\frac{dp(t)}{dt} = \frac{f}{m}. \tag{8.99}$$

a. Show that the general solution of (8.99), with initial conditions $p(t_0)$ and $\dot{p}(t_0)$ is given by

$$p(t) = p(t_0) + \dot{p}(t_0)(t - t_0) + \frac{1}{m} \int_{t_0}^{t} \int_{t_0}^{s} f(\tau) \, d\tau \, ds.$$

b. Define the state vector

$$x(t) = \begin{bmatrix} \dot{p}_1(t) \\ \dot{p}_2(t) \\ p_1(t) \\ p_2(t) \end{bmatrix}.$$

Find a state space representation of the system with measurements $p(t)$ and states $x(t)$, i.e. write the system in the form

$$\dot{x}(t) = \mathbf{F}x(t) + \mathbf{G}u(t),$$

$$p(t) = \mathbf{H}x(t)$$

and identify the components $\mathbf{F}, \mathbf{G}, \mathbf{H}$ and the input function $u(t)$.

7. Consider a RC electric circuit, consisting of a voltage source or battery (V), capacitor (C) and a resistor (R), all serially connected; for a circuit diagram see Fig. 8.14. Let $u(t)$ be the total voltage of the battery and $x(t)$ be the voltage of the capacitor, where R is the effective resistance of the combined load, source and component and C is the capacitance.

a. Use Kirchhoff's second law (total voltage) to derive the differential equation

$$\dot{x}(t) = -\frac{1}{CR}x(t) + \frac{1}{CR}u(t). \qquad (8.100)$$

b. If the output $y(t) = x(t)$ is the voltage of the capacitor, write down a state space representation of this system.
c. Show that this system is stable.
d. Suppose that the input voltage $u(t)$ follows the sinusoidal function

$$u(t) = A \sin \omega t,$$

where A is the amplitude and ω is the angular frequency.

Fig. 8.14 RC electric circuit. Shown are the voltage source (V), the capacitor (C) and the resistor (R), all serially connected

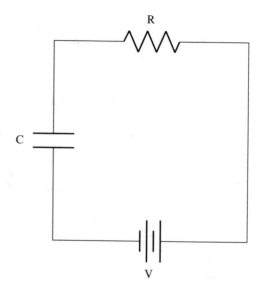

Show that the solution of (8.100) is

$$x(t) = \left[x(0) + \frac{ACR\omega}{1 + \omega^2 C^2 R^2} \right] \exp\left(-\frac{t}{CR} \right) + \frac{A(\sin \omega t - CR \cos \omega t)}{1 + \omega^2 C^2 R^2}.$$

Hence for large t, the output voltage of the capacitor is approximately equal to

$$x(t) \approx \frac{A(\sin \omega t - CR \cos \omega t)}{1 + \omega^2 C^2 R^2}.$$

8. Consider the linear system

$$\dot{x}(t) = \begin{bmatrix} 0 & -1 \\ 2 & -2 \end{bmatrix} x(t) \quad \text{and} \quad y(t) = [1, 1]x(t),$$

with

$$x(0) = \begin{bmatrix} x_1(0) \\ x_2(0) \end{bmatrix}.$$

a. Without making reference to part (b) below, show that this system is asymptotically stable.
b. Find the eigenvalues and eigenvectors of

$$F = \begin{bmatrix} 0 & -1 \\ 2 & -2 \end{bmatrix}$$

and use Eq. (8.33) to show

$$e^{Ft} = \begin{bmatrix} 2e^{-t} - e^{-2t} & -e^{-t} + e^{-2t} \\ 2e^{-t} - 2e^{-2t} & -e^{-t} + 2e^{-2t} \end{bmatrix}.$$

c. Use (b) to derive the solution of the state differential equation as $x_t = [x_1(t), x_2(t)]^T$, with

$$x_1(t) = x_1(0)(2e^{-t} - e^{-2t}) + x_2(0)(-e^{-t} + e^{-2t}),$$

$$x_2(t) = x_1(0)(2e^{-t} - 2e^{-2t}) + x_2(0)(-e^{-t} + 2e^{-2t}),$$

hence verify the asymptotic stability of part (a) by taking the limit of $x(t)$ when $t \to \infty$.

9. Consider the linear system

$$\dot{x}(t) = \begin{bmatrix} 1 & 0 \\ 1 & -1 \end{bmatrix} x(t) \quad \text{and} \quad y(t) = [1, 1]x(t),$$

with

$$x(0) = \begin{bmatrix} x_1(0) \\ x_2(0) \end{bmatrix}.$$

a. Without making reference to part (b) below, show that this system is unstable.

b. Follow a similar procedure as that of Exercise 9 to show that the solution of $x_t = [x_1(t), x_2(t)]^\top$ is

$$x_1(t) = x_1(0)e^t,$$

$$x_2(t) = \frac{x_1(0)}{2}e^t + \left(x_2(0) - \frac{x_1(0)}{2}\right)e^{-t}.$$

10. Consider the non-linear system of Example 8.12 (p. 433). Show this system is asymptotically stable using Lyapunov's indirect method.

11. Consider the Lorenz system (8.49a)–(8.49c) of Example 8.13, with $\sigma = 10$, $\rho = 0.9$ and $\beta = 8/3$. Using Lyapunov's indirect method show that the system is stable around the origin $(x, y, z) = (0, 0, 0)$.

12. Consider the system generated by the non-linear differential equation

$$\frac{d^2 y(t)}{dt^2} = \cos y(t) - \frac{3}{2\pi}y(t) - \alpha\frac{dy(t)}{dt},$$

for some constant α.

a. Find the state space representation of this system.

b. Show that the equilibrium point of this system is $x_e = [\pi/3, 0]^\top$, where $x = [x_1, x_2]^\top$ denotes the state vector of this system.

c. Identify the range of values of α, for which the system is asymptotically stable around x_e.

13. Consider the two-object system (8.18)–(8.19) of Example 8.6. Show that the system is stable, but not asymptotically stable.

14. Consider the system generated by the differential equations

$$\frac{d^2 y_1(t)}{dt^2} + \alpha_1\frac{dy_1(t)}{dt} + \alpha_2 y_1(t) = u_1 + \alpha_3 u_2$$

$$\frac{dy_2(t)}{dt} + \alpha_4 y_2(t) + \alpha_3\frac{y_1(t)}{dt} = \alpha_6 u_1,$$

where u_1, u_2 are inputs, $y_1(t), y_2(t)$ are the outputs and $\alpha_1, \alpha_2, \alpha_3, \alpha_4, \alpha_5$ and α_6 are known constants.

Define the state vector

$$x = \begin{bmatrix} y_1 \\ \dot{y}_1 \\ y_2 \end{bmatrix}.$$

a. With $y = [y_1, y_2]^\top$ the output vector, write down this system in state space form, or

$$\dot{x} = \mathbf{F}x + \mathbf{G}u,$$

$$y = \mathbf{H}x$$

and determine the matrices $\mathbf{F}, \mathbf{G}, \mathbf{H}$ and the input vector u.

b. If $\alpha_2 \neq 0$ and $\alpha_4 \neq 0$, show that the equilibrium point is

$$x_e = \left[\frac{u_1 + \alpha_3 u_2}{\alpha_2}, 0, -\frac{\alpha_6 u_1}{\alpha_4} \right]^\top.$$

c. If $u_1 = u_2 = 0$, $\alpha_4 \neq 0$, $\alpha_1, \alpha_2 > 0$, show that

 i. If $\alpha_4 < 0$, the system is asymptotically stable.
 ii. If $\alpha_4 > 0$, the system is unstable.

15. Consider Example 8.9 of the spring-mass attached on a wall. Assume that the position of the object attached to the spring observed subject to noise $\epsilon(t)$, which satisfies the conditions of (8.60).

a. If $k_2^2 - 4k_1 m < 0$, where m is the mass of the spring and k_1, k_2 are the friction coefficient and the spring constant, respectively, then show that the limit of the error covariance matrix (Kalman–Bucy filter) $\mathbf{P}(t)$ is the zero matrix, as $t \to \infty$. Hence, show that $\hat{x}(t)$, the estimate of $x(t)$ at the steady state satisfies

$$\hat{x}(t) = \sum_{n=0}^{\infty} \frac{(t - t_0)^n}{n!} \begin{bmatrix} 0 & 1 \\ -k_1 & -k_2 \end{bmatrix}^n x(t_0),$$

where $x(t_0)$ is the initial state vector.

b. If $t_0 = 0$, $k_1 = 2$, $k_2 = 3$ and $m = 2gr$, then show $\hat{x}(t) = [\hat{x}_1(t), \hat{x}_2(t)]^\top$ of part (a) has components

$$\hat{x}_1(t) = x_1(0)(2e^{-3t} - e^{-2t}) + x_2(0)(e^{-3t} - e^{-2t}),$$

$$\hat{x}_2(t) = x_1(0)(2e^{-2t} - 2e^{-3t}) + x_2(0)(2e^{-2t} - e^{-3t}),$$

where $x(0) = [x_1(0), x_2(0)]^\top$ is the initial state vector.

16. Consider Example 8.14 and assume that observations $y(t) = c$ are constant, for some c.

   a. Using $\mathbf{P} = \lim_{t \to \infty} \mathbf{P}(t)$ find K the limit of the Kalman gain $K(t)$ as $t \to \infty$.

   b. Using (a) show that the differential equation of $x(t)$ at steady state satisfies

   $$\dot{\hat{x}}(t) = (\mathbf{F} - KH)\hat{x}(t) + Kc.$$

   c. Write

   $$\begin{bmatrix} -1 & 3 \\ 0 & 1 \end{bmatrix}^n = \begin{bmatrix} f_{11}(n) & f_{12}(n) \\ f_{21}(n) & f_{22}(n) \end{bmatrix},$$

   for some sequences $f_{ij}(n)$. Show that the solution of the differential equation of part (b) is given as an infinite series

   $$\hat{x}(t) = -2c \sum_{n=0}^{\infty} \frac{t^n}{(n+1)n!} \begin{bmatrix} f_{11}(n) \\ f_{21}(n) \end{bmatrix}.$$

17. Consider the linear system

   $$\dot{x}(t) = -9x(t) + 3u(t), \tag{8.101}$$

   $$y(t) = 3x(t), \tag{8.102}$$

   for some input function $u(t)$, state $x(t)$ and output $y(t)$.

   a. Show that the transfer function of this system is

   $$H(s) = \frac{9}{9+s},$$

   where $s \in \mathbb{C}$.

   b. Suppose now that measurements $y(t)$ are observed subject to noise, hence Eq. (8.101) is replaced by

   $$y(t) = 3x(t) + \epsilon(t), \quad \epsilon(t) \sim N(0, 1), \tag{8.103}$$

   where $\epsilon(t)$ satisfies condition (8.60). At time $t_0 = 0$ let the initial distribution of $x(t_0)$ be $x(0) \sim N(1, 10)$ and assume that x is uncorrelated of $\epsilon(t)$, for all t. Use the Kalman–Bucy filter to answer the following.

   i. Show that the error variance $P(t)$ is

   $$P(t) = \frac{10}{6e^{18t} - 5}$$

and hence verify that $P = \lim_{t \to \infty} P(t) = 0$. Justify $P = 0$, without making reference of $P(t)$ given above.

ii. If the input function is a step function

$$u(t) = \begin{cases} k, & t_k \le t < \infty \\ 0, & \text{otherwise} \end{cases},$$

with $k > 0$, for some time $t_k > 0$. Show that at steady state the estimate of $x(t)$ is given by

$$\hat{x}(t) = e^{-9t} + \frac{k}{3}\left[1 - e^{9(t_k - t)}\right].$$

Hence show that as $t \to \infty$ the estimate $\hat{x}(t)$ of the state $x(t)$ converges to $k/3$.

18. Consider the linear system (8.101)–(8.103) of Exercise 17. The continuous-time system is discretised using $t_k = k\Delta t$, with $k = 0, 1, \ldots, N - 1$.

a. Simulate a path of x_k and y_k, for $N = 100$ and $\Delta t = 0.2$.
b. Using the discrete Kalman filter, provide the posterior distribution of x_k, given data $y_{1:k}$.
c. Plot the posterior mean $E(x_k \mid y_{1:k})$ against the simulated values x_k, for $k = 1, \ldots, 99$. Comment on the performance of this model based on this plot.

19. Consider the continuous-time time-invariant system of Example 8.14.

a. Using Eq. (8.79) show that the stead-state differential equation is

$$\dot{\hat{x}}(t) = \begin{bmatrix} -1 & 3 \\ 0 & 1 \end{bmatrix}\hat{x}(t) + \begin{bmatrix} 2 \\ 0 \end{bmatrix}y(t).$$

b. Diagonalise matrix

$$\begin{bmatrix} -1 & 3 \\ 0 & 1 \end{bmatrix}$$

and hence show

$$\exp\left\{\begin{bmatrix} -1 & 3 \\ 0 & 1 \end{bmatrix}t\right\} = \begin{bmatrix} e^{-t} & \frac{3}{2}e^{t} - \frac{3}{2}e^{-t} \\ 0 & e^{t} \end{bmatrix}.$$

c. Use part (b) to show that with $\hat{x}(t) = [\hat{x}_1(t), \hat{x}_2(t)]^T$, the solutions $\hat{x}_1(t)$ and $\hat{x}_2(t)$ of the differential equation of part (a) satisfy

$$\hat{x}_1(t) = e^{-t}\hat{x}_1(0) + \frac{3}{2}(e^t - e^{-t})\hat{x}_2(0) + 2\int_0^t e^{\tau - t} y(\tau)\, d\tau,$$

$$\hat{x}_2(t) = e^t \hat{x}_2(0),$$

where $\hat{x}_1(0)$ and $\hat{x}_2(0)$ are the initial estimates of $x_1(t)$ and $x_2(t)$ at $t = 0$, where $x(t) = [x_1(t), x_2(t)]^T$.

20. Consider the linear system

$$\dot{x}_1(t) = -x_1(t) + 6x_2(t),$$

$$\dot{x}_2(t) = -x_1(t) + 4x_2(t),$$

$$y(t) = x_1(t) + x_2(t) + \epsilon(t),$$

where $\epsilon(t) \sim N(0, 1)$ and it satisfied the error structure (8.60)–(8.61). Assume that the system is initialised at time $t_0 = 0$ with error covariance matrix $P(0) = I$.

a. Diagonalise the matrix

$$F = \begin{bmatrix} -1 & 6 \\ -1 & 4 \end{bmatrix}$$

and show that

$$e^{-Fw} = \begin{bmatrix} 3e^{-w} - 2e^{-2w} & -6e^{-w} + 6e^{-2w} \\ e^{-w} - e^{-2w} & -2e^{-w} + 3e^{-2w} \end{bmatrix}.$$

b. Using part (a) and the Kalman–Bucy filter determine the elements of $P^{-1}(t)$ (the inverse of the error covariance matrix of $x_t = [x_1(t), x_2(t)]^T$).

c. Using (b) or otherwise show that the limit P of $P(t)$ is

$$P = \lim_{t \to \infty} P(t) = \frac{1}{8}\begin{bmatrix} 17 & 11 \\ 11 & 9 \end{bmatrix}.$$

d. Show that the solution of the state differential equation (at steady state) is

$$\hat{x}_1(t) = \left(\frac{7}{2}e^{-2t} - \frac{5}{2}e^{-t}\right)\left[\hat{x}_1(0) + 6\int_0^t e^{-2\tau}y(\tau)\,d\tau - \frac{5}{2}\int_0^t e^{-\tau}y(\tau)\right]$$

$$+ \left(-\frac{5}{2}e^{-2t} + \frac{5}{2}e^{-t}\right)\left[\hat{x}_2(0) + 6\int_0^t e^{-2\tau}y(\tau)\,d\tau - \frac{7}{2}\int_0^t e^{-\tau}y(\tau)\right]$$

and

$$\hat{x}_2(t) = \left(\frac{7}{2}e^{-2t} - \frac{7}{2}e^{-t}\right)\left[\hat{x}_1(0) + 6\int_0^t e^{-2\tau}y(\tau)\,d\tau - \frac{5}{2}\int_0^t e^{-\tau}y(\tau)\right]$$

$$+ \left(-\frac{5}{2}e^{-2t} + \frac{7}{2}e^{-t}\right)\left[\hat{x}_2(0) + 6\int_0^t e^{-2\tau}y(\tau)\,d\tau - \frac{7}{2}\int_0^t e^{-\tau}y(\tau)\right],$$

where $x_t[x_1(t), x_2(t)]^\top$ and $\hat{x}(0) = [\hat{x}_1(0), \hat{x}_2(0)]^\top$ is the initial estimate of $x(t)$.

21. Consider the one-dimensional linear system

$$\dot{x}(t) = -cx(t) + \zeta(t), \quad \zeta(t) \sim N(0, \sigma_Z^2),$$

$$y(t) = \epsilon(t), \quad \epsilon(t) \sim N(0, \sigma_\epsilon^2),$$

where $c > 0$, the innovation functions $\zeta(t)$ and $\epsilon(t)$ satisfy assumptions (8.60)–(8.61) and $\sigma_Z^2, \sigma_\epsilon^2$ are known positive variances.

a. Show that this system is asymptotically stable.
b. Show that the error variance is equal to

$$P(t) = \frac{\sigma_Z^2}{2c} + \left[P(t_0) - \frac{\sigma_Z^2}{2c}\right]e^{-2c(t-t_0)},$$

where the system is initialised at t_0, with error variance $P(t_0) > 0$.
c. Using two different ways show that the limit P of $P(t)$ is

$$P = \lim_{t \to \infty} P(t) = \frac{\sigma_Z^2}{2c}.$$

d. Hence show that if $P(t_0) < \sigma_\epsilon^2/(2c)$, then $P(t)$ is increasing function; if $P(t_0) > \sigma_\epsilon^2/(2c)$, then $P(t)$ is decreasing function; if $P(t_0) = \sigma_\epsilon^2/(2c)$, then $P(t) = \sigma_\epsilon^2/(2c)$, for all $t \geq t_0$.

References

Adcock, C. J. (2007). Measuring portfolio performance using a modified measure of risk. *Journal of Asset Management, 7*(6), 388–403.

Adcock, C. J. (2010). Asset pricing and portfolio selection based on the multivariate extended skew-Student-t distribution. *Annals of Operations Research, 176*(1), 221–234.

Aguilar, O., & West, M. (2000). Bayesian dynamic factor models and portfolio allocation. *Journal of Business and Economic Statistics, 18*(3), 338–357.

Ahmad, S. M., Chipperfield, A. J., & Tokhi, M. O. (2001). Parametric modelling and dynamic characterization of a two-degree-of-freedom twin-rotor multi-input multi-output system. *Proceedings of the Institution of Mechanical Engineers, Part G: Journal of Aerospace Engineering, 215*(2), 63–78.

Ahmad, S. M., Chipperfield, A. J., & Tokhi, M. O. (2002). Dynamic modelling and open-loop control of a twin rotor multi-input multi-output system. *Proceedings of the Institution of Mechanical Engineers, Part I: Journal of Systems and Control Engineering, 216*(6), 477–496.

Aidala, V. I. (1979). Kalman filter behaviour in bearings-only tracking applications. *IEEE Transactions on Aerospace and Electronic Systems, 15*, 29–39.

Aldrich, J. (1998). Doing least squares: perspectives from Gauss and Yule. *International Statistical Review, 66*, 61–81.

Anand, D. K. (1984). *Introduction to control systems* (2nd ed.). London: Pergamon Press.

Anderson, B. D. O. (1971). Stability properties of Kalman-Bucy filters. *Journal of the Franklin Institute, 291*(2), 137–144.

Anderson, B. D. O., & Moore, J. B. (1979). *Optimal filtering*. Englewood Cliffs, NJ: Prentice Hall.

Anderson, O. D. (1976). *Time series analysis and forecasting: The Box-Jenkins approach.* Butterworths. Series R.

Andrieu, C., Doucet, A., & Holenstein, R. (2010). Particle Markov chain Monte Carlo methods. *Journal of the Royal Statistical Society Series B, 72*(3), 269–342.

Angelova, D., & Mihaylova, L. (2008). Extended object tracking using Monte carlo methods. *IEEE Transactions on Signal Processing, 56*(2), 825–832.

Applebaum, D. (2011). *Lévy processes and stochastic calculus* (2nd ed.). Cambridge: Cambridge University Press.

Arnold, B. C., & Groeneveld, R. A. (1995). Measuring skewness with respect to the mode. *The American Statistician, 49*, 34–38.

Asai, M., McAleer, M., & Yu, J. (2006). Multivariate stochastic volatility: a review. *Economtric Reviews, 25*(2-3), 145–175.

Ass, K., & Haff, I. H. (2006). The generalised hyperbolic skew Student's t-distribution. *Journal of Financial Econometrics, 4*, 275–309.

K. Triantafyllopoulos, *Bayesian Inference of State Space Models*, Springer Texts in Statistics, https://doi.org/10.1007/978-3-030-76124-0

Atkinson, A. C., Koopman, S. J., & Shephard, N. (1997). Detecting shocks: Outliers and breaks in time series. *Journal of Econometrics, 80*(2), 387–422.

Azzalini, A., & Capitanio, A. (2003). Distributions generated by perturbation of symmetry with emphasis on a multivariate skew t-distribution. *Journal of the Royal Statistical Society Series B, 65*(2), 367–389.

Bacciotti, A. (2019). *Stability and control of linear systems* (Vol. 185). New York: Springer.

Bakshi, G., Kapadia, N., & Madan, D. (2003). Stock return characteristics, skew laws and the differential pricing of individual equity options. *The Review of Financial Studies, 16*(1), 101–143.

Balakrishnan, A. V. (1984). *Kalman filtering theory*. New York: Optimization Software Inc.

Barndorff-Nielsen, O., & Schou, G. (1973). On the parameterization of autoregressive models by partial autocorrelations. *Journal of Multivariate Analysis, 3*, 408–419.

Bäuerle, N., & Li, Z. (2013). Optimal portfolios for financial markets with Wishart volatility. *Journal of Applied Probability, 50*(4), 1025–1043.

Bauwens, L., Laurent, S., & Rombouts, J. V. K. (2006). Multivariate GARCH models: a survey. *Journal of Applied Econometrics, 21*, 79–109.

Beach, C. M., & MacKinnon, J. G. (1978). A maximum likelihood procedure for regression with autocorrelated errors. *Econometrica, 46*, 51–58.

Beck, N. (1983). Time-varying parameter regression models. *American Journal of Political Science, 27*, 557–600.

Bersimis, S., Psarakis, S., & Panaretos, J. (2007). Multivariate statistical process control charts: an overview. *Quality and Reliability Engineering International, 23*(5), 517–543.

Bersimis, S., & Triantafyllopoulos, K. (2020). Dynamic non-parametric monitoring of air-quality. *Methodology and Computing in Applied Probability, 22*, 1457–1479.

Berzuini, C., & Gilks, W. (2001). Sequential Monte Carlo methods in practice. In A. Doucet, N. de Freitas, & N. Gordon (Eds.), *Statistics for engineering and information science* (pp. 117–138). New York: Springer.

Biernson, G. (1989). *Principles of feedback control: Feedback system design*. New York: Wiley.

Bingham, N. H., & Fry, J. M. (2010). *Regression: Linear models in statistics*. New York: Springer.

Bingham, N. H., & Kiesel, R. (2002). Semi-parametric modelling in finance: theoretical foundations. *Quantitative Finance, 2*, 241–250.

Bingham, N. H., & Kiesel, R. (2004). *Risk-neutral valuation: Pricing and hedging of financial derivatives* (2nd ed.). New York: Springer.

Bollerslev, T. (1986). Generalized autoregressive conditional heteroskedasticity. *Journal of Econometrics, 31*, 307–327.

Bollerslev, T. (1990). Modelling the coherence in short-run nominal exchange rates: A multivariate generalized Arch model. *The Review of Economics and Statistics, 72*(3), 498–505.

Box, G. E. P., Jenkins, G. M., & Reinsel, G. C. (2008). *Time series analysis: Forecasting and control* (4th ed.). New York: Wiley.

Box, G. E. P., Luceño, A., & del Carmen Paniagua-Quinones, M. (2009). *Statistical control by monitoring and adjustment* (2nd ed.). New York: Wiley.

Brandt, M. W., & Santa-Clara, P. (2006). Dynamic portfolio selection by augmenting the asset space. *The Journal of Finance, 61*, 2187–2217.

Brockwell, P. J., & Davis, R. A. (1991). *Time series: Theory and methods* (2nd ed.). New York: Springer.

Brockwell, P. J., & Davis, R. A. (2016). *Introduction to time series and forecasting* (3rd ed.). New York: Springer.

Brown, R. G. (1962). *Smoothing, forecasting and prediction of discrete time series*. Englewood Cliffs, NJ: Prentice-Hall.

Bru, M. (1991). Wishart processes. *Journal of Theoretical Probability, 4*, 725–751.

Büuhler, W. K. (1981). *Gauss: A biographical study*. New York: Springer.

Carlin, B. P., Polson, N. G., & Stoffer, D. S. (1992). A Monte Carlo approach to nonnormal and nonlinear state-space modeling. *Journal of the American Statistical Association, 87*, 493–500.

Carter, C. K., & Kohn, R. (1994). On Gibbs sampling for state space models. *Biometrika, 81*, 541–553.

Carter, M., & van Brunt, B. (2000). *The Lebesgue-Stieltjes integral: Practical introduction.* New York: Springer.

Catlin, D. E. (1989). *Estimation, control, and the discrete Kalman filter.* New York: Springer.

Chalupa, P., Přikryl, J., & Novák, J. (2015). Modelling of twin rotor mimo system. *Procedia Engineering, 100*, 249–258.

Chambers, J. M., Cleveland, W. S., Kleiner, B., & Tukey, P. A. (1983). *Graphical methods for data analysis.* Belmont, CA: Wadsworth.

Chan, H.-F., & Guo, L. (1991). *Identification and stochastic adaptive control.* Boston: Birkhäuser.

Chan, S. W., Goodwin, G. C., & Sin, K. S. (1984). Convergence properties of the Riccati difference equation in optimal filtering of nonstabilizable systems. *IEEE Transactions in Automatic Control, 29*, 10–18.

Chen, R., & Liu, J. S. (1996). Predictive updating methods with application to Bayesian classification. *Journal of the Royal Statistical Society Series B, 58*, 397–415.

Chib, S., Nardari, F., & Shephard, N. (2006). Analysis of high dimensional multivariate stochastic volatility models. *Journal of Econometrics, 134*, 341–371.

Chib, S., & Tiwari, R. C. (1994). Outlier detection in the state space model. *Statistics and Probability Letters, 20*(2), 143–148.

Chipman, J. S. (1950). The multisector multiplier. *Econometrica, 18*(4), 355–374.

Chiu, T. Y. M., Leonard, T., & Tsui, K.-W. (1996). The matrix-logarithmic covariance model. *Journal of the American Statistical Association, 91*(433), 198–210.

Chopin, N., & Papaspiliopoulos, O. (2020). *An introduction to sequential Monte Carlo.* New York: Springer.

Cobb, G. W. (1978). The problem of the Nile: conditional solution to a change-point problem. *Biometrika, 65*, 243–251.

Collett, D. (2003). *Modelling survival data in medical research* (2nd ed.). New York: Chapman and Hall.

Commandeur, J. J. F., & Koopman, S. J. (2007). *An introduction to state space time series analysis.* Oxford: Oxford University Press.

Cooper, J. D., & Harrison, P. J. (1997). A Bayesian approach to modelling the observed bovine spongiform encephalopathy epidemic. *Journal of Forecasting, 16*, 355–374.

Cootner, P. H. (1964). *The random character of stock market prices.* Cambridge, MA: MIT Press.

Cowan, C. F. N., & Grant, P. M. (1985). *Adaptive filters.* Englewood Cliffs, NJ: Prentice-Hall.

Cowles 3rd, A., & Jones, H. E. (1937). Some a posteriori probabilities in stock market action. *Econometrica, 5*, 280–294.

Cox, D. R. (1972). Regression models and life-tables (with discussion). *Journal of the Royal Statistical Society Series B, 34*, 187–220.

Cox, D. R. (1975). Partial likelihood. *Biometrika, 62*, 269–275.

Cox, D. R., & Oakes, D. (1984). *Analysis of survival data.* New York: Chapman and Hall.

Crowdy, D. G., & Luca, E. (2014). Solving Wiener-Hopf problems without kernel factorization. *Proceedings of the Royal Society A, 470*(2170).

Davis, H. T. (1941). *The analysis of economic time series, cowles commission monograph no. 6.* Bloomington, Indiana: Principia Press.

De Jong, P. (1989). Smoothing and interpolation with the state-space model. *Journal of the American Statistical Association, 84*, 1085–1088.

De Jong, P., & Penzer, J. R. (1998). Diagnosing shocks in time series. *Journal of the American Statistical Association, 93*, 796–806.

De Jong, P., & Penzer, J. R. (2004). The ARMA model in state space form. *Statistics and Probability Letters, 70*, 119–125.

Dempster, A. P., Laird, N. M., & Rubin, D. B. (1977). Maximum likelihood from incomplete data via the em algorithm. *Journal of the Royal Statistical Society Series B, 39*, 1–38.

Diaconis, P., & Ylvisaker, D. (1979). Conjugate priors for exponential families. *Annals of Statistics, 7*, 269–281.

Díaz-García, J. A., & Gutiérrez, J. R. (1997). Proof of the conjectures of H. Uhlig on the singular multivariate beta and the jacobian of a certain matrix transformation. *Annals of Statistics, 25,* 2018–2023.

Díaz-García, J. A., & Gutiérrez, J. R. (1998). Singular matrix beta distribution. *Journal of Multivariate Analysis, 99,* 637–648.

Dickey, D., & Fuller, W. (1979). Distribution of the estimators for autoregressive time series with a unit root. *Journal of the American Statistical Association, 74*(366), 427–431.

Ding, Z. (2013). *Nonlinear and adaptive control systems* (Vol. 84). IET Control Engineering Series.

Djeundje, V. B., & Crook, J. (2019). Dynamic survival models with varying coefficients for credit risks. *European Journal of Operational Research, 16*(1), 319–333.

Doob, J. L. (1955). *Stochastic processes.* New York: Wiley.

Douc, R., Cappe, O., & Moulines, E. (2005). Comparison of resampling schemes for particle filtering. In *Image and Signal Processing Analysis.*

Doucet, A., de Freitas, N., & Gordon, N. (2001). *Sequential Monte Carlo methods in practice.* New York: Springer.

Doucet, A., Godsill, S., & Andrieu, C. (2000). On sequential monte carlo sampling methods for Bayesian filtering. *Statistics and Computing, 10,* 197–208.

Duncan, D. B., & Horn, S. D. (1972). Linear dynamic recursive estimation from the viewpoint of regression analysis. *Journal of the American Statistical Association, 67,* 815–821.

Durbin, J. (2004). Introduction to state space time series analysis. In A. C. Harvey, S. J. Koopman, & N. Shephard (Eds.), *State space and unobserved componet models: Theory and applications* (pp. 3–25). Cambridge: Cambridge University Press.

Durbin, J., & Koopman, S. J. (2012). *Time series analysis by state space methods* (2nd ed.). Oxford: Oxford University Press.

Dyke, P. P. G. (1999). *An introduction to Laplace transforms and Fourier series.* New York: Springer.

Elliott, R. J., Hoek, J. V. D., & Malcolm, W. P. (2005). Pairs trading. *Quantitative Finance, 5,* 271–276.

Engle, R. (1982). Autoregressive conditional heteroscedasticity with estimates of the variance of united kingdom inflation. *Econometrica, 50,* 987–1007.

Engle, R., & Granger, C. (1987). Co-integration and error correction: representation, estimation and testing. *Econometrica, 55*(2), 251–276.

Engle, R. F. (2002). Dynamic conditional correlation: a simple class of multivariate generalized autoregressive conditional heteroskedasticity models. *Journal of Business and Economic Statistics, 20,* 339–350.

Engle, R. F., & Granger, C. W. J. (1987). Co-integration and error-correction: representation, estimation and testing. *Econometrica, 55,* 251–276.

Eubank, R. L. (2006). *A Kalman filter primer.* New York: Chapman and Hall.

Fahrmeir, L. (1992). Posterior mode estimation by extended Kalman filtering for multivariate generalised linear models. *Journal of the American Statistical Association, 87,* 501–509.

Fahrmeir, L. (1994). Dynamic modelling and penalized likelihood estimation for discrete time survival data. *Biometrika, 81*(2), 317–330.

Fahrmeir, L., & Tutz, G. (2001). *Multivariate statistical modelling based on generalized linear models.* New York: Springer.

Fama, E. F. (1965). Random walks in stock market prices. *Financial Analysts Journal, 21*(5), 55–59.

Fearnhead, P. (2002). Markov chain Monte Carlo, sufficient statistics, and particle filters. *Journal of Computational and Graphical Statistics, 11,* 848–862.

FeedbackInstruments. (1996). Twin Rotor MIMO System Manual 33–007–0 [Computer software manual]. Sussex, UK.

Fernandez, C., & Steel, M. F. J. (1998). On Bayesian modeling of fat tails and skewness. *Journal of the American Statistical Association, 93,* 359–371.

Fitzgerald, R. J. (1971). Divergence of the kalman filter. *IEEE Transactions in Automatic Control, AC-16,* 736–747.

Franklin, G. F., Powell, J. D., & Emami-Naeini, A. (2018). *Feedback control of dynamic systems* (7th ed.). Pearson.

Fruhwirth-Schnatter, S. (1994a). Applied state space modelling of non-Gaussian time series using integration-based Kalman filtering. *Statistics and Computing, 4,* 259–269.

Fruhwirth-Schnatter, S. (1994b). Data augmentation and dynamic linear models. *Journal of Time Series Analysis, 15,* 183–202.

Gamerman, D. (1991). Dynamic Bayesian models for survival data. *Journal of the Royal Statistical Society Series C, 40*(1), 63–79.

Gamerman, D. (1998). Markov chain Monte Carlo for dynamic generalised linear models. *Biometrika, 85,* 215–227.

Gamerman, D., & Lopes, H. F. (2006). *Markov chain Monte Carlo: Stochastic simulation for Bayesian inference* (2nd ed.). New York: Chapman and Hall.

Gamerman, D., dos Santos, T. R., & Franco, G. C. (2013). A non-Gaussian family of state space-models with exact marginal likelihood. *Journal of Time Series Analysis, 34,* 625–645.

Gamerman, D., & West, M. (1987). An application of dynamic survival models in unemployment studies. *The Statistician, 36,* 269–274.

Gatev, E., Goetzmann, W. N., & Rouwenhorst, K. G. (2006). Pairs trading: Performance of a relative-value arbitrage rule. *Review of Financial Studies, 19*(3), 797–827.

Gauss, C. F. (1809). *Theoria Motus Corporum Celestium (English translation by C.H. Davis (1857). Reprinted, 1963).* New York: Dover Publications.

Gauss, C. F. (1821/23/26). *Theoria Combinutionis Observurionum Erroribus Minirnus Obnoxiue, in two parts with a supplement. reprinted with an english translation and notes by G.W. Stewart, (1995).* Philadelphia: SIAM.

Geman, S., & Geman, D. (1984). Stochastic relaxation, gibbs distributions and the Bayesian restoration of images. *IEEE Transactions on Pattern Analysis and Machine Intelligence, 6,* 721–741.

Gilks, W. R., & Berzuini, C. (2001). Following a moving target – Monte Carlo inference for dynamic Bayesian models. *Journal of the Royal Statistical Society Series B, 63*(1), 127–146.

Godolphin, E. J., & Harrison, P. J. (1975). Equivalence theorems for polynomial projecting predictors. *Journal of the Royal Statistical Society Series B, 37,* 205–215.

Godolphin, E. J., & Johnson, S. E. (2003). Decomposition of time series dynamic linear models. *Journal of Time Series Analysis, 24,* 513–527.

Godolphin, E. J., & Stone, J. M. (1980). On the structural representation for polynomial-projecting predictor models based on the Kalman filter. *Journal of the Royal Statistical Society Series B, 42,* 35–45.

Godolphin, E. J., & Triantafyllopoulos, K. (2006). Decomposition of time series models in state-space form. *Computational Statistics and Data Analysis, 50,* 2232–2246.

Godsill, S., Doucet, A., & West, M. (2004). Monte Carlo smoothing for nonlinear time series. *Journal of the American Statistical Association, 99*(465), 156–168.

Gordon, N. J., Salmond, D. J., & Smith, A. F. M. (1993). Novel approach to nonlinear/non-Gaussian Bayesian state estimation. *IEE-Proceedings-F, 140,* 107–113.

Gourieroux, C., Jasiak, J., & Sufana, R. (2009). The Wishart autoregressive process of multivariate stochastic volatility. *Journal of Econometrics, 150,* 167–181.

Graf, J. (2016). *PID control fundamentals paperback.* Sinus Engineering.

Grewal, M. S., & Andrews, A. P. (2010). Applications of Kalman filtering in aerospace 1960 to the present: Historical perspectives. *IEEE Control Systems Magazine, 30*(3), 69–78.

Grewal, M. S., & Andrews, A. P. (2015). *Kalman filtering: Theory and practice using MATLAB* (4th ed.). New York: Wiley.

Grewal, M. S., Henderson, V., & Miyasako, R. (1991). Application of Kalman filtering to the calibration and alignment of inertial navigation systems. *IEEE Transactions in Automatic Control, 36*(1), 4–14.

Guo, S., & Han, L. (2018). *Stability and control of nonlinear time-varying systems.* New York: Springer.

Gupta, A. K., & Nagar, D. K. (1999). *Matrix variate distributions.* New York: Chapman and Hall.

Halcombe Laning Jr., J., & Battin, R. H. (1956). *Random processes in automatic control*. New York: McGraw-Hill.

Han, Y. (2006). Asset allocation with a high dimensional latent factor stochastic volatility model. *The Review of Financial Studies, 19*(1), 237–271.

Hannan, E. J., & Deistler, M. (1988). *The statistical theory of linear systems*. New York: Wiley.

Harrison, P. J. (1965). Short-term sales forecasting. *Applied Statistics, 15*, 102–139.

Harrison, P. J. (1967). Exponential smoothing and short-term forecasting. *Management Science, 13*, 821–842.

Harrison, P. J. (1997). Convergence and the constant dynamic linear model. *Journal of Forecasting, 16*, 287–292.

Harrison, P. J., & Stevens, C. (1971). A Bayesian approach to short-term forecasting. *Operations Research Quarterly, 22*, 341–362.

Harrison, P. J., & Stevens, C. (1976). Bayesian forecasting (with discussion). *Journal of the Royal Statistical Society Series B, 38*, 205–247.

Hartigan, J. A. (1969). Linear Bayesian methods. *Journal of the Royal Statistical Society Series B, 31*, 446–454.

Harvey, A. C. (1981). The Kalman filter and its applications in econometrics and time series analysis. *Methods of Operations Research, 44*, 3–18.

Harvey, A. C. (1984). A unified view of statistical forecasting procedures. *Journal of Forecasting, 3*, 245–275.

Harvey, A. C. (1986). Analysis and generalisation of a multivariate exponential smoothing model. *Management Science, 32*, 374–380.

Harvey, A. C. (1989). *Forecasting, structural time series and the Kalman filter*. Cambridge: Cambridge University Press.

Harvey, A. C. (2004). Tests for cycles. In A. C. Harvey, S. J. Koopman, & N. Shephard (Eds.), *State space and unobserved component models: Theory and applications* (pp. 102–119). Cambridge: Cambridge University Press.

Harvey, A. C., & Fernandes, C. (1989). Time series models for count or qualitative observations. *Business and Econmic Statistics, 7*, 407–417.

Harvey, A. C., Gardner, G., & Phillips, G. D. A. (1980). An algorithm for exact maximum likelihood estimation of autoregressive-moving average models by means of Kalman filtering. *Applied Statistics, 29*, 311–322.

Harvey, A. C., & Koopman, S. J. (1992). Diagnostic checking of unobserved-components time series models. *Journal of Business and Economic Statistics, 10*(4), 377–389.

Harvey, A. C., Koopman, S. J., & Shephard, N. (2004). *State space and unobserved component models: Theory and applications*. Cambridge: Cambridge University Press.

Harvey, A. C., Ruiz, E., & Shephard, N. (1994). Multivariate stochastic variance models. *Review of Economic Studies, 61*, 247–264.

Harville, D. A. (1997). *Matrix Algebra from a Statistician's perspective*. New York: Springer.

Hastings, W. K. (1970). Monte Carlo sampling methods using Markov chains and their applications. *Biometrika, 57*, 97–109.

Hata, H., & Sekine, J. (2013). Risk-sensitive asset management under a Wishart autoregressive factor model. *Journal of Mathematical Finance, 3*, 222-229.

Haykin, S. (2001). *Adaptive filter theory* (4th ed.). New Jersey: Prentice Hall.

He, J., McGee, D. L., & Niu, X. (2010). Application of the Bayesian dynamic survival model in medicine. *Statistics in Medicine, 29*, 347–360.

Hemming, K., & Shaw, J. E. H. (2002). A parametric dynamic survival model applied to breast cancer survival times. *Journal of the Royal Statistical Society Series C, 51*(4), 421–435.

Hemming, K., & Shaw, J. E. H. (2005). A class of parametric dynamic survival models. *Lifetime Data Analysis, 11*, 81–98.

Hilmer, S. C., & Tiao, G. C. (1982). An arima-model-based approach to seasonal adjustment. *Journal of the American Statistical Association, 77*, 63–70.

Hirsch, M. W., Smale, S., & Devaney, R. L. (2012). *Differential equations, dynamical systems, and an introduction to chaos*. Academic Press.

Horn, R. A., & Johnson, C. R. (2013). *Matrix analysis* (2nd ed.). Cambridge: Cambridge University Press.

Hue, C., Cadre, J. P. L., & Pérez, P. (2002). Sequential Monte Carlo methods for multiple target tracking and data fusion. *IEEE Transactions on Signal Processing, 50*(2), 309–325.

Iqbal, F., & Triantafyllopoulos, K. (2019). Bayesian inference of multivariate rotated GARCH models with skew returns. *Communications in Statistics - Simulation and Computation.*

Ishihara, T., Omori, Y., & Asai, M. (2016). Matrix exponential stochastic volatility with cross leverage. *Computational Statistics and Data Analysis, 100*, 331–350.

Jacobson, N. (1953). *Lectures in abstract Algebra.* New York: Van Nostrand.

Jacquier, E., Polson, N. G., & Rossi, P. E. (1994). Bayesian analysis of stochastic volatility models. *Journal of Business and Economic Statistics, 12*(4), 371–389.

Jazwinski, A. H. (1969). Adaptive filtering. *Automatica, 5*, 475–485.

Jazwinski, A. H. (1970). *Stochastic processes and filtering theory.* New York: Academic Press.

Johansen, S. (1991). Estimation and hypothesis testing of cointegration vectors in Gaussian vector autoregressive models. *Econometrica, 59*(6), 1551–1580.

Johnson, M. A., & Moradi, M. H. (2010). *PID control: New identification and design methods.* New York: Springer.

Johnson, R., & Núñez, C. (2014). The Kalman-Bucy filter revisted. *Discrete and Continuous Dynamical Systems, 34*(10), 4139–4153.

Jones, R. H. (1966). Exponential smoothing for multivariate time series. *Journal of the Royal Statistical Society Series B, 28*, 241–251.

Julier, S. J. (2002). The scaled unscented transformation. In *Proceedings of the 2002 American Control Conference.*

Julier, S. J., & Uhlmann, J. K. (1997). A new extension of the Kalman filter to nonlinear systems. In *Proceedings of AeroSense: The 11th International Symposium on Aerospace/Defence Sensing, Simulation and Controls.*

Julier, S. J., & Uhlmann, J. K. (2004). Unscented filtering and nonlinear estimation. In *Proceedings of the IEEE* (Vol. 92, pp. 401–422).

Julious, S. A., Campbell, M. J., Bianchi, S. M., & Murray-Thomas, T. (2011). Seasonality of medical contacts in school-aged children with asthma: association with school holidays. *Public Health, 125*, 769–776.

Kalaba, R., & Tesfatsion, L. (1988). The flexible least squares approach to time-varying linear regression. *Journal of Economic Dynamics and Control, 12*, 43–48.

Kalman, R. E. (1960). A new approach to linear filtering and prediction problems. *Journal of Basic Engineering, 82*, 35–45.

Kalman, R. E., & Bertram, J. E. (1960). Control system analysis and design via the second method of Lyapunov. I. Continuous-time systems. *Journal of Basic Engineering, 82*, 371–393.

Kalman, R. E., & Bucy, R. S. (1961). New results in linear filtering and prediction theory. *Journal of Basic Engineering, 83*(1), 95–108.

Kappl, J. J. (1971). Nonlinear estimation via Kalman filtering. *IEEE Transactions on Aerospace and Electronic Systems, AES-7*(1), 79–84.

Karlis, D., & Xekalaki, E. (2005). Mixed Poisson distribution. *International Statistical Review, 73*(1), 35–58.

Kearns, B., Stevenson, M. D., Triantafyllopoulos, K., & Manca, A. (2019). Generalized linear models for flexible parametric modeling of the hazard function. *Medical Decision Making, 39*, 867–878.

Kearns, B., Stevenson, M. D., Triantafyllopoulos, K., & Manca, A. (2021). The extrapolation performance of survival models for data with a cure fraction: a simulation study. *Value in Health* (in press). https://doi.org/10.1016/j.jval.2021.05.009

Kedem, B., & Fokianos, K. (2002). *Regression models for time series analysis.* New York: Wiley.

Khatri, C. G., & Pillai, K. C. S. (1965). Some results on the non-central multivariate beta distribution. *Annals of Mathematical Statistics, 36*, 1511–1520.

Kim, S., Shephard, N., & Chib, S. (1998). Stochastic volatility: Likelihood inference and comparison with ARCH models. *The Review of Economic Studies, 65*(3), 361–393.

Kitagawa, G. (1987). Non-Gaussian state-space modelling of nonstationary time series (with discussion). *Journal of the American Statistical Association, 82*, 1032–1063.

Kitagawa, G. (1998). A self-organizing state-space model. *Journal of the American Statistical Association, 93*, 1203–1215.

Kitagawa, G., & Gersch, W. (1996). *Smoothness priors analysis of time series.* New York: Springer.

Kolmogorov, A. N. (1941). Stationary sequences in Hilbert space (in Russian). *Moscow University Mathematics Bulletin, 2*(6), 228–271.

Konno, Y. (1988). Exact moments of the multivariate F and beta distributions. *Journal of the Japan Statistical Society, 18*, 123–130.

Koopman, S. J. (1993). Disturbance smoother for state space models. *Biometrika, 80*, 117–126.

Koopman, S. J. (1997). Exact initial Kalman filtering and smoothing for non-stationary time series models. *Journal of the American Statistical Association, 92*, 1630–1638.

Kulikov, G. Y., & Kulikova, M. V. (2018). Stability analysis of extended, cubature and unscented Kalman filters for estimating stiff continuous–discrete stochastic systems. *Automatica, 90*, 91–97.

Lang, S. (1987). *Calculus of several variables* (3rd ed.). New York: Springer.

Legendre, A. M. (1805). *Nouvelles Méthodes pour la Détermination des Orbites des Comètes.* Paris: F. Didot.

Leonard, T., & Hsu, J. S. J. (1999). *Bayesian methods.* Cambridge: Cambridge University Press.

Liesenfeld, R., & Richard, J.-F. (2003). Univariate and multivariate stochastic volatility models: estimation and diagnostics. *Journal of Empirical Finance, 10*, 505–531.

Lindsey, J. K. (2004). *Statistical analysis of stochastic processes in time.* Cambridge: Cambridge University Press.

Lintner, J. (1965). The valuation of risk assets and the selection of risky investments in stock portfolios and capital budgets. *Review of Economics and Statistics, 47*(1), 13–37.

Liu, J., & West, M. (2001). Sequential Monte Carlo Methods in practice. In D. A., de Freitas N., & G. N. (Eds.), chap. *Combined parameter and state estimation in simulation-based filtering.* New York: Springer.

Ljung, L. (1979). Asympotic behaviour of the extended Kalman filter as a parameter estimator for linear systems. *IEEE Transactions in Automatic Control, AC-24*, 36–50.

Loève, M. (1955). *Probability theory.* Van Nostrand Company Inc.

Longley, J. W. (1967). An appraisal of least-squares programs from the point of view of the user. *Journal of the American Statistical Association, 62*, 819–841.

Lorenz, E. N. (1963). Deterministic nonperiodic flow. *Journal of the Atmospheric Sciences, 20*, 130–141.

Lundbergh, S., Teräsvirta, T., & van Dijk, D. (2003). Time-varying smooth transition autoregressive models. *Journal of Business and Economic Statistics, 21*, 104–121.

Magnus, J. R., & Neudecker, H. (1988). *Matrix differential calculus with applications in statistics and econometrics.* New York: Wiley.

Makridakis, S., Wheelwright, S. C., & Hyndman, R. J. (1998). *Forecasting: methods and applications.* New York: Wiley.

Manley, G. (1974). Central England temperatures: monthly means 1659 to 1973. *Quarterly Journal of the Royal Meteorological Society, 100*, 389–405.

Markowitz, H. M. (1952). Portfolio selection. *The Journal of Finance, 7*(1), 77–91.

Markowitz, H. M. (1959). *Portfolio selection: Efficient diversification of investments.* New York: Wiley.

Martinussen, T., & Scheike, T. H. (2006). *Dynamic regression models for survival data.* New York: Springer.

McCullagh, P., & Nelder, J. A. (1989). *Generalised linear models* (2nd ed.). New York: Chapman and Hall.

McCulloch, R. E., & Tsay, R. S. (1993). Bayesian inference and prediction for mean and variance shifts in autoregressive time series. *Journal of the American Statistical Association, 88*(423), 968–978.

McGee, L. A., & Schmidt, S. F. (1985, November). *Discovery of the Kalman filter as a practical tool for aerospace and industry.* NASA Technical Memorandum 86847.

McGilchrist, C. A., & Sandland, R. L. (1979). Recursive estimation of the general linear model with dependent errors. *Journal of the Royal Statistical Society Series B, 41,* 65–68.

Meinhold, R. J., & Singpurwalla, N. D. (1983). Understanding the Kalman filter. *The American Statistician, 37,* 123–127.

Metropolis, N., Rosenbluth, A. W., Rosenbluth, M. N., Teller, A. H., & Teller, E. (1953). Equations of state calculations by fast computing machine. *Journal of Chemical Physics, 21,* 1087–1091.

Meucci, A. (2005). *Risk and asset allocation.* New York: Springer.

Mihaylova, L., Carmi, A. Y., Septier, F., & Gning, A. (2014). Overview of Bayesian sequential Monte Carlo methods for group andextended object tracking. *Digital Signal Processing, 25,* 1–16.

Minkler, G., & Minkler, J. (1993). *Theory and application of Kalman filtering.* Magellan Book Company.

Molinari, D. A. (2009). *Monitoring and adaptation in time series models, MSc thesis, School of Mathematics and Statistics, University of Sheffield.*

Montana, G., Triantafyllopoulos, K., & Tsagaris, T. (2009). Flexible least squares for temporal data mining and statistical arbitrage. *Expert Systems with Applications, 36,* 2819–2830.

Morris, A. S., & Sterling, M. J. H. (1979). Model tuning using the extended Kalman filter. *Electronics Letters, 15,* 201–202.

Morrison, G. W., & Pike, D. H. (1977). Kalman filtering applied to statistical forecasting. *Management Science, 23,* 768–774.

Muirhead, R. J. (1982). *Aspects of multivariate statistical theory.* New York: Wiley.

Muth, J. F. (1960). Optimal properties of exponentially weighted forecasts. *Journal of the American Statistical Association, 55,* 299–305.

Nelder, J. A., & Wedderburn, R. W. M. (1972). Generalised linear models. *Journal of the Royal Statistical Society Series A, 135,* 370–384.

Nolan, D., & Lang, D. T. (2015). *Data science in R: A case studies approach to computational reasoning and problem solving.* CRC Press.

Ogata, K. (1970). *Modern control engineering.* Englewood Cliffs, NJ: Prentice-Hall.

O'Hagan, A., & Forster, J. J. (2004). *Bayesian inference (Kendall's advanced theory of statistics: Volume 2B)* (2nd ed.). London: Arnold.

Okuguchi, K., & Irie, K. (1990). The Schur and Samuelson conditions for a cubic equation. *The Manchester School of Economic and Social Science, 58*(4), 414–418.

Oppenheim, A., & Willsky, A. (1983). *Signals and systems.* Englewood Cliffs, NJ: Prentice Hall.

Pakrashi, A., & Namee, B. M. (2019). Kalman filter-based heuristic ensemble (kfhe): A new perspective on multi-class ensemble classification using Kalman filters. *Information Sciences, 485,* 456–485.

Pan, X., & Jarrett, J. (2004). Applying state space to SPC: monitoring multivariate time series. *Journal of Applied Statistics, 31,* 397–418.

Pankratz, A. (1991). *Forecasting with dynamic regression models.* New York: Wiley.

Parisi, A., & Liseo, B. (2018). Objective Bayesian analysis for the multivariate skew-t model. *Statistical Methods and Applications, 27,* 277–295.

Parks, P. C. (1992). A. M. Lyapunov's stability theory - 100 years on. *IMA Journal of Mathematical Control and Information, 9*(4), 275–303.

Petris, G. (2010). An R package for dynamic linear models. *Journal of Statistical Software, 36,* 1–16.

Petris, G., Petrone, S., & Campagnoli, P. (2009). *Dynamic linear models with R.* New York: Springer.

Philipov, A., & Glickman, M. E. (2006). Multivariate stochastic volatility via Wishart processes. *Journal of Business and Economic Statistics, 24,* 313–328.

Phillips, P., & Perron, P. (1988). Testing for a unit root in time series regression. *Biometrika, 75*(2), 335–346.

Pitt, M. K., & Shephard, N. (1999). Filtering via simulation: auxiliary particle filters. *Journal of the American Statistical Association, 94*(446), 590–599.

Plackett, R. L. (1950). Some theorems in least squares. *Biometrika, 37*, 149–157.

Plackett, R. L. (1991). *Regression analysis*. Oxford: Oxford University Press.

Pole, A. (2007). *Statistical arbitrage. Algorithmic trading insights and techniques*. Wiley Finance.

Pole, A., West, M., & Harrison, P. J. (1994). *Applied Bayesian forecasting and time series analysis*. New York: Chapman and Hall.

Pollock, D. S. G. (2003). Recursive estimation in econometrics. *Computational Statistics and Data Analysis, 44*, 37–75.

Ponomareva, K., & Date, P. (2013). Higher order sigma point filter: a new heuristic for nonlinear time series filtering. *Applied Mathematics and Computation, 221*, 662–671.

Prado, R., & West, M. (2010). *Time series: Modeling, computation, and inference*. New York: Chapman and Hall.

Press, S. J. (1989). *Bayesian statistics: Principles, models, and applications*. New York: Wiley.

Priestley, M. B. (1981). *Spectral analysis and time series, volume 1: Univariate time series*. London: Academic Press.

Priestley, M. B., & Rao, T. S. (1975). The estimation of factor scores and Kalman filtering for discrete parameter stationary processes. *International Journal of Control, 21*, 971–975.

Punchihewa, Y. G., Vo, B.-T., Vo, B.-N., & Kim, D. Y. (2018). Multiple object tracking in unknown backgrounds with labeled random finite sets. *IEEE Transactions on Signal Processing, 66*(11), 3040–3055.

Quintana, J. M., & West, M. (1987). An analysis of international exchange rates using multivariate DLM. *The Statistician, 36*, 275–281.

Radhakrishnan, R., Yadav, A., Date, P., & Bhaumik, S. (2018). A new method for generating sigma points and weights for nonlinear filtering. *IEEE Control Systems Letters, 2*(3), 519–524.

Rao, M. J. M. (2000). Estimating time-varying parameters in linear regression models using a two-part decomposition of the optimal control formulation. *Sankhyā Series B, 62*, 433–447.

Raunch, H. E., Tung, F., & Streibel, C. T. (1965). Maximum likelihood estimators of linear dynamic systems. *American Institute of Aeronautics and Astronautics Journal, 3*, 1445–1450.

Robert, C. P. (2007). *The Bayesian choice: From decision-theoretic foundations to computational implementation* (2nd ed.). New York: Springer.

Robinson, R. C. (2012). *An introduction to dynamical systems: Continuous and discrete* (2nd ed.). Pearson Education Inc.

Rousseeuw, P. J., & Leroy, A. M. (1987). *Robust regression and outlier detection*. New York: Wiley.

Rudin, W. (1976). *Principles of mathematical analysis* (3rd ed.). New York: McGraw-Hill.

Saab, S. S. (2004). A heuristic Kalman filter for a class of nonlinear systems. *IEEE Transactions in Automatic Control, 49*(12), 2261–2265.

Salvador, M., & Gargallo, P. (2003). Automatic selective intervention in dynamic linear models. *Journal of Applied Statistics, 30*(10), 1161–1184.

Salvador, M., & Gargallo, P. (2004). Automatic monitoring and intervention in multivariate dynamic linear models. *Computational Statistics and Data Analysis, 47*(3), 401-431.

Samuelson, P. A. (1941). Conditions that the roots of a polynomial be less than unity in absolute value. *Annals of Mathematical Statistics, 12*, 360–364.

Särkkä, S. (2013). *Bayesian filtering and smoothing*. Cambridge: Cambridge University Press.

Sastry, S. (1999). *Nonlinear systems: Analysis, stability, and control* (Vol. 10). New York: Springer.

Schmidt, S. F. (1981). The Kalman filter: Its recognition and development for aerospace applications. *Journal of Guidance and Control, 4*(1), 4–7.

Schulmerich, M., Leporcher, Y.-M., & Eu, C.-H. (2015). *Applied asset and risk management*. New York: Springer.

Schweppe, F. C. (1965). Evaluation of likelihood signals for Gaussian signals. *IEEE Transactions on Information Theory, 11*, 61–70.

Schweppe, F. C. (1968). Recursive stateestimation:unknownbut bounded errors and system inputs. *IEEE Transactions in Automatic Control, AC-13*(1), 22–28.

Schweppe, F. C. (1973). *Uncertain dynamic systems*. Englewood Cliffs, NJ: Prentice Hall.

Seddon, J. M., & Newman, S. (2011). *Basic Helicopter aerodynamics* (3rd ed.). New York: Wiley.

Shephard, N. (1994a). Local scale models: state space alternative to integrated GARCH processes. *Journal of Econometrics, 60*, 181–202.

Shephard, N. (1994b). Partial non-Gaussian state space models. *Biometrika, 81*, 115–131.

Shephard, N., & Pitt, M. K. (1997). Likelihood analysis for non-Gaussian measurement time series. *Biometrika, 84*, 653–667.

Shumway, R. H., & Stoffer, D. S. (1982). An approach to time series smoothing and forecasting using the EM algorithm. *Journal of Time Series Analysis, 3*, 253–264.

Shumway, R. H., & Stoffer, D. S. (2017). *Time series analysis and its applications: With R examples* (4th ed.). New York: Springer.

Smirnov, V. I. (1992). Biography of A. M. Lyapunov. *International Journal of Control, 55*(3), 775–784.

Smith, J. Q. (1979). A generalisation of the Bayesian steady forecasting model. *Journal of the Royal Statistical Society Series B, 41*, 375–387.

Smith, J. Q. (1981). The multiparameter steady model. *Journal of the Royal Statistical Society Series B, 43*, 256–260.

Smith, R. L., & Miller, J. E. (1986). A non-Gaussian state space model with application to prediction of records. *Journal of the Royal Statistical Society Series B, 48*, 79–88.

Soyer, R., & Tanyeri, K. (2006). Bayesian portfolio selection with multi-variate random variance models. *Journal of Operational Research, 171*, 977–990.

Spivak, M. (1995). *Calculus* (3rd ed.). Cambridge: Cambridge University Press.

Stevens, B. L., Lewis, F. L., & Johnson, E. N. (2016). *Aircraft control and simulation: Dynamics, controls design, and autonomous systems*. New York: Wiley.

Stigler, S. M. (1986). *The history of statistics*. Cambridge, MA: Harvard University Press.

Storvik, G. (2002). Particle filters for state-space models with the presence of unknown static parameters. *IEEE Transactions on Signal Processing, 50*, 281–290.

Strachan, R., & Inder, B. (2004). Bayesian analysis of the error correction model. *Journal of Econometrics, 123*, 307–325.

Stroock, D. W. (2018). *Elements of stochastic calculus and analysis*. New York: Springer.

Svenson, A. (1981). On the goodness-of-fit test for the multiplicative poisson model. *Annals of Statistics, 9*, 697–704.

Tesfatsion, L., & Kalaba, R. (1989). Time-varying linear regression via flexible least squares. *Computers and Mathematics with Applications, 17*, 1215–1245.

Tokhi, O. M., & Azad, A. K. M. (2008). *Flexible robot manipulators: Modelling, simulation and control*. The Institution of Engineering and Technology.

Tong, H. (1996). *Non-linear time series*. Oxford: Clarendon Press.

Triantafyllopoulos, K. (2006a). Multivariate control charts based on Bayesian state space models. *Quality and Reliability Engineering International, 22*, 693–707.

Triantafyllopoulos, K. (2006b). Multivariate discount weighted regression and local level models. *Computational Statistics and Data Analysis, 50*, 3702–3720.

Triantafyllopoulos, K. (2007a). Convergence of discount time series dynamic linear models. *Communications in Statistics - Theory and Methods, 36*, 2117–2127.

Triantafyllopoulos, K. (2007b). Covariance estimation for multivariate conditionally Gaussian dynamic linear models. *Journal of Forecasting, 26*, 551–569.

Triantafyllopoulos, K. (2008a). Missing observation analysis for matrix-variate time series data. *Statistics and Probability Letters, 78*, 2647–2653.

Triantafyllopoulos, K. (2008b). Multivariate stochastic volatility with Bayesian dynamic linear models. *Journal of Statistical Planning and Inference, 138*, 1021–1037.

Triantafyllopoulos, K. (2009). Inference of dynamic generalised linear models: on-line computation and appraisal. *International Statistical Review, 77*, 439–450.

Triantafyllopoulos, K. (2011a). Real-time covariance estimation for the local level model. *Journal of Time Series Analysis, 32*, 93–107.

Triantafyllopoulos, K. (2011b). Time-varying vector autoregressive models with stochastic volatility. *Journal of Applied Statistics, 38*, 369–382.

Triantafyllopoulos, K. (2012). Multivariate stochastic volatility modelling using Wishart autoregressive processes. *Journal of Time Series Analysis, 33*, 48–60.

Triantafyllopoulos, K. (2014). Multivariate stochastic volatility estimation using particle filters. In M. Akritas, S. Lahiri, & D. Politis (Eds.), *Topics in nonparametric statistics. Springer proceedings in mathematics and statistics* (Vol. 74, pp. 335–345). New York: Springer.

Triantafyllopoulos, K., Aldebrez, F. M., Zinober, A. S. I., & Tokhi, M. O. (2009). Bayesian dynamic modelling and tracking control for flexible manoeuvring systems. In *2009 3rd International Conference on Signals, Circuits and Systems* (pp. 1–6).

Triantafyllopoulos, K., Godolphin, J. D., & Godolphin, E. J. (2005). Process improvement in the microelectronic industry by state-space modelling. *Quality and Reliability Engineering International, 21*, 465–475.

Triantafyllopoulos, K., & Han, S. (2013). Detecting mean reverted patterns in algorithmic pairs trading. In P. Latorre Carmona, J. S. Sánchez, & A. L. Fred (Eds.), *Mathematical methodologies in pattern recognition and machine learning* (pp. 127–147). New York: Springer.

Triantafyllopoulos, K., & Harrison, P. J. (2008). Posterior mean and variance approximation for regression and time series problems. *Statistics: A Journal of Theoretical and Applied Statistics, 42*, 329–350.

Triantafyllopoulos, K., & Montana, G. (2011). Dynamic modeling of mean-reverting spreads for statistical arbitrage. *Computational Management Science, 8*, 23–49.

Triantafyllopoulos, K., & Nason, G. P. (2007). A Bayesian analysis of moving average processes with time-varying parameters. *Computational Statistics and Data Analysis, 52*, 1025–1046.

Triantafyllopoulos, K., & Nason, G. P. (2009). A note on state-space representations of locally stationary wavelet time series. *Statistics and Probability Letters, 79*, 50–54.

Triantafyllopoulos, K., & Pikoulas, J. (2002). Multivariate Bayesian regression applied to the problem of network security. *Journal of Forecasting, 21*, 579–594.

Triantafyllopoulos, K., Shakandli, M., & Campbell, M. J. (2019). Count time series prediction using particle filters. *Quality and Reliability Engineering International, 35*, 1445–1459.

Trosset, M. W. (2009). *An introduction to statistical inference and its applications with R.* Chapman and Hall.

Tsay, R. S. (2002). *Analysis of financial time series.* New York: Wiley.

Uhlig, H. (1994). On singular Wishart and singular multivariate beta distributions. *Annals of Statistics, 22*, 395–405.

Uhlig, H. (1997). Bayesian vector autoregressions with stochastic volatility. *Econometrica, 65*, 59–73.

Van Der Merwe, R., Doucet, A., de Freitas, N., & Wan, E. A. (2001). The unscented particle filter. In T. G. D. Todd K. Leen & V. Tresp (Eds.), *Advances in neural information processing systems* (Vol. 13).

Van Der Weide, R. (2002). GO-GARCH: a multivariate generalised orthogonal GARCH model. *Journal of Applied Econometrics, 17*, 549–564.

van Houwelingen, H., & Putter, H. (2012). *Dynamic prediction in clinical survival analysis.* CRC Press.

Venables, W. N., & Ripley, B. D. (2002). *Modern applied statistics with S-PLUS* (4th ed.). New York: Springer.

Vidyamurthy, G. (2004). *Pairs trading.* Wiley Finance.

Virbickaite, A., Ausín, M. C., & Galeano, P. (2015). Bayesian inference methods for univariate and multivariate GARCH models: a survey. *Journal of Economic Surveys, 29*(1), 76–96.

Vrontos, I. D., Dellaportas, P., & Politis, D. N. (2003). A full-factor multivariate GARCH model. *Econometrics Journal, 6*(2), 312-334.

Wagner, H. (2011). Bayesian estimation and stochastic model specification search for dynamic survival models. *Statistics and Computing, 21*, 231–246.

Wan, E. A., & Van Der Merwe, R. (2000). The unscented Kalman filter for nonlinear estimation. In *Proceedings of the IEEE 2000 Adaptive Systems for Signal Processing, Communications, and Control Symposium.*

West, M. (1986). Bayesian model monitoring. *Journal of the Royal Statistical Society Series B, 48,* 70–78.

West, M. (1997). Time series decomposition. *Biometrika, 84,* 489–494.

West, M., & Harrison, P. J. (1986). Monitoring and adaptation in Bayesian forecasting models. *Journal of the American Statistical Association, 81,* 741–750.

West, M., & Harrison, P. J. (1997). *Bayesian forecasting and dynamic models* (2nd ed.). New York: Springer.

West, M., Harrison, P. J., & Migon, H. S. (1985). Dynamic generalised linear models and Bayesian forecasting (with discussion). *Journal of the American Statistical Association, 80,* 73–97.

West, M., Prado, R., & Krystal, A. D. (1999). Evaluation and comparison of EEG traces: latent structure in nonstationary time series. *Journal of the American Statistical Association, 94,* 375–387.

Whittle, P. (1984). *Prediction and Regulation by linear least-square methods* (2nd ed.). Oxford: Blackwell.

Wiener, N. (1949). *Extrapolation, Interpolation and smoothing of stationary time series with engineering applications.* Cambridge, MA: MIT Press.

Wiener, N., & Hopf, E. (1931). Über eine klasse singulärer integralgleichungen. *Sem-Ber Preuss Akad Wiss, 31,* 696–706.

Wilson, K. J., & Farrow, M. (2017). Bayes linear kinematics in a dynamic survival model. *International Journal of Approximate Reasoning, 80,* 239–256.

Wise, J. (1956). Stationarity conditions for stochastic processes of the autoregressive and moving-average type. *Biometrika, 48,* 216–219.

Yedavalli, R. K. K. (2016). *Robust control of uncertain dynamic systems: A linear state space approach.* New York: Springer.

Young, P. C. (1968). The use of linear regression and related procedures for the identification of dynamic processes. In *7th Symposium of Adaptive Processes.* IEEE.

Young, P. C. (1969). *The differential equation error method of process parameter estimation* (Unpublished doctoral dissertation). Cambridge, England: University of Cambridge.

Young, P. C. (2011). *Recursive estimation and time-series analysis: An introduction for the student and practitioner* (2nd ed.). New York: Springer.

Yu, J., & Meyer, R. (2006). Multivariate stochastic volatility models: Bayesian estimation and model comparison. *Economtric Reviews, 25,* 361–384.

Yu, P. L. H., Li, W. K., & Ng, F. C. (2017). The generalized conditional autoregressive Wishart model for multivariate realized volatility. *Journal of Business and Economic Statistics, 35*(4), 513–527.

Yule, G. U. (1927). On a method of investigating periodicities in disturbed series with special reference to Wolfer's sunspot numbers. *Philosophical Transactions, Royal Society, 226,* 267–298.

Zadeh, L., & Desoer, C. (1979). *Linear system theory.* Huntington, NY: Krieger Publishing Company.

Zadeh, L. A., & Desoer, C. A. (1963). *Linear system theory: The state space approach.* New York: McGraw Hill.

Zaritskii, V. S., Svetnik, V. B., & Shimelevich, L. I. (1975). Monte Carlo technique in problems of optimal data processing. *Automation and Remote Control, 12,* 95–103.

Zellner, A., & Tiao, G. C. (1964). Bayesian analysis of the regression model with autocorrelated errors. *Journal of the American Statistical Association, 59,* 763-778.

Zhang, H., & Zhang, Q. (2008). Trading a mean reverting asset: buy low and sell high. *Automatica, 44,* 1511–1518.

Zinober, A. S. I. (1994). Variable Structure and Lyapunov Control. In A. S. I. Zinober (Ed.), chap. *An introduction to sliding mode variable structure control* (Vol. 193). New York: Springer.

Zinober, A. S. I. (2001). *Nonlinear and adaptive control* (Vol. 281). New York: Springer.

Index

Printed in the United States
by Baker & Taylor Publisher Services